Phagocytosis of Dying Cells: From Molecular Mechanisms to Human Diseases

Phagocytosis of Dying Cells: From Molecular Mechanisms to Human Diseases

Dmitri V. Krysko • Peter Vandenabeele

Editors

Department for Molecular Biomedical Research,
VIB, Ghent, Belgium

Department of Molecular Biology,
Ghent University, Ghent, Belgium

 Springer

Editors
Dr. Dmitri V. Krysko, M.D., Ph.D.
VIB Department for Molecular Biomedical
Research, VIB and Ghent University,
Technologiepark 927
9052 Ghent, Belgium
dmitri.krysko@ugent.be

Prof. Dr. Peter Vandenabeele, Ph.D.
VIB Department for Molecular Biomedical
Research, VIB and Ghent University,
Technologiepark 927
9052 Ghent, Belgium
peter.vandenabeele@dmbr.ugent.be

ISBN 978-1-4020-9292-3   e-ISBN 978-1-4020-9293-0

Library of Congress Control Number 2008939363

*Cover illustration: Based on scanning electron micrograph of necrotic cell internalization by macro-
phage. Kindly provided by Prof. Dr. Katharina D'Herde, Ghent University, Ghent, Belgium.*

Printed on acid-free paper

9 8 7 6 5 4 3 2 1

springer.com

Preface

Phagocytosis has been at the forefront of cell biology for more than a century. Initially, phagocytosis, which comes from Greek words meaning "devouring cells," was discovered in the late 19th century by Ilya Metchnikoff, who was awarded, together with Paul Ehrlich, the Nobel Prize in Physiology and Medicine in 1908 "in recognition of their work on immunity." At that time Metchnikoff had already identified a function for phagocytes not only in host defense but also as scavengers of degenerating host cells during metamorphosis of tadpoles, thus providing one of the first descriptions of apoptotic cell clearance by macrophages (Kaufmann 2008). Since then, much has been learned about phagocytosis, and the previous several decades have witnessed outstanding progress in understanding the functions and the molecular mechanisms of phagocytosis. Two main types of targets are cleared by phagocytosis: microbial pathogens and dying cells. Rapid recognition and clearance of dying cells by phagocytes plays a pivotal role in development, maintenance of tissue homeostasis, control of immune responses, and resolution of inflammation. Clearance of dying cells can be divided into several stages, including sensing, recognition, binding and signaling, internalization, and immunological responses.

In this book, our contributors address these different stages of dead cell clearance and examine how impaired clearance of dying cells may lead to human diseases. We have attempted to provide sufficient cross-referencing and indexing to enable the reader to easily locate the ideas elaborated in the different chapters. These cross-references to other chapters in the book are indicated by [brackets] in the text. We have divided the book into two major parts: Part I "Molecular mechanisms of phagocytosis of dying cells" and Part II "Impairment of phagocytosis of dying cells and its role in the development of diseases". The first part addresses the issues of the attraction of phagocytes, the recognition of dying cells, genetic pathways, and immunological responses of professional and non-professional phagocytes. This part we begin with two introductory chapters to provide the reader with an overview of cell death types: apoptotic, autophagic and necrotic cell death. In the first chapter, Krysko, Kaczmarek and Vandenabeele provide a general overview of the molecular pathways responsible for the execution of apoptotic, necrotic and autophagy-associated cell death. We distinguish between accidental necrosis, which is due to direct physiochemical damage, and programmed necrosis, by some others also called necroptosis, which is mediated by Rip1-kinase dependent signaling. We also highlight the multiple roles of autophagy, such as in cell survival, death and clearance, and also its contribution to pathogenesis of human diseases. Diez-Fraile, Lammens and D'Herde continue with this subject by considering examples from embryonic development, tissue and organ remodeling, and age-related diseases in which apoptotic, necrotic and autophagic cell death play an important role. They also address the issue of cell death types in invertebrate and plant systems.

It is striking that about 500×10^9 cells in the human body die each day. These cells are either shed off directly from body surfaces or continuously removed by a remarkably efficient phagocytic system without causing inflammation or scar formation. This raises the question of how phagocytes, which are usually not close to the dying cells, manage to reach their targets in time. The answer to this question is provided by Peter, Wesselborg and Lauber, who review what is currently known about attraction and danger signals. These "eat me" signals, released from apoptotic and necrotic cells, direct phagocytes to the site of dead cells and contribute to their clearance as well as to the immunological response. Once phagocytes arrive at the site of cell death, they first have to distinguish dying cells from living cells, and then adhere to and bind them. These steps involve a complex and redundant array of ligands and receptors located on the surface of dying cells and phagocytes, respectively. Napirei and Mannherzs provide a global view of how phagocytes continually sense dead and dying cells by means of a set of cell surface receptors and ligands. (These themes are picked up later by Ucker; Gregory and Pound; Lucy-Hubert). Napirei and Mannherzs also emphasize the extracellular mechanisms involved in the degradation of dying cells with subsequent clearance of debris by phagocytes, and they elaborate on the role of nucleases and proteases in these processes. Recognition and binding of phagocytes to their dying targets is followed by signaling events leading to initiation of engulfment. Gronski and Ravichandran provide an in-depth comparative study of the biochemical pathways involved in clearance of dead cells in phylogenetically diverse multicellular organisms ranging from *C. elegans* and *D. melanogaster* to mammals. In the following two chapters, Ucker and Lacy-Hulbert address the issue of phagocyte responses to dying cells. Ucker emphasizes that apoptotic cells exert anti-inflammatory effects on phagocytes at the transcriptional level independently of engulfment. He also provides examples of how immunosuppressive pathways engaged by apoptotic cells are targeted by pathogens. In this discussion he also proposes two dimensions of immune recognition. In addition to the unidimensional view of immune discrimination as a self-versus-non-self phenomenon to signal danger, apoptotic "calm" (conserved apoptotic ligand for response modulation) determinants engage a second dimension of immune discrimination devoted to the maintenance of homeostasis and tolerance. Lacy-Hubert continues by dealing with the responses of different classes of professional and non-professional phagocytes to apoptotic cells. He emphasizes that their responses are more than waste disposal, and that they have immunomodulatory effects as well as promoting proliferation, angiogenesis and tissue regeneration. Aspects of the immunomodulatory properties of apoptotic cells are also touched upon by Divito and Morelli in the second part of this book. Finally, Trahtemberg and Mevorach close Part I with a chapter devoted to discussion of different methods and models that are widely used to study the clearance of dying cells in vitro and in vivo, together with their pros and cons and some practical advice. Researchers studying the clearance of dying cells can derive particular benefit from this chapter, especially in view of the potential bias to which these methods may lead.

Part II, "Impairment of phagocytosis of dying cells and its role in the development of diseases," turns to a different, more clinically relevant viewpoint. The authors examine whether dysregulation of dying cell clearance may alter immune responses in ways that contribute to human diseases. Gregory and Pound start this part with one of today's most intriguing themes: mouse knock-out models of clearance. In this well-detailed essay, there is again emphasis on the complexity of the array of molecules implicated in the removal of dying cells. They look at the association between defects in cell clearance and development of autoimmunity and inflammatory disorders in several knock-out mouse models, a matter that has been far from clear. Indeed, great benefit may be derived from knock-out models of clearance because they provide not only a basis for understanding clearance mechanisms, but also for identification of molecular targets for therapeutic manipulation of the clearance process. Mevorach continues by discussing mechanisms that might link the defective clearance of apoptotic cells with increased autoimmune responses in the development and acceleration of systemic lupus erythematosus (SLE), a multisystem autoimmune disease of unknown etiology, and drug-induced lupus. In this chapter, the following subjects are discussed: altered clearance of dying cells, accelerated leukocyte apoptosis, genetic or functional deficiencies of natural opsonins, the presence of autoantibodies, and dysfunction of phagocytes. Mevorach concludes that all these events could lead to an autoimmune response and persistence of inflammation.

We placed the chapter written by Divito and Morelli in the second part of the book because, besides discussing the mechanisms of down-regulation of adaptive immune responses by apoptotic cells, they focus on the effects of apoptotic cells on allograft survival in experimental models, as well as on the potential use of apoptotic cells in the prevention and treatment of graft rejection and autoimmune disorders. Bartunkova and Spisek overview the rapid developments in the use of dying cells as immunotherapy for cancer, as well as its advantages and disadvantages. The authors also outline the molecular markers of immunogenic cell death and delineate the future goals of immunotherapy. Martinet, Schrijvers and De Meyer focus on atherosclerosis and state that unstable rupture-prone plaques are characterized by the accumulation of macrophages and non-engulfed apoptotic cells, which points to inefficient removal of dying cells. They consider mechanisms contributing to the defective clearance of apoptotic cells in atherosclerosis and the consequences of this defect, and discuss therapeutic strategies that could be used to limit the detrimental effects of impaired phagocytosis of dying cells. However, it remains unclear whether the number of non-engulfed apoptotic cells in advanced human plaques is large enough to evoke the plaque-destabilizing events. Reynolds and Hodge critically examine the hypothesis of defective clearance of apoptotic material in the pathogenesis of chronic lung diseases, such as chronic obstructive pulmonary disease (COPD), asthma and cystic fibrosis. They emphasize that on one hand there is evidence for increased apoptosis, and on the other hand that the phagocytic capacity of macrophages derived from patients is deficient as a result of defects related to the expression levels of cell surface receptors. However, the link between defective clearance of apoptotic cells and lung disease requires a lot more work before its potential as a therapeutic target can be properly assessed. Dini and Vergallo continue the discussion of COPD but from another point of view, by assessing the relationship between cigarette smoke, one of the main risk factors in the development of chronic pulmonary diseases, and phagocytosis of dying cells. They follow by pointing out that other environmental factors, including static magnetic fields, also may affect the phagocytosis of dying cells and potentially lead to the diseases. This mostly uninvestigated area of research deserves more attention, especially in view of the increasing rate of environmental pollution.

Phagocytosis on its own and in relation to dying cells is a very important aspect of life, and gaining the ability to manipulate it will have far-reaching consequences in many fields of medicine. Future insights into the molecular mechanisms of cell death and phagocytosis and their immunomodulatory features will open new avenues for research on autoimmunity and cancer and should eventually facilitate the development of new classes of therapeutics and disease-modifying agents. Recognition of impaired phagocytosis as one of the mechanisms of certain human diseases may give rise to new therapeutic approaches based on manipulating this fundamental and highly conserved process. All these authors recognize the complexity of the link between defective clearance of dying cells and development of the above-mentioned human diseases. Elucidating this link is of great importance to biology and medicine, and our authors pose this challenging question to our readers. We hope that this book will stimulate future research on phagocytosis of dying cells, from its molecular and cellular basis up to human pathologies, and that it will inspire new and insightful experiments.

Sincere thanks are given to the outstanding contributors of this book for their time and effort. We would like to express our special acknowledgement to Dr. A. Bredan for editing the chapters and working over the proofs. We are also deeply thankful to BD Biosciences, whose generous support made it possible to publish all illustrations in color.

Ghent, Belgium
July 2008

Dmitri V. Krysko and Peter Vandenabeele
Editors

Reference

Kaufmann SH (2008) Immunology's foundation: the 100-year anniversary of the Nobel Prize to Paul Ehrlich and Elie Metchnikoff. Nat Immunol 9, 705–712

Contents

Part I
Molecular Mechanisms of Phagocytosis
of Dying Cells

Chapter 1
Molecular Pathways of Different Types of Cell Death: Many Roads to Death

Dmitri V. Krysko, Agnieszka Kaczmarek and Peter Vandenabeele

Abstract: Cell death is a fundamental cellular response that has a crucial role in shaping our bodies during development and in regulating tissue homeostasis by eliminating unwanted cells. Three major morphologies of cell death have been described: apoptosis (type I), cell death associated with autophagy (type II) and necrosis (type III). In mammalian cells, the apoptotic response is mediated by either an intrinsic or an extrinsic pathway, depending on the origin of the death stimuli, and is almost always caspase-dependent. For a long time necrosis has been considered to be an accidental and uncontrolled form of cell death. However, evidence is accumulating that necrotic cell death in some cases can be as well controlled and programmed as caspase-dependent apoptosis. Autophagy is foremost a survival mechanism that is activated in cells subjected to nutrient or obligate growth factor deprivation. When cellular stress continues, cell death may continue by autophagy alone, or else it often becomes associated with features of apoptotic or necrotic cell death, depending on the stimulus and cell type. It is debatable whether autophagic cell death is an alternative way of dying, different from apoptotic and necrotic cell death, or whether failure of autophagy to rescue the cell can lead to cell death by either pathway. The aim of this chapter is to provide a general overview of current knowledge on signalling events that result in apoptosis, necrosis and cell death associated with autophagy.

Keywords: Apoptosis • Necrosis • Autophagy • Caspases • Mitochondria

D. V. Krysko (✉) · P. Vandenabeele (✉) · A. Kaczmarek
Molecular Signaling and Cell Death Unit
Department for Molecular Biomedical Research
VIB, Technologiepark 927, 9052 Ghent
Belgium
Tel.: +32 9 3313764
Fax: +32 9 3313609
E-mail: dmitri.krysko@ugent.be; peter.vandenabeele@dmbr.ugent.be

D. V. Krysko · A. Kaczmarek · P. Vandenabeele
Department of Molecular Biology
Ghent University
Technologiepark 927, 9052 Ghent
Belgium

D.V. Krysko, P. Vandenabeele (eds.), *Phagocytosis of Dying Cells,*
DOI: 10.1007/978-1-4020-9292-3_2, © Springer Science+Business Media B.V. 2009

1.1 Introduction

Our life begins from a single cell that needs to divide into cells that are destined for different fates to form a complex organism. It is almost paradoxical that millions of cells die during development and later in life. Since ancient times it has been known that some structures of multicellular organisms, such as the fetal arterial duct, is committed to disappear, but the first description of cell death was introduced after the establishment of the cell theory by Carl Vogt in 1842 (Clarke et al. 1995). Over the following years, many scientists described naturally occurring cell death, yet most biologists were interested in understanding the life of the cell rather than its death (Clarke et al. 1996). The concept of programmed cell death (PCD) was introduced in 1965 by Lockshin and Williams (Lockshin et al. 1965) as a process that occurs in predictable places and at predictable times during embryogenesis, pointing to the fact that cells are somehow programmed to die during the development of the organism. Later on Kerr and co-authors (Kerr et al. 1972) described the morphological characteristics of cell death during development and tissue homeostasis and coined the term apoptosis (derived from the Greek word meaning "falling off", as of leaves from a tree) to distinguish this type of cell death from necrosis, the death of cells due to physico-chemical insult. Nowadays in mammals it is possible to discriminate about eleven types of cell death (Melino et al. 2005). In this chapter we will briefly discuss several of these death types, with particular focus on type I (apoptotic cell death), type II (autophagic cell death) and type III (necrotic cell death) (Schweichel et al. 1973).

1.2 Apoptosis

Apoptosis (type I cell death) occurs through a sequence of specific morphological changes in the dying cell: condensation of the cytoplasm and margination of the nuclear chromatin into one or several large masses, with subsequent formation of membrane-bound apoptotic bodies, containing a variety of cytoplasmic organelles and nuclear fragments, which are engulfed by neighboring cells and by macrophages (Kerr et al. 1972; Schweichel et al. 1973). Our understanding of the mechanisms involved in apoptosis in mammalian cells was gained from investigations of the programmed cell death that occurs during development of the nematode *C. elegans* (Horvitz 1999). In this organism, 1090 somatic cells are generated during the formation of the adult worm, of which 131 undergo apoptosis or "programmed cell death". These 131 cells die at the same place and time during development, demonstrating the remarkable accuracy of this system. Apoptosis occurs during normal development and ageing as a homeostatic mechanism to maintain cell populations in tissues (Diez-Fraile et al., Chap. 2, this Vol.). The mechanisms of apoptosis are highly complex, involving an energy-dependent cascade of molecular events. To date, research indicates that there are two main pathways to apoptosis. In mammalian cells, the apoptotic response is mediated by either the intrinsic or the extrinsic pathway, depending on the death stimulus. Both pathways finally converge on the proteolytic activation of downstream effector caspases. Apoptotic cells exhibit several biochemical modifications, such as protein cleavage, protein cross-linking, DNA breakdown, and phagocytic recognition and engulfment. All the other chapters in this book are dedicated mainly to discussing the latest information about the phagocytosis of apoptotic cells.

Although the concept of apoptosis has been introduced 30 years ago, the death pathways that regulate apoptosis have remained elusive until the last decades. In the past years it has become clear that a group of cysteinyl aspartate-specific proteases, named caspases, are central regulators and executioners of apoptotic, inflammatory signalling pathways, cell differentiation and proliferation (Lamkanfi et al. 2007). Capases are widely expressed as inactive proenzymes in most cells and once activated they can often activate other procaspases, allowing initiation of a protease cascade. This proteolytic cascade, in which one caspase can activate others, amplifies the apoptotic signal and thus leads to rapid cell death. Human caspases are broadly categorized into initiators (caspase-2, -8, -9, -10), effectors or executioners (caspase-3, -6, -7), and inflammatory caspases (caspase-1, -4, -5; Cohen 1997; Rai et al. 2005). Inflammatory caspases in mouse include caspase-1, -11 and -12 (Lamkanfi et al. 2002). Caspase-11, which is reported to regulate cytokine maturation during septic shock (Kang et al. 2002), caspase-12, which may mediate endoplasmic-specific apoptosis (discussed later in this chapter; Groenendyk et al. 2005; Nakagawa et al. 2000), and involved in modulation of septic shock (Saleh et al. 2004; Lamkanfi et al. 2005). Caspase-14 in mouse and human is specifically expressed in the skin and is involved in proper functioning of the cornified envelop of the skin (Denecker et al. 2007; Denecker et al. 2008).

1.2.1 Extrinsic Pathway of Caspase Activation

The extrinsic signalling pathways that initiate apoptosis involved transmembrane receptor-mediated interactions. It is triggered at the cell surface by the binding of an extracellular death ligand, such as FasL, tumour necrosis factor (TNF), Apo3-ligand, or TRAIL (TNF-related apoptosis inducing ligand/Apo-2 ligand), to its cell-surface death receptor, such as Fas receptor and tumour necrosis factor-receptor 1, death receptor (DR) 3 (APO-3/TRAMP), DR 4 (TRAIL-R1), DR 5 (TRAIL-R2/TRICK2) and DR 6 (Nagata 1999; Sheikh et al. 2000; Chen et al. 2002). Death receptors typically consist of an extracellular region containing varying numbers of cysteine-rich domains required for ligand binding, and an intracellular region with a death domain (DD) motif for homotypic protein-protein interactions. A typical example for the extrinsic pathway is Fas-induced apoptosis (Fig. 1.1). When Fas ligand or agonistic antibodies bind to the homotrimeric Fas receptor, an apoptotic signal is transduced by the death domain through homotypic interactions; these interactions recruit adaptor molecules also containing a DD, such as FADD (Aravind et al. 1999;

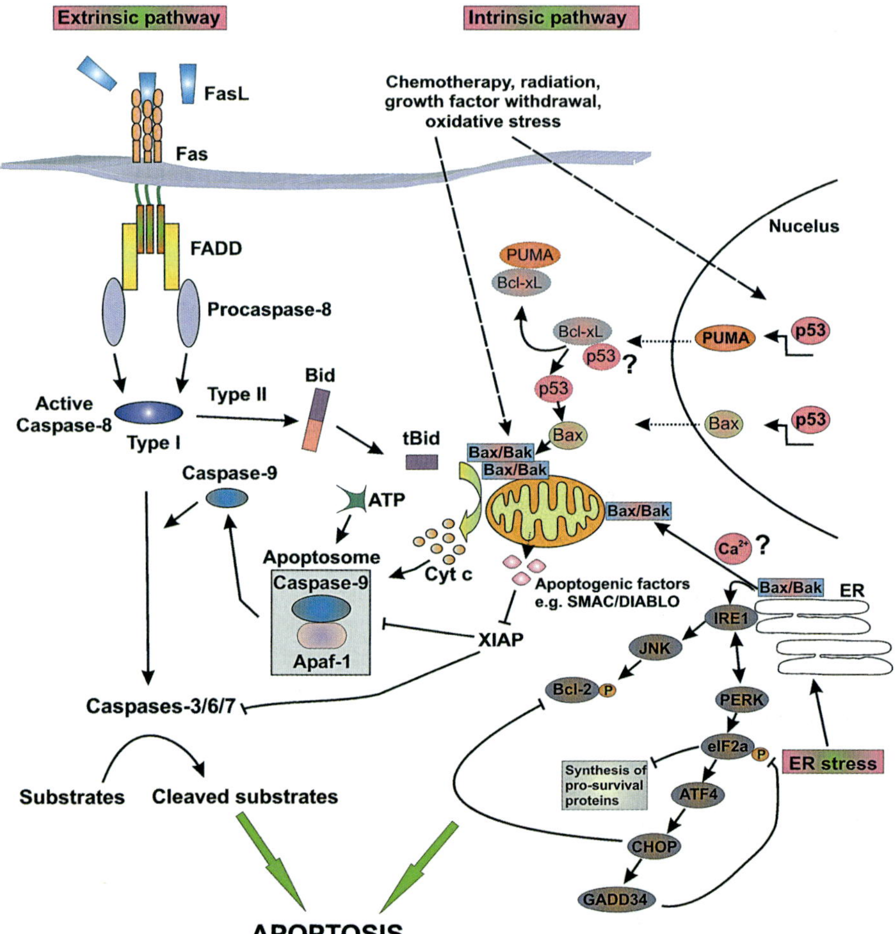

Fig. 1.1 An overview of apoptotic pathways. The extrinsic pathway involves oligomerization of death receptors by their ligands (e.g. FasL), resulting in recruitment and activation of caspase-8. Caspase-8 can execute apoptosis either directly, though cleavage of caspase-3 (type I), or indirectly (type II), by cleaving Bid, which then translocates to mitochondria to initiate the intrinsic pathway. As an example of the extrinsic pathway, Fas-mediated signalling is shown. The intrinsic pathway is activated when BH3-only proteins cause oligomerization of Bax or Bak, which triggers mitochondrial release of apoptogenic factors such as cytochrome c, SMAC/DIABLO, HtrA2/OMI, and EndoG into the cytosol. In the cytosol each of these proteins participate in a different way in the apoptotic process. Cytochrome c in the presence of ATP triggers apoptosome assembly and activation of the caspase cascade. SMAC/DIABLO and HtrA2/OMI contribute mainly to apoptosis by interfering with the caspase-inhibitory function of the IAP. In addition, to initially proposed role of p53 in the nucleus to trasnactivate genes such as puma to induce cell death, it was proposed that p53 activates the expression of PUMA, which then serves to release cytoplasmic p53 from the inhibitory interaction with Bcl-xL (Chipuk et al. 2004). However, the mechanism in which PUMA acts by liberating p53 from Bcl-xL in a way that it can bind and activate Bax on the mitochondria is controversial (Callus et al. 2008). p53 has been shown to activate the transcription of many genes, including *BAX* (Riley et al. 2008). Apoptosis by ER stress can occur through several pathways. Altered calcium homeostasis might contribute to the translocation of the death effectors Bax and

Hofmann 1999). The adaptor molecule FADD also contains a death effector domain (DED) that allows the homotypic recruitment of apoptotic initiator caspases containing DED, such as procaspase-8 in mice and procaspase-8 and -10 in humans (Muzio et al. 1996). These homotypic interactions lead to the formation of an oligomeric death-inducing signalling complex (DISC; Kischkel et al. 1995; Peter et al. 2003). The DISC forms a kind of molecular platform in which the initiator caspase-8 is conformationally activated, leading to autoproteolysis and proteolytic activation of the downstream effector caspases, such as caspase-3 and caspase-7. The extrinsic pathway can cross-talk with the intrinsic pathway through caspase-8-mediated cleavage of Bid (a BH3-only member of the Bcl-2 family proteins; Luo et al. 1998; Yang et al. 1998), which then triggers the release of mitochondrial proteins (Festjens et al. 2004). Two types of Fas receptor-induced apoptotic signalling have been discovered (reviewed in Krysko et al. 2007). Type I cells are characterized by high levels of DISC formation and increased amounts of active caspase-8 (Fig. 1.1). In this case, activated caspase-8 directly leads to the activation of downstream effector caspases without the need for a mitochondrial amplification loop. In type II cells, there are lower levels of DISC formation and, thus, lower levels of active caspase-8 (Scaffidi et al. 1998). In this case, signalling is dependent on the mitochondrial amplification loop, which involves the caspase-8-mediated cleavage of Bid to generate truncated (t) Bid, which induces the release of several mitochondrial factors (e.g. cytochrome c) that activate the mitochondrial apoptotic pathway (Fig. 1.1; Korsmeyer et al. 2000). Type II-induced apoptosis is blocked by Bcl-2 over-expression, whereas type I is not.

1.2.2 Intrinsic Pathway of Caspase Activation

The intrinsic signalling pathways that initiate apoptosis involve a diverse array of non-receptor-mediated stimuli that produce intracellular signals that act directly on targets within the cell and are mitochondrial-initiated events. The stimuli that initiate the intrinsic pathway produce either positive or negative intracellular signals. Negative signals involve the absence of certain growth factors, hormones and cytokines that can lead to failure to suppress death programs, thereby trig-

---

Bak from the ER to the mitochondria (Ron et al. 2007). Inositol-requiring protein-1 (IRE1)-mediated activation of Jun N-terminal kinase (JNK) might contribute to cell death by phosphorylating and inactivating the anti-apoptotic regulator Bcl-2. The pro-death proteins Bax and Bak can assistant in IRE1 activation. Protein kinase RNA (PKR)-like ER kinase (PERK)-mediated phosphorylation of eukaryotic translation factor-2α (eIF2α) can contribute to cell death by inhibiting the synthesis of pro-survival proteins such as the transcription factor CHOP, may repress Bcl-2 expression (Ron et al. 2007). ATF4, activating transcription factor-4; GADD34, growth arrest and DNA-damage-inducible protein-34; DIABLO, direct IAP binding protein with low PI; EndoG, endonuclease G; HtrA, high temperature requirement protein A; IAP, inhibitor of apoptosis proteins; SMAC, second mitochondria-derived activator of caspase; XIAP, X-linked inhibitor of apoptosis protein. This figure is adapted from Vousden (2005); Krysko et al. (2007); Meier et al. (2007); Ron et al. (2007).

gering apoptosis. Positive stimuli include DNA damage induced by irradiation, chemotherapeutics and ER stress. The intrinsic pathway (Fig. 1.1) is mediated by mitochondria (Wang 2001) and it is regulated by the Bcl-2 family of pro- and anti-apoptotic proteins. Bcl-2 was initially described in an acute B-cell leukemia cell line as a gene linked to the immunoglobulin heavy chain locus due to chromosomal translocation (Pegoraro et al. 1984). Pro-apoptotic Bcl-2 proteins function to permeabilize the mitochondrial outer-membrane, which is accompanied by release of several proteins from the mitochondrial intermembrane space into the cytoplasm in response to apoptotic stimuli (Wang 2001; Festjens et al. 2004; Saelens et al. 2004). It has also been shown that besides its role as transcription factor in the nucleus, p53 also possesses an extranuclear function in that it directly binds anti-apoptotic Bcl-2 proteins (Bcl-2 and Bcl-xL) and activates the pro-apoptotic multi-domain Bcl-2 proteins (Bax and Bak) and thus regulates mitochondrial outer-membrane permeabilization (Fig. 1.1; Chipuk et al. 2004; Chipuk et al. 2006). However, this is still a matter of ongoing debate (Callus et al. 2008). Some of the well-characterized proteins released from mitochondria include cytochrome c, SMAC, (second mitochondria-derived activator of caspases/DIABLO, a direct inhibitor of apoptosis [IAP]-binding protein with low pI), AIF (apoptosis-inducing factor), EndoG (endonuclease G) and OMI/HTRA2 (high-temperature-requirement protein A2; van Loo et al. 2002; Vande Walle et al. 2008). Probably the most important of these pro-apoptotic proteins is cytochrome c, which binds to and activates the protein Apaf-1 in the cytoplasm (Li et al. 1997). Cytochrome c, first described in 1930, is imported into the intermembrane space as an apoprotein and then converted to holocytochrome c by addition of a haem group by the enzyme cytochrome c haem lyase (Mayer et al. 1995). In viable cells, it functions in the respiratory chain as an electron carrier between complexes III and IV. However, under apoptotic conditions, membrane integrity is lost and cytochrome c leaks into the cytosol. Interestingly, we observed mitochondrial heterogeneity during apoptosis: in a model of granulosa explants cultured under serum-free conditions, a subset of respiring mitochondria retains cytochrome c function until the stage of chromatin condensation and nuclear fragmentation (D'Herde et al. 2000; Krysko et al. 2001). In agreement with this observation, it has been shown that cytochrome c is never complete released from the mitochondrial fraction of anti-Fas-induced apoptotic L929 cells (Denecker et al. 2001) and from mitochondria treated in vitro with tBid. These data indicate that dysfunction of a fraction of mitochondria can provoke activation of the apoptotic program while the other fraction of mitochondria, with canonically localized cytochrome c, continues cellular respiration and ATP production until the stage of chromatin condensation and fragmentation. However, by fluorescent labeling of cytochrome c with green fluorescent protein in situ (Goldstein et al. 2000) and tetracysteine-containing sequence (Goldstein et al. 2005) it was shown that all mitochondria within a single cell release all their cytochrome c within five minutes in several cell types (NCI-H1299 and HeLa treated either with UV or with actinomycin D and TNF plus cycloheximide). The authors suggested a model of a 'single step' release of cytochrome c. This difference in kinetics of the mitochondrial response to apoptosis induction can depend on whether the mitochondria are organized in a continuous (Amchenkova et al. 1988; Rizzuto et al. 1998; De Giorgi et al. 2000) or discontinuous (Collins et al. 2002; Collins et al. 2003) network in a cell. The different responses of mitochondria of the same cell type could be also related to culture conditions affecting the status of the mitochondrial networks (Egner et al. 2002) or to the strength of the cell death stimulus. Cytochrome c released into the cytoplasm acquires a cell death function because it can bind Apaf-1. This binding induces a conformational change that allows Apaf-1 to bind to ATP/dATP and to form the apoptosome complex (Jiang et al. 2000). The apoptosome functions as a platform for the recruitment of caspase-9 through homotypic CARD-CARD interactions, inducing a conformational change and activation of caspase-9 (Li et al. 1997; Rodriguez et al. 1999; Saleh et al. 1999; Zou et al. 1999). Activated caspase-9 converges again on the proteolytic activation of downstream effector caspases. Autoproteolysis of caspase-9 forms a negative feedback loop resulting in the release of caspase-9 from the apoptosome platform and its inactivation (Twiddy et al. 2004; Twiddy et al. 2006). In addition, Lakhani et al. (2006), using caspase-3 and caspase-7 deficient embryonic fibroblasts, provided evidence that these downstream caspases may amplify Bax translocation to mitochondria as well as cytochrome c release in response to ultraviolet radiation. This indicates that caspases-3 and -7 may participate in a feedback amplification loop to promote release of mitochondrial cytochrome c.

1.2.3 Organelles Other than Mitochondria Involved in Initiation of Apoptosis

Accumulating evidence points to other organelles, including the endoplasmic reticulum (ER), the Golgi apparatus and lysosomes, as major points of integration of pro-apoptotic and anti-apoptotic signalling or cellular damage sensing (Hicks et al. 2005; Malhotra et al. 2007).

1.2.3.1 The Endoplasmic Reticulum

The ER is a multifunctional organelle that has two main functions: it acts as the main cellular Ca^{2+} store and it controls synthesis, folding and post-translational modification of proteins and lipids (Groenendyk et al. 2005). The ER participates in initiating apoptosis by at least two different mechanisms, namely, Ca^{2+} signalling and the unfolded protein response. In support of the first mechanism is that the Bcl-2 family members localize not only in mitochondria but also in the ER. Thus, Bcl-2 proteins interrupt calcium homeostasis and lead to Ca^{2+} release from the ER, modulating cell survival and cell death signals. In addition prolonged ER stress can be responsible for activation of apoptosis (Nakamura et al. 2000; Groenendyk et al. 2005; Bernardi et al. 2007). Identification of caspase-12 on the cytoplasmic side of the ER and demonstration that caspase-12 is processed in cells treated with ER-stress agents favored the idea that it might be the initiator caspase in ER-stress-mediated apoptosis (Nakagawa et al. 2000). However, later on it was discovered that although caspase-12 is processed in ER-stress-mediated cell death in B16/B16 melanoma cells, the cells die to the same extent in the absence of caspase-12 (Kalai et al. 2003). Similarly, it was shown that murine cells lacking caspase-12 expression were not protected from apoptosis induced by ER stress agents (Obeng et al. 2005; Di Sano et al. 2006). Besides, over-expression of the anti-apoptotic Bcl-x$_L$ could provide protection for cells from ER stress-induced death (Obeng et al. 2005). These data suggest that although caspase-12 is processed in apoptosis mediated by ER stress, it may be dispensable for the execution of cell death prompted by ER stress (Lamkanfi et al. 2004). In addition, the existence of truncated human caspase-12 or enzymatically inactive caspase-12 (Lamkanfi et al. 2004) makes it unlikely that it would play a major role in neurodegenerative diseases such as Alzheimer's disease, a role that has been suggested on the basis of the reduced cytotoxicity of the β amyloid peptide in caspase-12-deficient mice (Nakagawa et al. 2000). In this regard, caspase-12 seems to be the cFLIP counterpart for regulating the inflammatory branch of the caspase cascade. In mice, caspase-12 deficiency confers resistance to sepsis and its presence exerts a dominant-negative suppressive effect on caspase-1, resulting in enhanced vulnerability to bacterial infection and septic mortality (Saleh et al. 2006). Several other ER-associated pro-apoptotic molecules have been reported, such as Bap31, a polytopic integral protein of the ER membrane that can bind caspase-8 and Scotin, which are involved in p53-mediated apoptosis (Lamkanfi et al. 2004). Other mechanisms the protein kinase RNA (PKR)-like ER kinase (PERK) / eukaryotic translation initiation factor-2α (eIF2α)—dependent transcription induction of the pro-apoptotic transcription factor CHOP; Bak/Bax-regulated Ca^{2+} release from the ER; inositol requiring protein 1 (IRE1)—mediated activation of apoptosis signal-regulating kinase 1 (ASK1) / Jun N-terminal kinase (JNK; Fig. 1.1; reviewed in detail by Malhotra et al. 2007; Ron et al. 2007).

1.2.3.2 The Golgi Apparatus

The Golgi apparatus consists of a series of parallel cisternae and vesicles that carry molecular "cargo". Its major function is to mediate protein and lipid modification, transport and storage. Recent studies suggest that the Golgi complex can also sense and transduce apoptotic signals. The discovery of a pool of caspase-2 at the cytoplasmic face of the Golgi complex indicates that caspase-2 may play a key role in apoptotic signalling at the Golgi complex. During apoptosis, the Golgi apparatus is disassembled due to caspase-mediated cleavage of golgins. These are proteins that maintain the structural and functional integrity of this organelle. Golgin-160 can be cleaved by caspase-2, -3, and -7. The cleavage of golgin-160 by caspase-2 occurs rapidly and precedes caspase-3 cleavage, indicating an early role in apoptosis for caspase-2 activation at the Golgi complex (Mancini et al. 2000). Furthermore, HeLa cells expressing a caspase-resistant mutant of golgin-160 were resistant to apoptosis induced by ligation of death receptors and by drugs that induce ER stress, but they were sensitive to other pro-apoptotic stimuli, including staurosporine, anisomycin, and etoposide (Maag et al. 2005). Several studies also indicate that another Golgi protein, namely p115, is also cleaved early in the apoptosis (Chiu et al. 2002; Mukherjee et al. 2007). Moreover, because the cytoskeleton maintains organization of the Golgi apparatus, it was also suggested to be involved either in Golgi apparatus disassembly or in disruption of the cytoskeleton, which could lead to onset of apoptosis. Mukherjee et al. (2007) showed that neither actin nor alpha-tubulin was cleaved in Fas-mediated apoptosis, which indicates that Golgi fragmentation precedes breakdown of the cytoskeleton. Taken together, these data indicate that some apoptotic signals may be sensed and integrated at the Golgi membranes. The Golgi complex may provide a link between ligation of death receptors and the ER stress response.

1.2.3.3 The Lysosomes

Lysosomes, which have been given the epithet 'suicide bags' (De Duve, Nobel prize speech, 1974), consist of numerous acid vesicles that contain many catabolic hydrolases received from the Golgi network. Their substrates are from inside the cell (autophagy) as well as from outside (heterophagy). These hydrolases contribute to type II (autophagic) cell death (discussed later). About 15 years ago, lysosomal membrane destabilization was for the first time considered as an early event in apoptosis. However, a recent study confirmed that lysosomal rupture may be an upstream event in some forms of apoptosis (Kurz et al. 2008). Although cathepsins may function at a late stage of apoptosis (Foghsgaard et al. 2001), they were also reported to translocate from lysosomes to the cytosol during early apoptosis, before cytochrome c release and caspase activation. Additionally, cathepsin-mediated activation of Bax and Bak indicates that cathepsins may induce apoptosis by the mitochondrial pathway (Deiss et al. 1996; Ferri et al. 2001; Blomgran et al. 2007). In conclusion, each organelle may possess sensors that detect specific alterations, locally activate signal transduction pathways, and emit signals that ensure inter-organelle cross-talk (Ferri et al. 2001).

Apoptosis is considered as a carefully regulated energy-dependent process, characterized by specific morphological and biochemical features in which caspase activation plays a central role. The importance of understanding the mechanistic machinery of apoptosis is vital because apoptotic cell death is a component of both health and disease, being initiated by various physiological and pathological stimuli. Disturbance of cell death regulation can be an important component of diseases, such as cancer, autoimmune lymphoproliferative syndrome, AIDS, ischemia, and neuron-generative diseases, e.g., Parkinson's disease, Alzheimer's diseases, Huntington's disease, and Amyotrophic Lateral Sclerosis. Some of these examples are discussed in more detail in Diez-Fraile et al. (Chap. 2, this Vol.). Notably, several human diseases can arise when apoptotic cells are not cleared sufficiently by phagocytes. In the second part of this book, the possible contribution of disturbed clearance of apoptotic cells to the pathogenesis of the following human diseases are discussed: systemic lupus erythematosus (Mevorach, Chap. 10, this Vol.), chronic lung diseases (cystic fibrosis, non-CF bronchiectasis, chronic obstructive pulmonary disease) and asthma (Reynolds and Hodge, Chap. 14, this Vol.), atherosclerosis (Martinet et al., Chap. 13, this Vol.), and cancer (Bartunkova and Spisek, Chap. 12, this Vol.). The potential for using apoptotic cells to treat or prevent graft rejection and autoimmune disorders is discussed by Divito and Morelli (Chap. 11, this Vol.).

1.3 Autophagy

In normal cells, two general mechanisms are involved in the degradation and recycling of the building blocks of organelles, proteins and other components of the cytoplasm. In the large-scale degradation of components of the cytoplasm, short-lived regulatory proteins are broken down to amino acids by the ubiquitin-proteasome system, and long-lived structures and proteins are targeted to the lysosome for hydrolysis by autophagy. Autophagy can be described as a process of cell recycling through degradation by lysosomes. It involves intracellular membrane reorganization to create auto-phagosomes, which sequester cytoplasm and organelles. After that they fuse with lysosomes and the cargo is degraded and recycled (Kelekar 2005; Hoyer-Hansen et al. 2008).

Several forms of autophagy have been described (Baehrecke 2005), including microautophagy and macroautophagy (hereafter referred to as autophagy) because of its association with type II cell death. In brief, several steps during autophagy can be distinguished: initiation, cargo packaging, maturation (docking and fusion) and breakdown (Fig. 1.2). First, a phagophore is generated; this is an isolated membrane that sequesters cytoplasm and organelles, or organelle fragments. The phagophore then expands to form an autophagosome, a double-membrane structure. During maturation, auto-phagosomes deliver their cargo to lysosomes by fusing with them to form auto-phagolysosomes. This compartment contains a range of hydrolases that can degrade proteins, lipids, nucleic acids and carbohydrates, which may lead to organelle degradation (Klionsky et al. 2000; Kelekar 2005; Lleo et al. 2007). Several origins have been proposed for the wrapping membrane structure. Among these are the ribosome-free regions of the rough endoplasmic reticulum and the *trans* Golgi network, but it is now believed that the autophagosome is formed mostly *de novo* from core membrane that expands through vesicular addition (Petiot et al. 2002; Klionsky 2007). Since the first description of autophagy (De Duve 1966, Ph.D. thesis), numerous studies have described it as a survival mechanism under poor nutritional conditions or birth-related starvation. Molecular mechanisms (Fig. 1.2) implicated in regulating autophagy in response to starvation were initially discovered in *Saccharomyces cerevisia*. Screens for yeast mutants that were starvation-sensitive or defective in the degradation of specific cytosolic proteins produced different mutants: *Apg* (autophagy-defective), *Aut* (autophagocytosis), and *Cvt* (cytoplasm-to-vacuole targeting) mutants; these partially overlap and are collectively designated as *Atg* genes (Tsukada et al. 1993; Thumm et al. 1994; Harding et al. 1995; Baba et al. 1997).

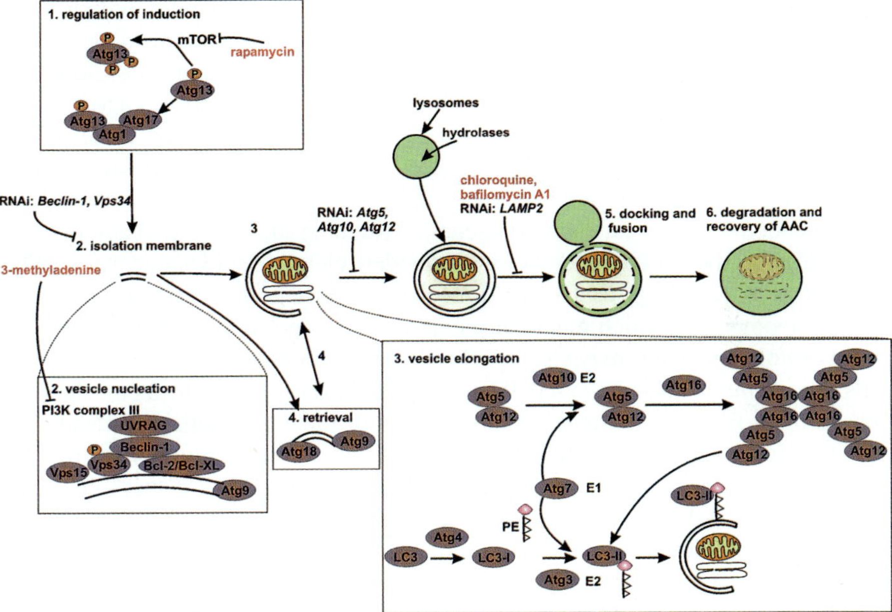

Fig. 1.2 Molecular mechanisms of autophagy. A portion of cytoplasm, including organelles, is enclosed by an expanding membrane sac, the phagophore (also called isolation membrane). Step first involves the de-repression of the mTOR Ser/Thr kinase, which inhibits autophagy by phosphorylating Atg13. This leads to the dissociation of Atg13 from a protein complex that contains Atg1 kinase and Atg17, and thus attenuates the Atg1 kinase activity. When mTOR is inhibited, re-association of de-phosphorylated Atg13 with Atg1 stimulates its catalytic activity and induces autophagy. Step two is vesicle nucleation starts from the activation of mammalian Vps34, a class III phosphatidylinositol 3-kinase (PI3K), to generate phosphatidylinositol-3-phosphate. Vps32 activation depends on the formation of a multiprotein complex in which Beclin-1, UVRAG and a myristylated kinase (Vps15) participate. The third step is auto-phagosomes formation (vesicle elongation) and involves two pathways: conjugation of Atg12 to Atg5, with the help of the E1-like enzyme Atg7 and the E2-like enzyme Atg10; conjugation of phosphatidylethanolamine (PE) to LC3 by the sequential action of the protease Atg4, the E1-like enzyme Atg7 and the E2-like enzyme Atg3. This leads to the conversion of the soluble form of LC3-I to the autophagic-vesicle-associated form LC3-II. The outer membrane of the auto-phagosomes subsequently fuses with a lysosome (forming autolysosomes), exposing the inner single membrane of the autophagosome to lysosomal hydrolases (step 5, docking and fusion). Finally the cargo-containing membrane compartment is lysed, and the contents are degraded or aminoacids (ACC) recovered and used for nutrients (step 6, degradation and recover of AAC). Chemical compounds are shown in red. E1, ubiquitin-activating enzyme; E2, ubiquitin-conjugating enzyme. This figure is adapted from Maiuri et al. (2007); Levine et al. (2008).

1.3.1 Autophagy as a Strategy for Cell Survival

The pro-survival function of autophagy is an ancient process conserved from yeast to mammals. In response to starvation, *C. elegans* larvae enter dauer, a latent developmental state, and RNAi-inactivation of autophagy genes (*bec-1*, *atg8* and *atg18*) disrupts normal dauer formation (Melendez et al. 2003). A recent study demonstrated that autophagic degradation and recovery of biomolecules within early embryos is essential for preimplantation in mammals (Tsukamoto et al. 2008) pointing to autophagy as a unique strategy for supporting mammalian development. These results are not inconsistent with the finding that conventional *Atg5*$^{-/-}$ mice survive preimplantation development (Kuma et al. 2004). However, these *Atg5*$^{-/-}$ mice were generated by mating *Atg5*$^{+/-}$ mice, so maternally inherited Atg5 protein in the cytoplasm of Atg5-null oocytes might have rescued the autophagy-deficient phenotype during early embryogenesis (Kuma et al. 2004; Tsukamoto et al. 2008). Rather, these two different mouse models clearly demonstrate the specific importance of autophagy as a survival mechanism during very early post-fertilization development, in which maternally inherited Atg5 protein remains in the cytoplasm. Autophagy genes may also be critical for maintaining cellular survival when cells are unable to take up external nutrients, e.g., during growth factor deprivation. Growth factor withdrawal usually results in rapid apoptotic cell death, but recent studies in apoptotic-deficient *bax*$^{-/-}$, *bak*$^{-/-}$ cells raveled an important role for autophagy genes in maintaining cellular survival following IL-3 deprivation (Lum et al. 2005). IL-3-deprived *bax*$^{-/-}$, *bak*$^{-/-}$ cells activate autophagy and ultimately die. Notably, the addition of growth factor anytime before death reverses the catabolic process and keeps the cells alive (Lum et al. 2005). Furthermore, RNAi against *beclin1*, *atg5*, *atg10*, and *atg12* enhances starvation-induced, but not staurosporine-induced, apoptotic cell death. This finding illustrates an interesting mechanism by which autophagy genes promote survival during nutrient deprivation, namely, suppression of the apoptotic death pathway (Boya et al. 2005). Autophagy occurs in almost all cells to help maintain homeostasis. It is also activated when cells need to generate energy during starvation or growth factor deficiency (Levine et al. 2008). Together, these data suggest that autophagy is a critical survival mechanism to overcome stress conditions.

1.3.2 Cell Death Associated with Autophagy

Paradoxically, the presence of autophagic structures in dying cells may also indicate the existence of type II autophagic cell death (Schweichel et al. 1973). Type II or autophagic cell death is a type of cell death accompanied by extensive autophagic vacuolization of the cytoplasm (Okada et al. 2004; Kroemer et al. 2005). In autophagic cell death, partial chromatin condensation occurs late if at all, and no DNA laddering is observed (Kroemer et al. 2005; Levine et al. 2005). It is important to emphasize that an optimal identification of autophagic cell death would require the use of transmission electron microscopy to distinguish auto-phagosomes from other types of vesicles such as endosomes, lysosomes and macropinosomes. Cell death associated with autophagy in most instances was found during the destruction of organs and large cell units in mouse and rat embryos and fetuses (Schweichel et al. 1973). Another example may be insect metamorphosis, in which cell death is typically autophagic, and blocking autophagy before pupation and die during metamorphosis (Juhasz et al. 2003). Autophagic cell death also has been observed in adult insects and vertebrates, including humans; it is often associated with the elimination of large secretory cells during the adjustment of sexual organs and ancillary tissues to seasonal reproduction (Bursch et al. 2000).

However, it is important to stress that whether type II cell death is a separate form of cell death remains unclear (Lockshin et al. 2004). Autophagosome formation during cell death does not necessarily means autophagic cell death. Pure autophagic cell death should be defined by the presence of auto-phagosomes simultaneously with the absence of apoptotic and necrotic hallmarks. Moreover, induction of autophagy (by starvation or rapamycin) should result in an increase in cell death and suppression of autophagy should lead to cell survival (White 2008). Yet, the literature contains strong evidence to support the notion that autophagic cell death may overlap either with caspase-dependent or caspase-independent cell death. Studies of steroid-triggered death of salivary gland cells during Drosophila development showed that caspases function in autophagic death of cells. Caspase-3 activation and *atg* gene transcription immediately precede autophagic cell death in salivary glands (Lee et al. 2001; Lee et al. 2003; Martin et al. 2004). Similarly, it has been shown that the pro-apoptotic signalling molecule, TNF-related apoptosis-inducing ligand (TRAIL), regulates autophagy in an in vitro model of mammary gland formation (Mills et al. 2004). In that study, the increased level of TRAIL was detected during morphogenesis of MCF-10A mammary epithelial cells in 3D basement-membrane cultures just before the death of luminal cells expressing active caspase-3, whereas these cells possess autophagic cell death morphology. Additionally, when lysosome-associated membrane protein-2 (LAMP2) was inactivated by RNAi or by homologous recombination in cells cultured in nutrient-deficient conditions, the cells accumulated autophagic vacuoles and then died with hallmarks of apoptosis, namely, caspase activation and chromatin condensation (Gonzalez-Polo et al. 2005). Other molecular links between autophagy and apoptosis may be the death-associated protein kinase (DAPk) and DAPk-related protein 1, which regulate membrane blebbing during apoptosis, but can also promote autophagic vacuole formation in dying cells (Inbal et al. 2002). It has been shown that Bcl-2 binds to Beclin 1 and disrupts its autophagic function, and that in the absence of Bcl-2 binding, *Beclin 1* mutants induce excessive autophagy and promote cell death (Pattingre et al. 2005). This study demonstrates the Beclin 1-Bcl-2 complex may function as a rheostat that ensures that autophagy levels remain within a physiological range rather than in a non-physiological range that triggers cell death (see Fig. 1.2 for an overview of molecular pathways). In addition, it has been demonstrated that Bcl-2 phosphorylation by the stress-activated signalling molecule, JNK1, is required for its starvation-induced dissociation from Beclin 1 and subsequent activation of autophagy (Wei et al. 2008). Taken together, these studies indicate that complex interrelationships may exist between autophagy and apoptotic cell death pathways.

In spite of the accumulating evidence for overlap of autophagic cell death with apoptosis, autophagy may also be related to caspase-independent cell death. It was shown that cell death induced by the human homologue of the Drosophila spin gene product (HSpin1; Gonzalez-Polo et al. 2005) is caspase-independent and autophagic (Yanagisawa et al. 2003). In support of this idea, RNAi directed against two autophagy genes, *atg7* and *beclin1*, severely affected cell death in mouse L929 cells treated with the caspase inhibitor zVAD-fmk (Yu et al. 2004). In this model, a new molecular pathway was defined: activation of Rip1 (a serine-threonine kinase) and Jun amino-terminal kinase induced cell death with the morphology of autophagy. Inhibition of caspase-8 induces cell death with autophagic morphology (Yu et al. 2004). Similarly, the autophagic cell death of $Bax^{-/-}Bak^{-/-}$ mouse embryonic fibroblasts treated with either etoposide or staurosporine was prevented by the knock-down of *Beclin1* and *Atg5* (Shimizu et al. 2004), suggesting that autophagic cell death occurs in a caspase-independent manner. In addition, these data might be interpreted to mean that either apoptosis effectors, such as caspases, Bax and Bak, actively suppress the autophagic components of cell death, or that the autophagic and apoptotic effector mechanisms constitute backup mechanisms that come into action when one or the other lethal pathway is inhibited (reviewed in Krysko et al. 2007).

It is now clear that autophagy has multiple roles. On one hand, autophagy is a mechanism for degrading long-lived proteins and damaged organelles through the auto-phagolysosomal pathway. On the other hand, it serves as a survival pathway preventing or delaying cell death. This cell survival mechanism is activated in cells undergoing different forms of cellular stress. On the other hand, when stress continues, autophagy may eventually become a cell death mechanism. Despite recent advances in the understanding of molecular mechanisms and biological functions of autophagic cell death, it is not clear whether it is an alternative way of dying that differs from apoptotic and necrotic cell death, or whether failure of autophagy to rescue the cell can lead to cell death by either pathway. It is also conceivable that the extent of autophagy determines the decision to survive or to die.

1.3.3 Autophagy and Clearance of Dying Cells

It was recently shown that cells dying by autophagy or by another mechanism associated with autophagy are engulfed by professional and non-professional phagocytes. Phosphatidylserine (PS) exposure mediates recognition and engulfment of cells dying by autophagy (Petrovski et al. 2007a). Phagocytic uptake of dying autophagic cells by macrophages leads to a pro-inflammatory response characterized by the production of IL-6, TNF and IL-8 (Petrovski et al. 2007b). However, there is only limited knowledge of the molecular mechanisms of recognition and phagocytic uptake of cells dying in association with autophagy and of its functional consequences; additional in depth studies are required.

Several studies indicate that autophagy may be involved in optimal clearance of apoptotic cells. Recently, data was obtained on embryonic cavitation, which is the earliest programmed cell death process in mammalian development. It was shown that inner ectodermal cells in embryonic bodies lacking the essential autophagic gene *Beclin1* or *Atg5* underwent apoptosis normally but were not engulfed by neighboring cells, because they fail to express PS and because they secrete lower levels of lysophosphatidylcholine (LPC; Qu et al. 2007). However, it is important to note that this phenomenon was observed for Beclin1$^{-/-}$ embryos in vitro and in vivo, while for Atg5$^{-/-}$ embryos only in vitro data were obtained. Further studies will resolve this contradiction. Another report showed that treatment of retinas with the autophagic inhibitor 3-methyladenine resulted in accumulation of numerous apoptotic cells (Mellen et al. 2008). This was accompanied by the absence of the 'eat me' PS signal and by deficiency of subsequent clearance of apoptotically dying cells. These studies demonstrated that autophagy, beyond its survival and cell death function, might also be essential in the exposure of 'eat-me' signals on apoptotic cells, which indicates that autophagosome formation during apoptosis may be one of the mechanisms used to expose phagocytosis signals on the surface of dying cells.

1.3.4 Role of Autophagy in Human Diseases

Autophagy is a highly conserved process enabling clearance of excess or aberrant organelles and long-lived proteins (discussed above). This process also plays important roles in embryogenesis, tissue remodelling, degradation of intracellular pathogens (Yorimitsu et al. 2005), and adaptation to changing environment conditions. Importantly, autophagy is also involved in both preventing and contributing to some types of human diseases (Yorimitsu et al. 2005; Levine et al. 2008). Several human pathologies are associated with increased auto-phagosomes formation, such as neurodegenerative diseases (Alzheimer's disease, Parkinson's disease, transmissible spongiform encephalopathies, and Huntington's disease). Recent studies indicate that activation of autophagy and accumulation of auto-phagosomes in patients' brains play a protective role because they enable clearance of aggregates of mutant proteins that are characteristic of neurodegenerative diseases (Kelekar 2005; Levine et al. 2008). For example, in Huntington's disease is found a mutated huntingtin protein (mHtt) with an abnormal number of glutamine residues (polyQ), mutant alpha-synucleins were observed in Parkinson's disease, and different mutations of tau proteins were found in temporal dementia (Williams et al. 2006). These potentially toxic oligomeric and aggregated proteins are inaccessible to proteasomes and so they are degraded through autophagy (Levine et al. 2008). Similarly, extensive accumulation of auto-phagosomes in the muscle and leading to cardiomyopathy is characteristic of a Danon disease. In this case, myopathy is a result of the lysosomes' inability to fuse with auto-phagosomes due to a mutation in LAMP-2 protein, which is involved in maturation of auto-phagosomes. Other myopathies associated with autophagic vacuole accumulation are X-linked myopathy and inclusion body myositis (Kelekar 2005).

In several experimental models it has been shown that autophagy could function as a back up mechanism when cells fail to execute apoptosis (Yu et al. 2004; Levine et al. 2005; Lum et al. 2005; Lefranc et al. 2007). It is well known that cancer cells are characterized by deregulated cell proliferation and suppression of apoptosis. Autophagy activation can have contradictory consequences for tumour development. On one hand, recent studies indicate that autophagy is important for tumour progression (Hippert et al. 2006). Since autophagy is known to be a survival mechanism that prevents cells from dying in stressful conditions (discussed above) it could also serve as a survival mechanism in solid tumours, where limited angiogenesis leads to nutrient depravation and reduced oxygen levels (Hippert et al. 2006; Hoyer-Hansen et al. 2008). On the other hand, it has been proposed that autophagy has an anti-cancer role as well. Strong evidence connecting autophagy with cancer appeared in the late 1990s, when it was discovered that the ATG gene *Beclin-1* may serve as a tumour suppressor (Qu et al. 2003; Klionsky 2007). It was proposed that autophagy can inhibit tumour development in the early stage. Later, other potential tumour suppressor genes were identified, including UV irradiation resistance-associated gene *(UVRAG)*, *Atg4C*, and damage-regulated autophagy modulator *(DRAM*; Hoyer-Hansen et al. 2008). Interestingly, p53 positively regulates autophagy, thereby corroborating the notion of tumour suppression by autophagy. In this regard, genotoxic stress caused by DNA-damaging agents induces p53-dependent autophagy (Feng et al. 2005; Zeng et al. 2007). Likewise, oncogenic activation, stimulated by forced expression of ARF or p53, induces autophagy in human cancer cells (Abida et al. 2008). Surprisingly, inhibition of p53 with pfithrin-α, knockdown of *p53* with siRNA, and genetic deletion of *p53* increases autophagy, whereas basal levels of p53 apparently inhibit autophagy (Tasdemir et al. 2008). The study by Tasdemir et al. (2008) revealed further complexity in the homeostatic regulation by showing that p53 and autophagy are interconnected in a hitherto unexpected bi-directional fashion. Although how autophagy inhibits tumour progression is still unclear, several hypotheses have been proposed to explain how this pro-survival mechanism can contribute to inhibition of cancer development. One hypothesis is that autophagy may suppress tumourigenesis by reducing necrotic cell lysis and therefore limiting tumour inflammation (Hoyer-Hansen et al. 2008). Another hypothesis is that autophagy suppresses tumour development by removing damaged mitochondria, thus reducing production of ROS, which is a potential source of DNA damage (Levine 2006; Mathew et al. 2007). Yet another hypothesis is that autophagy may influence cell cycle by degrading proteins responsible for regulation of cell growth and thereby leading to slower proliferation of tumour cell lines (Koneri et al. 2007; Levine et al. 2008). Taken together, these data indicate that autophagy may both stimulate and inhibit cancer, depending on the context.

The occurrence of autophagy in cancer cells may provide the grounds for using inhibitors or inducers of autophagy in the treatment of cancer. Autophagy can be pharmacologically inhibited by bafilomycin A1 (which interferes with phagosome-lysosome fusion) or by targeting 3-methyladenine to class III PI3K, which is involved in autophagosome formation (Levine et al. 2008). However, the autophagy inhibitors that are clinically relevant are rather few at the moment. These include microtubule-disrupting agents (vincristine and paclitaxel), which are used for treatment of various forms of cancer (Groth-Pedersen et al. 2007), and chloroquine, which is in a pre-clinical trial as a sensitizer to radiotherapy and chemotherapy (Amaravadi et al. 2007; Hoyer-Hansen et al. 2008). Inducers of autophagy include rapamycin, xestospongin B and lithium (Levine et al. 2008). Importantly, several trials have been started to examine the potential usefulness of mTOR inhibitors (Fig. 1.2; rampamycin, CCl-779), EB1089 (targets a vitamin D receptor) and angiogenesis inhibitors (anti-VEGF anibody, ADH-1, Kringle5) in treatment of numerous cancers, such as breast, prostate, brain, lung, ovary and cervix carcinomas (Noda et al. 1998; Ambalavanan et al. 2005; Nguyen et al. 2007). Moreover, combining autophagy inhibitors with other metabolic inhibitors may result in a synergic effect. Cancer cell death was reported when the lysosome-damaging drug siramesine was combined with blockade of autophagy by removal of Atg proteins or treatment with microtubule destabilizing drugs (vincristine, taxotere or 2-methoxyestradiol) (Escuin et al. 2005). On the other hand, some alkylating chemo-therapeutic agents (actinomycin D), gene therapy (p53), cytokines and radiation have been shown to induce cell death type II in various cancer cell lines (Lefranc et al. 2007; Hoyer-Hansen et al. 2008), but their potential effect still needs to be verified in vivo.

In conclusion, targeting autophagy may represent an alternative way for treatment of different diseases, including tumours resistant to radiotherapy and pro-apoptotic chemotherapy. Yet, the uses of pro- or anti-autophagic treatment still needs to be validated in vivo. Understanding autophagy may ultimately allow scientists and clinicians to harness this process for the purpose of improving human health.

1.4 Necrosis

Type III cell death, according to the classification of Schweichel and Merker, is characterized by swelling of the endoplasmic reticulum, mitochondria, and the cytoplasm, with subsequent collapse of the plasma membrane and lysis of the cells (Schweichel et al. 1973). The term oncosis was proposed to refer to the early stage of primary necrosis, during which cells committed to death pass through a pre-lethal process in which they swell (Majno et al. 1995). Notably, in the absence of phagocytosis, apoptotic cells may lose their cytoplasmic membrane integrity and proceed to a stage called secondary necrosis. Secondary necrotic cells resemble necrotic cells, but they have already gone through an apoptotic stage that can still be recognized at the ultrastructural level by the typical chromatin condensation pattern (reviewed in Krysko et al. 2007). It is important to distinguish necrosis from other forms of cell death, particularly because it is often associated with unwarranted loss of cells in human pathologies. However, there is no clear biochemical definition of necrotic cell death and consequently no positive biochemical markers that unambiguously discriminate necrosis from apoptosis. Therefore, we propose that the definition of cell death types should be based on three different criteria: morphological features, biochemical features, and interactions with phagocytes (Krysko et al. 2008).

Necrotic cell death was often considered to be a passive process, lacking underlying signalling events and occurring under extreme physico-chemical conditions, such as abrupt anoxia, sudden shortage of nutrients, heat and detergents. However, in disagreement with the classical textbook notion that it is merely an accidental consequence of non-physiological stress, necrosis might actually be programmed in terms of both its course and its occurrence. This idea is supported by several data. In several physiological and pathological conditions, necrosis is described as a form of caspase-independent cell death. For example, necrosis has been found to be a potential substitute for apoptosis during development. The loss of interdigital cells in the mouse embryo, a prototype of programmed cell death, still occurs by necrosis either upon inhibition of caspases by drugs, or in mice bearing a mutation in the *apaf-1* gene (Chautan et al. 1999). Caspase-independent cell death is also involved in processes such as the negative selection of lymphocytes (Smith et al. 1996; Doerfler et al. 2000; Jaattela et al. 2003), TNF-mediated liver injury (Kunstle et al. 1999) and the death of chondrocytes controlling the longitudinal growth of bones (Roach et al. 2000). Necrotic cell death is also instrumental during cellular turnover in the human large intestine (Barkla et al. 1999) and in response to viral infection (Chan et al. 2003). Susceptibility to necrotic death can be regulated by genetic and epigenetic factors. It has been shown that mouse strains vary in their susceptibility to brain ischemia, which was also dependent on the age of the mice and the brain regions (Mattson et al. 2006). In addition, necrotic cell death is involved in the pathogenesis of several human pathologies, such as neurodegeneration, ischemia-reperfusion and infection (reviewed in Vanlangenakker et al. 2008).

Next, we would like to consider some biochemical mechanisms implicated in programmed necrotic cell death. Depending on the cell line used, TNF can induce either apoptotic or necrotic cell death (Laster et al. 1988). In fibrosarcoma L929 cell line, TNF induces necrotic cell death without the involvement of caspases, even though stimulation of Fas leads to classical apoptosis in the same cells (Vercammen et al. 1998b). The necrotic pathway triggered in the L929 cell line by FasL can be detected only before the inactivation of the caspases by zVAD-fmk (Vercammen et al. 1998a). Thus, depending on the cellular context, both death domain-containing receptors (TNFR1 and Fas) can initiate apoptotic or necrotic cell death. This death receptor-induced caspase-independent pathway is not restricted to the L929 cell line. Similarly, in the U937 or Jurkat T lymphocyte cell lines, TNF, Fas and TRAIL trigger necrotic cell death when caspases are inhibited (Khwaja et al. 1999; Matsumura et al. 2000; Holler et al. 2000). Besides that, TNF in the presence of caspase inhibitors can induce caspase-independent cell death in murine embryonic fibroblasts (MEFs; Bernardi et al. 2006). FADD is an important adaptor molecule serving as a platform for the initiation of apoptotic as well as necrotic cell death. Over-expression of a FADD containing only DD (FADD-DD) leads to necrosis, while over-expression of FADD containing only DED (FADD-DED) kills the L929 cells by apoptosis; also, cell death can be redirected from apoptosis to necrosis in the presence of zVAD-fmk (Boone et al. 2000; Vanden Berghe et al. 2004). These

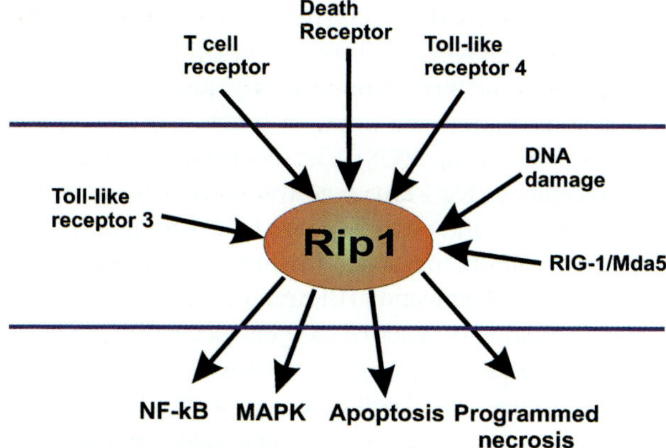

Fig. 1.3 Rip1 is a bifurcation point between cell death and cell live. Several signalling pathways converge on Rip1 such as activation of the T cell receptors, death receptors and TLR-3 and -4, and signalling pathways initiated upon detection of intracellular stress (dsRNA or DNA damage). The activation of NF-κB, MAPKs, apoptosis or necrosis is dependent on the cellular context. This figure is adapted from Festjens et al. (2007). dsRNA, double stranded RNA; MAPK, mitogen activated protein kinase; Rig1, retinoic acid inducible gene-1; Mda5, melanoma differentiation-associated gene 5.

data suggest that a bifurcation between apoptosis and necrosis might be situated at the level of the adaptor protein FADD. Binding of caspase-8 to DED of FADD leads to its activation and subsequent apoptosis, but if the recruitment or the enzymatic activity of caspase-8 is prevented, the presence of death domain of FADD in the receptor complex leads to necrosis (Boone et al. 2000; Vanden Berghe et al. 2004). Another molecule involved in the necrotic signalling cascade is Rip1 (Fig. 1.3). A study employing FADD- and Rip1-deficient Jurkat cells identified the kinase Rip1 as a crucial component of Fas- and TNF-mediated necrotic cell death (Holler et al. 2000). TNF signal is transduced by homotypic interaction of DD with RIP1, which allows Rip1 to bind directly to the TNFR1 or indirectly through TRADD (Stanger et al. 1995; Harper et al. 2003). It is interesting to note that Rip1, a key mediator in the necrotic cell death, can be cleaved by activated caspase-8 (Lin et al. 1999; Martinon et al. 2000), indicating that apoptotic and necrotic cell death pathways interfere with each other, and that the initiation of apoptosis might actively suppress the necrotic pathway. In addition, studies on the heat shock protein (Hsp) 90 (Kalai et al. 2002) have also revealed the importance of Rip1 in necrotic signalling. Necrosis induced by Fas or TNFR1 is inhibited by the Hsp90 inhibitors geldanamycin and radicicol, which are responsible for a strong down-regulation of Rip1 levels (Lesnefsky et al. 2001; Ma et al. 2005).

Degterev et al. (2005) characterized "necrostatins" as the first class inhibitors of necrotic cell death. To distinguish programmed necrosis that occurs in the absence of caspase activation from accidental necrosis, these authors named it "necroptosis". Later on the same group demonstrated that necrostatins inhibit Rip1 kinase activity, the key upstream kinase involved in the activation of programmed necrosis (Degterev et al. 2008). This discovery has a great potential in the treatment of human pathologies, such as myocardial infarction and cerebral ischemia, in which cell death exhibits necrotic features. Consequently, necrostatins have recently been shown to reduce histopathology and improve functional outcome after controlled cortical impact in mice (You et al. 2008). The mechanism that leads to the execution of necroptosis downstream of Rip1 kinase activation remains unclear and the subject of active research, especially the possible role of Rip1 ubiquitination and deubiquitination in this respect could be influence the biological outcome of Rip1 between apoptosis, necrosis or NF-κB activation (Bertrand et al. 2008). Ample evidence suggests that excessive formation of reactive oxygen species (ROS) is involved. Inhibition of caspases (which sensitizes to necrosis) results in increased formation of ROS. The addition of the scavenger butylated hydroxyanisole (BHA) protects L929 cells from TNF-induced necrosis, but not from Fas-mediated apoptosis (Vercammen et al. 1998b; Goossens et al. 1999), while necrosis induced by Fas is also blocked by BHA (Vercammen et al. 1998a). The strong protective effect of BHA may go beyond its oxygen radical scavenging activity. Festjens et al. (2006a) reported that BHA may directly inhibit several mechanisms that are implicated in necrotic cell death, such as complex I activity, PLA2 activation and lipoxygenase activity.

A connection between Rip1 and a non-mitochondrial source of ROS is the plasma membrane-associated NADPH oxidase-1 (Nox1). Knockdown of Nox1, which is responsible for TNF-induced generation of superoxide anions, delays necrosis in mouse fibroblasts. In these cells, TNF treatment induces Nox1 activation through a Rip1-dependent signalling complex containing TRADD, NOXO1 and the small GTPase Rac1 (Kim et al. 2007; Vanden Berghe et al. 2007).

Besides death receptor-induced necrosis, triggering pathogen recognition receptors (PRRs) such as Toll-like receptors (TLRs), the cytosolic NOD-like receptors (NLRs), and RIG-I-like receptors (RLRs) leads to initiation of inflammation and/or cell death (Fig. 1.3; Vanlangenakker et al. 2008). Synthetic dsRNA, via TLR3, induces necrotic cell death in human Jurkat cells and murine L929 fibrosarcoma cells in a caspase-8 and FADD-independent manner, and type I and II-interferons (IFNs) can sensitize for this necrosis (Dennis et al. 1991). Also, the gram-negative bacterium *Shigella flexneri* induces necrosis in neutrophils by activating the type III secretion systems, specific toxins, and actin polymerization (Francois et al. 2000). This bacterium causes mitochondrial damage with subsequent necrosis in human monocytes-derived macrophages (Koterski et al. 2005). Monocytes from patients carrying disease-associated mutations in the NLR member, cryopirin, exhibit excessive necrotic cell death. *Shigella flexneri* infection also causes cryopyrin-dependent macrophage necrosis with features similar to the death caused by mutant cryopirin. This necrotic death is independent of caspase-1 and IL-1β, and thus independent of the inflammasome (Willingham et al. 2007).

The picture described above indicates a complexity of death receptor-induced necrotic signalling networks that far exceeds that of the simple linear pathways originally indicated by the discovery of the receptor-triggered caspase cascade. This picture also supports the notion that, besides uncontrollable necrosis that may occur following mechanical damage or harsh chemical treatment of tissues, programmed necrosis may occur with certain molecular events (reviewed in detail Festjens et al. 2006b; Vanlangenakker et al. 2008).

The final fate of necrotic cells as well as apoptotic cells is engulfment by professional or non-professional phagocytes. In the following chapters several authors discuss the molecular mechanisms of necrotic cells engulfment: immunological factors released by necrotic cells (Peter et al., Chap. 3, this Vol.), surface molecules involved in engulfment (Napirei and Mannherz, Chap. 4, this Vol.), and the immunological consequences upon clearance of necrotic cells (Ucker, Chap. 6; Lacy-Hulbert, Chap. 7, this Vol.).

1.5 Conclusions

There is mounting interest in apoptotic, necrotic and autophagic cell death types, but a precise description of the morphological and biochemical events in these processes has not yet been derived, and the importance of this cell death type in embryology, ontogeny, physiology and pathophysiology is not yet known. As discussed above, accumulating evidence supports the concept that necrotic cell death is programmed. However, there is no clear biochemical definition of necrotic cell death and consequently no positive biochemical markers that unambiguously discriminate necrosis from apoptosis (Krysko et al. 2008). Unraveling the molecular players and defining the biochemical pathways in necrosis will provide us with powerful and specific methods for identifying necrotic cell and help us to distinguish it positively from other forms of cell death. Future studies aimed at understanding the molecular mechanisms of necrotic cell death will also create many opportunities for development of new therapeutic strategies to modulate necrotic cell death. In this regard, it has been shown that necrostatins, an inhibitor of necrotic cell death, prevent tissue damage in mouse models of cerebral ischemia and myocardial infarction (Degterev et al. 2005; Lim et al. 2007; Smith et al. 2007). We have also discussed the paradoxical roles of autophagy in cell survival and cell death and emphasized that it is not yet clear whether autophagy represents a separate type of cell death or whether it eventually results in apoptotic and necrotic cell death. Future investigations of the molecular mechanisms of all cell death types will yield better insight into the evolution of cell-death programs, their inter-relationships, and their potential for molecular targeting and manipulation in many diseases.

Acknowledgments   We thank Dr. A. Bredan for editing the manuscript. Dr. Dmitri V. Krysko is a postdoctoral fellow and Agnieszka Kaczmarek is a doctoral fellow, both paid by fellowships from FWO (Fonds Wetenschappelijk Onderzoek – Vlaanderen). This work has been supported by Flanders Institute for Biotechnology (VIB) and several Grants from the European Union (EC Marie Curie Training and Mobility Program, FP6, ApopTrain, MRTN-CT-035624; EC RTD Integrated Project, FP6, Epistem, LSHB-CT-2005-019067, APO-SYS, FP7, HEALTH-F4-2007-200767), the Interuniversity Poles of Attraction-Belgian Science Policy (IAP6/18), the Fonds voor Wetenschappelijk Onderzoek—Vlaanderen (3G.0218.06), and the Special Research Fund of Ghent University (Geconcerteerde Onderzoekstacties 12.0505.02).

References

Abida WM, Gu W (2008) p53-Dependent and p53-independent activation of autophagy by ARF. Cancer Res 68:352–357

Amaravadi RK, Yu D, Lum JJ et al (2007) Autophagy inhibition enhances therapy-induced apoptosis in a Myc-induced model of lymphoma. J Clin Invest 117:326–336

Ambalavanan N, Tyson JE, Kennedy KA et al (2005) Vitamin A supplementation for extremely low birth weight infants: outcome at 18 to 22 months. Pediatrics 115:E249–E254

Amchenkova AA, Bakeeva LE, Chentsov YS et al (1988) Coupling membranes as energy-transmitting cables. I. Filamentous mitochondria in fibroblasts and mitochondrial clusters in cardiomyocytes. J Cell Biol 107:481–495

Aravind L, Dixit VM, Koonin EV (1999) The domains of death: evolution of the apoptosis machinery. Trends Biochem Sci 24:47–53

Baba M, Osumi M, Scott SV et al (1997) Two distinct pathways for targeting proteins from the cytoplasm to the vacuole/lysosome. J Cell Biol 139:1687–1695

Baehrecke EH (2005) Autophagy: dual roles in life and death? Nat Rev Mol Cell Biol 6:505–510

Barkla DH, Gibson PR (1999) The fate of epithelial cells in the human large intestine. Pathology 31:230–238

Bernardi P, Krauskopf A, Basso E et al (2006) The mitochondrial permeability transition from in vitro artifact to disease target. FEBS J 273:2077–2099

Bernardi P, Rasola A (2007) Calcium and cell death: the mitochondrial connection. Subcell Biochem 45:481–506

Bertrand MJ, Milutinovic S, Dickson KM et al (2008) cIAP1 and cIAP2 facilitate cancer cell survival by functioning as E3 ligases that promote RIP1 ubiquitination. Mol Cell 30:689–700

Blomgran R, Zheng L, Stendahl O (2007) Cathepsin-cleaved Bid promotes apoptosis in human neutrophils via oxidative stress-induced lysosomal membrane permeabilization. J Leukocyte Biol 81:1213–1223

Boone E, Vanden Berghe T, Van Loo G et al (2000) Structure/Function analysis of p55 tumour necrosis factor receptor and fas-associated death domain. Effect on necrosis in L929sA cells. J Biol Chem 275:37596–37603

Boya P, Gonzalez-Polo RA, Casares N et al (2005) Inhibition of macroautophagy triggers apoptosis. Mol Cell Biol 25:1025–1040

Bursch W, Ellinger A, Gerner C et al (2000) Programmed cell death (PCD). Apoptosis, autophagic PCD, or others? Ann N Y Acad Sci 926:1–12

Callus BA, Ekert PG, Heraud JE et al (2008) Cytoplasmic p53 is not required for PUMA-induced apoptosis. Cell Death Differ 15:213–215 author reply 215–216

Chan FK, Shisler J, Bixby JG et al (2003) A role for tumour necrosis factor receptor-2 and receptor-interacting protein in programmed necrosis and antiviral responses. J Biol Chem 278:51613–51621

Chautan M, Chazal G, Cecconi F et al (1999) Interdigital cell death can occur through a necrotic and caspase-independent pathway. Curr Biol 9:967–970

Chen G, Goeddel DV (2002) TNF-R1 signalling: a beautiful pathway. Science 296:1634–1635

Chipuk JE, Green DR (2006) Dissecting p53-dependent apoptosis. Cell Death Differ 13:994–1002

Chipuk JE, Kuwana T, Bouchier-Hayes L et al (2004) Direct activation of Bax by p53 mediates mitochondrial membrane permeabilization and apoptosis. Science 303:1010–1014

Chiu R, Novikov L, Mukherjee S et al (2002) A caspase cleavage fragment of p115 induces fragmentation of the Golgi apparatus and apoptosis. J Cell Biol 159:637–648

Clarke PG, Clarke S (1995) Historic apoptosis. Nature 378:230

Clarke PG, Clarke S (1996) Nineteenth century research on naturally occurring cell death and related phenomena. Anat Embryol (Berl) 193:81–99

Cohen GM (1997) Caspases: the executioners of apoptosis. Biochem J 326(Pt 1) 1–16

Collins TJ, Berridge MJ, Lipp P et al (2002) Mitochondria are morphologically and functionally heterogeneous within cells. Embo J 21:1616–1627

Collins TJ, Bootman MD (2003) Mitochondria are morphologically heterogeneous within cells. J Exp Biol 206:1993–2000

De Giorgi F, Lartigue L, Ichas F (2000) Electrical coupling and plasticity of the mitochondrial network. Cell Calcium 28:365–70

Degterev A, Hitomi J, Germscheid M et al (2008) Identification of RIP1 kinase as a specific cellular target of necrostatins. Nat Chem BBiol 4:313–321

Degterev A, Huang Z, Boyce M et al (2005) Chemical inhibitor of non-apoptotic cell death with therapeutic potential for ischemic brain injury. Nat Chem Biol 1:112–119

Deiss LP, Galinka H, Berissi H et al (1996) Cathepsin D protease mediates programmed cell death induced by interferon-gamma, Fas/APO-1 and TNF-alpha. Embo J 15:3861–3870

Denecker G, Hoste E, Gilbert B et al (2007) Caspase-14 protects against epidermal UVB photodamage and water loss. Nat Cell Biol 9:666–674

Denecker G, Ovaere P, Vandenabeele P et al (2008) Caspase-14 reveals its secrets. J Cell Biol 180:451–458

Denecker G, Vercammen D, Steemans M et al (2001) Death receptor-induced apoptotic and necrotic cell death: differential role of caspases and mitochondria. Cell Death Differ 8:829–840

Dennis SC, Gevers W, Opie LH (1991) Protons in ischemia: where do they come from; where do they go to? J Mol Cell Cardiol 23:1077–1086

D'Herde K, De Prest B, Mussche S et al (2000) Ultrastructural localization of cytochrome c in apoptosis demonstrates mitochondrial heterogeneity. Cell Death Differ 7:331–337

Di Sano F, Ferraro E, Tufi R et al (2006) Endoplasmic reticulum stress induces apoptosis by an apoptosome-dependent but caspase 12-independent mechanism. J Biol Chem 281: 2693–2700

Doerfler P, Forbush KA, Perlmutter RM (2000) Caspase enzyme activity is not essential for apoptosis during thymocyte development. J Immunol 164: 4071-4079

Egner A, Jakobs S, Hell SW (2002) Fast 100-nm resolution three-dimensional microscope reveals structural plasticity of mitochondria in live yeast. Proc Natl Acad Sci U S A 99:3370–3375

Escuin D, Kline ER, Giannakakou P (2005) Both microtubule-stabilizing and microtubule-destabilizing drugs inhibit hypoxia-inducible factor-1alpha accumulation and activity by disrupting microtubule function. Cancer Res 65:9021–9028

Feng Z, Zhang H, Levine AJ et al (2005) The coordinate regulation of the p53 and mTOR pathways in cells. Proc Natl Acad Sci U S A 102:8204–8209

Ferri KF, Kroemer G (2001) Organelle-specific initiation of cell death pathways. Nat Cell Biol 3: E255–E263

Festjens N, Kalai M, Smet J et al (2006a) Butylated hydroxyanisole is more than a reactive oxygen species scavenger. Cell Death Differ 13:166–169

Festjens N, van Gurp M, van Loo G et al (2004) Bcl-2 family members as sentinels of cellular integrity and role of mitochondrial intermembrane space proteins in apoptotic cell death. Acta Haematol 111:7–27

Festjens N, Vanden Berghe T, Cornelis S et al (2007) RIP1, a kinase on the crossroads of a cell's decision to live or die. Cell Death Differ 14:400–410

Festjens N, Vanden Berghe T, Vandenabeele P (2006b) Necrosis, a well-orchestrated form of cell demise: signalling cascades, important mediators and concomitant immune response. Biochim Biophys Acta 1757:1371–1387

Foghsgaard L, Wissing D, Mauch D et al (2001) Cathepsin B acts as a dominant execution protease in tumour cell apoptosis induced by tumour necrosis factor. J Cell Biol 153:999–1010

Francois M, Le Cabec V, Dupont MA et al (2000) Induction of necrosis in human neutrophils by Shigella flexneri requires type III secretion, IpaB and IpaC invasins, and actin polymerization. Infect Immun 68:1289–1296

Goldstein JC, Munoz-Pinedo C, Ricci JE et al (2005) Cytochrome c is released in a single step during apoptosis. Cell Death Differ 12:453–462

Goldstein JC, Waterhouse NJ, Juin P et al (2000) The coordinate release of cytochrome c during apoptosis is rapid, complete and kinetically invariant. Nat Cell Biol 2:156–162

Gonzalez-Polo RA, Boya P, Pauleau AL et al (2005) The apoptosis/autophagy paradox: autophagic vacuolization before apoptotic death. J Cell Sci 118:3091–3102

Goossens V, De Vos K, Vercammen D et al (1999) Redox regulation of TNF signalling. Biofactors 10:145–156

Groenendyk J, Michalak M (2005) Endoplasmic reticulum quality control and apoptosis. Acta Biochim Pol 52:381–395

Groth-Pedersen L, Ostenfeld MS, Hoyer-Hansen M et al (2007) Vincristine induces dramatic lysosomal changes and sensitizes cancer cells to lysosome-destabilizing siramesine. Cancer Res 67:2217–2225

Harding TM, Morano KA, Scott SV et al (1995) Isolation and characterization of yeast mutants in the cytoplasm to vacuole protein targeting pathway. J Cell Biol 131:591–602

Harper N, Hughes M, MacFarlane M et al (2003) Fas-associated death domain protein and caspase-8 are not recruited to the tumour necrosis factor receptor 1 signalling complex during tumour necrosis factor-induced apoptosis. J Biol Chem 278:25534–25541

Hicks SW, Machamer CE (2005) Golgi structure in stress sensing and apoptosis. Biochim Biophys Acta 1744:406–414

Hippert MM, O'Toole PS, Thorburn A (2006) Autophagy in cancer: good, bad, or both? Cancer Res 66:9349–9351

Hofmann K (1999) The modular nature of apoptotic signalling proteins. Cell Mol Life Sci 55:1113–1128

Holler N, Zaru R, Micheau O et al (2000) Fas triggers an alternative, caspase-8-independent cell death pathway using the kinase RIP as effector molecule. Nat Immunol 1:489–495

Horvitz HR (1999) Genetic control of programmed cell death in the nematode Caenorhabditis elegans. Cancer Res 59:1701S–1706S

Hoyer-Hansen M, Jaattela M (2008) Autophagy—an emerging target for cancer therapy. Autophagy 4

Inbal B, Bialik S, Sabanay I et al (2002) DAP kinase and DRP-1 mediate membrane blebbing and the formation of autophagic vesicles during programmed cell death. J Cell Biol 157:455–68

Jaattela M, Tschopp J (2003) Caspase-independent cell death in T lymphocytes. Nat Immunol 4:416–23

Jiang X, Wang X (2000) Cytochrome c promotes caspase-9 activation by inducing nucleotide binding to Apaf-1. J Biol Chem 275:31199–31203

Juhasz G, Csikos G, Sinka R et al (2003) The Drosophila homolog of Aut1 is essential for autophagy and development. FEBS Lett 543:154–158

Kalai M, Lamkanfi M, Denecker G et al (2003) Regulation of the expression and processing of caspase-12. J Cell Biol 162:457–467

Kalai M, Van Loo G, Vanden Berghe T et al (2002) Tipping the balance between necrosis and apoptosis in human and murine cells treated with interferon and dsRNA. Cell Death Differ 9:981–994

Kang SJ, Wang S, Kuida K et al (2002) Distinct downstream pathways of caspase-11 in regulating apoptosis and cytokine maturation during septic shock response. Cell Death Differ 9:1115–1125

Kelekar A (2005) Autophagy. Ann N Y Acad Sci 1066:259–271

Kerr JF, Wyllie AH, Currie AR (1972) Apoptosis: a basic biological phenomenon with wide-ranging implications in tissue kinetics. Br J Cancer 26: 239–257

Khwaja A, Tatton L (1999) Resistance to the cytotoxic effects of tumour necrosis factor alpha can be overcome by inhibition of a FADD/caspase-dependent signalling pathway. J Biol Chem 274:36817–36823

Kim YS, Morgan MJ, Choksi S et al (2007) TNF-induced activation of the Nox1 NADPH oxidase and its role in the induction of necrotic cell death. Mol Cell 26:675–687

Kischkel FC, Hellbardt S, Behrmann I et al (1995) Cytotoxicity-dependent APO-1 (Fas/CD95)-associated proteins form a death-inducing signalling complex (DISC) with the receptor. Embo J 14:5579–5588

Klionsky DJ (2007) Autophagy: from phenomenology to molecular understanding in less than a decade. Nat Rev Mol Cell Biol 8:931–937

Klionsky DJ, Emr SD (2000) Autophagy as a regulated pathway of cellular degradation. Science 290:1717–1721

Koneri K, Goi T, Hirono Y et al (2007) Beclin 1 gene inhibits tumour growth in colon cancer cell lines. Anticancer Res 27:1453–1457

Korsmeyer SJ, Wei MC, Saito M et al (2000) Pro-apoptotic cascade activates BID, which oligomerizes BAK or BAX into pores that result in the release of cytochrome c. Cell Death Differ 7:1166–1173

Koterski JF, Nahvi M, Venkatesan MM et al (2005) Virulent Shigella flexneri causes damage to mitochondria and triggers necrosis in infected human monocyte-derived macrophages. Infect Immun 73:504–513

Kroemer G, El-Deiry WS, Golstein P et al (2005) Classification of cell death: recommendations of the Nomenclature Committee on Cell Death. Cell Death Differ 12 Suppl 2:1463–1467

Krysko DV, Roels F, Leybaert L et al (2001) Mitochondrial transmembrane potential changes support the concept of mitochondrial heterogeneity during apoptosis. J Histochem Cytochem 49:1277–1284

Krysko DV, Vanden Berghe T, D'Herde K et al (2008) Apoptosis and necrosis: Detection, discrimination and phagocytosis. Methods 44:205–221

Krysko DV, Vandenabeele P, D'Herde K (2007) Cell death at a glance. In: CR Kettleworth (ed) "Cell apoptosis research advances" pp. 1–21 Nova Science publishers, Inc

Kuma A, Hatano M, Matsui M et al (2004) The role of autophagy during the early neonatal starvation period. Nature 432:1032–1036

Kunstle G, Hentze H, Germann PG et al (1999) Concanavalin A hepatotoxicity in mice: tumour necrosis factor-mediated organ failure independent of caspase-3-like protease activation. Hepatology 30:1241–1251

Kurz T, Terman A, Gustafsson B et al (2008) Lysosomes in iron metabolism, ageing and apoptosis. Histochem Cell Biol 129:389–406

Lakhani SA, Masud A, Kuida K et al (2006) Caspases 3 and 7: key mediators of mitochondrial events of apoptosis. Science 311:847–851

Lamkanfi M, Declercq W, Kalai M et al (2002) Alice in caspase land. A phylogenetic analysis of caspases from worm to man. Cell Death Differ 9:358–361

Lamkanfi M, D'Hondt K, Vande Walle L et al (2005) A novel caspase-2 complex containing TRAF2 and RIP1. J Biol Chem 280:6923–6932

Lamkanfi M, Festjens N, Declercq W et al (2007) Caspases in cell survival, proliferation and differentiation. Cell Death Differ 14:44–55

Lamkanfi M, Kalai M, Vandenabeele P (2004) Caspase-12: an overview. Cell Death Differ 11:365–368

Laster SM, Wood JG, Gooding LR (1988) Tumour necrosis factor can induce both apoptic and necrotic forms of cell lysis. J Immunol 141:2629–2634

Lee CY, Baehrecke EH (2001) Steroid regulation of autophagic programmed cell death during development. Development 128:1443–1455

Lee CY, Clough EA, Yellon P et al (2003) Genome-wide analyses of steroid- and radiation-triggered programmed cell death in Drosophila. Curr Biol 13:350–357

Lefranc F, Facchini V, Kiss R (2007) Proautophagic drugs: a novel means to combat apoptosis-resistant cancers, with a special emphasis on glioblastomas. Oncologist 12:1395–1403

Lesnefsky EJ, Gudz TI, Moghaddas S et al (2001) Aging decreases electron transport complex III activity in heart interfibrillar mitochondria by alteration of the cytochrome c binding site. J Mol Cell Cardiol 33:37–47

Levine B (2006) Unraveling the role of autophagy in cancer. Autophagy 2:65–66

Levine B, Kroemer G (2008) Autophagy in the pathogenesis of disease. Cell 132:27–42

Levine B, Yuan J (2005) Autophagy in cell death: an innocent convict? J Clin Invest 115:2679–2688

Li P, Nijhawan D, Budihardjo I et al (1997) Cytochrome c and dATP-dependent formation of Apaf-1/caspase-9 complex initiates an apoptotic protease cascade. Cell 91:479–489

Lim SY, Davidson SM, Mocanu MM et al (2007) The cardioprotective effect of necrostatin requires the cyclophilin-D component of the mitochondrial permeability transition pore. Cardiovasc Drugs Ther 21:467–469

Lleo A, Invernizzi P, Selmi C et al (2007) Autophagy: highlighting a novel player in the autoimmunity scenario. J Autoimmun 29:61–68

Lockshin RA, William CM (1965) Programmed Cell Death. 3. Neural Control of the Breakdown of the Intersegmental Muscles of Silkmoths. J Insect Physiol 11:601–610

Lockshin RA, Zakeri Z (2004) Apoptosis, autophagy, and more. Int J Biochem Cell Biol 36:2405–2419

Lum JJ, Bauer DE, Kong M et al (2005) Growth factor regulation of autophagy and cell survival in the absence of apoptosis. Cell 120:237–248

Luo X, Budihardjo I, Zou H et al (1998) Bid, a Bcl2 interacting protein, mediates cytochrome c release from mitochondria in response to activation of cell surface death receptors. Cell 94:481–490

Ma Y, Temkin V, Liu H et al (2005) NF-kappaB protects macrophages from lipopolysaccharide-induced cell death: the role of caspase 8 and receptor-interacting protein. J Biol Chem 280:41827–41834

Maag RS, Mancini M, Rosen A et al (2005) Caspase-resistant Golgin-160 disrupts apoptosis induced by secretory pathway stress and ligation of death receptors. Mol Biol Cell 16:3019–3027

Maiuri MC, Zalckvar E, Kimchi A et al (2007) Self-eating and self-killing: crosstalk between autophagy and apoptosis. Nat Rev Mol Cell Biol 8:741–752

Majno G, Joris I (1995) Apoptosis, oncosis, and necrosis. An overview of cell death. Am J Pathol 146:3–15

Malhotra JD, Kaufman RJ (2007) The endoplasmic reticulum and the unfolded protein response. Semin Cell Dev Biol 18:716–731

Mancini M, Machamer CE, Roy S et al (2000) Caspase-2 is localized at the Golgi complex and cleaves golgin-160 during apoptosis. J Cell Biol 149:603–612

Martin DN, Baehrecke EH (2004) Caspases function in autophagic programmed cell death in Drosophila. Development 131:275–284

Mathew R, White E (2007) Why sick cells produce tumours: the protective role of autophagy. Autophagy 3:502–505

Matsumura H, Shimizu Y, Ohsawa Y et al (2000) Necrotic death pathway in Fas receptor signalling. J Cell Biol 151, 1247–1256

Mattson MP, Magnus T (2006) Ageing and neuronal vulnerability. Nat Rev Neurosci 7:278–294

Mayer A, Neupert W, Lill R (1995) Translocation of apocytochrome c across the outer membrane of mitochondria. J Biol Chem 270:12390–12397

Meier P, Vousden KH (2007) Lucifer's labyrinth--ten years of path finding in cell death. Mol Cell 28:746–754

Melendez A, Talloczy Z, Seaman M et al (2003) Autophagy genes are essential for dauer development and life-span extension in C. elegans. Science 301:1387–1391

Melino G, Knight RA, Nicotera P (2005) How many ways to die? How many different models of cell death? Cell Death Differ 12 Suppl 2:1457–1462

Mellen MA, de la Rosa EJ, Boya P (2008) The autophagic machinery is necessary for removal of cell corpses from the developing retinal neuroepithelium. Cell Death Differ

Mills KR, Reginato M, Debnath J et al (2004) Tumour necrosis factor-related apoptosis-inducing ligand (TRAIL) is required for induction of autophagy during lumen formation in vitro. Proc Natl Acad Sci U S A 101:3438–3443

Mukherjee S, Chiu R, Leung SM et al (2007) Fragmentation of the Golgi apparatus: an early apoptotic event independent of the cytoskeleton. Traffic 8:369–378

Nagata S (1999) Fas ligand-induced apoptosis. Annu Rev Genet 33:29–55

Nakagawa T, Zhu H, Morishima N et al (2000) Caspase-12 mediates endoplasmic-reticulum-specific apoptosis and cytotoxicity by amyloid-beta. Nature 403:98–103

Nakamura K, Bossy-Wetzel E, Burns K et al (2000) Changes in endoplasmic reticulum luminal environment affect cell sensitivity to apoptosis. J Cell Biol 150:731–740

Nguyen TM, Subramanian IV, Kelekar A et al (2007) Kringle 5 of human plasminogen, an angiogenesis inhibitor, induces both autophagy and apoptotic death in endothelial cells. Blood 109:4793–4802

Noda T, Ohsumi Y (1998) Tor, a phosphatidylinositol kinase homologue, controls autophagy in yeast. J Biol Chem 273:3963–3966

Obeng EA, Boise LH (2005) Caspase-12 and caspase-4 are not required for caspase-dependent endoplasmic reticulum stress-induced apoptosis. J Biol Chem 280:29578–29587

Okada H, Mak TW (2004) Pathways of apoptotic and non-apoptotic death in tumour cells. Nat Rev Cancer 4:592–603

Pattingre S, Tassa A, Qu X et al (2005) Bcl-2 antiapoptotic proteins inhibit Beclin 1-dependent autophagy. Cell 122:927–939

Pegoraro L, Palumbo A, Erikson J et al (1984) A 14;18 and an 8;14 chromosome translocation in a cell line derived from an acute B-cell leukemia. Proc Natl Acad Sci U S A 81:7166–7170

Peter ME, Krammer PH (2003) The CD95(APO-1/Fas) DISC and beyond. Cell Death Differ 10:26–35

Petiot A, Pattingre S, Arico S et al (2002) Diversity of signalling controls of macroautophagy in mammalian cells. Cell Struct Funct 27:431–441

Petrovski G, Zahuczky G, Katona K et al (2007a) Clearance of dying autophagic cells of different origin by professional and non-professional phagocytes. Cell Death Differ

Petrovski G, Zahuczky G, Majai G et al (2007b) Phagocytosis of cells dying through autophagy evokes a pro-inflammatory response in macrophages. Autophagy 3:509–511

Qu X, Yu J, Bhagat G et al (2003) Promotion of tumourigenesis by heterozygous disruption of the beclin 1 autophagy gene. J Clin Invest 112:1809–1820

Qu X, Zou Z, Sun Q et al (2007) Autophagy gene-dependent clearance of apoptotic cells during embryonic development. Cell 128:931–946

Rai NK, Tripathi K, Sharma D et al (2005) Apoptosis: a basic physiologic process in wound healing. Int J Low Extrem Wounds 4:138–144

Riley T, Sontag E, Chen P et al (2008) Transcriptional control of human p53-regulated genes. Nat Rev Mol Cell Biol 9:402–412

Rizzuto R, Pinton P, Carrington W et al (1998) Close contacts with the endoplasmic reticulum as determinants of mitochondrial Ca2+ responses. Science 280:1763–1766

Roach HI, Clarke NM (2000) Physiological cell death of chondrocytes in vivo is not confined to apoptosis. New observations on the mammalian growth plate. J Bone Joint Surg Br 82:601–613

Rodriguez J, Lazebnik Y (1999) Caspase-9 and APAF-1 form an active holoenzyme. Genes Dev 13:3179–3184

Ron D, Walter P (2007) Signal integration in the endoplasmic reticulum unfolded protein response. Nat Rev Mol Cell Biol 8:519–529

Saelens X, Festjens N, Vande Walle L et al (2004) Toxic proteins released from mitochondria in cell death. Oncogene 23:2861–2874

Saleh A, Srinivasula SM, Acharya S et al (1999) Cytochrome c and dATP-mediated oligomerization of Apaf-1 is a prerequisite for procaspase-9 activation. J Biol Chem 274:17941–17945

Saleh M, Mathison JC, Wolinski MK et al (2006) Enhanced bacterial clearance and sepsis resistance in caspase-12-deficient mice. Nature 440:1064–1068

Saleh M, Vaillancourt JP, Graham RK et al (2004) Differential modulation of endotoxin responsiveness by human caspase-12 polymorphisms. Nature 429:75–79

Scaffidi C, Fulda S, Srinivasan A et al (1998) Two CD95 (APO-1/Fas) signalling pathways. EMBO J 17:1675–1687

Schweichel JU, Merker HJ (1973) The morphology of various types of cell death in prenatal tissues. Teratology 7:253–266

Sheikh MS, Fornace AJ Jr (2000) Death and decoy receptors and p53-mediated apoptosis. Leukemia 14:1509–1513

Shimizu S, Kanaseki T, Mizushima N et al (2004) Role of Bcl-2 family proteins in a non-apoptotic programmed cell death dependent on autophagy genes. Nat Cell Biol 6:1221–1228

Smith CC, Davidson SM, Lim SY et al (2007) Necrostatin: a potentially novel cardioprotective agent? Cardiovasc Drugs Ther 21:227–233

Smith KG, Strasser A, Vaux DL (1996) CrmA expression in T lymphocytes of transgenic mice inhibits CD95 (Fas/APO-1)-transduced apoptosis, but does not cause lymphadenopathy or autoimmune disease. Embo J 15:5167–5176

Stanger BZ, Leder P, Lee TH et al (1995) RIP: a novel protein containing a death domain that interacts with Fas/APO-1 (CD95) in yeast and causes cell death. Cell 81:513–523

Tasdemir E, Maiuri MC, Galluzzi L et al (2008) Regulation of autophagy by cytoplasmic p53. Nat Cell Biol 10:676–687

Thumm M, Egner R, Koch B et al (1994) Isolation of autophagocytosis mutants of Saccharomyces cerevisiae. FEBS Lett 349:275–280

Tsukada M, Ohsumi Y (1993) Isolation and characterization of autophagy-defective mutants of Saccharomyces cerevisiae. FEBS Lett 333:169–174

Tsukamoto S, Kuma A, Murakami M et al (2008) Autophagy Is Essential for Preimplantation Development of Mouse Embryos. Science 321:117–120

Twiddy D, Brown DG, Adrain C et al (2004) Pro-apoptotic proteins released from the mitochondria regulate the protein composition and caspase-processing activity of the native Apaf-1/caspase-9 apoptosome complex. J Biol Chem 279:19665–19682

Twiddy D, Cohen GM, Macfarlane M et al (2006) Caspase-7 is directly activated by the approximately 700-kDa apoptosome complex and is released as a stable XIAP-caspase-7 approximately 200-kDa complex. J Biol Chem 281:3876–3888

van Loo G, van Gurp M, Depuydt B et al (2002) The serine protease Omi/HtrA2 is released from mitochondria during apoptosis. Omi interacts with caspase-inhibitor XIAP and induces enhanced caspase activity. Cell Death Differ 9:20–26

Vande Walle L, Lamkanfi M, Vandenabeele P (2008) The mitochondrial serine protease HtrA2/Omi: an overview. Cell Death Differ 15:453–460

Vanden Berghe T, Declercq W, Vandenabeele P (2007) NADPH oxidases: new players in TNF-induced necrotic cell death. Mol Cell 26:769–771

Vanden Berghe T, van Loo G, Saelens X et al (2004) Differential signalling to apoptotic and necrotic cell death by Fas-associated death domain protein FADD. J Biol Chem 279:7925–7933

Vanlangenakker N, Berghe TV, Krysko DV et al (2008) Molecular mechanisms and pathophysiology of necrotic cell death. Curr Mol Med 8:207–220

Vercammen D, Beyaert R, Denecker G et al (1998a) Inhibition of caspases increases the sensitivity of L929 cells to necrosis mediated by tumour necrosis factor. J Exp Med 187:1477–1485

Vercammen D, Brouckaert G, Denecker G et al (1998b) Dual signalling of the Fas receptor: initiation of both apoptotic and necrotic cell death pathways. J Exp Med 188:919–930

Vousden KH (2005) Apoptosis. p53 and PUMA: a deadly duo. Science 309:1685–1686

Wang X (2001) The expanding role of mitochondria in apoptosis. Genes Dev 15:2922–2933

Wei Y, Pattingre S, Sinha S et al (2008) JNK1-mediated phosphorylation of Bcl-2 regulates starvation-induced autophagy. Mol Cell 30:678–688

White E (2008) Autophagic cell death unraveled: Pharmacological inhibition of apoptosis and autophagy enables necrosis. Autophagy 4:399–401

Williams A, Jahreiss L, Sarkar S et al (2006) Aggregate-prone proteins are cleared from the cytosol by autophagy: therapeutic implications. Curr Top Dev Biol 76:89–101

Willingham SB, Bergstralh DT, O'Connor W et al (2007) Microbial pathogen-induced necrotic cell death mediated by the inflammasome components CIAS1/cryopyrin/NLRP3 and ASC. Cell Host Microbe 2:147–159

Yanagisawa H, Miyashita T, Nakano Y et al (2003) HSpin1, a transmembrane protein interacting with Bcl-2/Bcl-xL, induces a caspase-independent autophagic cell death. Cell Death Differ 10:798–807

Yang X, Chang HY, Baltimore D (1998) Essential role of CED-4 oligomerization in CED-3 activation and apoptosis. Science 281:1355–1357

Yorimitsu T, Klionsky DJ (2005) Autophagy: molecular machinery for self-eating. Cell Death Differ 12 Suppl 2:1542–1552

You Z, Savitz SI, Yang J et al (2008) Necrostatin-1 reduces histopathology and improves functional outcome after controlled cortical impact in mice. J Cereb Blood Flow Metab
Yu L, Alva A, Su H et al (2004) Regulation of an ATG7-beclin 1 program of autophagic cell death by caspase-8. Science 304:1500–1502
Zeng X, Yan T, Schupp JE et al (2007) DNA mismatch repair initiates 6-thioguanine--induced autophagy through p53 activation in human tumour cells. Clin Cancer Res 13:1315–1321
Zou H, Li Y, Liu X et al (1999) An APAF-1.cytochrome c multimeric complex is a functional apoptosome that activates procaspase-9. J Biol Chem 274:11549–11556

Chapter 2
Apoptotic, Autophagic and Necrotic Cell Death Types in Pathophysiological Conditions: Morphological and Histological Aspects

Araceli Diez-Fraile, Tim Lammens and Katharina D'Herde

Abstract: This chapter is intended as an assembly of a minimal atlas of how cells in animal and plant tissues can die in a controlled way. In accordance with the recommendations of the Nomenclature Committee on Cell Death, we recognize three types of programmed cell death (PCD) based on morphological features: type 1 (apoptosis), type 2 (autophagic cell death) and type 3 (necrotic cell death). We present evidence on the inter-relation or simultaneous occurrence of the different PCD modes, which poses difficulties for the study of cell death in organized tissues. We also address particular examples of cell death, such as mitotic catastrophe, entosis, cornification, and formation of lens fibers and erythrocytes, which do not fit in the three-part classification.

Keywords: Apoptosis • Autophagic cell death • Entosis • Mitotic catastrophe • Necrosis

2.1   Introduction

The concept of programmed cell death (PCD) was initially introduced by Lockshin and Williams (1964) to describe cell death that occurs in predictable places and at predictable times during development. This concept emphasizes that cells are

K. D'Herde (✉) · A. Diez-Fraile
Department of Basic Medical Sciences
Campus Heymans, 4B3
University Hospital, De Pintelaan 185, B-9000 Ghent, Belgium
Tel.: +32 (0) 93325244; Fax: +32 (0) 93323809
E-mail: katharina.dherde@ugent.be

T. Lammens
Department of Plant Systems Biology
Flanders Institute for Biotechnology (VIB)
9052 Ghent, Belgium

Department of Molecular Genetics
Ghent University, 9052 Ghent, Belgium

D.V. Krysko, P. Vandenabeele (eds.), *Phagocytosis of Dying Cells,*
DOI: 10.1007/978-1-4020-9292-3_2, © Springer Science+Business Media B.V. 2009

programmed to die during the execution of the developmental plan of the organism (Lockshin and Williams 1964, 1965; Lockshin 1969). The first evidence that physiological cell death is genetically programmed came much later from developmental studies in *Caenorhabditis elegans* (*C. elegans*; Ellis and Horvitz 1986; Horvitz et al. 1982; Sulston and Horvitz 1977). Still later, in the 90s, it was found that the *C. elegans* genes involved in cell death had mammalian counterparts (Hengartner and Horvitz 1994; Yuan et al. 1993; Zou et al. 1997). In 1971, Kerr induced atrophy in the rat liver by ligating a major branch of the portal vein without compromising arterial flow. This intervention resulted in hepatocyte death with a peculiar morphology, initially called shrinkage necrosis. One year later Kerr, Wyllie and Currie (1972) published a landmark paper on the morphological features of a type of cell death that occurs during development and tissue homeostasis and is characterized by nuclear and cytoplasmic shrinkage. They named this form of cell death "apoptosis", a word derived from the Greek word for "falling off", as in autumn leaves falling from a tree.

Initially, there was general acceptance of a dichotomous view of cell death: all types of PCD have apoptotic morphology, while accidental, non-physiological passive cell death displays features of what was called cell necrosis (from the Greek word for "dying"). Despite its simplicity, this initial classification of cell death made a crucial contribution, promoted research on PCD, and provided clear-cut morphological clues for predicting the presence or absence of regulatory mechanisms behind cell death.

However, it became apparent that necrosis also occurs during normal cell physiology and development (Chu-Wang and Oppenheim 1978; Chautan et al. 1999; Kitanaka and Kuchino 1999; Dahmoun et al. 1999). In addition, in vitro findings revealed that necrosis too is regulated by an intrinsic cell death program (Krysko et al., Chap. 1, this Vol.; Leist and Jäättelä 2001; Festjens et al. 2006). Another milestone paper, neglected for a long time, was that of Schweichel and Merker (1973). In their ultrastructural study of rat and mouse embryos and fetuses, they proposed three main types of cell death in developing tissues based on the role of lysosomes. It was Clarke's review in 1990 that brought the report of Schweichel and Merker again to the attention of researchers in the field. Clarke's important contribution also included the analysis of many papers dealing with cell death morphologies during development. There is now general agreement that three types of PCD exist: type 1 (apoptosis), type 2 (autophagic cell death), and type 3 (necrosis) (Clarke et al. 1990; Zakeri et al. 1995; D'Herde et al. 1996; Bursch et al. 1997; Krysko et al., Chap. 1, this Vol.; Bursch 2001). In type 1 cell death (apoptosis) the contents of the dying cell are destroyed by heterophagocytosis. In type 2 cell death, lysosomal activity increases remarkably within the dying cell to enable it to digest its own contents. For that reason, type 2 cell death is also called self-cannibalism. By contrast, in type 3 cell death lysosomes do not participate in the cell lysis process. Thus, the terms apoptosis and PCD should not be used synonymously, but unfortunately, this practice continues. Moreover, necrotic cell death should be considered a heterogeneous phenomenon encompassing both programmed and accidental cell death. Majno and Joris (1995) proposed using the term *oncosis* (from the Greek word for swelling) for any cell death characterized by marked cellular swelling, and the term *necrosis* for the features that appear after the cell has died (Trump et al. 1997). However, this terminology has not been widely accepted. In the following paragraphs we will follow the proposal of the Nomenclature Committee for Cell Death classification (Kroemer et al. 2005) as we discuss the morphological features of the three main types of programmed cell death: apoptosis, autophagic cell death, and necrosis. Indeed, these cell death modalities are still defined by morphological rather than by biochemical criteria (Galluzzi et al. 2007).

2.2 Classification of Programmed Cell Death by Morphology

2.2.1 *Apoptosis*

During apoptosis, the cell loses contact with neighbouring cells or extracellular matrix, its volume decreases as it condenses and rounds up, and microvilli and other specialized surface features are lost. Some vesiculation of the endoplasmic reticulum may occur. After fusion with the plasma membrane, these vesicles give rise to a characteristic blebbed appearance in scanning electron micrographs. Membrane blebbing is an early feature of apoptosis and is responsible for the "boiling" morphology also designated as zeiosis. The chromatin coalesces into characteristic crescent-shaped masses lying against the nuclear membrane. The nucleus may become markedly indented and eventually it fragments. A process designated as "budding", effectuated by the sealing of surface protuberances, produces so-called apoptotic bodies containing cytoplasmic material, nuclear fragments, or both. These acidophilic membrane-bound structures of varying size remain viable and have an intact plasma membrane, as shown by their initial exclusion of vital dyes. Usually, apoptotic bodies are rapidly removed by professional phagocytes or neighbouring phagocytic cells, making this mode of cell death often inconspicuous in tissues (Ucker, Chap. 6; Lacy-Hulbert, Chap. 7; Gregory and Pound, Chap. 9, this Vol.). Besides general cytoplasmic condensation, all organelles, including the mitochondria, are fairly well preserved at the ultrastructural level. The interval between commitment to cell death and the appearance of the first characteristic cellular features varies according to the cell type and the type of lethal stimulus. But there is general agreement that it may take a few hours from the first structural changes until uptake within a phagosome (Bursch et al. 1990). Because of the short "washout time", during which apoptotic cells are recognized, low rates of apoptosis can still be responsible for major reductions in the total cell number in a tissue. Apoptotic cells that are not recognized by phagocytes undergo secondary necrosis (Wyllie 1981). In most cases the suicidal cell is phagocytosed, but the macrophage may play an active role in the cell death process (e.g. during development of the pupillary membrane in the mouse eye (Lang and Bishop 1993). Before pathologists recognized that apoptotic bodies are the final stage of a cell death programme, they gave them different names, such as sunburn cells and civatte bodies in the skin, coucilman bodies in the liver, and tingible bodies in lymphoid germinal centers (Wyllie 1981).

2.2.2 Autophagic Cell Death

Although the term 'autophagic cell death' seems to imply that cell death is *due to* autophagy, it actually refers to cell death in the presence of autophagy. Indeed, so far there is no in vivo evidence that the conserved genes required for autophagy (ATG genes) promote physiological PCD in vivo, for instance during development. Moreover, it must be emphasized that under most circumstances, autophagy is a stress adaptation pathway that promotes cell survival (Krysko et al., Chap. 1, this Vol.; Levine and Kroemer 2008). In some cases, however, cell death is mediated by autophagy, because inhibiting autophagy abolishes cell death. In other models, inhibiting autophagy does not change the fate of the cells but only changes the morphology by a switch from autophagic cell death to apoptosis.

Identification of autophagic cell death relies on the presence of extensive autophagocytosis, revealed by numerous autophagosomes in which organelles are recognizable. These sequestered organelles then fuse with primary lysosomes, converting the original autophagosome into an autophagolysosome. Cells undergoing this type of cell death often display an enlarged Golgi apparatus. In contrast to apoptosis and necrosis, an initial increase of endocytosis is frequently seen in autophagic cell death (Clarke 1990). Nuclear changes in the form of chromatin condensation occur late or not at all (D'Herde et al. 1996). Some authors report that this chromatin condensation pattern can be distinguished from apoptotic condensation because the condensed chromatin mass is centrally placed and so it does not lie next to the nuclear membrane (Bursch et al. 1997; D'Herde et al. 1996). Autophagosomes, by definition, have a double-membrane and contain cytoplasmic organelles or cytosol. Thus, autophagosomes are distinguishable by electron microscopy from other types of vesicles such as endosomes, lysosomes, and apoptotic blebs. During the late stages of autophagic cell death, the vacuoles increase in size and number, and many of them contain myelin figures; the connection between autophagy and myelin figures has been reported (Clarke 1990). Autophagocytosis seems to depend on an intact cytoskeleton, but this structure undergoes changes during the initiation phase of apoptosis (Bursch et al. 2000). The late autophagic debris is frequently removed by heterophagy, but this tends to occur late and seems less conspicuous than the clearance of apoptotic bodies (Clarke 1990).

2.2.3 Necrosis

It is important to keep in mind that programmed necrosis cannot be distinguished at the ultrastructural level from accidental necrosis (Krysko et al., Chap. 1, this Vol.; Degterev and Yuan 2008). However, images of programmed necrosis in in vitro models are often taken at an early stage and, therefore, do not reveal the full picture of 'classical' necrosis. Necrosis defines a situation in which a cell swells but retains contact with neighbouring cells through specialized junctions, which initially remain intact. The cellular swelling explains the decrease of electron density observed at the ultrastructural level. Unlike the boiling morphology of apoptotic cells, necrosis is characterized by small surface evaginations in the form of bubbles or blisters that can be seen by phase contrast or scanning electron microscopy (Rello et al. 2005). Transmission electron microscopy shows that these bubbles contain few or no organelles (Trump et al. 1997).

The nucleus of the necrotic cell has a mottled appearance caused by clumped but loosely packed chromatin. Furthermore, the nuclear membrane is detached and ruptured. All organelles in the cytoplasm, including mitochondria, are generally dilated. Disruption of membrane integrity is accompanied by osmotic swelling and finally lysis of the cells. By light microscopy, these necrotic cells are less conspicuous than apoptotic cells. Therefore, necrosis is often defined in a negative fashion: a type of cell death that involves rupture of the plasma membrane without the hallmarks of apoptosis and without massive autophagic vacuolization. Phagocytes eventually ingest the necrotic debris by a macropinocytosis, in contrast to the zipper-like mechanism of phagocytosis of apoptotic bodies (Krysko et al. 2003). Conventional textbooks indicate that apoptosis is a clean and silent non-inflammatory modality of cell death, whereas necrosis leads to spilling of the intracellular contents into the intercellular space, triggering an inflammatory response. However, it has become clear that this view is not valid (Galluzzi 2007).

2.2.4 *Nuclear Changes in Dying Cells*

Terms describing changes in the nucleus have been used since the 19th century, but back then they were not related to specific modes of cell death. Karyorhexis refers to the disintegration of the nucleus, while karyolysis describes the disappearance or fading of the nuclear material. The terms pyknosis and chromatin margination were introduced by Arnheim (1890). The term chromatolysis (the disappearance of nuclear staining) was introduced by Flemming (1885) in his description of the nuclear changes of granulosa cells during follicular atresia. Linking these descriptive terms to the cell death mode is not straightforward.

Karyolysis and chromatolysis, the disappearance of the nucleus and the disappearance of nuclear staining, respectively, is the ultimate fate of nuclei in all types of cell death. Karyorrhexis, falling apart of the nucleus, indicates nuclear fragmentation during apoptosis, but this term too is not specific because it is also used to describe nuclear changes in necrosis (Wyllie 1981). The term pyknosis (i.e. condensed hyperchromatic nuclei), which is used in light microscopy, is not specific for apoptotic nuclei, as it is also seen during necrosis and autophagic cell death.

Hence, the pattern of chromatin margination seen in apoptosis as sharply delineated geometric masses, detectable even by light microscopy, remains the only specific nuclear feature pointing to one specific cell death mode.

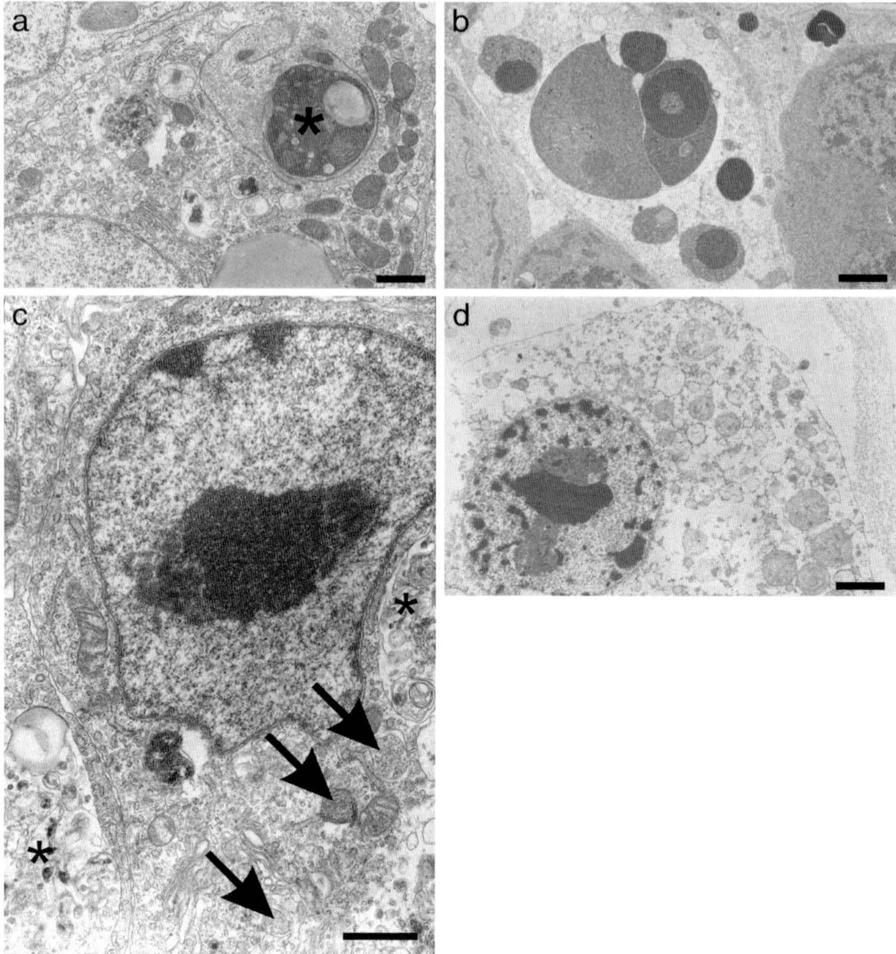

Fig. 2.1 a. Transmission electron microscopy of quail granulosa. An apoptotic body (*) with condensed cytoplasm being engulfed by a neighbouring viable cell. b. Transmission electron micrograph showing apoptosis of nurse cells in the ovary of the invertebrate *Artemia salina*. Several apoptotic bodies, some of them containing a mass of condensed chromatin, are being engulfed by a somatic cell undergoing necrosis. c. Autophagic cell death as one of the three cell death modes during follicular atresia in the quail ovary. Autophagic cell death is identified by double-membraned vacuoles containing recognizable cytoplasmic material (arrows). The nucleus has an indented nuclear membrane and a central mass of condensed chromatin. Notice the area marked with an asterisk were autophagic vacuoles fuse and degradation of cytoplasm is more extensive. d. In follicular atresia, apoptotic cell death, autophagic cell death and necrotic cell death are found together in an in vivo model (D'Herde et al. 1996). Transmission electron micrograph of a necrotic granulosa cell as one of the 3 subtypes of PCD in follicular atresia. The pattern of condensed ill-defined chromatin masses is apparent. Notice the swollen mitochondria, the dilated endoplasmic reticulum, and the detached ribosomes. Scale bars: 1 μm.

2.3   PCD in Vertebrate Tissues

2.3.1   PCD in Normal Prenatal Development

Large numbers of dying cells are seen in nearly all parts of the embryo. Dying cells may be few and isolated, or they may be so numerous in certain areas that whole organs or tissue regions are broken down (Nussbaum 1901). The extent to which cell death during embryonic and fetal development is used to mold the shape of organs had been apparent to developmental biologists before the concept of PCD was generally accepted (see reviews of Glucksmann 1951; Saunders 1966). Glucksmann reviewed 76 examples of cell death occurring as part of vertebrate ontogeny. Cell death is not only important for the shaping of organs, but also for fusing tissues from different origins and for establishing organ systems, such as the nervous system, the immune system and the reproductive system.

From a functional standpoint, developmental PCD can have three rationales, as initially proposed by Glucksmann (1951). First, cells that had served a purpose earlier in evolution die because they no longer have a function. The degeneration of the pronephros, a vestigial organ, is an example of cell death as component of phylogenesis. Another example concerns cells that differentiate and assume a required, but transient function and are then removed when the life style of the organism changes. A well-documented example is the disappearance of the tadpole tail (Tata 1968; Kerr et al. 1974). Second, cell death is instrumental in morphogenesis as it may modify the tissue to permit differentiation into its final form, best exemplified by the cell death occurring in the interdigital tissue (Zakeri and Ahuja 1997). Third, cell death plays a role in histogenesis as is illustrated in the adjustment of the number of neurons in the central nervous system to the number of targets, such as striated muscle fibers.

Which modalities of cell death function during normal development? The three types of PCD during prenatal development were well illustrated in the provocative paper of Schweichel and Merker (1973). A role of autophagic cell death in the regulation of vertebrate development has long been proposed by studies reporting morphological features of autophagy during embryogenesis (Schweichel and Merker 1973; Clarke 1990; Baehrecke 2002; Lockshin and Zakeri 2004). However, more recently it was shown that functional deficiency of *Ambra-1*, the *Beclin-1*-regulating gene involved in autophagy, leads to severe neural tube defects associated -with autophagy impairment and excessive apoptosis, illustrating that autophagy has a pro-survival role in neural development (Fimia et al. 2007). Knock-out of many genes influencing cell death results in no specific phenotype at birth, reinforcing the idea that cell death processes are functionally redundant in the embryo (for a list of developmental phenotypes in mice with targeted deletions of apoptotic proteins see Mirkes 2008). In an often-cited paper, Pierre Golstein and colleagues showed that PCD associated with loss of inter-digital tissue and formation of the digits is not exclusively due to apoptosis, and that the necrotic mode of cell death participates in some of the interdigital cell death in wild type mice (Chautan et al. 1999).

However, other examples of necrotic cell death during undisturbed prenatal development remain to the best of our knowledge scanty. The findings of Schweichel and Merker (1973) on necrosis during normal mineralization of vacuolated cartilage were confirmed by Roach and Clarke (2000). Furthermore, some ultrastructural studies clearly document necrosis in various neuronal populations during development (Pilar and Landmesser 1976; Chu-Wang and Oppenheim 1978).

2.3.2 *PCD from Birth to Adulthood in Health and in Age-related Diseases*

Faithful control of cell proliferation and PCD is critical for the harmonized functions of different cell populations in sophisticated multicellular organisms. All types of PCD have been morphologically observed in healthy humans. While apoptosis plays a predominant role in neonatal tissue morphogenesis and in proliferative tissue homeostasis throughout life, autophagy has been mainly observed early after birth in long-lived post-mitotic cells and in hormone-dependent tissues. To our knowledge, necrosis has only been reported in healthy females two days after menstruation, a physiological, intense, ischemic event. Changes in the extent of PCD lead to the decline of tissue-specific function, a typical characteristic of normal aging. In addition, homeostatic defences can be overwhelmed, leading to the development of diseases. This section focuses on the morphological evidence supporting the existence of different types of PCD under physiological conditions from birth to adulthood, during genetic disorders that resemble premature aging, and in relevant age-dependent diseases, i.e. neurodegenerative and cardiovascular diseases.

2.3.2.1 PCD in the Neonate

In general, apoptosis in humans seems to be down-regulated during early post-natal life as compared to fetal life (Malamitsi-Puchner et al. 2001; Sarandakou et al. 2003). Nonetheless, apoptosis seems to make important contributions to the orderly perinatal morphogenesis of certain organs and tissues. These include the cardiovascular system (James 1998), blood vessels that regress early after birth (Cho et al. 1995; Slomp et al. 1997; Kim et al. 1998; Meeson et al. 1999; Kim et al. 2000), some neuronal populations of the spinal cord (Coggeshall et al. 1994; Lawson et al. 1997), development of the gut (Godlewski et al. 2007), thymocyte maturation, thymic epithelium involution (Gray et al. 2006), and elimination of excess or abnormal germ cells (Rodriguez et al. 1997; Matova and Cooley 2001; Jahnukainen et al. 2004). In contrast to apoptosis, autophagy is scarce during embryogenesis, but seems to play a crucial role in neonate survival during the severe starvation at birth caused by the sudden interruption of the trans-placental nutrient supply (Kuma et al. 2004).

Apoptosis in Long-lived Post-mitotic Cells

Cardiovascular Tissue Remodelling

Post-natal involution of the heart right ventricle, including the atrioventricular node and the His bundle, is mediated by apoptosis (James 1998). Myocytes of the His bundle exhibit apoptotic nuclei, and scavenging macrophages are evenly spaced (James 1998; Tran et al. 2002). Early after birth, the ductus arteriosus, a muscular artery that connects the elastic aorta to the pulmonary artery during fetal life, regresses. Apoptosis of the smooth muscle cells in the tunica media (the muscular middle layer of an artery) of areas with evident macroscopic signs of degeneration has been implicated in the remodelling of the ductus arteriosus in human neonates (Slomp et al. 1997). A high rate of apoptosis was described in the post-natal intra-abdominal umbilical artery and abdominal aorta in sheep (Cho et al. 1995). Also in humans, apoptosis has been reported to be one of the mechanisms of closure of the umbilical vessel early after birth. The proportion of TUNEL positive cells was 80% in the intima, 40% in the media, and 80% in connective tissue of the umbilical cord (Kim et al. 1998). A later study in humans confirmed that apoptosis is involved in the closure and regression of the human ductus arteriosus and umbilical vessels and in the remodelling of the great arteries that branch off the aortic arch. In addition, they pointed to a critical involvement of Bcl-2 family members in apoptosis during human neonatal vascular remodelling (Kim et al. 2000). In the rodent neonatal eye, regression of capillaries of the pupillary membrane is thought to be partly caused by macrophage-induced apoptosis in endothelial cells (Meeson et al. 1996; Meeson et al. 1999; Lang and Bishop 1993).

Neural Tissue Morphogenesis

Four to five days after the birth of rats, apoptosis peaks in interneurons (Lawson et al. 1997) and to a lesser extent in dorsal root ganglion cells (Coggeshall et al. 1994), but almost no apoptosis could be detected in spinal motor neurons (Harris and McCaig 1984). Death of these neuronal populations in neonates seems to be necessary for the elimination of aberrant neural connections and for adjustment of the size of neuronal projections to their synaptic targets.

Apoptosis in Proliferative Tissues

Gut Tissue Development

It has been reported that enhanced mitosis in the gut of neonatal piglets occurs in parallel with a significant decrease in apoptosis during the first two days after birth. This phenomenon contributes substantially to the enlargement of the gut mucosa. Thereafter, the mitosis rate normalizes until the weaning period, when apoptosis increases (Godlewski et al. 2004). The local lack of growth factors, cytokines, and hormones in milk seem to be related to the apoptosis-inducing mechanism during the physiological development of the gut epithelium (Blum and Baumrucker 2002; Woliński et al. 2003; Godlewski et al. 2005). In the rapidly developing gut of piglets, in contrast to the gut of adult animals, apoptotic cells can be found along the entire length of the villi, including their lower half (Godlewski et al. 2004).

Thymic Involution

Stromal cell numbers change markedly during thymic involution, maintaining a stable ratio with thymocytes after birth (Gray et al. 2006). The development of the T cell repertoire continues through adolescence and up to early adulthood. The cortex is dominated by double-positive (CD4$^+$ CD8$^+$) T cell precursors. Continued migration toward the medulla brings thymocytes in contact with DCs and medullary thymic epithelial cells that, in a process termed negative selection, induce apoptosis or anergy of potentially autoreactive thymocytes possessing self determinants (Anderson and Jenkinson 2001). The surviving single-positive mature thymocytes are found within the medulla. Thus, the thymus is an organ of selective apoptosis with substantial cell death (10–15%) taking place under physiological conditions in vivo (Tadakuma et al. 1990; Gavrieli et al. 1992).

Apoptosis in Hormone-dependent Tissues

In the female, germline attrition in the ovary begins five months after fertilization and continues until puberty. Shortly after birth, all oocytes (about 2,000,000 in humans) are arrested in the dictyate stage of late prophase in the developing follicles. A small number of follicles (fewer than 300,000) reaches puberty, while the rest are lost by apoptosis (Matova and Cooley 2001). Between birth and the first weeks of post-natal life, a wave of apoptosis also takes place in germ cells of male rats and mice. Cell death affects mainly spermatogonia and early spermatocytes, and apoptosis is thought to be required for the proper development of functional mature sperm (Rodríguez et al. 1997; Jahnukainen et al. 2004).

Autophagy in Neonates

After natural birth, high levels of autophagy can be detected, particularly in the heart muscle, diaphragm, alveolar cells, and skin in mice (Kuma et al. 2004). The presence of autophagic cells has also been confirmed in the developing gut of neonate piglets and coexistence of autophagy with apoptosis has been suggested (Godlewski et al. 2004; Godlewski et al. 2007). It has been proposed that autophagy contributes to the maintenance of energy homeostasis at birth: mice deficient in *Atg5*, a gene essential for autophagosome formation, appear normal at birth but die within one day (Kuma et al. 2004).

2.3.2.2 PCD During Adulthood

PCD is rarely observed in non-proliferative tissue of healthy adults. Nevertheless, PCD exerts a homeostatic function in long-lived post-mitotic cells as well as in cells undergoing continuous or cyclic tissue renewal. The presence of PCD in adults is outlined below for several types of tissues, including long-lived post-mitotic, proliferative, and hormone-dependent tissues.

PCD in Long-lived Post-mitotic Cells

Apoptosis is rarely observed in terminally differentiated cells, such as neurons and cardiomyocytes. If these essential post-mitotic cells, which cannot be replaced, undergo apoptosis, it may lead to pathology. On the other hand, autophagy is considered a beneficial physiological response that protects organisms against diverse neurodegenerative and cardiovascular diseases (reviewed by Levine and Kroemer 2008).

PCD in Proliferative Tissues

PCD in the Hair Follicle

The hair follicle is a cutaneous organ that goes through cyclic activity comprised of periods of active hair growth, apoptosis-driven involution, hair shedding, and relative resting. A spatio-temporal distribution of TUNEL$^+$ cells in the hair follicle during hair involution may be viewed as a wave starting from the melanogenic area, propagating to the hair matrix, and then to the root sheaths and hair shafts. Most follicular epithelial cells and melanocytes are susceptible to apoptosis, whereas dermal papilla fibroblasts and some of the keratinocytes and melanocytes selected for survival are resistant to apoptosis. Apoptosis of the dermal papilla of involuting hair follicles in humans and mice is never seen under physiological conditions (reviewed by Botchkareva et al. 2006). Bcl-2 specifically protects melanocyte stem cells in the hair follicle from apoptosis, and Bcl-2 deficiency results in their elimination by apoptosis and consequent premature hair graying (Veis et al. 1993; Nishimura et al. 2005).

PCD in the Gut Epithelium

The lifespan of cells in the gut epithelium seldom exceeds 48 hours, and the quick turnover is based on the dynamic equilibrium between cell mitosis and PCD. Enterocytes undergoing apoptotic and/or autophagic cell death have been localized in the top of the villi of adult humans and rodents, and in the upper third of the villi in monkeys (Iwanaga 1995; Shibahara et al. 1995). Cells undergoing apoptosis are not only desquamated, some of them are phagocytosed by macrophages present in the gut mucosa of guinea pig and monkey.

Attenuated Form of Cell Death During Terminal Differentiation

Certain cell types, including epithelial keratinocytes, lens epithelial cells, and mammalian red blood cells, undergo a highly specialized process of terminal differentiation. These cells have been shown to share pathways with cells undergoing apoptosis, and it has been suggested that terminal differentiation represents an attenuated form of PCD. Indeed, during terminal differentiation and apoptosis, nuclei and organelles are eliminated, cell shape changes, and transglutaminases and proteases are activated. However, these terminally differentiated cells are not phagocytosed like cells undergoing apoptosis. Instead, they assume an important physiological function (Dahm 1999). For example, epidermal keratinocytes differentiate as part of the stratum corneum, nail plate, or sebaceous glands. In human keratinocytes in culture, over-expression of a *Bcl-2* transgene inhibits terminal differentiation (Nataraj et al. 1994). Moreover, it has been shown that terminal differentiation of epidermal keratinocytes is associated with the expression of keratinocyte-specific enzymes that are related to classical pro-apoptotic factors, i.e. caspase-14 and/or DNase1L2 (Jäger et al. 2007; Eckhart et al. 2000). In lens epithelial cells, loss of cytoplasmic organelles seems to be crucial for lens transparency (Zimmermann and Font 1966). Similarities and differences between key features of cells undergoing apoptosis and those of fiber cell differentiation have been previously reviewed (Dahm 1999). Effective terminal maturation of erythrocytes depends on intracellular signalling networks that regulate cell growth and apoptosis. In particular, regulated expression of the antiapoptotic genes *Bcl-xL* and *Nix* are thought to mediate erythroid maturation (Gregory et al. 1999; Aerbajinai et al. 2003; Opferman 2007).

PCD in Hormone-dependent Tissues

PCD in the Ovary, Uterus and Mammary Gland in the Adult Female

During the normal menstrual cycle, follicles in the ovary and endometrium (the lining of the uterus) follow a precisely programmed series of morphologic and physiologic events characterized by growth, differentiation, and in the absence of conception, degeneration and regeneration. A single dominant follicle is generally selected for ovulation in human, while the cohort of antral follicles undergoes atresia (McGee and Hsueh 2000). Although apoptosis initiated within the granulosa layer of the antral follicle (Hsueh et al. 1994; Morita and Tilly 1999; Tilly 2001) is the basic mechanism of follicular atresia in vertebrates (Depalo et al. 2003), autophagic and necrotic cell death have been documented in the granulosa of starvation-induced atretic follicles of Japanese quail (Fig. 2.1; D'Herde et al. 1996). Furthermore, PCD is also involved in ovulation (Murdoch 1995) as well as during regression of the mammalian corpus luteum (Dharmarajan et al. 2004). After the corpus luteum ceases to produce progesterone, it disappears from the ovary as a result of both apoptosis and autophagocytosis of luteal cells (Gaytán et al. 1998; Vaskivuo et al. 2002; Quatacker 1971; Fraser et al. 1999). The endometrium changes dramatically during the estrous cycle, pregnancy, and post-partum involution. It seems that uterine homeostasis is controlled mainly by apoptosis during the estrous cycle and during early pregnancy in several species, and that it involves various hormones and cytokines. Manifestations of apoptosis, before the term came into use, were first observed in the endometrium by Bartelmez (1933). Since then, apoptosis has been observed in human endometrium, mainly in the luteal phase and during the menstruation (Dahmoun et al. 1999; Kokawa et al. 1996; Verma 1983). Necrotic areas have also been observed in the endometrial stroma during the first two days of menstruation (Dahmoun et al. 1999). Apoptotic features during the embryo implantation window and early pregnancy have been reported in several species (Schlafke and Enders 1975; Parr et al. 1987; Welsh and Enders 1993; Piacentini and Autuori 1994; Akcali et al. 1996; Galán et al. 2000; Wang et al. 2003; Okano et al. 2007). An increase in apoptosis during late pregnancy has been reported in the rat uterus (Leppert 1998). After parturition, uterine involution requires massive remodelling of the extracellular matrix in association with cell proliferation and apoptosis as the uterus returns to the pre-pregnancy state (Takamoto et al. 1998). Additionally, autophagocytosis was noted in smooth muscle cells, and in some instances extensive autophagy appeared to lead to cell death in rat uterus following parturition. It was proposed that this autophagy is a means for reducing the size of smooth muscle cells (Henell et al. 1983). It was also proposed that absence of biologically significant estrogenic stimulation in the post-menopausal years leads to progressive endometrial involution, mainly by apoptosis: in the mid-50s the endometrium becomes inactive, and it atrophies in the late-60s. This hypothesis is supported by the induction of apoptosis in the uterine endometrium of hamster, rabbit, and monkey in response to steroid hormone withdrawal (Rotello et al. 1992). Nevertheless, the post-menopausal endometrium rather than being atrophic seems to be in a quiescent state, since it can respond to adequate hormonal stimulation in post-menopausal women (Cicinelli et al. 1993).

Although morphologic evidence of apoptosis has been detected in normal breast tissue during ovulatory cycles (Ferguson and Anderson 1981), PCD is particularly prominent during withdrawal of maternal breast-feeding. During mammary gland involution, the extracellular matrix and the alveolar basement membrane are degraded. The alveoli lose their structural integrity and death of mammary epithelial cells is extensive. Regulation of bovine mammary gland remodelling during the lactation cycle includes increased intensity of both apoptosis and autophagy, as evidenced by characteristic biochemical and morphological features (Zarzyńska et al. 2007). Apoptotic cell death has also been described during involution of lactating breast in human (Walker et al. 1989). In addition, evidence indicates that autophagic cell death can be also observed in human mammary epithelial cells (Mills et al. 2004).

PCD in Male Reproductive Organs

Spermatogenesis involves a series of mitoses and meioses and results in the production of up to 200 million spermatozoa daily. It is estimated that a substantial level of cell death, up to 75% of the spermatogonia, occurs during normal spermatogenesis in mammals (Huckins and Oakberg 1978). It has become apparent that degeneration during normal spermatogenesis occurs mainly by apoptosis (Blanco-Rodríguez and

Martínez-García 1996; Blanco-Rodríguez 1998; Rodríguez et al. 1997; Jahnukainen et al. 2004). Two putative roles have been proposed for the death of such a large proportion of potential spermatozoids during normal spermatogenesis: limitation of the germ cell population to numbers that can be supported by the Sertoli cells (Lee et al. 1997), and selective depletion of abnormal spermatozoa (Blanco-Rodríguez et al. 2003).

Like in other hormone-dependent tissues, apoptosis has been observed in prostate epithelium upon castration and to a lesser extent in healthy prostate tissue (Kerr and Searle 1973). Columnar epithelium, which ultrastructurally appears more differentiated than basal epithelium, lacked detectable Bcl-2, but this protein could be observed at the basal cuboidal epithelium of the prostate gland (Hockenbery et al. 1991).

2.3.2.3　PCD in Progeroid Syndromes

Progeroid syndromes, also called segmental aging syndromes, constitute a group of genetic disorders that clinically resemble premature aging but do not display all of the traits of natural aging. Among them, Werner syndrome (WS) and Hutchinson-Gilford progeria syndrome (HGPS) have been the most extensively studied, as they closely mimic natural aging. Both syndromes lead to accelerated aging of certain tissues, but remarkable differences between these syndromes have been described. Cellular aging in WS is characterized by premature loss of proliferating cells by apoptosis, and cultures of fibroblast from WS patients are rich in senescent cells (de Magalhães et al. 2004). By contrast, cellular aging of HGPS fibroblasts is characterized by a period of hyperproliferation, and it terminates with a substantial increase in the rate of apoptosis (Bridger and Kill 2004). The increased rate of senescence and apoptosis in HGPS does not seem to be related to an enhanced p53 response (O'Neil et al. 2003), whereas the ability of fibroblasts from WS individuals to undergo p53-mediated apoptosis is attenuated (Spillare et al. 1999). In HGPS, the occurrence of cells with nuclear abnormalities seems to result from cell division, because the proportion of these abnormalities increases with passages in cell culture (Brigder and Kill 2004). Initially, minor architectural defects may appear, such as herniations in the nuclear periphery, folds, and crevices. As the cellular age advances, nuclear fragments and lobules, including micronuclei, are observed in cultured fibroblasts (Bridger and Kill 2004). Other typical characteristics of aging fibroblasts in HGPS, such as heterochromatin loss, nuclear pore clustering, and reduced telomere length, can also be observed (Allsopp et al. 1992; Spillare et al. 1999). To our knowledge, structural abnormalities in WS nuclei have not been adequately characterized.

2.3.2.4　PCD in Age-related Diseases

Senescence or aging are associated with altered gene expression due to damage of nuclear and mitochondrial DNA in proliferating tissues, which may result in the development of malignant neoplasms. The diminishing ability of long-lived cells to remove biological waste material has also been implicated in the aging process (Martinez-Vicente and Cuervo 2007). It is clear that accumulation of lesions in long-lived cells, such as neurons or cardiomyocytes, is more detrimental because it leads to the most severe age-related changes (Strehler 1977). PCD seems to be instrumental in the progress of both neurodegenerative disorders and cardiovascular diseases in aged humans.

PCD in Tumours

Resistance to cancer drugs is a major obstacle limiting the efficacy of cancer chemotherapy. Many lines of clinical and experimental evidence have demonstrated that a defect in the apoptotic machinery (e.g. in the p53 status of blood cell tumours) is the most frequent cause of cancer drug resistance. This has two fundamental bases. First, the neoplastic process is driven by oncogenic mutations that increase tumour cell number by activating the cell cycle and/or by inhibiting the normal apoptotic process. This means that cancer cells are genetically pre-disposed to apoptotic resistance. Second, conventional anticancer agents, regardless of their targets and mechanisms, are mostly apoptogenic, and so development of drug resistance via anti-apoptotic mechanisms seems to be often inevitable.

Nevertheless, apoptosis has been observed in vivo in pre-neoplastic cell foci, in untreated tumours, in cells killed by cytotoxic T lymphocytes and natural killer cells, and in tumours sensitive to chemotherapeutic agents and radiation (Searle et al. 1973; Wyllie 1981; Sarraf and Bowen 1988; Schulte-Hermann et al. 1995). In solid tumours, the picture is complicated by various sources of cell loss, such as necrosis due to ischemia in the center of the tumour or migration and subsequent exfoliation of tumour cells (Kerr and Harmon 1994). Several anticancer treatments induce autophagy or self-cannibalism. However, it is unclear when this response is a mechanism for promoting cell survival and when it is a mechanism promoting cell death. In other words, the question is in which tumours is induction of autophagy beneficial for the cancer patient and in which tumours is it detrimental (Levine 2007). A few examples of anticancer drugs inducing autophagic death of tumour cells are already known. In this context, one of the best-studied models is tamoxifen-induced autophagy in mammary gland carcinoma cells (Schulte-Hermann et al. 1995). There is recent evidence that proautophagic chemotherapy can overcome apoptosis resistance in cancer cells. Temozolomide, a proautophagic cytotoxic drug, is therapeutically beneficial for glioblastoma patients, and it is now in clinical trials for several types of apoptosis-resistant cancers (Lefranc et al. 2007).

Besides apoptosis, autophagic cell death and necrosis, mitotic catastrophe has been observed in tumours treated by ionizing radiation (Cohen-Jonathan et al. 1999) or after treatment with certain chemotherapeutic agents. Mitotic catastrophe is nowadays considered as either a prestage to necrosis or apoptosis or a survival mechanism for tumours (Vakifahmetoglu et al. 2008) Mitotic catastrophe as such can be differentiated morphologically from apoptosis, despite reports stating the contrary. Indeed, cells undergoing mitotic catastrophe are very large, multinucle-ated, and contain one to several micronuclei that do not contain the typical pattern of chromatin condensation seen in apoptosis. Furthermore, these cells do not show an increase in cytoplasmic density.

Permanent growth arrest, known as senescence, is also considered a type of cell death in the context of cancer therapy. Different classes of chemotherapeutic agents and irradiation induce senescence in human cancer cell lines in vitro and in mouse xenografts. Senescent cells in culture are large, flattened, frequently vacuolated, and have a distinct heterochromatic structure (Ricci and Zong 2006; Okada and Mak 2004). Finally, an interesting finding in human cancers is the cell-in-cell feature, called entosis, after the Greek word *'entos'*, which means inside or within. In entosis, a living cell detached from its matrix is taken up by another living cell and eventu-ally either destroyed or released. Several hypotheses can be proposed to explain how tumour growth can be affected by this form of cannibalism (Overholtzer et al. 2007).

In conclusion, mounting evidence shows that the response of tumour cells to chemotherapy and radiation is not confined to apoptosis; it also includes other modes of death (Brown and Attardi, 2005). In view of this evidence, we should redirect our efforts to improve cancer therapy by using combinations of drugs that attack tumours by mechanisms other than just by inducing canonical apoptosis.

PCD in Neurodegenerative Disorders

Neuronal death in specific regions of the brain and loss of synapses underlie the symptoms of many human age-dependent and progressive neurodegenerative dis-orders, including Alzheimer's, Parkinson's and Huntington's diseases, and amyo-trophic lateral sclerosis. Apoptosis is intimately involved in the loss of neurons during neurodegenerative diseases (Thompson 1995; Bredesen 1995; Kostic et al. 1997; Ekshyyan and Aw 2004). Nevertheless, it is very difficult to convincingly demonstrate apoptosis in the brains of patients (Lossi and Gambino 2008), and the role of apoptosis as the principal death mediator in neurodegenerative disease has been intensely questioned (Dauer and Przedborski 2003; Jellinger 2006). Besides neural death and loss of neuronal connectivity, a common sign of many neurode-generative diseases is the accumulation and deposits of misfolded proteins, which affects various cell-signalling systems (Bence et al. 2001; Soto and Estrada 2008). Mounting evidence supports the notion that intracellular accumulation of these toxic misfolded proteins is a consequence of an age-related decline in autophagic and lys-osomal activity (Brunk and Terman 2002, Ravikumar et al. 2002; Nixon et al. 2005; Williams et al. 2006; Martinez-Vicente and Cuervo 2007), which eventually leads to apoptosis (Cataldo et al. 1996; Anglade et al. 1997).

Neuronal cell death in Parkinson's disease is more complex. In addition to the accumulation of aberrant autophagosome-like structures, there is apopto-sis and necrosis in response to oxidative damage (Jenner and Olanow 1996; Anglade et al. 1997; Tompkins et al. 1997; Stefanis 2005).

PCD in Age-associated Cardiovascular Diseases

Different types of cells undergo cell death in the diseased cardiovascular system. Loss of endothelial and smooth muscle cells is implicated in atherosclerosis, a major age-related disease in humans (Galis et al. 1994). Indeed, activation of the cellular suicide pathway leading to apoptosis of the endothelial cells represents an initial step in the development of atherosclerotic lesions (Bennett 1999; Kavurma et al. 2005). Furthermore, disease progression is defined by the extent of vascular smooth muscle cell apoptosis, which is responsible for the plaque instability that leads to heart attacks (Clarke and Bennett 2006). The complex micro-environment of the atherosclerotic plaques represents a challenge for the phagocytes trying to clear the apoptotic bodies. Indeed, the presence of secondary necrotic cells due to inadequate phagocytosis explains the persistence of inflammation in this pathology (Clarke et al. 2007). Besides atherosclerosis, death of endothelial cells and smooth muscle cells are also reported to contribute to the formation of aneurysms. However, it remains unclear whether apoptosis drives pathology or is just an unfortunate bystander effect of hemodynamic stress (Clarke et al. 2007).

The loss of the post-mitotic cardiomyocytes is associated with myocardial infarction and ischemic and dilated cardiomyopathies (Clarke et al. 2007). It is assumed that prevention of apoptosis may decrease the incidence of cardiac failure and improve the survival of endothelial and smooth muscle cells in the elderly (Duque 2000). In end stage heart failure, apoptotic rates of < 0.5% are found in combination with a seven-fold higher rate of necrosis. The elimination of cardiomyocytes does not only involve apoptosis and necrosis but includes extensive autophagocytosis, as has been documented in the cardiomyocytes of patients with dilated cardiomyopathy (Clarke et al. 2007; Levine and Kroemer 2008). As in other tissues, autophagy in the cardiovascular system is not invariably linked to a cell death mechanism (Levine and Kroemer 2008). Interestingly, heart-specific *atg5*-knockout in adult mice results in cardiac hypertrophy and contractile dysfunction, indicating the need for autophagy in normal heart physiology (Kuma et al. 2004). Finally, acute myocardial infarction, initially described as a paradigm of necrotic cell death due to breakdown of cellular energy metabolism, is also an example of apoptosis. Indeed, a central area of necrosis in the ischemic zone is bordered by apoptosis in the hypoperfused zone (Clarke et al. 2007). How reperfusion intervention influences the relative contributions of apoptosis and necrosis has not been fully elucidated.

2.4   Programmed Cell Death in Invertebrate Tissues

In the small nematode *C. elegans*, the 131 cells that die during development show nuclear chromatin aggregation, cytoplasmic condensation, and fragmentation of the cell into membrane-bound fragments (Sulston and Horvitz 1977). Non-apoptotic cell death has been reported only in rare cases, such as the developmental cell death of a linker cell (Abraham et al. 2007) with ultrastructural features comparable to

PCD type 3 or to necrosis in vertebrates (Clarke 1990). Moreover, Driscoll and colleagues demonstrated that mutations in mechanosensory genes in *C. elegans* result in necrotic death of neurons, which could mean that necrotic cell death is under genetic control in invertebrates as well (Lints and Driscoll 1996).During the development of *Drosophila melanogaster* (*D. melanogaster*) and later during its metamorphosis, large numbers of cells undergo death controlled by many different signals, yet most of them exhibit common morphological and biochemical changes that are characteristic of apoptosis in vertebrates (Abrams et al. 1993). However, activation of autophagy during *D. melanogaster* metamorphosis has been observed as well (Berry and Baehrecke 2007). Patterns of apoptosis can be typically visualized in live preparations using the vital dye, acridine orange.

The large tobacco hornworm moth, *Manduca sexta*, has been the subject of a number of studies on neuronal cell death and on muscle degeneration, both of which occur during metamorphosis. During degeneration of the intersegmental muscles of *Manducca sexta* there is an early lysosomal involvement, in the form of autophagy, and the characteristic morphology of apoptosis and the typical oligonucleosomal DNA fragmentation are absent (Zakeri et al. 1993).

Cell death in the salivary glands during metamorphosis of the blowfly, *Calliphora vomitoria*, is typically non-apoptotic, with vacuolation and enlargement of the cells, followed by their disintegration, which is indicative of necrosis (Bowen et al. 1993). Martin and Baehrecke (2004) provided evidence for a mixed cell death process in the salivary glands of *D. melanogaster*, with autophagic vacuoles, blebbing, and fragmentation of cells, but without the typical nuclear chromatin condensation and margination indicative of apoptosis.

2.5 Programmed Cell Death in Plants

As in animals, PCD in plants plays an indispensable role in development and in disease (Thomas and Franklin-Tong 2004; Bonke et al. 2003; Mea et al. 2007; Morel and Dangl 1997; Vercammen et al. 2007). Evidence for the occurrence of PCD *in planta* (as opposed to 'accidental' cell death) is provided by the existence of mutants that spontaneously mimic cell death lesions (Dietrich et al. 1994; Greenberg et al. 1994; Hanaoka et al. 2002). Studies on PCD in plants are much less advanced than in animals, and little initiative has been taken to classify and characterize plant PCD forms. Thus, it has not been determined whether processes such as leaf senescence, fruit ripening, and hypersensitivity responses overlap or are true separate PCD events. Although detailed molecular and ultrastructural analysis of how plant cells die is still underway, recent studies have provided indications that two animal PCD categories, namely apoptosis and autophagy, are more or less conserved (van Doorn and Woltering 2005).

The presence of apoptotic PCD *in planta* remains a subject of considerable debate. Morphological and biochemical hallmarks of apoptosis were demonstrated in plant PCD: cytoplasmic shrinkage, nuclear condensation, exposure of phosphatidylserine (PS), caspase-like activity, and DNA fragmentation down to nucleosomal ladders

(Greenberg 1996; Thomas and Franklin-Tong 2004). Most problematic in trying to decide whether or not apoptosis occurs in plants is that apoptotic bodies are seen only infrequently (Levine et al. 1996; Wang et al. 1996; Asai et al. 2000) and are never engulfed (see special features of plant PCD).

Autophagy in plants, which has been clearly documented, shows the typical characteristic massive vacuolization described in animal cells (Kroemer et al. 2005). Transmission electron microscopy shows that this process usually starts with disappearance of the endoplasmic reticulum and associated ribosomes, and it ends with degradation of the mitochondria and the nucleus. This is followed by collapse of the tonoplast and the plasma membrane. The final stage usually includes the disappearance of the rigid cell wall. Many authors reported autophagy during normal development, and during biotic and/or abiotic challenge of plants (Sect. 2.6.2 and Sect. 2.6.3).

2.5.1 Special Features of Plant PCD

Clearly, many of the morphological and biochemical features of animal PCD (apoptosis and autophagy) are conserved in the plant kingdom. However, in view of the differences in cellular architecture and evolutionary divergence of plants and animals, it is not surprising that plant PCD involves some plant-specific features while some known animal PCD features are absent. In plant tissue, the cell and its nucleus are not always fragmented into separate bodies, and the cell and its fragments are never engulfed by other cells. This has been postulated to be a consequence of the plants' rigid cell walls. Interestingly, the externalization of PS, which serves in animal PCD as an "eat me" signal, has also been observed in plants (Madeo et al. 2002). Did plant biologists 'miss' observing the engulfment of plant cells? Have plants lost the ability to recognize this signal? Does this molecule have another function?

Some features of plant PCD are different from those in animals due to the presence of plant-specific compartments; the chloroplast and even the cell wall assist in the induction and execution of plant PCD (Quirino et al. 2000; Lam et al. 2001). Several authors have shown that during PCD in plant cells the conversion of the remaining plastids into functional chloroplast rescues those plant cells from PCD. This implies that the point of no return for plant PCD seems to be more clearly defined than in animal cells (Kroemer et al. 2005; van Doorn 2005): in plants it is when chloroplasts begin to participate actively in PCD.

2.5.2 Plant Developmental PCD

As in animals, PCD is an essential part of the life cycle of plants and occurs from germination to seed production. Little is known about the mechanisms and initial signalling, but hormones and phytoregulators (ethylene, cytokinins, ABA, jasmonate, polyamines, etc.) seem to play very important roles.

PCD has been observed in many different phases of plant reproduction. During pollination, plants enforce self-incompatibility as an important strategy to avoid self-fertilization. Research on *Papaver rhoeas* has revealed that proteins in the pistil on which the pollen lands interact with the pollen and trigger PCD in incompatible (self) pollen. Typical apoptotic features were observed, such as DNA fragmentation into nucleosomal ladders, chromatin condensation, caspase-like activities, and cytochrome C release (Thomas and Franklin-Tong 2004). However, there was no mention of apoptotic bodies. In the tapetum, the cell layer that surrounds the developing pollen grains in the anther, cells undergo apoptosis-like PCD with consequent release of their contents, which then serve as nutrients for the pollen (Balk and Leaver 2001). Filonova et al. (2002) reported that PCD is the major mechanism responsible for elimination of subordinate embryos in polyembryonic seeds, a reproductive strategy commonly found in higher plants. Transmission electron micrographs clearly show engulfment of organelles and growing vacuoles, hallmarks of phagocytosis. Subsequent embryonic growth and germination are accompanied by the death of the suspensor cell of the embryo and parts of the seed (the endosperm), with formation of apoptotic bodies and DNA fragmentation (Young and Gallie 2000).

In animals, PCD is known for its essential role in morphogenesis (Penaloza et al. 2008). By contrast, morphogenesis in plants typically occurs by differential cell/tissue growth and differentiation (Beemster et al. 2006; De Veylder et al. 2007), without cell migration as seen in animal morphogenesis (Friedl et al. 2004). Only a few examples of plants in which cell death plays a role in the generation of leaf shape are known. In the genus *Monstera*, patches of cells die during the early stages of development of the leaf blade, generating holes or slits, which results in mature leaves containing a series of perforations or lobes (Kaplan 1984).

Finally, also the death of a whole plant organ is controlled at the level of PCD, which in this context is often termed senescence. A good example is leaf senescence, which is manifested as the yellowing of the leaf. Leaf senescence is important for converting cell material accumulated during growth of the leaf into nutrients, which are then utilized by the developing seeds and other growing organs (Lim et al. 2007). Also here, organelle disintegration and vacuolization are typical morphological characteristics, but in contrast to other autophagocytic PCD in plants, chloroplasts are involved first (Nam 1997; Otegui et al. 2005). These authors also reported the occurrence of chromatin condensation, DNA fragmentation and caspase-like activity in the senescent leaf (Cao et al. 2003), implying that senescent PCD could be a mixture of autophagy and apoptotic-like PCD.

Many more examples of PCD during plant development have been observed, such as vascular system development, pollen release and ovum development (Greenberg 1996).

2.5.3   *Plant Biotic and Abiotic PCD*

Plants, due to their sessile lifestyle, mount a whole battery of reactions to protect against a diverse set of biotic and abiotic threats, and PCD is in many cases an obligatory part in their survival-response.

Abiotic stress, such as wounding, salinity, drought, cold, heat, UV-B radiation, and oxidative stress, has been shown to induce a PCD-dependent defence response in plant tissue. A body of evidence indicates that PCD in response to abiotic stress can exhibit some apoptotic features (Li and Dickman 2004). However, it has been proposed that the cell death type that occurs following abiotic stress depends on the intensity (level and duration) of the stress (Li and Dickman 2004).

In response to organisms such as fungi, bacteria and viruses, plants mount the Hypersensitive Response (HR), which includes localized PCD, production of reactive oxygen intermediates, and rise in nitric oxide levels. HR-PCD induced by some pathogens and elicitors (molecules secreted by pathogens) seems morphologically similar to apoptosis in animals. Apoptotic features such as nuclear and cytoplasmic condensation, internucleosomal DNA cleavage and sometimes even apoptotic bodies are present in some cells undergoing the HR (Asai et al. 2000; Levine et al. 1996; Ryerson and Heath 1996; Wang et al. 1996).

Importantly, a pro-survival function for autophagy during stress responses has been reported. It has been demonstrated that autophagy promotes cell-survival during stress responses in plants (Liu et al. 2005) as well as in mammals (White 2008), which implies that there is intimate crosstalk between autophagy and apoptosis during stress responses.

2.6   Conclusions

PCD constitutes an integral part of plant and animal development. Moreover, PCD has been proven crucial in maintaining health and determining the outcome of disease. In general, we must recognize that much progress still has to be made in studying PCD at the level of organized tissues. The short duration of visible apoptotic stages and the co-existence of different PCD modes constrain advanced studies at the tissue level. We note that the phenomenon of autophagic cell death remains a puzzle. Morphological evaluation does not seem to be decisive in determining whether autophagy is instrumental or is an epiphenomenon of cell death. In-depth studies of pathogenesis in the complex environment of diseased tissue will be crucial for evaluating the types of PCD involved. Together with molecular and biochemical data, this should eventually generate therapeutic options for diseases such as cancer and neurodegenerative and cardiovascular disorders.

References

Abraham MC, Lu Y, Shaham S (2007) A morphologically conserved nonapoptotic program promotes linker cell death in *Caenorhabditis elegans*. Dev Cell 12:73–86

Abrams JM, White K, Fessler LI et al (1993) Programmed cell death during *Drosophila* embryogenesis. Development 117:29–43

Aerbajinai W, Giattina M, Lee YT et al (2003) The proapoptotic factor Nix is coexpressed with bcl-x_L during terminal erythroid differentiation. Blood 102:712–717

Akcali KC, Khan SA, Moulton BC (1996) Effect of decidualization on the expression of bax and bcl-2 in the rat uterine endometrium. Endocrinology 137:3123–3130

Allsopp RC, Vaziri H, Patterson C et al (1992) Telomere length predicts replicative capacity of human fibroblasts. Proc Natl Acad Sci USA 89:10114–10118

Anderson G, Jenkinson EJ (2001) Lymphostromal interactions in thymic development and function. Nat Rev Immunol 1:31–40

Anglade P, Vyas S, Hirsch EC et al (1997) Apoptosis in dopaminergic neurons of the human substantia nigra during normal aging. Histol Histopathol 12:603–610

Arnheim G (1890) Coagulationsnecrose und Kernschwund. Virchows Arch Pathol Anat 120:367–383

Asai T, Stone JM, Heard JE et al (2000) Fumonisin B1-induced cell death in *Arabidopsis* protoplasts requies jasmonate-, ethylene-, and salicylate-dependent signaling pathways. Plant Cell 12:1823–1836

Baehrecke EH (2002) How death shapes life during development. Nat Rev Mol Cell Biol 3:779–787

Balk J, Leaver CJ (2001) The PET1-CMS mitochondrial mutation in sunflower is associated with premature programmed cell death and cytochrome c release. Plant Cell 13:1803–1818

Bartelmez GW (1933) Histological studies on the menstruating mucous membrane of the human uterus. Contrib Embryol 24:141–186

Beemster GT, Vercruysse S, De Veylder L et al (2006) The *Arabidopsis* leaf as a model system for investigating the role of cell cycle regulation in organ growth. J Plant Res 119:43–50

Bence NF, Sampat RM, Kopito RR (2001) Impairment of the ubiquitin-proteasome system by protein aggregation. Science 292:1552–1555

Bennett MR (1999) Apoptosis of vascular smooth muscle cells in vascular remodelling and atherosclerotic plaque rupture. Cardiovasc Res 41:361–368

Berry DL, Baehrecke EH (2007) Growth arrest and autophagy are required for salivary gland cell degradation in *Drosophila*. Cell 131:1137–1148

Blanco-Rodríguez J (1998) A matter of death and life: the significance of germ cell death during spermatogenesis. Int J Androl 21:236–248

Blanco-Rodríguez J, Martínez-García C (1996) Induction of apoptotic cell death in the seminiferous tubule of the adult rat testis: assessment of the germ cell types that exhibit the ability to enter apoptosis after hormone suppression by oestradiol treatment. Int J Androl 19:237–247

Blanco-Rodríguez J, Martínez-García C, Porras A (2003) Correlation between DNA synthesis in the second, third and fourth generations of spermatogonia and the occurrence of apotosis in both spermatogonia and spermatocytes. Reproduction 126:661–668

Blum JW, Baumrucker CR (2002) Colostral and milk insulin-like growth factors and related substances: mammary gland and neonatal (intestinal and systemic) targets. Domest Anim Endocrinol 23:101–110

Bonke M, Thitamadee S, Mähönen AP et al (2003) APL regulates vascular tissue identity in *Arabidopsis*. Nature 426:181–186

Botchkareva NV, Ahluwalia G, Shander D (2006) Apoptosis in the hair follicle. J Invest Dermatol 126:258–264

Bowen ID, Morgan SM, Mullarkey K (1993) Cell death in the salivary glands of metamorphosing *Calliphora vomitoria*. Cell Biol Int 17:13–33

Bredesen DE (1995) Neural apoptosis. Ann Neurol 38:839–851

Bridger JM, Kill IR (2004) Aging of Hutchinson-Gilford progeria syndrome fibroblasts is characterised by hyperproliferation and increased apoptosis. Exp Gerontol 39:717–724

Brown JM, Attardi LD (2005) The role of apoptosis in cancer development and treatment response. Nat Rev Cancer 5:231–237

Brunk UT, Terman A (2002) The mitochondrial-lysosomal axis theory of aging: accumulation of damaged mitochondria as a result of imperfect autophagocytosis. Eur J Biochem 269:1996-2002

Bursch W (2001) The autophagosomal-lysosomal compartment in programmed cell death. Cell Death Differ 8:569–581

Bursch W, Ellinger A, Torok L et al (1997) In vitro studies on subtypes and regulation of active cell death. Toxicol In Vitro 11:579–588

Bursch W, Hochegger K, Torok L et al (2000) Autophagic and apoptotic types of programmed cell death exhibit different fates of cytoskeletal filaments. J Cell Sci 113:1189–1198

Bursch W, Paffe S, Putz B et al (1990) Determination of the length of the histological stages of apoptosis in normal liver and in altered hepatic foci of rats.Carcinogenesis 11:847–853

Cao J, Jiang F, Sodmergen et al (2003) Time-course of programmed cell death during leaf senescence in *Eucommia ulmoides*. J Plant Res 116:7–12

Cataldo AM, Hamilton DJ, Barnett JL et al (1996) Properties of the endosomal-lysosomal system in the human central nervous system: disturbances mark most neurons in populations at risk to degenerate in Alzheimer's disease. J Neurosci 16:186–199

Chautan M, Chazal G, Cecconi F et al (1999) Interdigital cell death can occur through a necrotic and caspase-independent pathway. Curr Biol 9:967–970

Cho A, Courtman DW, Langille BL (1995) Apoptosis (programmed cell death) in arteries of the neonatal lamb. Circ Res 76:168–175

Chu-Wang IW, Oppenheim RW (1978) Cell death of motoneurons in the chick embryo spinal cord. II. A quantitative and qualitative analysis of degeneration in the ventral root, including evidence for axon outgrowth and limb innervation prior to cell death. J Comp Neurol 177:59–85

Cicinelli E, Cignarelli M, Resta L et al (1993) Effects of the repetitive administration of progesterone by nasal spray in postmenopausal women. Fertil Steril 60:1020–1024

Clarke M, Bennett M (2006) Defining the role of vascular smooth muscle cell apoptosis in atherosclerosis. Cell Cycle 5:2329–2331

Clarke M, Bennett M, Littlewood T (2007) Cell death in the cardiovascular system. Heart 93:659–664

Clarke PGH (1990) Developmental cell death: morphological diversity and multiple mechanisms. Anat Embryol (Berl) 181:195–213

Coggeshall RE, Pover CM, Fitzgerald M (1994) Dorsal root ganglion cell death and surviving cell numbers in relation to the development of sensory innervation in the rat hindlimb. Brain Res Dev Brain Res 82:193–212

Cohen-Jonathan E, Bernhard EJ, McKenna WG (1999) How does radiation kill cells? Curr Opin Chem Biol 3:77–83

D'Herde K, De Prest B, Roels F (1996) Subtypes of active cell death in the granulosa of ovarian atretic follicles in the quail (*Coturnix coturnix japonica*). Reprod Nutr Dev 36:175–189

Dahm R (1999) Lens fibre cell differentiation. A link with apoptosis? Ophthalmic Res 31:163–183

Dahmoun M, Boman K, Cajander S et al (1999) Apoptosis, proliferation, and sex hormone receptors in superficial parts of human endometrium at the end of the secretory phase. J Clin Endocrinol Metab 84:1737–1743

Dauer W, Przedborski S (2003) Parkinson's disease: mechanisms and models. Neuron 39:889–909

de Magalhães JP, Migeot V, Mainfroid V et al (2004) No increase in senescence-associated β-galactosidase activity in Werner syndrome fibroblasts after exposure to H_2O_2. Ann N Y Acad Sci 1019:375–378

De Veylder L, Beeckman T, Inzé D (2007) The ins and outs of the plant cell cycle. Nat Rev Mol Cell Biol 8:655–665

Degterev A, Yuan J (2008) Expansion and evolution of cell death programmes. Nat Rev Mol Cell Biol 9:378–390

Depalo R, Nappi L, Loverro G et al (2003) Evidence of apoptosis in human primordial and primary follicles. Hum Reprod 18:2678–2682

Dharmarajan AM, Goodman SB, Atiya N et al (2004) Role of apoptosis in functional luteolysis in the pregnant rabbit corpus luteum: evidence of a role for placental-derived factors in promoting luteal cell survival. Apoptosis 9:807–814

Dietrich RA, Delaney TP, Uknes SJ et al (1994) *Arabidopsis* mutants simulating disease resistance response. Cell 20:565–577

Duque G (2000) Apoptosis in cardiovascular aging research: future directions. Am J Geriatr Cardiol 9:263–264

Eckhart L, Declercq W, Ban J et al (2000) Terminal differentiation of human keratinocytes and stratum corneum formation is associated with caspase-14 activation. J Invest Dermatol 115:1148–1151

Ekshyyan O, Aw TY (2004) Apoptosis: a key in neurodegenerative disorders. Curr Neurovasc Res 1:355–371

Ellis HM, Horvitz HR (1986) Genetic control of programmed cell death in the nematode *C. elegans*. Cell 44:817–829

Ferguson DJP, Anderson TJ (1981) Ultrastructural observations on cell death by apoptosis in the "resting" human breast. Virchows Arch 393:193–203

Festjens N, Vanden Berghe T, Vandenabeele P (2006) Necrosis, a well-orchestrated form of cell demise: signalling cascades, important mediators and concomitant immune response. Biochim Biophys Acta 1757:1371–1387

Filonova LH, von Arnold S, Daniel G et al (2002) Programmed cell death eliminates all but one embryo in a polyembryonic plant seed. Cell Death Differ 9:1057–1062

Fimia GM, Stoykova A, Romagnoli A et al (2007) Ambra1 regulates autophagy and development of the nervous system. Nature 447:1121–1125

Flemming W (1885) Über die Bildung van Richtungsfiguren in Säugetiereiern beim Untergang Graaf'scher Follikel. Arch Anat Entwickelungsgeschichte 221–224

Fraser HM, Lunn SF, Harrison DJ et al (1999) Luteal regression in the primate: different forms of cell death during natural and gonadotropin-releasing hormone antagonist or prostaglandin analogue-induced luteolysis. Biol Reprod 61:1468–1479

Friedl P, Hegerfeldt Y, Tusch M (2004) Collective cell migration in morphogenesis and cancer. Int J Dev Biol 48:441–449

Galán A, O'Connor JE, Valbuena D et al (2000) The human blastocyst regulates endometrial apoptosis in embryonic adhesion. Biol Reprod 63:430–439

Galis ZS, Sukhova GK, Lark MW et al (1994) Increased expression of matrix metalloproteinases and matrix degrading activity in vulnerable regions of human atherosclerotic plaques. J Clin Invest 94:2493–2503

Galluzzi L, Maiuri MC, Vitale I et al (2007) Cell death modalities: classification and pathophysiological implications. Cell Death Differ 14:1237–1243

Gavrieli Y, Sherman Y, Ben-Sasson SA (1992) Identification of programmed cell death in situ via specific labeling of nuclear DNA fragmentation. J Cell Biol 119:493–501

Gaytán F, Morales C, García-Pardo L et al (1998) Macrophages, cell proliferation, and cell death in the human menstrual corpus luteum. Biol Reprod 59:417–425

Glucksmann A (1951) Cell deaths in normal vertebrate ontogeny. Biol Rev 26:59–86

Godlewski MM, Biernat M, Woliński J et al (2004) Epithelial cell apoptosis in the gut of pig neonates and chicken - new application for laser scanning cytometry. In Diederich M (ed): Proceedings of the Signal Transduction. Signal transduction pathways as therapeutic targets. Luxembourg 2004, p 133

Godlewski MM, Hallay N, Bierla JB et al (2007) Molecular mechanism of programmed cell death in the gut epithelium of neonatal piglets. J Physiol Pharmacol 58:97–113

Godlewski MM, Słupecka M, Woliński J et al (2005) Into the unknown-The death pathways in the neonatal gut epithelium. J Physiol Pharmacol 56:7–24

Gray DHD, Seach N, Ueno T et al (2006) Developmental kinetics, turnover, and stimulatory capacity of thymic epithelial cells. Blood 108:3777–3785

Greenberg JT (1996) Programmed cell death: a way of life for plants. Proc Natl Acad Sci USA 93:12094–12097

Greenberg JT, Guo A, Klessig DF et al (1994) Programmed cell death in plants: a pathogen-triggered response activated coordinately with multiple defense functions. Cell 77:551–563

Gregory T, Yu C, Ma A et al (1999) GATA-1 and erythropoietin cooperate to promote erythroid cell survival by regulating bcl-x$_L$ expression. Blood 94:87–96

Hanaoka H, Noda T, Shirano Y et al (2002) Leaf senescence and starvation-induced chlorosis are accelerated by the disruption of an *Arabidopsis* autophagy gene. Plant Physiol 129:1181–1193

Harris AJ, McCaig CD (1984) Motoneuron death and motor unit size during embryonic development of the rat. J Neurosci 4:13–24

Henell F, Ericsson JL, Glaumann H (1983) An electron microscopic study of the post-partum involution of the rat uterus. With a note on apparent crinophagy of collagen. Virchows Arch B Cell Pathol Incl Mol Pathol 42:271–287

Hengartner MO, Horvitz HR (1994) Programmed cell death in *Caenorhabditis elegans*. Curr Opin Genet Dev 4:581–586

Hockenbery DM, Zutter M, Hickey W et al (1991) Bcl-2 protein is topographically restricted in tissues characterized by apoptotic cell death. Proc Natl Acad Sci USA 88:6961–6965

Horvitz HR, Chalfie M, Trent C et al (1982) Serotonin and octopamine in the nematode *Caenorhabditis elegans*. Science 216:1012–1014

Hsueh AJ, Billig H, Tsafriri A (1994) Ovarian follicle atresia: a hormonally controlled apoptotic process. Endocr Rev 15:707–724

Huckins C, Oakberg EF (1978) Morphological and quantitative analysis of spermatogonia in mouse testes using whole mounted seminiferous tubules. I. The normal testes. Anat Rec 192:519-528

Iwanaga T (1995) The involvement of macrophages and lymphocytes in the apoptosis of enterocytes. Arch Histol Cytol 58:151–159

Jäger K, Fisher H, Tschachler E et al (2007) Terminal differentiation of nail matrix keratinocytes involves up-regulation of DNase1L2 but is independent of caspase-14 expression. Differentiation 75:939–946

Jahnukainen K, Chrysis D, Hou M et al (2004) Increased apoptosis occurring during the first wave of spermatogenesis is stage-specific and primarily affects midpachytene spermatocytes in the rat testis. Biol Reprod 70:290–296

James TN (1998) Normal and abnormal consequences of apoptosis in the human heart. Annu Rev Physiol 60:309–325

Jellinger KA (2006) Challenges in neuronal apoptosis. Curr Alzheimer Res 3:377–391

Jenner P, Olanow CW (1996). Oxidative stress and the pathogenesis of Parkinson's disease. Neurology 47:S161–S170

Kaplan DR (1984) Alternative modes of organogenesis in higher plants In: White RA, Dickenson WC (eds) Contemporary problems in plant anatomy. Academic Press, New York, pp 261–300

Kavurma MM, Bhindi R, Lowe HC et al (2005) Vessel wall apoptosis and atherosclerotic plaque instability. J Thromb Haemost 3:465–472

Kerr JFR (1971) Shrinkage necrosis: a distinct mode of cellular death. J Pathol 105:13–20

Kerr JFR, Harmon BV, Searle J (1974) An electron-microscope study of cell deletion in the anuran tadpole tail during spontaneous metamorphosis with special reference to apoptosis of striated muscle fibers. J Cell Sci 14:571–585

Kerr JFR, Searle J (1973) Deletion of cells by apoptosis during castration-induced involution of the rat prostate. Virchows Arch B Cell Pathol 13:87–102

Kerr JFR, Winterford CM, Harmon BV (1994) Apoptosis. Its significance in cancer and cancer therapy. Cancer 73:2013-2026

Kerr JFR, Wyllie AH, Currie AR (1972) Apoptosis: a basic biological phenomenon with wide-ranging implications in tissue kinetics. Br J Cancer 26:239–257

Kim HS, Hwang KK, Seo JW et al (2000). Apoptosis and regulation of bax and bcl-x proteins during human neonatal vascular remodeling. Artheroscler Thromb Vasc Biol 20:957–963

Kim HS, Seo JU, Hwang KK et al (1998) Role of apoptosis in the closure of human umbilical vessel at birth and its mechanism. Atherosclerosis 136:77

Kitanaka C, Kuchino Y (1999) Caspase-independent programmed cell death with necrotic morphology. Cell Death Differ 6:508–515

Kokawa K, Shikone T, Nakano R (1996) Apoptosis in the human uterine endometrium during the menstrual cycle. J Clin Endocrinol Metab 81:4144–4147

Kostic V, Jackson-Lewis V, de Bilbao F et al (1997) Bcl-2: prolonging life in a transgenic mouse model of familial amyotrophic lateral sclerosis. Science 277:559–562

Kroemer G, El-Deiry WS, Golstein P et al (2005) Classification of cell death: recommendations of the Nomenclature Committee on Cell Death. Cell Death Differ 12:1463–1467

Krysko DV, Brouckaert G, Kalai M et al (2003) Mechanisms of internalization of apoptotic and necrotic L929 cells by a macrophage cell line studied by electron microscopy. J Morphol 258:336–345

Kuma A, Hatano M, Matsui M et al (2004) The role of autophagy during the early neonatal starvation period. Nature 432:1032–1036

Lam E, Kato N, Lawton M (2001) Programmed cell death, mitochondria and the plant hypersensitive response. Nature 411:848–853

Lang RA, Bishop JM (1993) Macrophages are required for cell death and tissue remodeling in the developing mouse eye. Cell 74:453–462

Lawson SJ, Davies HJ, Bennett JP et al (1997) Evidence that spinal interneurons undergo programmed cell death postnatally in the rat. Eur J Neurosci 9:794–799

Lee J, Richburg JH, Younkin SC et al (1997) The Fas system is a key regulator of germ cell apoptosis in the testis. Endocrinology 138:2081–2088

Lefranc F, Facchini V, Kiss R (2007) Proautophagic drugs: a novel means to combat apoptosis-resistant cancers, with a special emphasis on glioblastomas. Oncologist 12:1395–1403

Leist M, Jäättelä M (2001) Four deaths and a funeral: from caspases to alternative mechanisms. Nat Rev Mol Cell Biol 2:589–598

Leppert PC (1998) Proliferation and apoptosis of fibroblasts and smooth muscle cells in rat uterine cervix throughout gestation and the effect of the antiprogesterone onapristone. Am J Obstet Gynecol 178:713–725

Levine A, Pennell RI, Alvarez ME et al (1996) Calcium-mediated apoptosis in plant hypersensitive disease resistance response. Curr Biol 6:427–437

Levine B (2007) Cell biology: autophagy and cancer. Nature 446:745–747

Levine B, Kroemer G (2008) Autophagy in the pathogenesis of disease. Cell 132:27–42

Li W, Dickman MB (2004) Abiotic stress induces apoptotic-like features in tobacco that is inhibited by expression of human bcl-2. Biotechnol Lett 26:87–95

Lim PO, Kim HJ, Nam HG (2007) Leaf senescence. Annu Rev Plant Biol 58:115–136

Lints R, Driscoll M (1996) Programmed and pathological cell death in *Caenorhabditis elegans*. In: Holbrook N, Martin GR, Lockshin RA (eds) Cellular aging and cell death. Wiley, New York pp. 235–253

Liu Y, Schiff M, Czymmek K et al (2005) Autophagy regulates programmed cell death during the plant innate immune response. Cell 121:567–577

Lockshin RA (1969) Programmed cell death. Activation of lysis by a mechanism involving the synthesis of protein. J Insect Physiol 15:1505–1516

Lockshin RA, Willams CM (1964) Programmed cell death. II. Endocrine potentiation of the breakdown of the intersegmental muscles of silkmoths. J Insect Physiol 10:643–649

Lockshin RA, Williams CM (1965) Programmed cell death. V. Cytolytic enzymes in relation to the breakdown of the intersegmental muscles of silkmoths. J Insect Physiol 11:831–844

Lockshin RA, Zakeri Z (2004) Apoptosis, autophagy, and more. Int J Biochem Cell Biol 36:2405–2419

Lossi L, Gambino G (2008) Apoptosis of the cerebellar neurons. Histol Histopathol 23:367–380

Madeo F, Herker E, Maldener C et al (2002) A caspase-related protease regulates apoptosis in yeast. Mol Cell 9:911–917

Majno G, Joris I (1995) Apoptosis, oncosis and necrosis. An overview of cell death. Am J Pathol 146:3–15

Malamitsi-Puchner A, Sarandakou A, Tziotis J et al (2001) Evidence for a suppression of apoptosis in early postnatal life. Acta Obstet Gynecol Scand 80:994–997

Martin DN, Baehrecke EH (2004) Caspases function in autophagic programmed cell death in *Drosophila*. Development 131:275–284

Martinez-Vicente M, Cuervo AM (2007) Autophagy and neurodegeneration: when the cleaning crew goes on strike. Lancet Neurol 6:352–361

Matova N, Cooley L (2001) Comparative aspects of animal oogenesis. Dev Biol 231:291–320

McGee EA, Hsueh AJ (2000) Initial and cyclic recruitment of ovarian follicles. Endocr Rev 21:200–214

Mea MD, Serafini-Fracassini D, Duca SD (2007) Programmed cell death: similarities and differences in animals and plants. A flower paradigm. Amino Acids 33:395–404

Meeson A, Palmer M, Calfon M et al (1996) A relationship between apoptosis and flow during programmed capillary regression is revealed by vital analysis. Development 122:3929–3938

Meeson AP, Argilla M, Ko K et al (1999) VEGF deprivation-induced apoptosis is a component of programmed capillary regression. Development 126:1407–1415

Mills KR, Reginato M, Debnath J et al (2004) Tumor necrosis factor-related apoptosis-inducing ligand (TRAIL) is required for induction of autophagy during lumen formation in vitro. Proc Natl Acad Sci USA 101:3438–3443

Mirkes PE (2008) Cell death in normal and abnormal development. Congenit Anom (Kyoto) 48:7–17

Morel JB, Dangl JL (1997) The hypersensitive response and the induction of cell death in plants. Cell Death Differ 4:671–683

Morita Y, Tilly JL (1999) Oocyte apoptosis: like sand through an hourglass. Dev Biol 213:1–17

Murdoch WJ (1995) Programmed cell death in preovulatory ovine follicles. Biol Reprod 53:8–12

Nam HG (1997) The molecular genetic analysis of leaf senescence. Curr Opin Biotechol 8:200–207

Nataraj A, Pathak S, Hopwood VL et al (1994) Bcl-2 oncogene blocks differentiation and extends viability but does not immortalize normal human keratinocytes. Int J Oncology 4:1211–1218

Navarro CL, Cau P, Lévy N (2006) Molecular bases of progeroid syndromes. Human Mol Genet 15:R151–R161

Nishimura EK, Granter SR, Fisher DE (2005) Mechanisms of hair graying: incomplete melanocyte stem cell maintenance in the niche. Science 307:720–724

Nixon RA, Wegiel J, Kumar A et al (2005) Extensive involvement of autophagy in Alzheimer disease: an immuno-electron microscopy study. J Neuropathol Exp Neurol 64:113–122

Nussbaum M (1901) Zur Rückbildung embryonaler Anlagen. Arch Mikrosk Anat 57:676–705

O'Neill M, Núñez F, Melton DW (2003) p53 and a human premature ageing disorder. Mech Ageing Dev 124:599–603

Okada H, Mak TW (2004) Pathways of apoptotic and non-apoptotic death in tumour cells. Nat Rev Cancer 4:592–603

Okano A, Ogawa H, Takahashi H et al (2007) Apoptosis in the porcine uterine endometrium during the estrous cycle, early pregnancy and post partum. J Reprod Dev 53:923–930

Opferman JT (2007) Life and death during hematopoietic differentiation. Curr Opin Immunol 19:497–502

Otegui MS, Noh YS, Martínez DE et al (2005) Senescence-associated vacuoles with intense proteolytic activity develop in leaves of *Arabidopsis* and soybean. Plant J 41:831–844

Overholtzer M, Mailleux AA, Mouneimne G et al (2007) A nonapoptotic cell death process, entosis, that occurs by cell-in-cell invasion. Cell 131:966–979

Parr EL, Tung HN, Parr MB (1987) Apoptosis as the mode of uterine epithelial cell death during embryo implantation in mice and rats. Biol Reprod 36:211–225

Piacentini M, Autuori F (1994) Immunohistochemical localization of tissue transglutaminase and bcl-2 in rat uterine tissues during embryo implantation and post-partum involution. Differentiation 57:51–61

Pilar G, Landmesser L (1976) Ultrastructural differences during embryonic cell death in normal and peripherally deprived ciliary ganglia. J Cell Biol 68:339–356

Quatacker JR (1971) Formation of autophagic vacuoles during human corpus luteum involution. Z Zellforsch Mikrosk Anat 122:479–487

Quirino BF, Noh YS, Himelblau E et al (2000) Molecular aspects of leaf senescence. Trends Plant Sci 5:278–282

Ravikumar B, Duden R, Rubinsztein DC (2002) Aggregate-prone proteins with polyglutamine and polyalanine expansions are degraded by autophagy. Hum Mol Genet 11:1107–1117

Rello S, Stockert JC, Moreno V et al (2005) Morphological criteria to distinguish cell death induced by apoptotic and necrotic treatments. Apoptosis 10:201–208

Ricci MS, Zong WX (2006) Chemotherapeutic approaches for targeting cell death pathways. Oncologist 11:342–357

Roach HI, Clarke NM (2000) Physiological cell death of chondrocytes in vivo is not confined to apoptosis. New observations on the mammalian growth plate. J Bone Joint Surg Br 82:601–613

Rodríguez I, Ody C, Araki K et al (1997) An early and massive wave of germinal cell apoptosis is required for the development of functional spermatogenesis. EMBO J 16:2262–2270

Rotello RJ, Lieberman RC, Lepoff RB et al (1992) Characterization of uterine epithelium apoptotic cell death kinetics and regulation by progesterone and RU486. Am J Pathol 140:449–456

Ryerson DE, Heath MC (1996) Cleavage of nuclear DNA into oligonucleosomal fragments during cell death induced by fungal infection or by abiotic treatments. Plant Cell 8:393–402

Sarandakou A, Protonotariou E, Rizos D et al (2003) Soluble Fas antigen and soluble Fas ligand in early neonatal life. Early Hum Dev 75:1–7

Sarraf CE, Bowen ID (1988) Proportions of mitotic and apoptotic cells in a range of untreated experimental tumours. Cell Tissue Kinet 21:45–49

Saunders JW Jr (1966) Death in embryonic systems. Science 154:604–612

Schlafke S, Enders AC (1975) Cellular basis of interaction between trophoblast and uterus at implantation. Biol Reprod 12:41–65

Schulte-Hermann R, Bursch W, Grasl-Kraupp B et al (1995) Role of active cell death (apoptosis) in multistage carcinogenesis. Toxicol Lett 82/83:143–148

Schweichel JU, Merker HJ (1973) The morphology of various types of cell death in prenatal tissues. Teratology 7:253–266

Searle J, Collins DJ, Harmon B et al (1973) The spontaneous occurrence of apoptosis in squamous carcinomas of the uterine cervix. Pathology 5:163–169

Shibahara T, Sato N, Waguri S et al (1995) The fate of effete epithelial cells at the villus tips of the human small intestine. Arch Histol Cytol 58:205–219

Slomp J, Gittenberger-de Groot AC, Glukhova MA et al (1997) Differentiation, dedifferentiation, and apoptosis of smooth muscle cells during the development of the human ductus arteriosus. Arterioscler Thromb Vasc Biol 17:1003–1009

Soto C, Estrada LD (2008) Protein misfolding and neurodegeneration. Arch Neurol 65:184–189

Spillare EA, Robles AI, Wang XW et al (1999) p53-mediated apoptosis is attenuated in Werner syndrome cells. Genes Dev 13:1355–1360

Stefanis L (2005) Caspase-dependent and -independent neuronal death: two distinct pathways to neuronal injury. Neuroscientist 11:50–62

Strehler BL (ed) (1977) Time, cells, and aging, 2nd edn. Academic Press, New York

Sulston JE, Horvitz HR (1977) Postembryonic cell lineages of the nematode *Caenorhabditis elegans*. Dev Biol 56:110–156

Tadakuma T, Kizaki H, Odaka C et al (1990) CD4+CD8+ thymocytes are susceptible to DNA fragmentation induced by phorbol ester, calcium ionophore and anti-CD3 antibody. Eur J Immunol 20:779–784

Takamoto N, Leppert PC, Yu SY (1998) Cell death and proliferation and its relation to collagen degradation in uterine involution of rat. Connect Tissue Res 37:163–175

Tata JR (1968) Early metamorphic competence of *Xenopus* larvae. Dev Biol 18:415–440

Thomas GS, Franklin-Tong VE (2004) Self-incompatibility triggers programmed cell death in *Papaver* pollen. Nature 429: 305–309

Thompson CB (1995) Apoptosis in the pathogenesis and treatment of disease. Science 267:1456–1462

Tilly JL (2001) Emerging technologies to control oocyte apoptosis are finally treading on fertile ground. Scientific World Journal 1:181–183

Tompkins MM, Basgall EJ, Zamrini E et al (1997) Apoptotic-like changes in Lewy-body-associated disorders and normal aging in substantia nigral neurons. Am J Pathol 150:119–131

Tran HB, Ohlsson M, Beroukas D et al (2002) Subcellular redistribution of La/SSB autoantigen during physiologic apoptosis in the fetal mouse heart and conduction system. A clue to the pathogenesis of congenital heart block. Arthritis Rheum 46:202–208

Trump BF, Berezesky IK, Chang SH et al (1997) The pathways of cell death: oncosis, apoptosis, and necrosis. Toxicol Pathol 25:82–88

Vakifahmetoglu H, Olsson M, Zhivotovsky B (2008) Death through a tragedy: mitotic catastrophe. Cell Death Differ 15:1153–1162

van Doorn WG (2005) Plant programmed cell death and the point of no return. Trends Plant Sci 10:478–483

van Doorn WG, Woltering EJ (2005) Many ways to exit? Cell death categories in plants. Trends Plant Sci 10:117–122

Vaskivuo TE, Ottander U, Oduwole O et al (2002) Role of apoptosis , apoptosis-related factors and 17β-hydroxysteroid dehydrogenases in human corpus luteum regression. Mol Cell Endocrinol 194:191–200

Veis DJ, Sorenson CM, Shutter JR et al (1993) Bcl-2-deficient mice demonstrate fulminant lymphoid apoptosis, polycystic kidneys, and hypopigmented hair. Cell 75:229–240

Vercammen D, Declercq W, Vandenabeele P et al (2007) Are metacaspases caspases? J Cell Biol 179:375–380

Verma V (1983) Ultrastructural changes in human endometrium at different phases of the menstrual cycle and their functional significance. Gynecol Obstet Invest 15:193–212

Walker NI, Bennett RE, Kerr JFR (1989) Cell death by apoptosis during involution of the lactating breast in mice and rats. Am J Anat 185:19–32

Wang B, Xiao C, Goff AK (2003) Progesterone-modulated induction of apoptosis by interferon-tau in cultured epithelial cells of bovine endometrium. Biol Reprod 68:673–679

Wang H, Li J, Bostock RM et al (1996) Apoptosis: a functional paradigm for programmed plant cell death induced by a host-selective phytotoxin and invoked during development. Plant Cell 8:375–391

Welsh AO, Enders AC (1993) Chorioallantoic placenta formation in the rat. III. Granulated cells invade the uterine luminal epithelium at the time of epithelial cell death. Biol Reprod 49:38–57

White E (2008) Autophagic cell death unraveled. Pharmacological inhibition of apoptosis and autophagy enables necrosis. Autophagy 4:399–401

Williams A, Jahreiss L, Sarkar S et al (2006) Aggregate-prone proteins are cleared from the cytosol by autophagy: therapeutic implications. Curr Top Dev Biol 76:89–101

Woliński J, Biernat M, Guilloteau P et al (2003) Exogenous leptin controls the development of the small intestine in neonatal piglets. J Endocrinol 177:215–222

Wyllie AH (1981) Cell death: a new classification separating apoptosis from necrosis. In: Bowen ID, Lockshin RA (eds) Cell death in biology and pathology, pp. 9–34. Chapman and Hall, London.

Young TE, Gallie DR (2000) Programmed cell death during endosperm development. Plant Mol Biol 44:283–301

Yuan J, Shaham S, Ledoux S et al (1993) The *C. elegans* cell death gene ced-3 encodes a protein similar to mammalian interleukin-1β-converting enzyme. Cell 75:641–652

Zakeri Z, Bursch W, Tenniswood M et al (1995) Cell death: programmed apoptosis, necrosis, or other? Cell Death Differ 2:87–96

Zakeri Z, Penaloza C, Orlanski S et al (2008) Cell death in mammalian development. Curr Pharm Des 14:184–196

Zakeri ZF, Ahuja HS (1997) Cell death/apoptosis: normal, chemically induced, and teratogenic effect. Mutat Res 396:149–161

Zakeri ZF, Quaglino D, Latham T et al (1993) Delayed internucleosomal DNA fragmentation in programmed cell death. FASEB J 7:470–478

Zarzyńska J, Gajkowska B, Wojewódzka U et al (2007) Apoptosis and autophagy in involuting bovine mammary gland is accompanied by up-regulation of TGF-β1 and suppression of somatotropic pathway. Pol J Vet Sci 10:1–9

Zimmermann LE, Font RL (1966) Congenital malformations of the eye. JAMA 196:96–104

Zou H, Henzel WJ, Liu X et al (1997) Apaf-1, a human protein homologous to *C. elegans* CED-4, participates in cytochrome c-dependent activation of caspase-3. Cell 90:405–413

Chapter 3
Role of Attraction and Danger Signals in the Uptake of Apoptotic and Necrotic Cells and its Immunological Outcome

Christoph Peter, Sebastian Wesselborg and Kirsten Lauber

Abstract: In healthy multicellular organisms dying cells—apoptotic as well as necrotic ones—are swiftly engulfed either by phagocytosis-competent neighbouring cells or by professional phagocytes. The process of dying cell removal with special regard to the key players and the molecular mechanisms involved in this scenario as well as the postprandial reactions of the phagocyte and the immunological outcome is a rapidly evolving field of scientific interest. During the last years numerous studies have led to a detailed understanding of the interaction site between the dying cell and the phagocyte, which today is called the phagocytic synapse, and the current concept that apoptotic cell removal leads to an anti-inflammatory whereas necrotic cell removal stimulates a pro-inflammatory phagocyte reaction. Conversely, our knowledge about the soluble factors released from dying cells is very limited, although meanwhile it is generally accepted that not only the dying cell itself but also the substances, which are liberated during cell death, contribute to the process of dying cell removal as well as to its immunological outcome. Here, we intend to summarize the current knowledge about attraction and danger signals of apoptotic and necrotic cells, their function as chemoattractants in phagocyte recruitment, additional effects on the immune system, and the receptors, which are engaged in this scenario.

Keywords: Apoptosis • Necrosis • Phagocytosis • Engulfment • Migration • Chemotaxis • 'Find-me' signals • Attraction signals • Danger signals

K. Lauber (✉) · C. Peter · S. Wesselborg
Department of Internal Medicine I
Eberhard Karls University
Otfried-Mueller-Str. 10
D-72076 Tuebingen, Germany
Tel.: +49-7071-29-83137
Fax: +49-7071-294680
E-mail: kirsten.lauber@med.uni-tuebingen.de

D.V. Krysko, P. Vandenabeele (eds.), *Phagocytosis of Dying Cells,*
DOI: 10.1007/978-1-4020-9292-3_3, © Springer Science+Business Media B.V. 2009

3.1 Introduction

One of the paradoxes of life is that cell death is crucially required for the survival and homeostasis of multicellular organisms. In the course of daily cell turnover and tissue regeneration billions of cells die by apoptosis giving rise to a considerable amount of dead cell mass that has to be removed. Phagocytosis of these apoptotic cells is fundamentally important throughout life, because non-cleared apoptotic cells become secondary necrotic and release intracellular contents, thus instigating inflammatory and autoimmune responses (Gaipl et al. 2004; Lauber et al. 2004). Additionally, although considered as unphysiological form of cell death, necrosis also produces dead cells. These necrotic cells have to be removed even more rapidly, since due to their leaky plasma membrane they can and do liberate putatively cytotoxic, (auto)-antigenic and pro-inflammatory intracellular constituents. To this end, in healthy multicellular organisms dying cells—apoptotic as well as necrotic—are swiftly engulfed either by phagocytosis-competent neighbouring cells or by professional phagocytes (Krysko et al. 2006). In the latter case the question arises about how the phagocyte reaches its prey in time, since usually the phagocyte is not located right next to the dying cell. One solution to this dilemma would be that the dying cell secretes soluble mediators, which guide the scavenger to their source. Several studies provide support for this notion. Particularly in the context of apoptotic cell removal diverse attraction or 'find-me' signals have been characterized in the supernatants of different apoptotic cells. Amongst them the lysophospholipid lysophosphatidylcholine (LPC), thrombospondin-1 (TSP-1) and its fragments, a cross linked dimer of S19 ribosomal protein (dRP S19), endothelial monocyte-activating polypeptide II (EMAP II), cleavage products of human tyrosyl-tRNA synthetase (TyrRS) and even apoptotic micro-blebs have been described to actively stimulate phagocyte chemotaxis. Importantly, the chemoattractants produced by apoptotic cells do not only recruit macrophages to the site of cell death, they also modulate the activation and differentiation state of their target cells and can thereby influence the consecutive immune response. The effects on phagocyte activation and/or differentiation and the subsequent immune response are especially well studied for soluble factors released by necrotic cells, such as heat-shock proteins (HSPs), uric acid (MSU), high mobility group box 1 protein (HMGB-1), S100 proteins, ATP, and nucleic acids (RNA and DNA), whereas in this case the process of phagocyte recruitment is only poorly understood. This book chapter, therefore, is intended to discuss, which soluble attraction or 'find-me' signals are released by dying (apoptotic as well as necrotic) cells, how they are sensed by the phagocyte, and which other effects apart from phagocyte recruitment these factors can instigate.

3.2 'Find-me' Signals of Apoptotic Cells

As mentioned above, 'find-me' signals released from apoptotic cells play a crucial role in the timely removal of the apoptotic cell before it undergoes secondary necrosis. In the following, the different attraction signals from apoptotic cells are introduced (Fig. 3.4).

3.2.1 Dimer of Ribosomal Protein S19 (dRP S19)

Historically the first attraction signal of apoptotic cells that has been identified was a covalent dimer of ribosomal protein S19 (dRP S19) in 1998 (Horino et al. 1998; Fig. 3.1). Originally characterized as a part of the small subunit of ribosomes, RP S19 later had been described to act as the essential chemoattractant for monocytes in the extracts of rheumatoid arthritis-synovial lesions (Nishiura et al. 1996). In their report Horino and coworkers could observe cross-linking and release of dRP S19 24 hours after apoptosis induction in HL-60 cells upon heat treatment (43°C for 60 minutes). When the so treated HL-60 cells were injected into guinea pig skin or rabbit footpads, a strong monocyte/macrophage infiltration and subsequent phagocytic removal of the apoptotic HL-60 cells could be detected. Since in vitro the cross-linking of RP S19 and the release of the chemotactic activity were preceded by an increase in transglutaminase 2 activity and, furthermore, depletion with an anti-isopeptide-bond antibody could remove the attraction signal from apoptotic culture supernatants, the authors hypothesized that transglutaminase 2 could be responsible for covalent dimerization of RP S19—a notion, which was confirmed by in vitro transglutamination of purified RP S19 and subsequent mapping of the isopeptide-bond to Gln 137 (Nishimura et al. 2001). It should be emphasized that only the cross-linked dimer (dRP S19) and not the monomers of RP S19 could exert the chemoattractant effect and that chemotaxis induction was specific for monocytes and macrophages, whereas neutrophils were not recruited (Nishimura et al. 2001; Umeda et al. 2004). This was an intriguing finding, since additional studies utilizing neutralizing antibodies and receptor antagonists could identify the C5a receptor CD88, although expressed on monocytes as well as on neutrophils, to be the central G-protein coupled receptor (GPCR) in dRP S19 stimulated chemotaxis (Nishiura et al. 1998). Detailed mutational analyses could resolve this inconsistency by mapping the receptor binding sites of dRP S19 to KHK[43] and LDR[133] (Shibuya et al. 2001) and identifying a so-called switch moiety at the ultimate C-terminus (IAGQVAAANKK[144]), which determines between agonistic stimulation of CD88 in

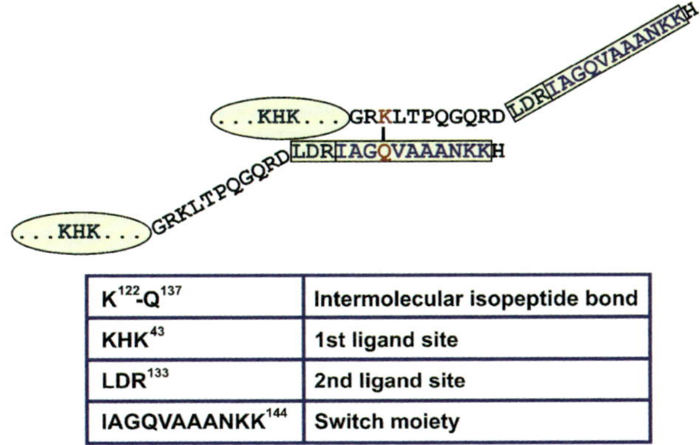

$K^{122}\text{-}Q^{137}$	Intermolecular isopeptide bond
KHK[43]	1st ligand site
LDR[133]	2nd ligand site
IAGQVAAANKK[144]	Switch moiety

Fig. 3.1 Schematic structure of ribosomal protein S19 (RP S19). Sites for ligand binding and transglutamination are indicated. Modified from Yamamoto (2007).

monocytes and, contrarily, antagonistic blocking of CD88 in neutrophils (Shrestha et al. 2003). Thus, dRP S19 can stimulate CD88-dependent monocyte migration, whereas it blocks CD88-mediated migration of neutrophils.

Apart from its chemoattractant effects on monocytes and its chemorepellant effects on neutrophils, dRP S19 has been reported to induce a slight respiratory burst reaction in monocytes and to augment spontaneous apoptosis in neutrophils. The latter observation might suggest that dRP S19 could contribute to the meanwhile accepted anti-inflammatory character of apoptotic cell removal (Yamamoto 2007). Although the mechanisms of dRP S19 formation during apoptosis and its effects on monocytes and neutrophils have been quite exhaustively studied, it remains to be clarified, how dRP S19 is secreted by apoptotic cells, if other cell types than the already tested ones (HL-60, AsPC-1 and NIH 3T3 cells) secrete dRP S19 during apoptosis and most importantly if this secretion can be observed when classic apoptotic stimuli, such as death receptor ligation, and not heat treatment, are applied.

3.2.2 *Endothelial Monocyte-activating Polypeptide II (EMAP II)*

Shortly after Horino et al. had described dimerized RP S19 as the first apoptotic 'find-me' signal, a report by Knies et al. (1998) identified endothelial monocyte-activating polypeptide II (EMAP II) as another attraction signal, which is released by apoptotic cells (Fig. 3.2). EMAP II was originally purified from the supernatant of methylcholanthrene A (meth A) transformed murine fibrosarcoma cells on the basis of its ability to induce tissue factor expression on entothelial cells in vitro (Kao et al. 1992). Later on, it became obvious by in vitro studies that it acts as a chemoattractant for polymorphonuclear granulocytes as well as for monocytes (Kao et al. 1994b). In their initial study Knies et al. observed a profound abundance of EMAP II mRNA at sites of active tissue remodelling where many apoptotic cells could be found as detected by dUTP end labelling during mouse embryogenesis. Co-localizing with the EMAP II mRNA signal Knies et al. (1998) could spot an accumulation of F4/80-positive macrophages. Consequently, the authors hypothesized that EMAP II might be responsible for the observed phagocyte accumulation and examined EMAP II production in murine meth A cells during apoptosis in further detail. The molecular mechanisms of EMAP II biosynthesis closely resemble those of IL-1β, since translation of the EMAP II mRNA results in a precursor protein with a molecular mass of 43 kDa, which lacks a conventional secretion signal and which has to be proteolytically processed in order to yield and release the mature and biologically active 23 kDa protein. This cleavage, again in accordance to IL-1β biosynthesis, occurs at an aspartate residue (within the peptide motif ASTD) during apoptosis but not during necrosis (Knies et al. 1998). A follow-up study could identify the executioner caspases-3 and -7 as being responsible for pro-EMAP II processing and thereby controlling the crucial regulatory step in EMAP II biogenesis (Behrensdorf et al. 2000). Pro-EMAP II p43 today is known to be identical to the p43 subunit of the mammalian tRNA multi-synthetase complex and is assumed to function as

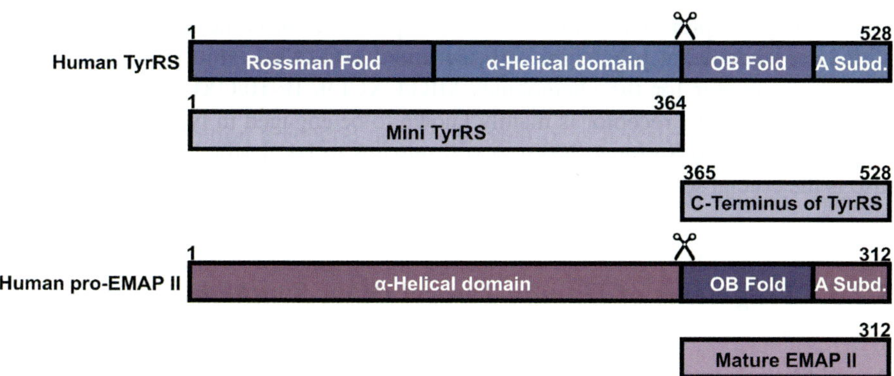

Fig. 3.2 Schematic domain structure of human endothelial monocyte-activating polypeptide II (EMAP II) and tyrosyl tRNA synthetase (TyrRS). Scissors indicate the sites of proteolytic processing. Modified from Ivakhno and Kornelyuk (2004).

a co-factor for the tRNA synthetases in the complex by stabilizing the interaction between these enzymes and the corresponding tRNAs (Quevillon et al. 1997). Yet, it has to be mentioned, that pro-EMAP II processing and release of mature EMAP II p23 have been observed to be rather late apoptotic events. In comparison to poly (ADP-ribose) polymerase (PARP) cleavage, which could be detected as early as 2 hours after stimulation of meth A transformed murine fibrosarcoma cells with tumour necrosis factor α (TNF-α) and cycloheximide (CHX), pro-EMAP II cleavage could only be observed after 9–13 hours of stimulation. Unfortunately, the authors did not examine, whether mature EMAP II p23 was released from truly apoptotic cells with an intact plasma membrane or rather by secondary necrotic cells, which have already lost their plasma membrane integrity. However, since in most in vivo settings, apoptotic cells are phagocytosed at an early stage before pro-EMAP II processing occurs, Behrensdorf and co-workers suggested that mature EMAP II p23 might only contribute to dead cell removal by recruitment of professional scavengers at sites of extensive cell death, where neighbouring cells and nearby phagocytes fail to clear all of the accruing apoptotic cells, so that some of them reach the late phase of apoptosis including pro-EMAP II processing and release. Thus, the additionally recruited mononuclear phagocytes could subsequently assist in the clearance process. This model would also take into account the formerly described pro-inflammatory effects of EMAP II p23 (myeloperoxidase release in neutrophils and stimulation of TNF-α production in monocytes; Kao et al. 1994a), and of pro-EMAP II p43 (induction of TNF-α synthesis, ICAM-1 adhesion molecule and several other cytokines) and chemokines (IL-8/CXCL8, MCP-1/CCL2, MIP-1α/CCL3, MIP-1β/CCL4, MIP-2α/CXCL2, RANTES/CCL5, and IL-1β in THP-1 cells; Ko et al. 2001), which otherwise would contradict the current dogma that apoptotic cell removal is an anti-inflammatory process, whereas clearance of secondary necrotic cells is rather pro-inflammatory. Taken together, these findings may suggest a function of EMAP II as an attraction and danger signal of secondary necrotic cells rather than of apoptotic cells.

                                                                 C. Peter et al.

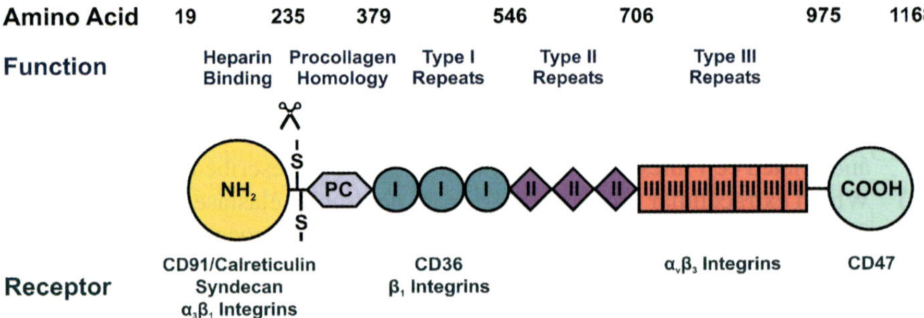

Fig. 3.3 Schematic domain structure of human thrombospondin-1 (TSP-1). Domain functions and the corresponding receptors are indicated. Scissors mark the site of proteolytic processing. Modified from Bornstein (1995).

Apart from its putative role as apoptotic 'find-me' signal Krispin and colleagues could demonstrate a tolerizing effect of TSP-1 on immature dendritic cells (iDCs), strengthening the concept of the anti-inflammatory nature of apoptotic cell removal (Krispin et al. 2006). The authors showed that lipopolysaccharide (LPS)-induced up-regulation of MHC class II and CD86 surface expression in iDCs could be efficiently reduced by pre-treatment with apoptotic monocytes in combination with purified TSP-1. A similar effect could also be observed by pre-incubation of iDCs with purified TSP-1 in the absence of apoptotic monocytes. Furthermore, the T cell activating capacity of iDCs (as measured in a mixed leukocyte reaction; MLR) treated with LPS following pre-incubation with TSP-1 was visibly decreased in comparison to iDCs stimulated with LPS alone. In this setting the authors failed to identify a single phagocyte receptor mediating the tolerizing TSP-1 effect but rather observed a simultaneous participation of CD29 (integrin β_1), CD36, CD47, CD51 (integrin α_v), and CD91, respectively (Krispin et al. 2006).

3.2.5 Lysophosphatidylcholine (LPC)

Aside from the previously reported chemoattractants on protein basis in 2003 we could present evidence that apoptotic cells also do secrete a lipid chemotactic signal that induces attraction of monocytic cell lines and primary human macrophages (Lauber et al. 2003). A detailed biochemical characterization identified this lipid 'find-me' signal as the phospholipid lysophosphatidylcholine (LPC). We could demonstrate that LPC was produced by the calcium-independent, cytosolic phospholipase $A_2\beta$ (iPLA$_2\beta$), which during apoptosis was cleaved and thereby activated in a caspase-3-dependent manner. In this context it should be noted that LPC was already described by Kim et al. as an 'eat-me' signal on the apoptotic cell surface, which was recognized by naturally occurring IgM antibodies (Kim et al. 2002). These findings support the intriguing concept that LPC might fulfill a dual role as a membrane bound 'eat-me' signal and as a soluble 'find-me' signal.

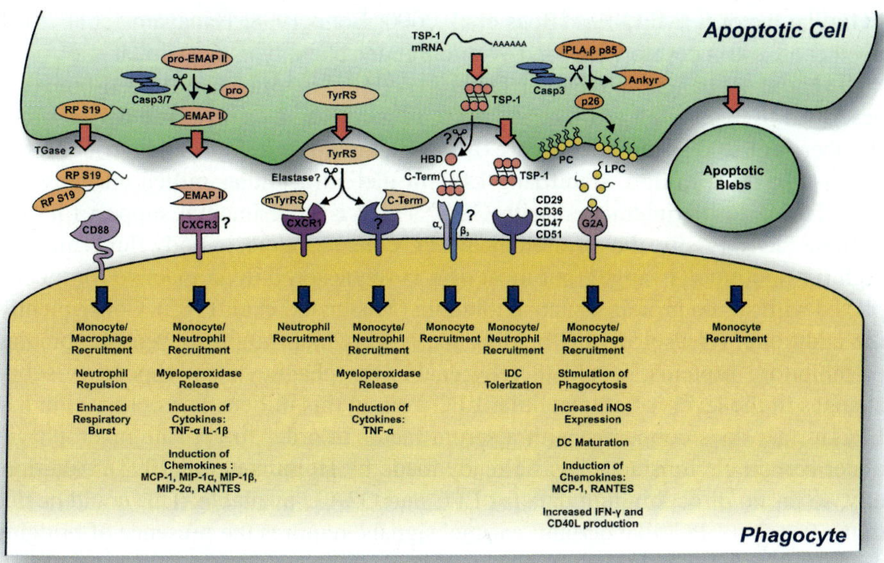

Fig. 3.4 Attraction signal synapse between an apoptotic cell and a phagocyte *Casp* Caspase, *EMAP II* endothelial monocyte-activating polypeptide II, *G2A* G protein-coupled receptor G2A, *HBD* heparin-binding domain, *iPLA2β* calcium-independent phospholipase A₂β, *LPC* lysophosphatidylcholine, *PC* phosphatidylcholine, *RP S19* ribosomal protein S19, *TGase 2* transglutaminase 2, *TSP-1* thrombospondin-1, *TyrRS* tyrosyl tRNA synthetase.

Although LPC had been known to be a very potent macrophage and lymphocyte attracting factor for years (McMurray et al. 1993; Quinn et al. 1988; Ryborg et al. 1994), from our initial observations we could not rule out that other lysophospholipids might act as chemoattractants released by apoptotic cells as well. Furthermore, it was possible that not LPC itself would be the actual phagocyte-attracting agent but that it would rather first have to be converted into the finally active form. A recent follow-up study revealed that among the lysophospholipids, which putatively can be produced by iPLA₂β during apoptosis, only LPC exhibited a chemotaxis-stimulating effect on monocytic cell lines. Moreover, we could provide evidence that LPC and none of its metabolic derivatives was responsible for phagocyte recruitment (Peter et al. 2007). In this report we could also identify the G-protein coupled receptor G2A as the crucial mediator engaged in the phagocyte's chemotactic response to LPC. This observation together with the phenotype of the corresponding G2A knock-out mice strongly supports the current notion that defects in the removal of apoptotic cells can lead to chronic inflammation and autoimmunity (Gaipl et al. 2004; Munoz et al. 2005). Thus, G2A knock-out mice have been reported to develop the typical autoimmune phenotype of a late-onset, multi-organ inflammation, closely related to human systemic lupus erythematosus (Le et al. 2001). Mice that are deficient for the expression of certain 'eat-me' signals or the corresponding receptors in the phagocytic synapse, such as complement protein C1q, receptor tyrosine kinase Mer, IgM, or milk fat globule

EGF-like factor 8 (MFG-E8) (Boes et al. 2000; Botto 1998; Hanayama et al. 2004; Scott et al. 2001) develop this kind of autoimmune phenotype in a similar way.

It should not be ignored that the factor(s) determining if LPC stays cell-bound, as described by Kim et al. (Kim et al. 2002), or is released, as observed by our group (Lauber et al. 2003), remain obscure. Is there a kind of transporter or carrier protein facilitating its secretion, stabilizing LPC in the extracellular milieu and possibly contributing to the interaction with G2A? There is experimental support for both of these ideas. Firstly, the physiological LPC concentrations in body fluids are very high (in micromolar range), but most of it is sequestered in an inactive form complexed with serum proteins, such as albumin (Mochizuki et al. 1982). Consequently, the additional release of LPC-binding factor(s), which might prevent the binding of inhibitory proteins, would thereby enable the phagocyte to respond to subtle changes in the levels of extracellular LPC even in this milieu. The observation that LPC in fact does cooperate with a serum factor in order to activate macrophages under certain circumstances was already made by Homma et al. (1993). Additionally, so far no direct binding data for LPC and G2A is available. This could be due to the fact that LPC, like certain 'eat-me' signals, requires the presence of bridging proteins in order to exert its function.

Besides its 'find-me' signal function diverse immunomodulatory effects of LPC have been reported. Murugesan and coworkers observed an LPC-mediated induction of classical monocyte attracting chemokines, such as MCP-1/CCL2 and RANTES/CCL5, in human vascular endothelial cells (Murugesan et al. 2003). This finding could be temptingly interpreted as an amplification loop of phagocyte recruitment, in which LPC would not only act as chemoattractant itself but also stimulate the release of additional attraction signals.

Furthermore, a phagocytosis promoting effect of LPC has been reported in diverse studies. In this context in vitro treatment of mouse peritoneal macrophages with lysophosphatidylcholine was described to result in a greatly enhanced Fc-receptor-mediated phagocytic activity (Homma et al. 1993; Ngwenya and Yamamoto 1985). If this finding also holds true for non-Fc-receptor-mediated phagocytic removal of apoptotic cells is a highly interesting question and remains to be clarified. Yan et al. described in a mouse model for experimental sepsis thatin vivo administration of LPC displayed a profound therapeutic effect, since treatment with LPC markedly enhanced the clearance of intraperitoneal bacteria as well as the bactericidal activity of neutrophils (Yan et al. 2004). Subsequent work of this group showed an anti-inflammatory effect of LPC based on the observation that LPC significantly suppressed the endotoxin-induced release of high-mobility group box 1 (HMGB-1) protein (a danger signal of necrotic cells and activated immune cells that will be discussed later on) from monocytes and macrophages in endotoxemia and sepsis (Chen et al. 2005). Concluding from these findings the authors suggested that LPC might confer protection against lethal experimental sepsis partly by facilitating the elimination of the invading pathogens and partly by inhibiting endotoxin-induced release of the pro-inflammatory cytokine HMGB-1.

Yet, on the contrary it should be mentioned that pro-inflammatory effects of LPC have also been described by several groups. Amongst them were stimulation of phospholipase D activity in murine peritoneal macrophages (Gomez-Munoz et al. 1999), augmentation of IL-1β-stimulated inducible NO synthase (iNOS) expression (Tani-uchi et al. 1999), promotion of dendritic cell maturation (up-regulation MHC class II-, CD83-, CD86-, and CD40-surface expression and T cell activating capacity in MLR; Coutant et al. 2002), and increase of anti-CD3 stimulated interferon-γ (IFN-γ) production and CD40 ligand expression in CD4$^+$ T cells (Sakata-Kaneko et al. 1998). Future studies have to clarify how these observations fit into the current concept that the removal of apoptotic cells basically constitutes an anti-inflammatory process.

3.2.6 *Apoptotic Micro-blebs*

All the apoptotic 'find-me' signals, which thus far have been described here, repre-sent clearly characterized protein or lipid molecules. Yet, also another attraction sig-nal with a rather undefined molecular character and particle nature has been reported in 1999 by Segundo et al. (1999) apoptotic micro-blebs. In this report the authors described that membranous particles approximately 0.2 μm in size were actively released by apoptotic human tonsil germinal center (GC) B cells and exhibited chem-oattractive activity on monocytes. The formation of these micro-blebs was dependent on active metabolism and Ca^{2+}-presence and was largely independent on protein syn-thesis, microfilament integrity and PKC-activation. After depletion of the micro-blebs by ultrafiltration or ultracentrifugation, it could be observed that the chemoattractant activity in the supernatants of apoptotic GC B cells was nearly completely removed. Thus, the authors concluded that the crucial chemotactic activity was associated with the micro-blebs. Yet, in how far phagocyte recruitment by apoptotic micro-blebs con-tributes to apoptotic cell removal in vivo still requires further investigations, since so far nothing is known about the distance, over which apoptotic micro-blebs can 'diffuse' in tissues. Furthermore, the phagocyte receptor(s), which is (are) engaged in micro-bleb-stimulated migration are completely unknown.

3.3 Danger Signals of Necrotic Cells

For a long time necrosis has been believed to be an accidental, uncontrolled form of cell death. However, recent data provide evidence that necrosis displays several features of a well controlled type of cell death and that it occurs not only under pathophysiological conditions but also in a physiological context during develop-ment (e.g. the death of chondrocytes controlling the longitudinal growth of bones) and adult tissue homeostasis (e.g. the death of intestinal epithelial cells; Festjens et al. 2006; Golstein and Kroemer 2007). In contrast to apoptosis necrosis has been described to be associated with inflammation. In general, exposed or released intra-cellular components are potentially cytotoxic (like lysosomal hydrolases) or can act as danger signals in order to 'inform' the immune system about tissue damage (Gallucci and Matzinger 2001). However, although most of the intracellular content

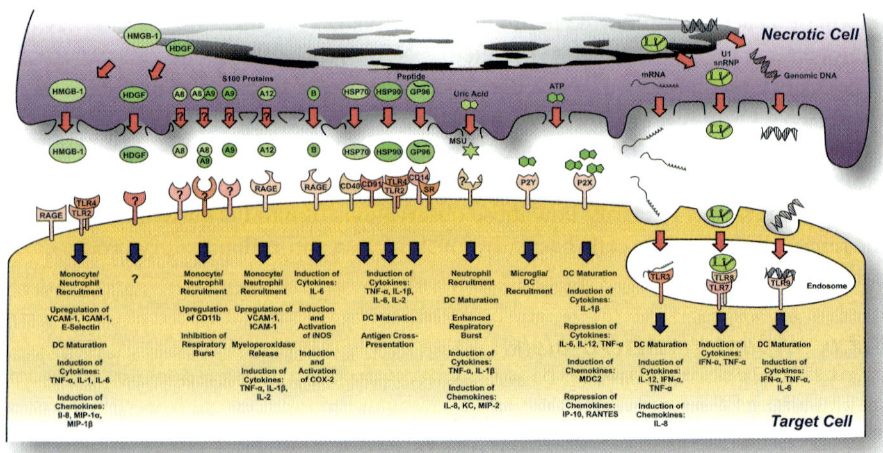

Fig. 3.5 Danger signal synapse between a necrotic cell and a target cell. *COX-2* cyclooxygenase-2, *GP* glycoprotein, *HDGF* hepatoma-derived growth factor, *HMGB-1* high mobility group box 1 protein, *HSP* heat-shock protein, *ICAM* intercellular adhesion molecule, *iNOS* inducible NO-synthase, *MSU* monosodium urate, *P2X* purinergic receptor X, *P2Y* purinergic receptor Y, *RAGE* receptor for advanced glycation end products, *snRNP* small nuclear ribonucleoprotein, *SR* scavenger receptor, *TLR* toll-like receptor, *VCAM* vascular cell adhesion molecule.

is released during necrosis, only a handful molecules have been identified to date that can elicit necrosis-induced immune signalling (Fig. 3.5).

3.3.1 High Mobility Group Box 1 Protein (HMGB-1)

High mobility group box 1 protein (HMGB-1; also known as amphoterin or HMG1) is a 30 kDa nuclear, non-histone protein, which is ubiquitously expressed in nuclei containing cells and is loosely bound to chromatin. Originally it has been described as a DNA-binding protein that stabilizes nucleosome formation and facilitates transcription (Bustin et al. 1978; Javaherian et al. 1978; Park et al. 2003), but recent studies have demonstrated an additional cytokine-like role of HMGB-1 in the context of tissue injury and inflammation (Bartunkova and Spisek, Chap. 12, this Vol.; Scaffidi et al. 2002; Semino et al. 2005; Wang et al. 2001).

The initial report showing that HMGB-1 is released by necrotic cells came from Scaffidi et al. (2002). Yet, so far the detailed mechanisms of nuclear export and subsequent release of HMGB-1 are not understood. Importantly however, HMGB-1 secretion did not occur when the cells were forced to undergo apoptosis, presumably because then HMGB-1 was bound to cruciform DNA, which during apoptosis is formed after internucleosomal DNA cleavage, or to hypoacetylated proteins within the apoptotic cell nucleus. This clear difference between apoptotic and necrotic cells in terms of HMGB-1 release might contribute to the vital distinction in the activation of a pro- or rather anti-inflammatory response to dying cells.

It should be mentioned that apart from its passive liberation during necrosis HMGB-1 can also be expressed on the cell surface and actively secreted by monocytes, macrophages, and dendritic cells in response to endogenous and exogenous inflammatory stimuli, such as IL-1, TNF-α, IFN-γ, or pathogen-derived LPS (Bonaldi et al. 2003; Gardella et al. 2002). In this context it exhibits the characteristics of a leaderless cytokine, which is not translocated from the Golgi apparatus to the cell membrane directly after synthesis (like IL-2 for example), but rather requires specialized means to be secreted in organelles that belong to the endolysosomal compartment, thus sharing properties with the IL-1 family (Andrei et al. 1999). A crucially required step in this scenario has been reported to be hyperacetylation of HMGB-1 at diverse of its 43 lysine residues, preventing it from interacting with the nuclear-import machinery and consequently leading to HMGB-1 accumulation in the cytosol. Acetylated, cytosolic HMGB-1 subsequently migrated to cytoplasmic secretory vesicles, waiting to be released into the extracellular space (Bonaldi et al. 2003). Interestingly, Chen and coworkers observed that pre-treatment with the apoptotic 'find-me' signal LPC interfered with this process and inhibited activation induced HMGB-1 release, thus, suggesting a kind of cross-talk between apoptotic 'find-me' and necrotic danger signals (Chen et al. 2005). Although the detailed signalling mechanisms, which regulate activation-induced HMGB-1 secretion, remain elusive, inhibitor studies have identified the central mediators as TLR4, CD14, members of the mitogen activated protein kinase (MAPK) family (like p38, MAPK/extracellular signal regulated kinases (MEK) 1 and 2), protein kinases B (PKB) and C (PKC), and phosphatidylinositol 3 kinase (PI3K). As described below, extracellular HMGB-1 has been reported to induce the activation of the MAPK and NF-κB pathway itself and, therefore, could be required for sustaining the signalling cascade to monocyte/macrophage activation and dendritic cell maturation. Thus, it is tempting to speculate that monocytes, macrophages, and dendritic cells can 'simulate' or keep up the pro-inflammatory milieu, which is generated when massive cell necrosis occurs, by actively secreting HMGB-1.

Once released into the extracellular space HMGB-1 has been shown to incite a plethora of immune cell reactions. In this context a strong stimulation of neutrophil recruitment could be observed in vitro as well as in vivo (Orlova et al. 2007). Although HMGB-1 in vitro exerted a direct chemotactic effect on neutrophils, in vivo the transendothelial migration of neutrophils presumably was also mediated in part by an additional activation of endothelial cells with concomitant increased expression of vascular cell adhesion molecule 1 (VCAM-1), intercellular adhesion molecule 1 (ICAM-1) and endothelial cell selectin (E-selectin), which facilitate neutrophil adhesion, a central pre-requisite for the initiation of leukodiapedesis (Treutiger et al. 2003). Aside from its role in neutrophil recruitment, an autocrine function of surface bound HMGB-1 in transendothelial monocyte migration has been described by Rouhiainen et al. (2004). However, the authors did not detect a direct chemotactic effect of soluble HMGB-1 on monocytes (Rouhiainen et al. 2004). In concert with IL-1 and IL-2 family members HMGB-1 has been reported to promote the interactions between natural killer cells and monocytes or iDCs, respectively, and to enhance subsequent DC maturation as measured by up-regulation of CD83, CD54, CD80, CD40, CD58, and MHC class II surface expression and consecutive T cell stimulating capacity in the MLR (Messmer et al. 2004; Semino et al. 2005; Semino et al. 2007).

Apart from affecting cellular functions, like migration, activation, and maturation, HMGB-1 has also been reported to modulate cytokine expression. In this context Andersson and co-workers described that in vitro culture of monocytes with purified HMGB-1 resulted in the release of pro-inflammatory cytokines, such as TNF-α, IL-1α, IL-1β, IL-6, IL-8/CXCL8, MIP-1α/CCL3, and MIP-1β/CCL4 (Andersson et al. 2000). An increase in TNF-α production was also observed when monocytes were challenged with necrotic HMGB-1$^{+/+}$ but not with HMGB-1$^{-/-}$ fibroblasts (Scaffidi et al. 2002). Finally, in vivo administration of purified HMGB-1 has been reported to lead to an increase in TNF-α serum levels. However, the direct effect of HMGB-1 on cytokine production still is under debate, since a follow-up study by Rouhiainen and co-workers showed that highly purified recombinant HMGB-1 has only a very weak direct pro-inflammatory activity in terms of cytokine production (Rouhiainen et al. 2007). The authors, therefore, suggested that HMGB-1 might need a kind of bound cofactor, like lipopolysaccharide or single stranded DNA, which is essentially required for stimulation of cytokine production. Future studies have to validate this hypothesis and have to identify the putative co-factor.

HMGB-1 has been described to transduce cellular signals via at least three different multi-ligand receptors, which otherwise are involved in the recognition of pathogen associated molecular patterns (PAMPS), such as lipopolysaccharide: The receptor for advanced glycation end products (RAGE) and the toll-like receptors (TLR) 2 and 4, respectively. The common downstream signalling for RAGE, TLR2, and TLR4 has been shown to involve a myeloid differentiation primary response protein 88 (MyD88)-dependent pathway that ultimately leads to NF-κB-activation and target gene transcription (Dumitriu et al. 2005; Parket al. 2004; Yu et al. 2006). TLR2 and TLR4 ligation by HMGB-1 additionally activates a signalling pathway comprising extracellular signal regulated kinases (ERK) 1 and/or 2, and the MAPK p38 (Park et al. 2003). Yet, the detailed molecular mechanisms of the latter so far remain unknown.

Increased extracellular HMGB-1 expression in serum and synovial fluid has been involved in a number of pathologic conditions, such as sepsis (Yang et al. 2004) or arthritis (Kokkola et al. 2002; Taniguchi et al. 2003) suggesting that HMGB-1 might be an intriguing molecular target for future therapies of these and other inflammatory disorders.

3.3.2 *Hepatoma-derived Growth Factor (HDGF)*

A new candidate for a necrotic danger signal very similar to HMGB-1 is Hepatoma-derived growth factor (HDGF). HDGF is a ubiquitously expressed protein with a bipartite nuclear localization signal lacking a classical signal peptide for secretion and sharing 32% amino acid sequence homology with HMGB-1 (Nakamura et al. 1994). HDGF has been reported to be mainly localized in the nucleus but also to be passively released by necrotic cells or actively secreted by healthy cells (although in very little amounts; Zhou et al. 2004). Importantly, Zhou and colleagues could not observe HDGF liberation from apoptotic cells, since there it tightly was retained in the nucleus. Further studies are necessary to explore the immunological consequences of HDGF secretion during necrosis and, thus, its putative role as danger signal.

3.3.3 S100 Proteins

The S100 protein or calgranulin family comprises more than 20 related, small (10–20 kDa), acidic, homo- and hetero-dimerizing, calcium-binding proteins of the helix-loop-helix ('EF-hand type') conformation. The name 'S100' was derived from the fact that theses proteins are Soluble in 100% ammonium sulfate solution at neutral pH. They are tissue specifically expressed and have been described to fulfil a variety of intracellular functions in the regulation of protein phosphorylation, enzyme activities, the dynamics of cytoskeleton components, transcription factors, Ca^{2+} homeostasis, and cell proliferation/differentiation (Donato 2003). However, at least for the family members S100A8 (MRP-8, CP-10), S100A9 (MRP-14), S100A12 (originally named EN-RAGE), which are predominantly expressed in granulocytes, monocytes, and early differentiation stages of macrophages, and S100B, which is mainly found in astrocytes and other brain cells, additional extracellular danger signal functions have been reported (Odink et al. 1987; Rothermundt et al. 2003; Yang et al. 2001).

For purified S100A8 a potent chemoattractant effect has been observed in murine and human neutrophils in vitro (Lackmann et al. 1992) and in vivo when injected into mouse footpads (Devery et al. 1994). Furthermore, the S100A8/S100A9 heterodimer and S100A9 alone, but not S100A8, have been described to play a role in monocyte recruitment (Eue et al. 2000). In this context Eue and colleagues observed high levels of endothelium associated S100A8 and S100A9 in inflamed tissues, particularly in the vicinity of transmigrating leukocytes. The authors suggested a causative role of S100A8 and S100A9 in leukocyte recruitment, since treatment with S100A9 or S100A8/S100A9 heterodimer led to an increased CD11b expression in human monocytes and strengthened the interaction between monocytes and endothelial cells. Finally, blocking antibodies against S100A9 alone or the S100A8/S100A9 heterodimer strongly inhibited transendothelial migration of monocytes (Eue et al. 2000). A role of the S100A8/S100A9 heterodimer in the propagation of inflammation has also been suggested, as it has been shown to take part in arachidonic acid and eicosanoid metabolism, secretion and extracellular transport (Kerkhoff et al. 1999). Despite the previously mentioned studies demonstrating a pro-inflammatory effect of S100A8 and S100A9 alone or in the form of a heterodimer, there are also reports showing a rather anti-inflammatory effect on macrophage and lymphocyte activation. Thus, the S100A8/S100A9 heterodimer has been described to inhibit macrophage activation as measured by the respiratory burst capacity (Aguiar-Passeti et al. 1997) and to inhibit immunoglobulin synthesis in lymphocytes (Brun et al. 1994).

Another S100 family member, which has been described to exert extracellular functions in the context of inflammation, is S100A12, originally named EN-RAGE (Hofmann et al. 1999). In extensive studies Hofmann et al. demonstrated that S100A12 displays a variety of different effects on endothelial cells as well as on monocytes/macrophages and periperal blood mononuclear cells (PBMCs) resulting in a strong inflammatory response. In vitro the authors observed that human umbilical vein endothelial cells (HUVECs) respond to S100A12 treatment with a strong up-regulation of VCAM-1 and ICAM-1 expression, which was accompanied by enhanced adhesion of mononuclear and putatively also polymorphonuclear cells—an initial prerequisite for extravasation. Additionally, incubation with purified S100A12 stimulated a strong chemotactic response in human peripheral monocytes and induction of IL-1β and TNF-α secretion in murine macrophages, whereas PBMCs responded with an increased IL-2 production. Some of these findings could also be observed in different in vivo models. Thus, systemic infusion of S100A12 led to an increase of VCAM-1 expression in the lung. Furthermore, local injection of S100A12 into mouse footpads stimulated a profound inflammatory response as measured on the basis of leukocyte infiltration and subcutaneous edema. All of the described effects of S100A12 were at least in part mediated by the receptor RAGE as could be confirmed by competition studies utilizing neutralizing antibodies and the soluble extracellular domain of RAGE (Hofmann et al. 1999). So far RAGE is the only receptor, which has been described for S100 proteins. Future studies, therefore, are required to validate if S100A8 and S100A9 also utilize this receptor or if additional receptors are involved. Moreover, the intracellular signalling cascade, which is instigated after RAGE ligation by S100A12, aside from the already described NF-κB-activation needs further clarification.

Apart from S100A8, S100A9, and S100A12, whose expression is mainly limited to cells of the myeloid lineage, S100B, a protein, which is basically expressed in brain cells, has been reported to exert pro-inflammatory effects on diverse cell types. In neurons treatment with S100B has been shown to induce NF-κB-dependent induction of IL-6 mRNA (Li et al. 2000). Furthermore, Hu and colleagues described a potent induction and activation of iNOS in astrocytes (Hu et al. 1996). Treatment of microglia cells with nanomolar to micromolar amounts of S100B resulted in increased cyclooxygenase-2 (COX-2) expression (Bianchi et al. 2007), whereas induction of iNOS mRNA expression as well as nitric oxide secretion by microglia cells could only be observed in combination of S100B with IFN-γ treatment (Adami et al. 2001). The microglial receptor engaged in this scenario has been identified to be RAGE (Adami et al. 2004). Yet, since at least some of the described effects were only dependent on the presence of the extracellular but not the signal transducing domain of RAGE, the involvement of RAGE in S100B-mediated microglia activation as well as the underlying signalling mechanisms remain largely obscure.

One of the crucial open questions concerning the role of S100 proteins as danger signals of necrotic cells is if and how they are released during necrosis, since so far only for S100B a damage-associated release from rat neonatal neurons, astrocytes, and microglia has been described in an in vitro stretch injury model (Ellis et al. 2007; Willoughby et al. 2004). However, due to their high expression levels

(S100A8 and S100A9 have been described to comprise up to 30% of the cytosolic proteins) and small molecular weight, it is feasible that also other S100 proteins might be released during necrosis.

Apart from this passive release an active secretion of S100A8, S100A9, and S100A12 comparable to the activation-induced secretion o HMGB-1 or HDGF has been observed in human monocytes subsequent to their interaction with TNF-α-activated endothelial cells (Frosch et al. 2000). Again in analogy to HMGB-1 or HDF, S100 proteins lack a conventional leader signal and have been described to be secreted via a non-classical pathway involving an intact tubulin network and protein kinase C activity (Rammes et al. 1997). For S100B a constitutive secretion by astroglia has been described, which could further be augmented by stimulation with a number of agents (Ciccarelli et al. 1999; Pinto et al. 2000; Whitaker-Azmitia et al. 1990).

Since high concentrations of S100 proteins are found in synovial fluid, sputum, stool and blood plasma/serum during inflammation, both the S100A8/S100A9 complex and S100A12 have been proven to be useful diagnostic markers of inflammation—especially in non-infectious inflammatory diseases, such as arthritis, chronic inflammatory lung and bowel disease (Foell et al. 2004). They indicate phagocyte activation more sensitively than conventional parameters of inflammation, exhibit a strong correlation to the status of inflammation in various acute and chronic disorders, and, thus, are sensitive parameters for the monitoring of disease activity and response to treatment in individual patients (Frosch et al. 2000). Future studies will have to clarify, whether S100 proteins aside from their role as biomarkers of inflammation might also represent therapeutic targets for the treatment of inflammatory diseases.

3.3.4 *Heat-shock Proteins (HSPs)*

Heat-shock proteins (HSPs) were first discovered in 1962 as a set of highly conserved proteins abundantly present in both prokaryotes and eukaryotes (Bukau et al. 2006). Mammalian HSPs have been classified into five families, according to their molecular size: HSP100, HSP90, HSP70, HSP60, and the small HSPs. Each family is composed of diverse members expressed either constitutively (like HSC70) or inductively (like HSP70) exhibiting different subcellular localizations. HSP inducing stimuli comprise various types of cellular insult, such as heat-shock or oxidative stress. HSPs play an essential role as molecular chaperones by assisting the correct folding of nascent and stress-accumulated misfolded proteins, and preventing their aggregation (Schmitt et al. 2007). Apart from this HSPs also have been reported to exert different extracellular, immunomodulatory functions (Bartunkova and Spisek, Chap. 12, this Vol.). In this context their presence in human serum has been shown to be associated with stress conditions, including inflammation, bacterial, and viral infections (Njemini et al. 2003; Xiao et al. 2005).

The passive release of HSPs by necrotic cells with compromised plasma membrane integrity has been described by different groups for HSP70, HSP90, GP98, and calreticulin (Basu et al. 2000; Mambula and Calderwood 2006a). Importantly, the HSPs were not liberated during apoptotic cell death. Thus, HSP release represents a clear difference between necrotically and apoptotically dying cells, comparable to the release of HMGB-1 or HDGF.

Apart from being passively released during necrosis, HSPs, again like HMGB-1, HDGF or S100 proteins, have been reported to be actively secreted by different cell types in response to a variety of stress/activation signals. For example, active release of HSC70 has been reported in K562 erythroleukemic cells in response to treatment with the pro-inflammatory cytokine interferon-gamma (IFN-γ; Barreto et al. 2003). So far, the mechanisms of HSP transport to the plasma membrane, and HSP export in this scenario remain enigmatic. Cytosolic HSPs, like other danger signals described above, do not contain leader peptides for secretion via the classic secretory pathway. However, as already mentioned a number of non-canonical pathways for the release of 'leaderless' proteins exist. Thus, HSP27, HSP70, HSC70, and HSP90 have been observed to be released within the lumen of 'exosomes' through such a pathway (Clayton et al. 2005; Lancaster and Febbraio 2005). Another alternative pathway of protein secretion, originally described for the IL-1 family (Andrei et al. 1999), involves the entry of the 'leaderless' protein into secretory lysosomal endosomes, migration of these organelles to the cell surface and release of the contents of the endolysosomes into the extracellular space. Indeed, HSP70 has recently been shown to be secreted from tumour cells and macrophages by this pathway (Mambula and Calderwood 2006b). For GP96, an endoplasmic reticulum-residing member of the HSP90 family, it was speculated that transport to the plasma membrane might be enabled by masking of the endoplasmic reticulum (ER)-retention sequence KDEL (Altmeyer et al. 1996). In conclusion, the knowledge about these externalization processes is still incomplete and further studies are required to determine the favored pathways for active secretion of HSPs as well as of HMGB-1, HDGF, and S100 proteins.

Extracellular HSPs have powerful effects on the immune response and can interact with the immune response in a number of contexts (Srivastava 2000). Thus, HSP70, HSP90, and GP96 have been reported to stimulate the production of the pro-inflammatory cytokines TNF-α, IL-1β, IL-6, and IL-12 in human monocytes and DCs (Asea et al. 2000; Asea et al. 2002; Basu et al. 2000). Furthermore, experiments by Singh-Jasuja and co-workers have shown that treatment with exogenous GP96 strongly induces maturation of mouse and human DCs as measured by up-regulation of CD86, production of IL-12 and TNF-α, and T cell activating capacity (Singh-Jasuja et al. 2000a).

Apart from their direct function on immune cells, HSPs can exert an indirect effect as peptide carriers. Due to their inherent chaperone activity HSPs (especially members of the HSP70 and HSP90 family) can potently bind intracellular peptides, including putatively antigenic peptides (for example from tumour cells) and piggyback them outside the cell. Such extracellular HSP-peptide complexes have been described to exert a potent effect on T cell activation by antigen presenting cells (APCs), like DCs. In this regard, HSP-chaperoned peptides have been reported to be swiftly internalized by DCs and processed via the endosomal pathway (Singh-

Jasuja et al. 2000b) followed by subsequent presentation of the originally HSP-bound peptide on MHC class I molecules on the DC surface (Arnold-Schild et al. 1999)—a process called cross-presentation. After cross-presentation an antigen-specific CD8[+]-T cell response was initiated in vitro and in tumour mouse models in vivo (Doody et al. 2004). HSPs have been found to be important players in the cross-presentation process of tumour-derived, antigenic peptides, and, thus, might represent putatively powerful vaccination agents for tumour immunotherapy.

It should not be ignored that HSPs also have been ascribed anti-inflammatory effects, which can be specifically noted in inflammatory diseases. Disorders, such as rheumatoid arthritis, can be triggered by cross-reactive T cells, which recognize common epitopes in mammalian and highly immunogenic prokaryotic HSPs (van Eden et al. 2005). Interestingly, however, administration of the corresponding mammalian HSP has been observed to suppress the pro-inflammatory responses to bacterial HSP epitopes and led to remission of inflammation (Kingston et al. 1996). Additionally, in vitro HSP72 has been reported to attenuate LPS- or TNF-α-induced HMGB-1 release from activated macrophages (Tang et al. 2007) Thus, depending on the context, HSPs can obviously be profoundly pro- or anti-inflammatory.

Different receptors have been observed to be engaged in the cellular responses to extracellular HSPs. One of them is CD91, the low density lipoprotein receptor-related protein 1 (LRP-1), which can be found on APCs and on other cells and has been proposed as receptor for all immunogenic HSPs, including HSP60, HSP70, GP96, and calreticulin (Basu et al. 2001; Binder et al. 2000). However, since Theriault et al. (2005) examined the ability of HSP70 to bind to cells with or without CD91 expression and observed minimal differences, its role as a direct high affinity HSP receptor is still under debate. Additionally, over-expression of the substrate binding domain of CD91 in cells deficient for HSP70 binding could not restore the latter (Theriault et al. 2005). Thus, HSP70 binding to CD91 might involve low affinity interactions or be rather indirect. Another surface molecule, which has been suggested to act as an HSP70 receptor, is CD40, a member of the tumour necrosis factor receptor family involved in the maturation of antigen presenting cells (Becker et al. 2002). Yet, Theriault et al. (2005) also tested CD40 in their experimental system and observed no increase in HSP70 binding capacity when CD40 was over-expressed in previously null cells. Consequently, the exact role of CD40 in HSP70 binding still remains obscure. Like in the case of HMGB-1, members of the PAMP-recognizing TLR family have also been reported to be involved in HSP sensing. In this context the heterodimeric TLR2/TLR4 cluster in combination with the lipopolysaccharide (LPS) receptor CD14 were identified as relevant receptor complexes coupling the binding of HSP60, HSP70, and GP96 to NF-κB-activation in part via the MyD88/IRAK pathway (Asea et al. 2000; Asea et al. 2002; Vabulas et al. 2001; Vabulas et al. 2002). Again, however, the study by Theriault and co-workers could not support the previous reports of direct HSP70 binding to TLR2/TLR4 and CD14, since cells stably expressing CD14, TLR2 or TLR4 did not bind avidly to HSP70 (Theriault et al. 2005). Another receptor family usually engaged in PAMP-recognition, which has been reported to bind to HSPs, is the scavenger receptor family (Murphy et al. 2005). In contrast to HSP90 and HSP60, which have been found to bind to

LOX-1 (Delneste et al. 2002), HSP70 has been shown additionally to bind to and to be internalized by scavenger receptor class F member 1 (SREC-I), and fasciclin egf-like (FEEL-1)/common lymphatic endothelial and vascular endothelial receptor-1 (CLEVER-1; Delneste et al. 2002; Theriault et al. 2006; Theriault et al. 2005). Moreover, GP96 and calreticulin have been reported to exhibit significant affinity to and to be internalized by scavenger receptor A (SR-A) and SREC-I (Berwin et al. 2003). Obviously, members of the scavenger receptor family, which are known to be expressed in a wide range of different cell types, might play an important role in HSP binding and internalization.

This confusingly large group of suggested HSP receptors might reflect the large and heterogeneous group of HSPs instigating different cellular effects. It is likely that different HSPs recognize different receptors on different cell types and that the multiplicity of receptors might indicate a certain specialization for individual functions: receptors like TLR2, TLR4, and CD40 might be adapted for transmembrane signalling, while CD91 and the scavenger receptors might be more important for the internalization of HSPs in the context of cross-presentation.

3.3.5 Uric Acid

In 2003 uric acid, the end product of purine metabolism in uricotelic mammals, was described as one of the principal endogenous immunological danger or danger signals of dying cells by Shi and colleagues (Shi et al. 2003). They identified uric acid on the basis of its adjuvant activity on the priming of CD8$^+$-T cell responses when coinjected with antigen into mice. Apart from its role in purine metabolism uric acid is a crucial intracellular antioxidant and, thus, is constitutively present in high intracellular concentrations (up to 4 mg/ml in liver cell cytosol). Furthermore, its concentration has been described to increase during cell injury, probably due to the augmented degradation of cellular RNA and DNA and subsequent metabolization of the liberated purines (Shi et al. 2003). Once released it has been reported to enhance not only CD8$^+$-T cell responses (Shi et al. 2000) but also CD4$^+$-T cell responses towards particulate antigens (Shi and Rock 2002). From these observations the authors concluded that uric acid must be affecting a step that is common to the generation of helper and cytotoxic T cell responses and examined the effects of uric acid treatment on antigen presenting cells (APCs). In vitro they could observe that uric acid treatment of murine bone marrow-derived dendritic cells (DCs) led to maturation as measured in form of increased CD86 and to a lesser extent also CD80 surface expression (Shi et al. 2003). Very recently, uric acid has also been reported to enhance antibody immunity and to improve inhibition of tumour growth when applied as adjuvant together with dying tumour cells in a tumour vaccination model (Behrens et al. 2008), thus again strengthening its role as endogenous adjuvans or danger signal.

Interestingly, uric acid is only soluble inside cells but precipitates and readily forms monosodium urate (MSU) microcrystals in its extracellular form, particularly in body fluids. Thus, Shi and co-workers concluded, that MSU crystals might be the biologically active form—a notion, which was supported by the observation that preformed MSU crystals strongly elicited a CD8$^+$-T cell response when administered together with antigen in vivo. Furthermore, in vitro treatment of DCs with MSU but not with allopurinol or aluminium oxide crystals led to up-regulation of CD86 expression. Eliminating uric acid in vivo by administration of allopurinol (a uric acid analogue, which inhibits its production) or uricase (an enzyme, which degrades uric acid to allantoin) inhibited the immune response to antigens associated with transplanted syngeneic cells and the proliferation of autoreactive T cells in a transgenic diabetes model (Shi et al. 2006). Again, this was due to the influence on DC maturation, since, uric acid depletion did not reduce the stimulation of T cells by mature, activated DCs.

Extracellular MSU crystals have also been reported to display major inflammatory properties, most evident when they accumulate and precipitate in joints and periarticular tissues, hence causing gout (Dalbeth and Haskard 2005). Although the molecular mechanisms of MSU crystal-induced gout are far from being completely understood, treatment with MSU crystals has been described to augment respiratory burst capacity in polymorphonuclear granulocytes (Abramson et al. 1982) and macrophages (Chen et al. 2004), and to induce the production of pro-inflammatory cytokines (TNF-α in human blood monocytes synovial cells, di Giovine et al. 1991; IL-1β in human blood monocytes and synovial fluid mononuclear cells, di Giovine et al. 1987; and IL-6 in human monocytes and synoviocytes, Guerne et al. 1989). In addition, increased levels of neutrophil chemotactic factors, such as the CXC chemokines IL-8/CXCL8 (in humans; Terkeltaub et al. 1991), KC/CLCL11 and MIP-2 (in rodents; Murakami et al. 2003; Murakami et al. 2002) and, not surprisingly, neutrophil infiltration have been observed (Terkeltaub et al. 1998). Yet, it should be mentioned, that a crucial role in neutrophil attraction could also be attributed to additionally released danger signals, like S100A8/A9 (Ryckman et al. 2004) and S100A12 (Rouleau et al. 2003).

In search of a receptor for MSU crystals, Chen and colleagues analyzed neutrophil infiltration after i.p. injection of MSU crystals in mice deficient for different TLRs and adaptor proteins with a Toll/IL-1 receptor (TIR) domain (Chen et al. 2006). Very interestingly, none of the TLR deficient mice (TLR1, -2, -4, -2/4, -6, -7, -9, or -11 have been tested, TLR5 and TLR8 deficient mice so far are not available) showed a reduction in MSU crystal-induced intraperitoneal neutrophil recruitment, although TLR2 and TLR4 had previously been described to play a role in a mouse air-pouch model of MSU crystal-induced neutrophil recruitment (Liu-Bryan et al. 2005). The study of Chen and co-workers could only detect a significant decrease in neutrophil recruitment when mice lacking the TIR domain protein MyD88 were employed. As upstream activator of MyD88 the authors could identify the IL-1 receptor. Since MSU crystals previously had been described to engage the caspase-1-activating, IL-1 producing, NALP3 inflammasome, resulting in the production of active IL-1β and IL-18, the authors have concluded, that MSU crystals initially might stimulate the inflammasome-dependent IL-1β production in macrophages, which subsequently leads to IL-1R- and MyD88-mediated neutrophil attraction. This notion is strongly supported by the observations that macrophages from mice deficient in various components of the inflammasome, such as caspase-1, ASC and NALP3, have been observed to be defective in MSU crystal-induced IL-1β-activation and that inflammasome- or IL-1R-deficient mice exhibit impaired neutrophil infiltration in an in vivo model of MSU crystal-induced peritonitis (Martinon et al. 2006). These findings provide first insight into the molecular processes underlying the inflammatory conditions of gout, and further support a pivotal role of the endogenous danger signal uric acid in this scenario.

3.3.6 *Adenosine-5'-triphosphate (ATP)*

The intracellular role of ATP in energy metabolism and its extracellular function in neurotransmission have been well known for decades. However, meanwhile a large number of studies have established another extracellular function of ATP: it can act as a danger signal in the context of tissue damage (di Virgilio 2000). Under normal conditions levels of intracellular ATP are high (5–10 millimolar), whereas in the extracellular compartment, its concentration is only in the nanomolar range. However, transient or permanent damage of cell membranes (for example during shear stress) has been described to massively increase the extracellular ATP concentration (Burnstock 2006; Grierson and Meldolesi 1995). Apart from this passive, cell damage-associated release, ATP, like other danger signals, can also actively be secreted by different cells, such as astrocytes and macrophages (Ferrari et al. 1997b). Although here the detailed molecular mechanisms are far from being understood, there are data available suggesting that it might be liberated by exocytosis, both when there is complete fusion between vesicle and cell membrane, and also when fusion is only transient, the so-called 'kiss-and-run' release (MacDonald et al. 2006). Additionally, ATP might be released through connexin hemichannels (Eltzschig et al. 2006).

Once in the extracellular space, ATP exerts a plethora of effects on different cell types. Thus, extracellular ATP has been reported to induce rat microglia cell chemotaxis in vitro (Honda et al. 2001). This finding could also be supported in vivo, since a very plastic model employing time-lapse confocal imaging of adult mouse brain slice preparations showed that upon traumatic brain injury, microglial branches rapidly and autonomously converged on the site of injury. This rapid chemotactic response could be mimicked by local injection of ATP and could be inhibited by addition of the ATP-hydrolyzing enzyme apyrase (Davalos et al., 2005; Wu et al. 2007). Thus, extracellular ATP obviously can regulate microglial branch dynamics in the intact brain, and its release from the damaged tissue and/or activated surrounding astrocytes mediates a rapid microglial response towards cellular insult. Apart from its chemotactic effect on microglia, ATP has also been reported to act as chemoattractant for immature DCs in vitro (Idzko et al. 2002).

Besides its chemokine function extracellular ATP has been described to modulate cytokine production and immune cell activation. In this regard, stimulation with ATP has been observed to trigger IL-1β release by human macrophages (Ferrari et al. 1997a) and to promote DC maturation (la Sala et al. 2002). In the latter study ATP-treated immature DCs have been observed to upregulate expression of CXCR4 and CCR7, the chemokine receptors for SDF-1α/CXCL12 and ELC/CCL19, and to down-regulate expression of CCR5, the receptor for MIP-1β/CCL4. This was paralleled by enhanced DC migration towards SDF-1α/CXCL12 and ELC/CCL19 and decreased migration towards MIP-1β/CCL4- the chemotactic fingerprint of mature DCs setting them for enhanced lymph node localization. At the same time, ATP-treatment augmented constitutive MDC/CCL22 production and inhibited LPS-induced IP-10/CXCL10 and RANTES/CCL5 production, leading to a strong chemotaxis induction in Th2 but not in Th1 cells by DC supernatants. Since in previous studies it had been observed that DCs matured by classic stimuli, such as LPS and CD40L, in the presence of ATP failed to induce production of IL-6, IL-1-α, IL-1β, TNF-α, and IL-12, whereas the up-regulation of membrane molecules, such as CD80 and CD86, and of the anti-inflammatory cytokines IL-10 and IL-1 receptor antagonist remained unaffected (Hasko et al. 2000; la Sala et al. 2001), la Sala and colleagues concluded that ATP fosters DC maturation in terms of a Th2-polarized immune response. This is also strengthened by a recent observation by Bulanova et al., who reported a rather Th2-like cytokine pattern (IL-4, IL-6, IL-13, and TNF-α) being released by ATP-stimulated bone marrow derived mast cells (Bulanova et al. 2005).

Extracellular nucleotides, like ATP, are recognized by P2 purinergic receptors (P2Rs), which are ubiquitously expressed on the surface of a variety of cells. P2Rs are divided into the P2XR and the P2YR subfamilies, the former identified as multimeric ligand-gated plasma membrane ion channels and the latter as seven transmembrane domain spanning G protein-coupled receptors (Burnstock 2006). Interestingly, these different subfamilies seem to transduce different physiological effects in response to ATP ligation. Whereas the molecular sensors, which have been described to mediate migratory responses towards extracellular ATP, obviously belong to the G protein-coupled receptor subfamily P2YR of (Davalos et al. 2005; Honda et al. 2001; Wu et al. 2007), the modulatory effects of ATP on cytokine production and DC activation/maturation have been linked to the ion channel-like subfamily P2XR (Ferrari et al. 1997a; la Sala et al. 2001). The activation of P2XRs by extracellular ATP has been described to lead to an increase in plasma membrane permeability to ions (Na^+, K^+, and Ca^{2+}). P2X7 (originally named P2Z) differs from the other P2XR family members due to its ability to undergo a progressive increase in size upon sustained ligand binding, thus, leading to the formation of a non-selective, reversible membrane pore, permeable for low molecular weight hydrophilic solutes. As downstream targets of P2XR signalling the transcription factors NF-κB and NFAT have been identified to be involved in target gene transcription (Ferrari et al. 1999; Ferrari et al. 1997c). Conversely, P2YR ligation has been reported to induce phospholipase C-mediated inositol trisphosphate generation, Ca^{2+} release from intracellular stores, or adenylate cyclase stimulation/inhibition by utilization of different G proteins (Gi/o or Gq/11; Burnstock 2006). Intriguingly, P2R-mediated responses are obviously regulated by the local ATP concentration: The EC_{50}-values of P2XRs have been determined to be in the low micromolar range (for P2 X7 in the hundred micromolar range), whereas P2YRs displayed higher affinity responding already to nanomolar concentrations of ATP (Burnstock 2006). These observations neatly fit into the above mentioned different biological functions of the two subfamilies. For chemoattraction, particularly at sites several rows of cells away from the ATP releasing source, where the local ATP concentration supposedly is rather low, receptors with higher affinity and, thus, sensitivity, like the P2YRs, are required. Conversely, cytokine production and immune cell activation/maturation should only take place at the site of the cellular injury, where the ATP-concentration is rather high. Hence, lower affinity receptors, like the P2XRs, should be engaged. Another factor that modulates P2R-mediated responses is the extracellular nucleotide metabolism: ATP can be hydrolyzed to adenosine-5'-diphosphate (ADP) by ecto-ATP/ADPase (CD39; Robson et al. 2005) and NTPDase 2 (CD73; Picher et al. 2003). CD39, which can also hydrolyze ADP to adenosine-5'-monophosphate (AMP), is expressed on the membrane of a wide variety of cells, such as Langerhans cells and DCs, macrophages, natural killer (NK) cells, activated B lymphocytes, and endothelial cells. In addition, AMP is the substrate for 5'-ectonucleotidase (CD73), an enzyme which also is present on the surface of many different cell types, leading to the generation of adenosine. Adenosine itself has been reported to exert a variety of effects on different immune cells and it remains to be clarified in how far metabolization of ATP might contribute to its function as danger signal.

A recent report by Idzko et al. (2007) could identify a crucial role of extracellular ATP in pathologic conditions, such as allergic reactions. Utilizing an experimentally induced mouse asthma model they have shown that intrapulmonal allergen challenge caused an acute accumulation of ATP in the airways—a finding, which could also be observed in asthmatic humans. Intriguingly, all cardinal features of asthma, including eosinophilic airway inflammation, Th2 cytokine production and bronchial hyper-reactivity, were abrogated when lung ATP levels were locally neutralized using apyrase or when mice were treated with broad-spectrum P2R antagonists. Furthermore, Th2 sensitization to inhaled antigen was enhanced by endogenous or exogenous ATP. These adjuvant effects of ATP were due to the recruitment and activation of lung myeloid DCs, which induced Th2 responses in the mediastinal lymph nodes. Taken together these data show that the danger signal ATP and subsequent purinergic signalling events exert a key role in allergen-driven lung inflammation, thus, rendering it an attractive target for therapeutic intervention.

3.3.7 mRNA, Small Nuclear Ribonucleoproteins (snRNPs) and Genomic DNA

The immunostimulatory effects of non-host dsRNA (for example viral RNA) and non-host DNA (for example non-methylated CpG DNA) are well established (Kaisho and Akira 2006). Yet, there is also accruing experimental evidence that self-RNA and self-DNA can stimulate pro-inflammatory reactions. Necrotic cells have been reported to release cytosolic mRNA (Ni et al. 2002). Intriguingly, Ni and co-workers could observe that incubation with extracellular mRNA led to an activation and maturation of DCs in vitro. As measured by an up-regulation of MHC class II molecules and the co-activating molecules CD80 and CD86, extracellular mRNA exhibited a DC stimulating potency comparable to that of classic DC activators, such as LPS, dsRNA and CD40L. Application of extracellular mRNA also induced production of the pro-inflammatory cytokines IL-12, IFN-α, IL-8, and TNF-α by DCs. Furthermore, the expression pattern of chemokine receptors on the DC surface changed in response to mRNA-treatment, as CCR5 and CCR6 were down- and CXCR4 was up-regulated, thus, preparing the maturing DCs for homing to the lymph nodes. Intriguingly, not only in vitro prepared mRNA, but also endogenous RNA released from or associated with necrotic cells also stimulated DCs, leading to IFN-α secretion, which could be abolished by pre-treatment of necrotic cells with RNase (Kariko et al. 2004). Since Poly(A) RNA could mimick at least some of the observed mRNA effects on DCs and these effects could be reduced by pre-incubation with pertussis toxin or by ATP-mediated desensitization, Ni et al. (2002) suggested a possible involvement of the purinergic receptor subfamily P2YR in extracellular mRNA-signalling. Yet, it should be mentioned that also in vitro transcribed RNA lacking a Poly(A)-signal could induce DC maturation although to a lesser extent than Poly(A)-containing mRNA. Thus, other sensors apart from the P2YR family have to be involved. One receptor involved in this context was identified in a follow-up study by the same group as TLR3, the toll-like receptor known to bind viral dsRNA (Kariko et al. 2004). In this study the authors presented evidence that heterologous RNA released from or associated with necrotic cells or generated by in vitro transcription stimulated TLR3-dependent immune activation in terms of DC maturation (MHC II and CD83 up-regulation) and pro-inflammatory cytokine (IL-8, IL-12 and IFN-α) expression. Employing neutralizing anti-TLR3 antibodies the authors could show that TLR3 and subsequent NF-κB-activation were crucially involved in extracellular mRNA-stimulated IL-8/CXCL8 expression and DC maturation. These results demonstrate that RNA, likely through the formation of secondary structures, is a potent host-derived activator of TLR3 (Kariko et al. 2004). However, it should be kept in mind that the danger signal function of mRNA seems to be limited by its inherent instability due to the ubiquitiuos presence of RNases in the extracellular space. Additionally, TLR3 is known to be located in the endosomal compartment. Therefore, future studies are required in order to determine if mRNA released by necrotic cells occasionally might be complexed by stabilizing compounds, like membrane fragments for example, in order to provide it with enhanced longevity and target it to endosomes, a vital pre-requisite for its function as a TLR3-triggering danger signal of damaged cells.

Aside from encoding mRNA also non-encoding RNA has been reported to be released by necrotic cells and to stimulate a pro-inflammatory response, particularly when complexed in small nuclear ribonucleoproteins, like U1 snRNP (Vollmer et al. 2005). Employing cells from TLR7 knock-out mice Vollmer et al. could observe that U1 snRNP stimulated TLR7 signalling leading to IFN-α secretion from plasmacytoid DCs. Ectopically TLR8-expressing human embryonic kidney 293 cells revealed that U1 snRNP-induced TNF-α secretion was dependent on TLR8 signalling. These stimulatory effects required the RNA component of the particle, since they could be diminished by RNase but not proteinase treatment, and could be reproduced by synthetic oligoribonucleotides as short as nine bases in length. Notably, stimulation required transfection of U1 snRNP or the formation of immune complexes comprised of anti-RNP antibodies from SLE patients and U1 snRNP, which plasmacytoid DCs appeared to take up through the Fcγ-receptor II. 'Antimalarial' small molecules inhibiting the endosomal pathway were shown to block the immunostimulatory effects of U1 snRNP implying that in order to activate endosomally located TLR7 or TLR8 U1 snRNP has to be taken up and directed to the endosomal compartment.

In addition to mRNA and snRNA, double stranded genomic dsDNA released from necrotic cells has been shown to exert immunostimulatory effects on murine macrophages and bone marrow-derived murine DCs (Ishii et al. 2001). In this report the authors observed that genomic dsDNA but not denatured ssDNA triggered APCs to up-regulate expression of MHC I/II and co-stimulatory molecules (CD40, CD54, CD69). Subsequent reports could also show induction of pro-inflammatory cytokines, such as IFN-α, TNF-α, and IL-6, by DCs in response to extracellular DNA (Yasuda et al. 2005a; Yasuda et al. 2006; Yasuda et al. 2005b). Intriguingly, transfection of dsDNA strongly augmented its DC maturating capacity in comparison to incubation with naked dsDNA. Moreover, inhibitors of endocytosis and endosomal acidification substantially reduced DC maturation and cytokine production in response to dsDNA (Yasuda et al. 2005a; Yasuda et al. 2005b; Yasuda et al. 2006) suggesting that, like in the case of U1 sn RNP (see above), the endosomal pathway might be crucially involved in this setting. Functionally, dsDNA enhanced APC function in vitro as measured by the generation of antigen-specific T cells. This finding could also be observed in vivo, since significantly higher OVA-specific antibody titers were induced by immunization of mice with the combination of OVA antigen plus dsDNA in comparison to administration of OVA alone. Additionally, co-administering OVA plus dsDNA also led to profoundly increased OVA-specific cytotoxic T cell activity. Consecutive studies by different groups identified activation of endosomally located TLR9, the toll-like receptor formerly known to be responsible for bacterial CpG-DNA recognition, and subsequent MyD88 signalling to be vital elements in the context of genomic dsDNA-stimulated DC activation (Decker et al. 2005; Yasuda et al. 2005a; Yasuda et al. 2005b; Yasuda et al. 2006).

3.3.8 IL-6

Interestingly, necrotic cells might not only induce the production of pro-inflammatory cytokines in target cells by releasing danger signals as described above, but instead produce and release pro-inflammatory cytokines, like IL-6, themselves (Vanden Berghe et al. 2006). In this study the authors observed that necrotic cell death induced by several stimuli coincided with NF-κB-and p38 MAPK-mediated up-regulation and secretion of IL-6. Production of IL-6 was profoundly reduced or absent in conditions of apoptotic cell death. Thus, besides the capacity of necrotic cells to induce an inflammatory response due to leakage of intracellular contents, necrotically dying cells themselves are involved in the expression and secretion of inflammatory cytokines.

3.4 Danger Signals of Secondary Necrotic Cells

Physiologically apoptotic cells are rapidly engulfed by semi-professional or professional phagocytes. However, if the process of apoptotic cell clearance fails, primary apoptotic cells become secondary necrotic. This is characterized by a loss of plasma membrane integrity and release of intracellular contents into the extracellular space. From a time course point of view, it could be argued that the repertoire of soluble mediators released by secondary necrotic cells should be a sum of the factors released by apoptotic and necrotic cells. However, it has to be kept in mind that apoptotic factors, like LPC, dRP S19, EMAP-II or TyRS, are released much earlier and that due to the ubiquitous presence of hydrolases (proteases, lipases, etc.) in the extracellular space the stability of these compounds is likely to be limited. Hence, it is questionable if these mediators are still present in reasonable concentrations at the time point, when the apoptotically dying cell starts undergoing secondary necrosis. Unfortunately, so far no experimental data is available addressing this question. Furthermore, especially the release of HMGB-1 and HDGF by secondary necrotic cells should be very limited or even not detectable, since HMGB-1 and HDGF have been reported to be sequestered within the nuclei of apoptotic cells (Scaffidi et al. 2002; Zhou et al. 2004). Thus, they should also be retained when these cells acquire the secondary necrotic state with a compromised plasma membrane. Conversely, during apoptosis small DNA-fragments are formed, which can readily leave the nucleus and then can also pass the disintegrated plasma membrane (see also below) putatively carrying HMGB-1 and HDGF piggyback outside the secondary necrotic cell. Initial experimental data provide some support for this notion (Bell et al. 2006), yet, require further validation.

Another danger factor, which probably might not be released by secondary instead of primary necrotic cells, is ATP. The requirement for and the consumption of ATP during apoptosis have been reported by many different studies (Chiarugi 2005). Thus, the intracellular ATP-level of an apoptotic cell undergoing secondary necrosis supposedly is much lower than that of an originally healthy cell undergoing primary necrosis. Consequently, much less ATP can be liberated into the surroundings. However, again critical experimental data are missing to support this hypothesis.

3.4.1 Nucleosomes

A crucial danger signal, which is released by secondary necrotic cells in larger quantities than by primary necrotic cells, is genomic DNA in form of nucleosomes. Whereas during apoptosis endonucleases are activated, which lead to internucleosomal DNA cleavage and, thus, production of free nucleosomes containing DNA fragments of 180 bp or multiples of 180 bp in length, during necrosis unspecific cleavage leads to large-scale DNA fragmentation and cleavage products of many kbp in length. The small nucleosomes produced during apoptosis in contrast to the large DNA fragments generated during necrosis can readily leave the nucleus via

of pro- or anti-inflammation. As summarized here, the repertoire of released substances is dependent on the form of cell death; thus, apoptotic, primary and secondary necrotic cells obviously 'smell' different. Although we have already gained first insight into the plethora of effects, which can be exerted by substances, like dRP S19, TyRS, TSP-1, EMAP-II, LPC, HMGB-1, HPSs, S100 proteins, uric acid, ATP, genomic DNA and others, future studies are required in order to fully understand, why and how the immune system responds differently to the 'interstitial fingerprint' of apoptotic and necrotic cells. This is of particular interest, since many inflammatory and autoimmune diseases have been linked to a de-regulated appearance and/or function of these factors. Hence, liberated attraction and danger signals might represent suitable markers for disease monitoring and possibly also putative targets for future therapies. Time, and more research, will tell.

Acknowledgments We thank the members of the Wesselborg lab for stimulating discussions and apologize to all colleagues in the field of engulfment and inflammation, whose work could not be cited here owing to space constraints. This work was supported by the DFG We 1801/2-4 (KL, SW), GRK1302 (KL, SW), SFB 685 (KL, SW), the Interdisciplinary Center of Clinical Research Tübingen (IZKF; Fö. 01KS9602; KL, SW), the Wilhelm Sander-Stiftung (2004.099.1; SW), the Landesforschungsschwerpunktprogramm of the Ministry of Science, Research and Arts of the Land Baden-Wuerttemberg (1423-98101; SW) and the Fortune Program of the University of Tübingen (No. 1250-0-0; KL).

References

Abramson S, Hoffstein ST, Weissmann G (1982) Superoxide anion generation by human neutrophils exposed to monosodium urate. Arthritis Rheum 25:174–180

Adami C, Bianchi R, Pula G et al (2004) S100B-stimulated NO production by BV-2 microglia is independent of RAGE transducing activity but dependent on RAGE extracellular domain. Biochim Biophys Acta 1742:169–177

Adami C, Sorci G, Blasi E et al (2001) S100B expression in and effects on microglia. Glia 33:131–142

Adams J C (2001) Thrombospondins: multifunctional regulators of cell interactions. Annu Rev Cell Dev Biol 17:25–51

Aguiar-Passeti T, Postol E, Sorg C et al (1997) Epithelioid cells from foreign-body granuloma selectively express the calcium-binding protein MRP-14, a novel down-regulatory molecule of macrophage activation. J Leukocyte Biol 62:852–858

Altmeyer A, Maki RG, Feldweg AM et al (1996) Tumour-specific cell surface expression of the KDEL containing, endoplasmic reticular heat shock protein gp96. Int J Cancer 69:340–349

Amoura Z, Piette JC, Chabre H et al (1997) Circulating plasma levels of nucleosomes in patients with systemic lupus erythematosus: correlation with serum antinucleosome antibody titers and absence of clear association with disease activity. Arthritis Rheum 40:2217–2225

Andersson U, Wang H, Palmblad K et al (2000) High mobility group 1 protein (HMG-1) stimulates pro-inflammatory cytokine synthesis in human monocytes. J Exp Med 192:565–570

Andrei C, Dazzi C, Lotti L et al (1999) The secretory route of the leaderless protein interleukin 1beta involves exocytosis of endolysosome-related vesicles. Mol Biol Cell 10:1463–1475

Arnold-Schild D, Hanau D, Spehner D et al (1999) Cutting edge: receptor-mediated endocytosis of heat shock proteins by professional antigen-presenting cells. J Immunol 162:3757–3760

Asea A, Kraeft SK, Kurt-Jones EA et al (2000) HSP70 stimulates cytokine production through a CD14-dependant pathway, demonstrating its dual role as a chaperone and cytokine. Nat Med 6:435–442

Asea A, Rehli M, Kabingu E et al (2002) Novel signal transduction pathway utilized by extracellular HSP70: role of toll-like receptor (TLR) 2 and TLR4. J Biol Chem 277:15028–15034

Barreto A, Gonzalez JM, Kabingu E et al (2003) Stress-induced release of HSC70 from human tumours. Cell Immunol 222:97–104

Basu S, Binder RJ, Ramalingam T et al (2001) CD91 is a common receptor for heat shock proteins gp96, hsp90, hsp70, calreticulin. Immunity 14:303–313

Basu S, Binder RJ, Suto R et al (2000) Necrotic but not apoptotic cell death releases heat shock proteins, which deliver a partial maturation signal to dendritic cells and activate the NF-kappa B pathway. Int Immunol 12:1539–1546

Becker T, Hartl FU, Wieland F (2002) CD40, an extracellular receptor for binding and uptake of Hsp70-peptide complexes. J Cell Biol 158:1277–1285

Behrens MD, Wagner WM, Krco CJ et al (2008) The endogenous danger signal, crystalline uric acid, signals for enhanced antibody immunity. Blood 111:1472–1479

Behrensdorf HA, van de Craen M, Knies UE et al (2000) The endothelial monocyte-activating polypeptide II (EMAP II) is a substrate for caspase-7. FEBS Lett 466:143–147

Bell CW, Jiang W, Reich CF 3rd et al (2006) The extracellular release of HMGB1 during apoptotic cell death. Am J Physiol Cell Physiol 291:C1318–C1325

Bell DA, Morrison B (1991) The spontaneous apoptotic cell death of normal human lymphocytes in vitro: the release of, immunoproliferative response to, nucleosomes in vitro. Clin Immunol Immunopathol 60:13–26

Berg CP, Stein GM, Keppeler H et al (2008) Apoptosis-associated antigens recognized by autoantibodies in patients with the autoimmune liver disease primary biliary cirrhosis. Apoptosis 13:63–75

Berwin B, Hart JP, Rice S et al (2003) Scavenger receptor-A mediates gp96/GRP94 and calreticulin internalization by antigen-presenting cells. Embo J 22:6127–6136

Bianchi R, Adami C, Giambanco I et al (2007) S100B binding to RAGE in microglia stimulates COX-2 expression. J Leukocyte Biol 81:108–118

Binder RJ, Han DK, Srivastava PK (2000) CD91: a receptor for heat shock protein gp96. Nat Immunol 1:151–155

Boes M, Schmidt T, Linkemann K et al (2000) Accelerated development of IgG autoantibodies and autoimmune disease in the absence of secreted IgM. Proc Natl Acad Sci U S A 97:1184–1189

Bonaldi T, Talamo F, Scaffidi P et al (2003) Monocytic cells hyperacetylate chromatin protein HMGB1 to redirect it towards secretion. EMBO J 22:5551–5560

Bornstein P (1995) Diversity of function is inherent in matricellular proteins: an appraisal of thrombospondin 1. J Cell Biol 130:503–506

Botto M (1998) C1q knock-out mice for the study of complement deficiency in autoimmune disease. Exp Clin Immunogenet 15:231–234

Brun JG, Ulvestad E, Fagerhol MK et al (1994) Effects of human calprotectin (L1) on in vitro immunoglobulin synthesis. Scand J Immunol 40:675–680

Bukau B, Weissman J, Horwich A (2006) Molecular chaperones and protein quality control. Cell 125:443–451

Bulanova E, Budagian V, Orinska Z et al (2005) Extracellular ATP induces cytokine expression and apoptosis through P2′7 receptor in murine mast cells. J Immunol 174:3880–3890

Burnstock G (2006) Pathophysiology and therapeutic potential of purinergic signalling. Pharmacol Rev 58:58–86

Bustin M, Hopkins RB, Isenberg I (1978) Immunological relatedness of high mobility group chromosomal proteins from calf thymus. J Biol Chem 253:1694–1699

Cascieri MA, Springer MS (2000) The chemokine/chemokine-receptor family: potential and progress for therapeutic intervention. Curr Opin Chem Biol 4:420–427

Casciola-Rosen LA, Anhalt G, Rosen A (1994) Auto-antigens targeted in systemic lupus erythematosus are clustered in two populations of surface structures on apoptotic keratinocytes. J Exp Med 179:1317–1330

Chen CJ, Shi Y, Hearn A et al (2006) MyD88-dependent IL-1 receptor signalling is essential for gouty inflammation stimulated by monosodium urate crystals. J Clin Invest 116:2262–2271

Chen G, Li J, Qiang X et al (2005) Suppression of HMGB1 release by stearoyl lysophosphatidylcholine:an additional mechanism for its therapeutic effects in experimental sepsis. J Lipid Res 46:623–627

Chen L, Hsieh MS, Ho HC et al (2004) Stimulation of inducible nitric oxide synthase by monosodium urate crystals in macrophages and expression of iNOS in gouty arthritis. Nitric Oxide 11:228–236

Chiarugi A (2005) "Simple but not simpler": toward a unified picture of energy requirements in cell death. FASEB J 19:1783–1788

Ciccarelli R, Di Iorio P, Bruno V et al (1999) Activation of A(1) adenosine or mGlu3 metabotropic glutamate receptors enhances the release of nerve growth factor and S-100beta protein from cultured astrocytes. Glia 27:275–281

Clayton A, Turkes A, Navabi H et al (2005) Induction of heat shock proteins in B-cell exosomes. J Cell Sci 118:3631–3638

Coutant F, Perrin-Cocon L, Agaugue S et al (2002) Mature dendritic cell generation promoted by lysophosphatidylcholine. J Immunol 169:1688–1695

Dalbeth N, Haskard DO (2005) Mechanisms of inflammation in gout. Rheumatology (Oxford) 44:1090–1096

Davalos D, Grutzendler J, Yang G et al (2005) ATP mediates rapid microglial response to local brain injury in vivo. Nat Neurosci 8:752–758

Decker P, Singh-Jasuja H, Haager S et al (2005) Nucleosome, the main auto-antigen in systemic lupus erythematosus, induces direct dendritic cell activation via a MyD88-independent pathway: consequences on inflammation. J Immunol 174:3326–3334

Delneste Y, Magistrelli G, Gauchat J et al (2002) Involvement of LOX-1 in dendritic cell-mediated antigen cross-presentation. Immunity 17:353–362

Devery JM, King NJ, Geczy CL (1994) Acute inflammatory activity of the S100 protein CP-10. Activation of neutrophils in vivo and in vitro. J Immunol 152:1888–1897

di Giovine FS, Malawista SE, Nuki G et al (1987) Interleukin 1 (IL 1) as a mediator of crystal arthritis. Stimulation of T cell and synovial fibroblast mitogenesis by urate crystal-induced IL 1. J Immunol 138:3213–3218

di Giovine FS, Malawista SE, Thornton E et al (1991) Urate crystals stimulate production of tumour necrosis factor alpha from human blood monocytes and synovial cells. Cytokine mRNA and protein kinetics, cellular distribution. J Clin Invest 87:1375–1381

di Virgilio F (2000) Dr. Jekyll/Mr. Hyde: the dual role of extracellular ATP. J Autonom Nerv Syst 81:59–63

Donato R (2003) Intracellular and extracellular roles of S100 proteins. Microsc Res Techniq 60: 540-551

Doody AD, Kovalchin JT, Mihalyo MA et al (2004) Glycoprotein 96 can chaperone both MHC class I- and class II-restricted epitopes for in vivo presentation, but selectively primes CD8+ T cell effector function. J Immunol 172:6087–6092

Dumitriu IE, Baruah P, Bianchi ME et al (2005) Requirement of HMGB1 and RAGE for the maturation of human plasmacytoid dendritic cells. Eur J Immunol 35:2184–2190

Ellis EF, Willoughby KA, Sparks SA et al (2007) S100B protein is released from rat neonatal neurons, astrocytes, microglia by in vitro trauma and anti-S100 increases trauma-induced delayed neuronal injury and negates the protective effect of exogenous S100B on neurons. J Neurochem 101:1463–1470

Eltzschig HK, Eckle T, Mager A et al (2006) ATP release from activated neutrophils occurs via connexin 43 and modulates adenosine-dependent endothelial cell function. Circ Res 99:1100–1108

Eue I, Pietz B, Storck J et al (2000) Transendothelial migration of 27E10+ human monocytes. Int Immunol 12:1593–1604

Ferrari D, Chiozzi P, Falzoni S et al (1997a) Extracellular ATP triggers IL-1 beta release by activating the purinergic P2Z receptor of human macrophages. J Immunol 159:1451–1458

Ferrari D, Chiozzi P, Falzoni S et al (1997b) Purinergic modulation of interleukin-1 beta release from microglial cells stimulated with bacterial endotoxin. J Exp Med 185:579–582

Ferrari D, Stroh C, Schulze-Osthoff K (1999) P2´7/P2Z purinoreceptor-mediated activation of transcription factor NFAT in microglial cells. J Biol Chem 274:13205–13210

Ferrari D, Wesselborg S, Bauer MK et al (1997c) Extracellular ATP activates transcription factor NF-kappaB through the P2Z purinoreceptor by selectively targeting NF-kappaB p65. J Cell Biol 139:1635–1643

Festjens N, Vanden Berghe T, Vandenabeele P (2006) Necrosis, a well-orchestrated form of cell demise: signalling cascades, important mediators and concomitant immune response. Biochim Biophys Acta 1757:1371–1387

Foell D, Frosch M, Sorg C et al (2004) Phagocyte-specific calcium-binding S100 proteins as clinical laboratory markers of inflammation. Clin Chim Acta 344:37–51

Frosch M, Strey A, Vogl T et al (2000) Myeloid-related proteins 8 and 14 are specifically secreted during interaction of phagocytes and activated endothelium and are useful markers for monitoring disease activity in pauciarticular-onset juvenile rheumatoid arthritis. Arthritis Rheum 43:628–637

Gaipl US, Franz S, Voll RE et al (2004) Defects in the disposal of dying cells lead to autoimmunity. Curr Rheumatol Rep 6:401–407

Gallucci S, Matzinger P (2001) Danger signals: SOS to the immune system. Curr Opin Immunol 13:114–119

Gardella S, rei C, Ferrera D et al (2002) The nuclear protein HMGB1 is secreted by monocytes via a non-classical, vesicle-mediated secretory pathway. EMBO Rep 3:995–1001

Golstein P, Kroemer G (2007) Cell death by necrosis: towards a molecular definition. Trends Biochem Sci 32:37–43

Gomez-Munoz A, O'Brien L, Hundal R et al (1999) Lysophosphatidylcholine stimulates phospholipase D activity in mouse peritoneal macrophages. J Lipid Res 40:988–993

Grierson JP, Meldolesi J (1995) Shear stress-induced [Ca2+]i transients and oscillations in mouse fibroblasts are mediated by endogenously released ATP. J Biol Chem 270:4451–4456

Guerne PA, Terkeltaub R, Zuraw B et al (1989) Inflammatory microcrystals stimulate interleukin-6 production and secretion by human monocytes and synoviocytes. Arthritis Rheum 32:1443–1452

Hanayama R, Tanaka M, Miyasaka K et al (2004) Autoimmune disease and impaired uptake of apoptotic cells in MFG-E8-deficient mice. Science 304:1147–1150

Hasko G, Kuhel DG, Salzman AL et al (2000) ATP suppression of interleukin-12 and tumour necrosis factor-alpha release from macrophages. Br J Pharmacol 129:909–914

Herrmann M, Voll RE, Zoller OM et al (1998) Impaired phagocytosis of apoptotic cell material by monocyte-derived macrophages from patients with systemic lupus erythematosus. Arthritis Rheum 41:1241–1250

Hofmann MA, Drury S, Fu C et al (1999) RAGE mediates a novel pro-inflammatory axis: a central cell surface receptor for S100/calgranulin polypeptides. Cell 97:889–901

Homma S, Yamamoto M and Yamamoto N (1993) Vitamin D-binding protein (group-specific component) is the sole serum protein required for macrophage activation after treatment of peritoneal cells with lysophosphatidylcholine. Immunol Cell Biol 71 (Pt 4):249–257

Honda S, Sasaki Y, Ohsawa K et al (2001) Extracellular ATP or ADP induce chemotaxis of cultured microglia through Gi/o-coupled P2Y receptors. J Neurosci 21:1975–1982

Horino K, Nishiura H, Ohsako T et al (1998) A monocyte chemotactic factor, S19 ribosomal protein dimer, in phagocytic clearance of apoptotic cells. Lab Invest 78:603–617

Hou Y, Plett PA, Ingram DA et al (2006) Endothelial-monocyte-activating polypeptide II induces migration of endothelial progenitor cells via the chemokine receptor CXCR3. Exp Hematol 34:1125–1132

Hu J, Castets F, Guevara JL et al (1996) S100 beta stimulates inducible nitric oxide synthase activity and mRNA levels in rat cortical astrocytes. J Biol Chem 271:2543–2547

Idzko M, Dichmann S, Ferrari D et al (2002) Nucleotides induce chemotaxis and actin polymerization in immature but not mature human dendritic cells via activation of pertussis toxin-sensitive P2y receptors. Blood 100:925–932

Idzko M, Hammad H, van Nimwegen M et al (2007) Extracellular ATP triggers and maintains asthmatic airway inflammation by activating dendritic cells. Nat Med 13:913–919

Ishii KJ, Suzuki K, Coban C et al (2001) Genomic DNA released by dying cells induces the maturation of APCs. J Immunol 167:2602–2607

Ivakhno SS, Kornelyuk AI (2004) Cytokine-like activities of some aminoacyl-tRNA synthetases and auxiliary p43 cofactor of aminoacylation reaction and their role in oncogenesis. Exp Oncol 26:250–255

Javaherian K, Liu JF, Wang JC (1978) Non-histone proteins HMG1 and HMG2 change the DNA helical structure. Science 199:1345–1346

Kaisho T, Akira S (2006) Toll-like receptor function and signalling. J Allergy Clin Immunol 117:979–987; quiz 988

Kao J, Fan YG, Haehnel I et al (1994a) A peptide derived from the amino terminus of endothelial-monocyte-activating polypeptide II modulates mononuclear and polymorphonuclear leukocyte functions, defines an apparently novel cellular interaction site, induces an acute inflammatory response. J Biol Chem 269:9774–9782

Kao J, Houck K, Fan Y et al (1994b) Characterization of a novel tumour-derived cytokine. Endothelial-monocyte activating polypeptide II. J Biol Chem 269:25106–25119

Kao J, Ryan J, Brett G et al (1992) Endothelial monocyte-activating polypeptide II. A novel tumour-derived polypeptide that activates host-response mechanisms. J Biol Chem 267:20239–20247

Kariko K, Ni H, Capodici J et al (2004) mRNA is an endogenous ligand for Toll-like receptor 3. J Biol Chem 279:12542–12550

Kerkhoff C, Klempt M, Kaever V et al (1999) The two calcium-binding proteins, S100A8 and S100A9, are involved in the metabolism of arachidonic acid in human neutrophils. J Biol Chem 274:32672–32679

Kim SJ, Gershov D, Ma X et al (2002) I-PLA(2) activation during apoptosis promotes the exposure of membrane lysophosphatidylcholine leading to binding by natural immunoglobulin M antibodies and complement activation. J Exp Med 196:655–665

Kingston AE, Hicks CA, Colston MJ et al (1996) A 71-kD heat shock protein (hsp) from Mycobacterium tuberculosis has modulatory effects on experimental rat arthritis. Clin Exp Immunol 103:77–82

Knies UE, Behrensdorf HA, Mitchell CA et al (1998) Regulation of endothelial monocyte-activating polypeptide II release by apoptosis. Proc Natl Acad Sci U S A 95:12322–12327

Ko YG, Park H, Kim T et al (2001) A cofactor of tRNA synthetase, p43, is secreted to up-regulate pro-inflammatory genes. J Biol Chem 276:23028–23033

Kokkola R, Sundberg E, Ulfgren AK et al (2002) High mobility group box chromosomal protein 1: a novel pro-inflammatory mediator in synovitis. Arthritis Rheum 46:2598–2603

Krispin A, Bledi Y, Atallah M et al (2006) Apoptotic cell thrombospondin-1 and heparin-binding domain lead to dendritic-cell phagocytic and tolerizing states. Blood 108:3580–3589

Krysko DV, Denecker G, Festjens N et al (2006) Macrophages use different internalization mechanisms to clear apoptotic and necrotic cells. Cell Death Differ 13:2011–2022

la Sala A, Ferrari D, Corinti S et al (2001) Extracellular ATP induces a distorted maturation of dendritic cells and inhibits their capacity to initiate Th1 responses. J Immunol 166:1611–1617

la Sala A, Sebastiani S, Ferrari D et al (2002) Dendritic cells exposed to extracellular adenosine triphosphate acquire the migratory properties of mature cells and show a reduced capacity to attract type 1 T lymphocytes. Blood 99:1715–1722

Lackmann M, Cornish CJ, Simpson RJ et al (1992) Purification and structural analysis of a murine chemotactic cytokine (CP-10) with sequence homology to S100 proteins. J Biol Chem 267:7499–7504

Lancaster GI, Febbraio MA (2005) Exosome-dependent trafficking of HSP70: a novel secretory pathway for cellular stress proteins. J Biol Chem 280:23349–23355

Lauber K, Blumenthal SG, Waibel M et al (2004) Clearance of apoptotic cells; getting rid of the corpses. Mol Cell 14:277–287

Lauber K, Bohn E, Krober SM et al (2003) Apoptotic cells induce migration of phagocytes via caspase-3-mediated release of a lipid attraction signal. Cell 113:717–730

Le LQ, Kabarowski JH, Weng Z et al (2001) Mice lacking the orphan G protein-coupled receptor G2A develop a late-onset autoimmune syndrome. Immunity 14:561–571

Li Y, Barger SW, Liu L et al (2000) S100beta induction of the pro-inflammatory cytokine interleukin-6 in neurons. J Neurochem 74:143–150

Liu-Bryan R, Scott P, Sydlaske A et al (2005) Innate immunity conferred by Toll-like receptors 2 and 4 and myeloid differentiation factor 88 expression is pivotal to monosodium urate monohydrate crystal-induced inflammation. Arthritis Rheum 52:2936–2946

Lymn JS, Patel MK, Clunn GF et al (2002) Thrombospondin-1 differentially induces chemotaxis and DNA synthesis of human venous smooth muscle cells at the receptor-binding level. J Cell Sci 115:4353–4360

MacDonald PE, Braun M, Galvanovskis J et al (2006) Release of small transmitters through kiss-and-run fusion pores in rat pancreatic beta cells. Cell Metab 4:283–290

Mambula SS, Calderwood SK (2006a) Heat induced release of Hsp70 from prostate carcinoma cells involves both active secretion and passive release from necrotic cells. Int J Hyperthermia 22:575–585

Mambula SS, Calderwood SK (2006b) Heat shock protein 70 is secreted from tumour cells by a non-classical pathway involving lysosomal endosomes. J Immunol 177:7849–7857

Mansfield PJ, Boxer LA, Suchard SJ (1990) Thrombospondin stimulates motility of human neutrophils. J Cell Biol 111:3077–3086

Mansfield PJ, Suchard SJ (1994) Thrombospondin promotes chemotaxis and haptotaxis of human peripheral blood monocytes. J Immunol 153:4219–4229

Martinon F, Petrilli V, Mayor A et al (2006) Gout-associated uric acid crystals activate the NALP3 inflammasome. Nature 440:237–241

McMurray HF, Parthasarathy S, Steinberg D (1993) Oxidatively modified low density lipoprotein is a chemoattractant for human T lymphocytes. J Clin Invest 92:1004–1008

Messmer D, Yang H, Telusma G et al (2004) High mobility group box protein 1: an endogenous signal for dendritic cell maturation and Th1 polarization. J Immunol 173:307–313

Mochizuki M, Zigler JS Jr, Russell P et al (1982) Serum proteins neutralize the toxic effect of lysophosphatidyl choline. Curr Eye Res 2:621–624

Munoz LE, Gaipl US, Franz S et al (2005) SLE--a disease of clearance deficiency? Rheumatology (Oxford) 44:1101–1107

Murakami Y, Akahoshi T, Hayashi I et al (2003) Inhibition of monosodium urate monohydrate crystal-induced acute inflammation by retrovirally transfected prostaglandin D synthase. Arthritis Rheum 48:2931–2941

Murakami Y, Akahoshi T, Kawai S et al (2002) Anti-inflammatory effect of retrovirally transfected interleukin-10 on monosodium urate monohydrate crystal-induced acute inflammation in murine air pouches. Arthritis Rheum 46:2504–2513

Murphy JE, Tedbury PR, Homer-Vanniasinkam S et al (2005) Biochemistry and cell biology of mammalian scavenger receptors. Atherosclerosis 182:1–15

Murugesan G, Sandhya Rani MR, Gerber CE et al (2003) Lysophosphatidylcholine regulates human microvascular endothelial cell expression of chemokines. J Mol Cell Cardiol 35:1375–1384

Nakamura H, Izumoto Y, Kambe H et al (1994) Molecular cloning of complementary DNA for a novel human hepatoma-derived growth factor. Its homology with high mobility group-1 protein. J Biol Chem 269:25143–25149

Ngwenya BZ, Yamamoto N (1985) Activation of peritoneal macrophages by lysophosphatidylcholine. Biochim Biophys Acta 839:9–15

Ni H, Capodici J, Cannon G et al (2002) Extracellular mRNA induces dendritic cell activation by stimulating tumour necrosis factor-alpha secretion and signalling through a nucleotide receptor. J Biol Chem 277:12689–12696

Nishimura T, Horino K, Nishiura H et al (2001) Apoptotic cells of an epithelial cell line, AsPC-1, release monocyte chemotactic S19 ribosomal protein dimer. J Biochem 129:445–454

Nishiura H, Shibuya Y, Matsubara S et al (1996) Monocyte chemotactic factor in rheumatoid arthritis synovial tissue. Probably a cross-linked derivative of S19 ribosomal protein. J Biol Chem 271:878–882

Nishiura H, Shibuya Y, Yamamoto T (1998) S19 ribosomal protein cross-linked dimer causes monocyte-predominant infiltration by means of molecular mimicry to complement C5a. Lab Invest 78:1615–1623

Njemini R, Lambert M, Demanet C et al (2003) Elevated serum heat-shock protein 70 levels in patients with acute infection: use of an optimized enzyme-linked immunosorbent assay. Scand J Immunol 58:664–669

Odink K, Cerletti N, Bruggen J et al (1987) Two calcium-binding proteins in infiltrate macrophages of rheumatoid arthritis. Nature 330:80–82

Orlova VV, Choi EY, Xie C et al (2007) A novel pathway of HMGB1-mediated inflammatory cell recruitment that requires Mac-1-integrin. EMBO J 26:1129–1139

Park JS, Arcaroli J, Yum HK et al (2003) Activation of gene expression in human neutrophils by high mobility group box 1 protein. Am J Physiol Cell Physiol 284:C870–C879

Park JS, Svetkauskaite D, He Q et al (2004) Involvement of toll-like receptors 2 and 4 in cellular activation by high mobility group box 1 protein. J Biol Chem 279:7370–7377

Peter C, Waibel M, Radu CG et al (2007) Migration to apoptotic 'Find-Me' signals is mediated via the phagocyte receptor G2A. J Biol Chem

Picher M, Burch LH, Hirsh AJ et al (2003) Ecto 5'-nucleotidase and non-specific alkaline phosphatase. Two AMP-hydrolyzing ectoenzymes with distinct roles in human airways. J Biol Chem 278:13468–13479

Pinto SS, Gottfried C, Mendez A et al (2000) Immunocontent and secretion of S100B in astrocyte cultures from different brain regions in relation to morphology. FEBS Lett 486:203–207

Prasad S, Soldatenkov VA, Srinivasarao G et al (1998) Identification of keratins 18, 19 and heat-shock protein 90 beta as candidate substrates of proteolysis during ionizing radiation-induced apoptosis of estrogen-receptor negative breast tumour cells. Int J Oncol 13:757–764

Quevillon S, Agou F, Robinson JC et al (1997) The p43 component of the mammalian multi-synthetase complex is likely to be the precursor of the endothelial monocyte-activating polypeptide II cytokine. J Biol Chem 272:32573–32579

Quinn MT, Parthasarathy S, Steinberg D (1988) Lysophosphatidylcholine: a chemotactic factor for human monocytes and its potential role in atherogenesis. Proc Natl Acad Sci U S A 85:2805–2809

Rammes A, Roth J, Goebeler M et al (1997) Myeloid-related protein (MRP) 8 and MRP14, calcium-binding proteins of the S100 family, are secreted by activated monocytes via a novel, tubulin-dependent pathway. J Biol Chem 272:9496–9502

Robson SC, Wu Y, Sun X et al (2005) Ectonucleotidases of CD39 family modulate vascular inflammation and thrombosis in transplantation. Semin Thromb Hemost 31:217–233

Ronnefarth VM, Erbacher AI, Lamkemeyer T et al (2006) TLR2/TLR4-independent neutrophil activation and recruitment upon endocytosis of nucleosomes reveals a new pathway of innate immunity in systemic lupus erythematosus. J Immunol 177:7740–7749

Rosen A, Casciola-Rosen L (1999) Auto-antigens as substrates for apoptotic proteases: implications for the pathogenesis of systemic autoimmune disease. Cell Death Differ 6:6–12

Rothermundt M, Peters M, Prehn JH et al (2003) S100B in brain damage and neurodegeneration. Microsc Res Techniq 60:614–632

Rouhiainen A, Kuja-Panula J, Wilkman E et al (2004) Regulation of monocyte migration by amphoterin (HMGB1). Blood 104:1174–1182

Rouhiainen A, Tumova S, Valmu L et al (2007) Pivotal advance: analysis of pro-inflammatory activity of highly purified eukaryotic recombinant HMGB1 (amphoterin). J Leukocyte Biol 81:49–58

Rouleau P, Vandal K, Ryckman C et al (2003) The calcium-binding protein S100A12 induces neutrophil adhesion, migration, release from bone marrow in mouse at concentrations similar to those found in human inflammatory arthritis. Clin Immunol 107:46–54

Ryborg AK, Deleuran B, Thestrup-Pedersen K et al (1994) Lysophosphatidylcholine: a chemoattractant to human T lymphocytes. Arch Dermatol Res 286:462–465

Ryckman C, Gilbert C, de Medicis, R et al (2004) Monosodium urate monohydrate crystals induce the release of the pro-inflammatory protein S100A8/A9 from neutrophils. J Leukocyte Biol 76:433–440

Sakata-Kaneko S, Wakatsuki Y, Usui T et al (1998) Lysophosphatidylcholine upregulates CD40 ligand expression in newly activated human CD4+ T cells. FEBS Lett 433:161–165

Savill J, Hogg N, Ren Y et al (1992) Thrombospondin cooperates with CD36 and the vitronectin receptor in macrophage recognition of neutrophils undergoing apoptosis. J Clin Invest 90:1513–1522

Scaffidi P, Misteli T, Bianchi ME (2002) Release of chromatin protein HMGB1 by necrotic cells triggers inflammation. Nature 418:191–195

Schmitt E, Gehrmann M, Brunet M et al (2007) Intracellular and extracellular functions of heat shock proteins: repercussions in cancer therapy. J Leukocyte Biol 81:15–27

Scott RS, McMahon EJ, Pop SM et al (2001) Phagocytosis and clearance of apoptotic cells is mediated by MER. Nature 411:207–211

Segundo C, Medina F, Rodriguez C et al (1999) Surface molecule loss and bleb formation by human germinal center B cells undergoing apoptosis: role of apoptotic blebs in monocyte chemotaxis. Blood 94:1012–1020

Semino C, Angelini G, Poggi A et al (2005) NK/iDC interaction results in IL-18 secretion by DCs at the synaptic cleft followed by NK cell activation and release of the DC maturation factor HMGB1. Blood 106:609–616

Semino C, Ceccarelli J, Lotti LV et al (2007) The maturation potential of NK cell clones toward autologous dendritic cells correlates with HMGB1 secretion. J Leukocyte Biol 81:92–99

Shi Y, Evans, JE, Rock KL (2003) Molecular identification of a danger signal that alerts the immune system to dying cells. Nature 425:516–521

Shi Y, Galusha SA, Rock KL (2006) Cutting edge: elimination of an endogenous adjuvant reduces the activation of CD8 T lymphocytes to transplanted cells and in an autoimmune diabetes model. J Immunol 176:3905–3908

Shi Y, Rock KL (2002) Cell death releases endogenous adjuvants that selectively enhance immune surveillance of particulate antigens. Eur J Immunol 32:155–162

Shi Y, Zheng W, Rock KL (2000) Cell injury releases endogenous adjuvants that stimulate cytotoxic T cell responses. Proc Natl Acad Sci U S A 97:14590–14595

Shibuya Y, Shiokawa M, Nishiura H et al (2001) Identification of receptor-binding sites of monocyte chemotactic S19 ribosomal protein dimer. Am J Pathol 159:2293–2301

Shrestha A, Shiokawa M, Nishimura T et al (2003) Switch moiety in agonist/antagonist dual effect of S19 ribosomal protein dimer on leukocyte chemotactic C5a receptor. Am J Pathol 162:1381–1388

Singh-Jasuja H, Scherer HU, Hilf N et al (2000a) The heat shock protein gp96 induces maturation of dendritic cells and down-regulation of its receptor. Eur J Immunol 30:2211–2215

Singh-Jasuja H, Toes RE, Spee P et al (2000b) Cross-presentation of glycoprotein 96-associated antigens on major histocompatibility complex class I molecules requires receptor-mediated endocytosis. J Exp Med 191:1965–1974

Srivastava PK (2000) Heat shock protein-based novel immunotherapies. Drug News Perspect 13:517–522

Tang D, Kang R, Xiao W et al (2007) The anti-inflammatory effects of heat shock protein 72 involve inhibition of high-mobility-group box 1 release and pro-inflammatory function in macrophages. J Immunol 179:1236–1244

Taniguchi N, Kawahara K, Yone K et al (2003) High mobility group box chromosomal protein 1 plays a role in the pathogenesis of rheumatoid arthritis as a novel cytokine. Arthritis Rheum 48:971–981

Taniuchi M, Otani H, Kodama N et al (1999) Lysophosphatidylcholine up-regulates IL-1 beta-induced iNOS expression in rat mesangial cells. Kidney Int Suppl 71:S156–S158

Terkeltaub R, Baird S, Sears P et al (1998) The murine homolog of the interleukin-8 receptor CXCR-2 is essential for the occurrence of neutrophilic inflammation in the air pouch model of acute urate crystal-induced gouty synovitis. Arthritis Rheum 41:900–909

Terkeltaub R, Zachariae C, Santoro D et al (1991) Monocyte-derived neutrophil chemotactic factor/interleukin-8 is a potential mediator of crystal-induced inflammation. Arthritis Rheum 34:894–903

Theriault JR, Adachi H, Calderwood SK (2006) Role of scavenger receptors in the binding and internalization of heat shock protein 70. J Immunol 177:8604–8611

Theriault JR, Mambula SS, Sawamura T et al (2005) Extracellular HSP70 binding to surface receptors present on antigen presenting cells and endothelial/epithelial cells. FEBS Lett 579:1951–1960

Treutiger CJ, Mullins GE, Johansson AS et al (2003) High mobility group 1 B-box mediates activation of human endothelium. J Intern Med 254:375–385Umeda Y, Shibuya Y, Semba U, et al (2004) Guinea pig S19 ribosomal protein as precursor of C5a receptor-directed monocyte-selective leukocyte chemotactic factor. Inflamm Res 53:623–630

Vabulas RM, Ahmad-Nejad P, da Costa, C et al (2001) Endocytosed HSP60s use toll-like receptor 2 (TLR2) and TLR4 to activate the toll/interleukin-1 receptor signalling pathway in innate immune cells. J Biol Chem 276:31332–31339

Vabulas RM, Braedel S, Hilf N et al (2002) The endoplasmic reticulum-resident heat shock protein Gp96 activates dendritic cells via the Toll-like receptor 2/4 pathway. J Biol Chem 277:20847–20853

van Eden W, Van Der Zee R, Prakken B (2005) Heat-shock proteins induce T-cell regulation of chronic inflammation. Nat Rev Immunol 5:318–330

van Nieuwenhuijze AE, van Lopik T, Smeenk RJ et al (2003) Time between onset of apoptosis and release of nucleosomes from apoptotic cells: putative implications for systemic lupus erythematosus. Ann Rheum Dis 62:10–14

Vanden Berghe T, Kalai M, Denecker G et al (2006) Necrosis is associated with IL-6 production but apoptosis is not. Cell Signal 18:328–335

Vollmer J, Tluk S, Schmitz C et al (2005) Immune stimulation mediated by auto-antigen binding sites within small nuclear RNAs involves Toll-like receptors 7 and 8. J Exp Med 202:1575–1585

Wakasugi K, Schimmel P (1999) Two distinct cytokines released from a human aminoacyl-tRNA synthetase. Science 284:147–151

Wang H, Yang H, Czura CJ et al (2001) HMGB1 as a late mediator of lethal systemic inflammation. Am J Respir Crit Care Med 164:1768–1773

Whitaker-Azmitia PM, Murphy R, Azmitia EC (1990) Stimulation of astroglial 5-HT1A receptors releases the serotonergic growth factor, protein S-100, alters astroglial morphology. Brain Res 528:155–158

Williams RC Jr, Malone CC, Meyers C et al (2001) Detection of nucleosome particles in serum and plasma from patients with systemic lupus erythematosus using monoclonal antibody 4H7. J Rheumatol 28:81–94

Willoughby KA, Kleindienst A, Muller C et al (2004) S100B protein is released by in vitro trauma and reduces delayed neuronal injury. J Neurochem 91:1284–1291

Wu LJ, Vadakkan KI, Zhuo M (2007) ATP-induced chemotaxis of microglial processes requires P2Y receptor-activated initiation of outward potassium currents. Glia 55:810–821

Wu X, Molinaro C, Johnson N et al (2001) Secondary necrosis is a source of proteolytically modified forms of specific intracellular auto-antigens: implications for systemic autoimmunity. Arthritis Rheum 44:2642–2652

Xiao Q, Mandal K, Schett G et al (2005) Association of serum-soluble heat shock protein 60 with carotid atherosclerosis: clinical significance determined in a follow-up study. Stroke 36:2571–2576

Yamamoto T (2007) Roles of the ribosomal protein S19 dimer and the C5a receptor in pathophysiological functions of phagocytic leukocytes. Pathol Int 57:1–11

Yan JJ, Jung JS, Lee JE et al (2004) Therapeutic effects of lysophosphatidylcholine in experimental sepsis. Nat Med 10:161–167

Yang H, Ochani M, Li J et al (2004) Reversing established sepsis with antagonists of endogenous high-mobility group box 1. Proc Natl Acad Sci U S A 101:296–301

Yang Z, Tao T, Raftery MJ et al (2001) Pro-inflammatory properties of the human S100 protein S100A12. J Leukocyte Biol 69:986–994

Yasuda K, Ogawa Y, Yamane I et al (2005a) Macrophage activation by a DNA/cationic liposome complex requires endosomal acidification and TLR9-dependent and -independent pathways. J Leukocyte Biol 77:71–79

Yasuda K, Rutz M, Schlatter B et al (2006) CpG motif-independent activation of TLR9 upon endosomal translocation of "natural" phosphodiester DNA. Eur J Immunol 36:431–436

Yasuda K, Yu P, Kirschning CJ et al (2005b) Endosomal translocation of vertebrate DNA activates dendritic cells via TLR9-dependent and -independent pathways. J Immunol 174:6129–6136

Yu M, Wang H, Ding A et al (2006) HMGB1 signals through toll-like receptor (TLR) 4 and TLR2. Shock 26:174–179

Zhou Z, Yamamoto Y, Sugai F et al (2004) Hepatoma-derived growth factor is a neurotrophic factor harbored in the nucleus. J Biol Chem 279:27320–27326

Chapter 4
Molecules Involved in Recognition and Clearance of Apoptotic/Necrotic Cells and Cell Debris

Markus Napirei and Hans Georg Mannherz

Abstract: Disposal of apoptotic or necrotic cells or cell remnants is of paramount importance for the survival of multicellular organisms. Plasma membrane rupture of accumulating dying cells would lead to a deluge of cellular components into the extracellular spaces and could lead to impairment of tissue organisation, occlusion of vessels and the induction of inflammatory and autoimmune reactions. Therefore, efficient mechanisms exist, that secure safe disposal of dying cells and their contents. Neighbouring cells and professional phagocytes like macrophages and immature dendritic cells are the main effector cells of these processes. The mechanism of engulfment of dying cells and the signalling pathways leading to phagocytosis are described. In addition, extracellular mechanisms seem to exist that are activated for the disposal of necrotic cells and their remnants.

Keywords: Apoptosis • Necrosis • Phagocytosis • Phagocyte receptors

4.1 Programmed Versus Accidental Cell Death, in Simple Apoptosis Versus Necrosis

Cell death is a characteristic feature in the lifespan of eukaryotic cells and essential for the survival of multicellular organisms (for reviews see Majno and Joris 1995; Zakeri et al. 1995; Fiers et al. 1999; Van Cruchten and Van Den Broeck 2002; Fink and Cookson 2005). Coordinated cell demise is necessary for sculpturing tissues during embryogenesis (organogenesis), for preserving a constant organ cell number

M. Napirei (✉) · H. G. Mannherz
Department of Anatomy and Embryology
Medical Faculty
Ruhr-University Bochum
Universitätsstraße 150, D-44801 Bochum, Germany
Tel.: +49 2343223164
Fax: +49 2343214474
E-mail: markus.napirei@rub.de; hans.g.mannherz@rub.de

D.V. Krysko, P. Vandenabeele (eds.), *Phagocytosis of Dying Cells,*
DOI: 10.1007/978-1-4020-9292-3_4, © Springer Science+Business Media B.V. 2009

by balancing continuous renewal with elimination of aged or non-functional cells (tissue homeostasis), and in the immune system during the processes that lead to optimal antigen recognition or for maintaining self-tolerance by eliminating lymphocytes with autoimmune activity. During these physiological situations cell death occurs by genetically determined pathways and has therefore been termed programmed cell death (PCD). However, there are different modes of PCD like apoptosis and autophagy, type I and type II cell death, respectively (Krysko et al., Chap. 1; Diez-Fraile et al., Chap. 2, this Vol.). The former being the most frequently occurring form of PCD. The morphological features of apoptosis were described in detail by Kerr and colleagues (Kerr et al. 1972). Characteristically apoptotic cells detach from neighbouring cells and the extracellular matrix, exhibit cytoplasmic shrinkage, pyknosis, i.e. nuclear shrinkage and chromatin condensation at the inner nuclear lamina, nuclear fragmentation (karyorrhexis), and subsequently they disintegrate into membrane-surrounded cell fragments (apoptotic bodies; Fig. 4.1a–d and Fig. 4.2c). The morphological features of apoptosis are caused by the selective cleavage of essential cellular proteins of the nuclear scaffold, focal adhesions, cytoskeleton and the DNA-repair machinery after activation of a group of aspartate-specific cysteine proteases called caspases. Caspase activation results from the transmission of death signals by cell death receptors (extrinsic pathway) and/or release of pro-apoptotic factors from mitochondria (intrinsic pathway). However, also under pathological conditions (infections, irradiation, intoxication, and mutagenic events) PCD is initiated with the aim to protect the organism against neoplasia, invading micro-organisms and to overcome inflammatory reactions.

In contrast to PCD, accidental cell death (ACD), also better known as necrosis or type III cell death, is usually regarded as the consequence of non-physiological insults. It can occur when an organism is exposed to extreme physical forces and/or to aggressive chemicals and in particular under pathological conditions such as anoxia caused by ischemia, mechanical trauma, heat, chill, irradiation, and/or drug intoxication. In a number of instances necrosis is initiated by suppression of

---

Fig. 4.1 Apoptosis and necrosis in MCF-7 cells. a–d: Human mamma adenocarcinoma MCF-7 cells treated with 2 μM staurosporin exhibit typical morphological signs of apoptosis. Early after induction of apoptosis (2 hours) the cells shrink (b) and expose PS on the plasma membrane (b'', positive plasma membrane staining with annexin V-FLUOS). After 4 hours, the cells display chromatin condensation (c'), however, the plasma membrane remains intact (c'', negative nuclear staining with propidium iodide, PI). Secondary (apoptotic) necrosis with osmotic swelling (d) due to plasma membrane damage occurs after 24 hours (d'', positive nuclear staining with PI). e–h: MCF-7 cells treated with 1 mM H_2O_2 and 1 mm sodium azide exhibit typical signs of primary necrosis, like pyknosis and osmotic cell swelling (oncosis). Early after induction of necrosis the cells display membrane rupture and PS exposure in parallel (e'' and f'', positive nuclear and plasma membrane staining with PI and annexin V-FLUOS, respectively). In the presence of murine wild-type serum, chromatin breakdown occurs with ongoing time (16 h) and DNA-fragments diffuse from the nucleus into the cytoplasm (g' and g''). Chromatin breakdown depends on DNASE1 as revealed by its lack in the presence of serum derived from *Dnase1* KO mice (h' and h''). Parts of this figure were already published in (Napirei et al. 2004). a–h: phase contrast microscopy. a'–h': DNA-staining by the plasma membrane-permeable dye Hoechst 33258. a'' and c''–h'': DNA-staining with the plasma membrane-impermeable dye PI. b'' and f'': staining of PS by annexin V-FLUOS. Bars: 20 μm.

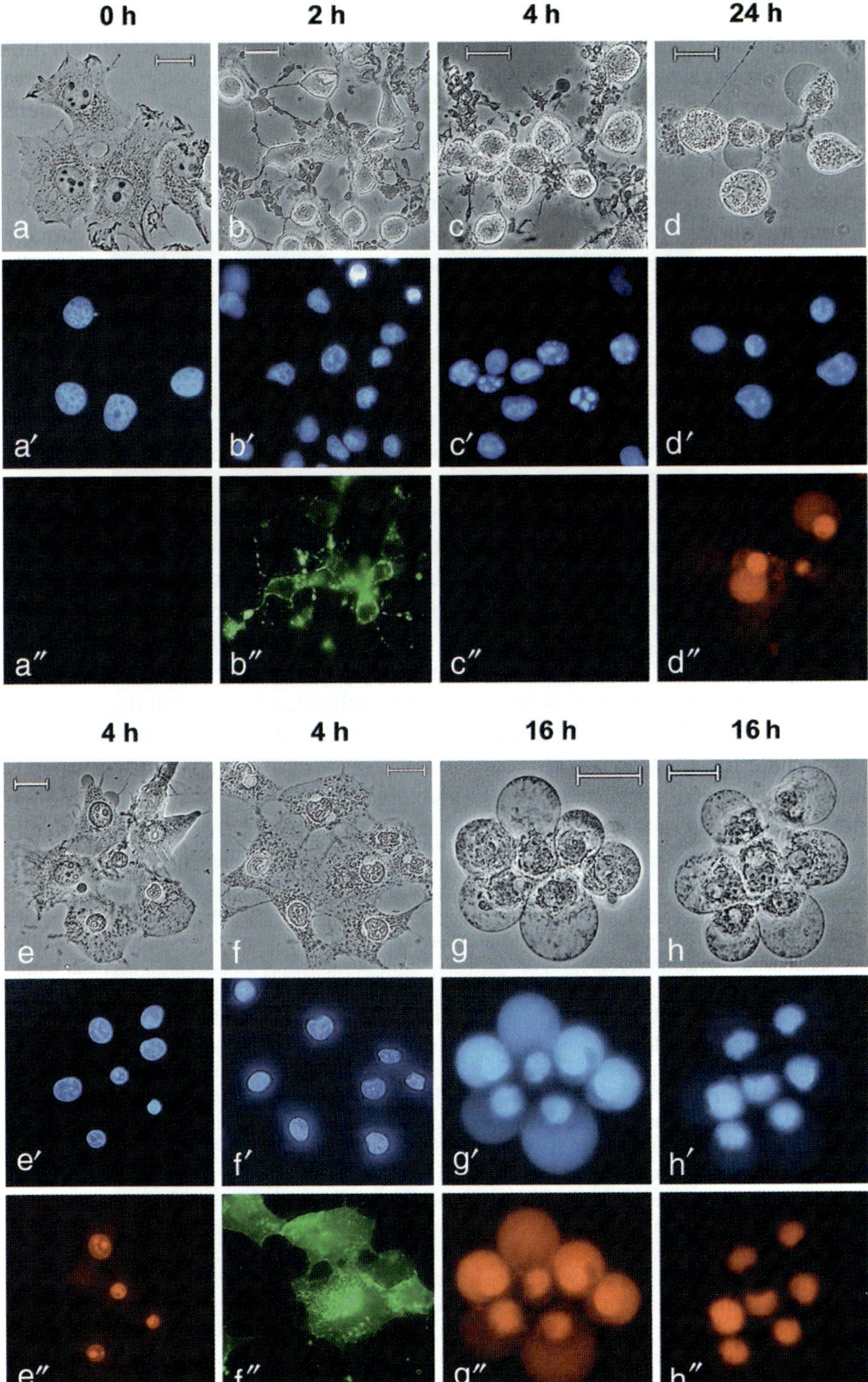

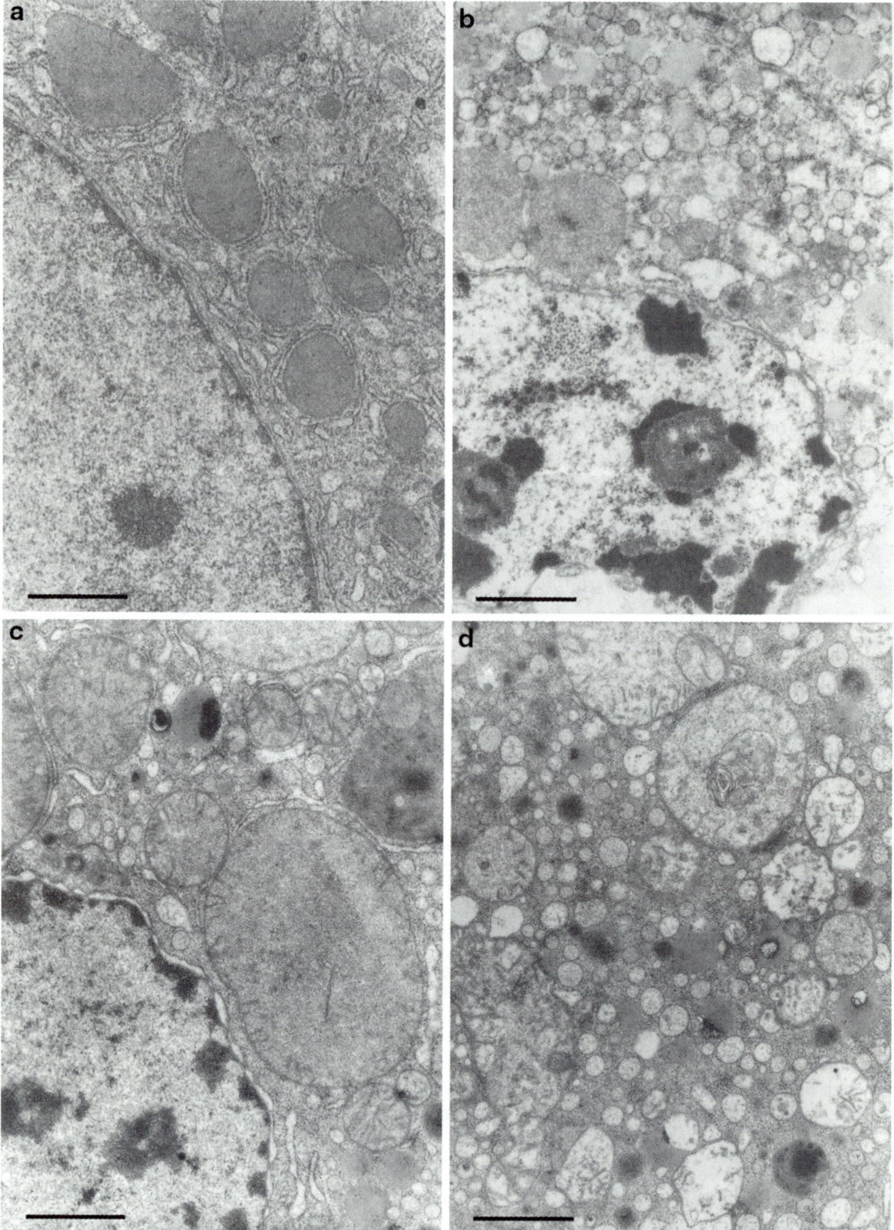

Fig. 4.3 Ultrastructural features of necrotic hepatocytes. a–d: Electron microscopy of murine liver sections derived from male CD-1 mice. a: Healthy hepatocyte of an untreated control mouse showing regular shaped rough ER, mitochondria and a section of the nucleus with one nucleolus. b–d: Ultrastructure of hepatocytes of mice which were exposed to a treatment of 800 mg/kg acetaminophen (paracetamol) for 18 hours (Jacob et al. 2007). b: Necrotic hepatocyte displaying nuclear pyknosis characterized by uniformly distributed condensed chromatin clods. Two nucleoli

after insults such as ischemia (Jaeschke and Lemasters 2003; Eum et al. 2007; Wang et al. 2007) and intoxications (Gujral et al. 2002; Gujral et al. 2003).

The main differences between PCD and ACD are: **First**, in PCD specialized hydrolytic enzymes are activated in a strictly regulated manner, whose primary function is the execution of cell death. In contrast, ACD is characterized by the collapse of the internal cellular milieu and loss of cytoplasmic components due to rupture of the plasma membrane that may also allow the inward diffusion of extracellular proteases and nucleases. **Secondly**, PCD requires energy, whereas ACD predominantly results from energy depletion. **Third**, PCD generates small "eupeptic" cell corpses for rapid and efficient engulfment by professional phagocytes or neighbouring cells. Thus, PCD and subsequent phagocytosis (efferocytosis, deCathelineau and Henson 2003; heterophagy, Zakeri et al. 1995) are two consecutive steps essential to complete cell elimination. This scenario can be supplemented by a process called autophagy, which means that cells are able of "self-eating" in addition to "self-killing" (for review see Maiuri et al. 2007). In contrast to PCD, ACD is characterized by cell swelling, which most likely represents a steric obstacle for effective engulfment and therefore leads to a delay of phagocytosis. Extracellular mechanisms most likely assist the complete removal of necrotic cell debris, especially when the cells have ruptured and spilled out their contents into the extracellular space. **Fourth**, PCD is predominantly anti-inflammatory, whereas ACD is pro-inflammatory by releasing soluble "danger signals" like heat shock proteins (e.g. Hsp60, Hsp70, gp96, calreticulin), the nuclear factor high mobility group box 1 (HMGB1), uric acid, DNA, mRNA, or fucose (Peter et al., Chap. 3, this Vol.; Proskuryakov et al. 2005; Krysko et al. 2006a).

4.1.1 Elimination of Apoptotic Cells

The elucidation of the molecular mechanism of cell demise by apoptosis has dominated the last decade of cell death research. In contrast, only little emphasis has been laid on the mechanisms of the disposal of the apoptotic cell corpses. Apoptotic cells or bodies are initially enclosed by an intact plasma membrane. If they are not eliminated within a short time period, they undergo secondary necrosis and release potentially immunogenic intracellular components into the extracellular

---

are visible. The perinuclear space reveals dilatation and the rough ER has more or less completely transformed to vacuoles. The mitochondria remain a regular structure. c: Necrotic hepatocyte displaying pyknosis with similar nuclear shrinkage, but less severe chromatin condensation in comparison to the nucleus shown in b. Two nucleoli are apparent. The perinuclear space is dilatated and vesiculation of the rough and smooth ER is visible. In addition, some of the mitochondria appear to be swollen. d: Necrotic hepatocyte displaying massive vacuolization and ongoing destruction of the inner mitochondrial structure. Most of the vacuoles appeared to be derived from the smooth ER or from the rough ER after displacement of the ribosomes. Bars: 1 μm. The photographs were kindly provided by Dr. Monika Jacob, Department of Anatomy and Embryology, Ruhr-University Bochum.

Table 4.1 Molecules involved in the recognition and clearance of apoptotic cells. Molecules involved in the recognition and clearance of apoptotic cells are listed and divided into those involved in the elimination of early and late apoptotic cells. Co-receptors, their corresponding bridging molecules and ligands are depicted in italics, whereas the phagocytic receptors and their bridging molecules and ligands are depicted in normal font. Synonyms of molecules are presented in brackets [ ].The presented data are partially a composition of results obtained from independent and separate experimental approaches performed by different groups (see text). Some of the presented data are speculative(?), however are conceivable due to the entire published literature concerning this issue.

Elimination of early apoptotic cells

Phagocyte receptor		Co-receptor	Bridging molecule(s)	Ligand(s)	Phagocyte(s)
PS receptors:	TIM1			PS	kidney tubule cells
	TIM4			PS	non-elicited peritoneal Mφ, thymic Mφ
TAM receptors:	MER	*αvβ5 integrin*	GAS6, protein S / *MFG-E8*	PS, oxPS / *PS, oxPS*	RPE cells, iDCs
	MER		GAS6, protein S	PS, oxPS	BMD-Mφ, thymic/ splenic Mφ
LPS receptor:	CD14			modified ICAM-3	MD-Mφ, BMD- Mφ, non-elicited peritoneal Mφ phagocytes: thymus, spleen, lung, liver, gut
Scavenger receptors:					
SR-A				oxLDL like sites	thymic Mφ, peritoneal Mφ
CD36		*αvβ3 integrin*	TSP	TSP binding sites	MD-Mφ
CD36		?	?	oxPS	elicited peritoneal Mφ
SR-BI				PS, PI	Sertoli cells
LOX-1				oxLDL like sites, PS	BAE cells
Integrin receptors:					
αvβ3		*CD36*	TSP	TSP binding sites	MD-Mφ
αvβ3		?	TSP	TSP binding sites	glomerular mesangial cells
αvβ3 ?		?	MFG-E8	PS, oxPS	mammary gland Mφ and ECs tingible body Mφ, elicited peritoneal Mφ

Table 4.1 (continued)

αvβ5	CD36	MFG-E8 / TSP ?	PS, oxPS / TSP binding sites ?	RPE cells, iDCs
αvβ5	MER	MFG-E8 / GAS6, protein S	PS, oxPS / PS, oxPS	RPE cells, iDCs
αmβ2 [CR3]			deposited C3b/bi	iDCs, MD-Mφ
αxβ2 [CR4]			deposited C3b/bi	iDCs, MD-Mφ
annexin II ?		β2GPI	PS	THP-1 derived Mφ, elicited peritoneal Mφ
Elimination of late apoptotic cells				
CD91 [LRP]	CRT [cC1qR]	MBL	carbohydrates, DNA	MD-Mφ, non-elicited peritoneal Mφ
CD91 [LRP]	CRT [cC1qR]	SP-A	carbohydrates, DNA	alveolar Mφ, MD-Mφ
CD91 [LRP]	CRT [cC1qR]	SP-D	carbohydrates, DNA	alveolar Mφ, MD-Mφ
CD91 [LRP]	CRT [cC1qR]	C1q	C1q binding sites	not described
CD91 [LRP]	CRT [cC1qR]	C1q – IgM	LPC	BMD- Mφ
CD91 [LRP]	CRT [cC1qR]	C1q – SAP	PE	MD-Mφ
CD91 [LRP]	CRT [cC1qR]	C1q – CRP	PC, LPC	MD-Mφ
CD91 [LRP]	CRT [cC1qR]	C1q – PTX3	PTX3 binding sites	inhibits phagocytosis by iDCs / Mφ
CD91 [LRP]			CRT	J774 Mφ
Fcγ receptors:				
FcγRI		SAP / CRP	PE / PC, LPC	J774 Mφ, peritoneal Mφ, leukocytes
FcγRI		HRG1	HRG1 binding sites	MD-Mφ
FcγRIIa/b		CRP	PC, LPC	Leukocytes
FcγRIII		SAP / CRP	PE / PC, LPC	J774 Mφ, peritoneal Mφ

BAE cells = bovine aortic endothelial cells; β2GPI = β2-glycoprotein I; BMD Mφ = bone-marrow derived Mφ; CRT = calreticulin; ECs = epithelial cells; iDCs = immature dendritic cells; LPC = lysophosphatidylcholine; LRP = low-density lipoprotein receptor-related protein; MD-Mφ = monocyte derived Mφ; MFG-E8 = milk fat globule protein E8; PC = phosphatidylcholine; PE = phosphatidylethanolamine; PI = phosphatidylinositol; PS = phosphatidylserine; TSP = thrombospondin; RPE cells = retinal pigment epithelial cells.

space. Within the human body tissues with high cellular turnover are epithelia and the hematopoietic and immune system. Usually dying epithelial cells are shed from their surface like for instance into the lumen of the small intestine, therefore in these tissues phagocytosis rarely occurs. In solid organs dying cells are in most instances phagocytosed by neighbouring cells (Ucker, Chap. 6; Lacy-Hulbert, Chap. 7, this Vol.). Professional phagocytes; however, engulf apoptotic white blood cells like neutrophils or lymphocytes. In most cases macrophages are involved and they accumulate also when large numbers of apoptotic cells are generated due to tissue damage by mild ischemia or infection.

Phagocytosis necessitates the selective recognition of apoptotic cells by the engulfing cell followed by firm interaction and subsequent engulfment by the phagocyte. These processes are presently understood only in vague outlines. There is general agreement that the apoptotic cell itself signals its ongoing demise by exposing a number of different "eat-me" signals on its surface. This implies alterations in the composition or of individual components of the plasma membrane. Apoptotic cells shrink and detach from the extracellular matrix like for instance the basal membrane and from neighbouring cells. During these processes junctional complexes like tight and adherens junctions unclench and separate and gap junctions loose their communicating capability (Theiss et al. 2007). Cell singling necessarily leads to exposure of epitopes previously buried in cell-cell and ECM contacts. However, very little is known about their possible contribution to recognition processes by phagocytic cells.

In contrast, it is well-established that during apoptosis new molecules appear on the cell surface (for reviews see Kagan et al. 2003; Fadeel 2004; Gardai et al. 2006; Table 4.1). The best known is phosphatidylserine (PS) that is translocated from the inner to the outer leaflet of the plasma membrane (Fadok et al. 1992; Martin et al. 1995; Fig. 4.1b"). In viable cells the majority of negatively charged aminophospholipids (PS and phosphatidylethanolamine, PE) is mainly confined to the inner leaflet of the plasma membrane, whereas choline-containing phospholipids (phosphatidylcholine (PC) and sphingomyelin) are localized mainly in its outer leaflet (Bevers et al. 1999; Kagan et al. 2003). This asymmetry is maintained by aminophospholipid translocase, an ATP-dependent "flippase" that transports PS and PE from the outer to the inner leaflet of the membrane, which counteracts the activity of phospholipid scramblases facilitating the bi-directional movement of all classes of phospholipids across the lipid bilayer (Bevers et al. 1999). During apoptosis the aminophospholipid translocase is inactivated by caspase-3 cleavage (Mandal et al. 2005) and furthermore the activity of phospholipid scramblases is supposed to be enhanced leading to the accumulation of PS and PE on the outer leaflet of the membrane (Kagan et al. 2003). In addition, the ATP-binding-cassette transporter 1 (ABCA1) is implicated in the distribution of lipids across the membrane. ABCA1, which is supposed to be the structural orthologue of *Caenorhabditis elegans* CED-7 (Wu and Horvitz 1998), was shown to be required for optimal engulfment of apoptotic cells by macrophages (Luciani and Chimini 1996). *Abca1* KO mice display a reduced clearance of interdigital apoptotic cells within the limb buds and a reduced clearance of i.p. injected apoptotic wild-type thymocytes by peritoneal macrophages possibly due to a reduced exposure of PS by macrophages (Hamon et al. 2000). Indeed, ABCA1 was not only shown to be a carrier facilitating the efflux of cholesterol and choline-containing phospholipids to acceptor proteins like apolipoprotein A-I during the formation of lipoproteins, but also to be involved in a calcium-dependent externalization of PS, i.e. it appears to function as a phospholipid "floppase" (Hamon et al. 2000). A recent study indicates that not ABCA1, but its homologue ABCA7 is the structural orthologue of CED-7 (Jehle et al. 2006). In conclusion, efficient phagocytosis of apoptotic cells necessitates PS exposure on both, the apoptotic plasma membrane as well as on the surface of macrophages. Indeed, masking of PS on the cell surface of macrophages by annexin V inhibits phagocytosis of lymphocytes as effectively as masking PS on the target cell surface (Callahan et al. 2000). It is supposed that annexin I exposed on the cell surface of apoptotic T lymphocytes interacts with PS on the surface of macrophages and that annexin I and II exposed on the surface of macrophages interacts with PS on the surface of apoptotic cells vice versa (Fan et al. 2004). In addition to PS externalization, a fraction of the PS within the inner leaflet of the plasma membrane is oxidized (oxPS) either directly by reactive oxygen species resulting from disruption of the mitochondrial electron transport or indirectly by oxidized cytochrome c (Kagan et al. 2003).

The importance of PS for recognition and phagocytosis by macrophages was first shown for red blood cells (Schroit et al. 1985) and indeed, elimination of aged erythrocytes seems to be analogous to apoptotic cells and dependent on a Fas/caspase-8/caspase-3-dependent signalling leading to PS externalization (Mandal et al. 2005). Apart from PS, other modified phospholipids appear on the surface of apoptotic cells, in particular lyso-phospholipids that are generated by the increased activity of the calcium-independent phospholipase A_2 (iPLA$_2$) due to caspase 3 cleavage (Atsumi et al. 2000) or by the action of secreted phospholipase A_2 (sPLA$_2$) (Atsumi et al. 1997). In addition, cytosolic annexin I (Arur et al. 2003; Fan et al. 2004), modified cell-surface ICAM-3 (Moffatt et al. 1999), increased levels of calreticulin from the endoplasmic reticulum (Gardai et al. 2005), and even DNA (Casciola-Rosen et al. 1994; Palaniyar et al. 2004) were reported to appear on the surface of or are released from apoptotic cells. Further modifications affect the composition of the glycocalix. Thus, an increased level of α-D-mannose and β-D-galactose-rich plasma membrane glycoproteins is accompanied by a decrease in the level of α-2,3-sialic-acid-containing glycoproteins in apoptotic plasma membranes (Bilyy and Stoika 2007). Furthermore, N-acetylglucosamine and fucose are increasingly expressed during apoptosis (Duvall et al. 1985; Russell et al. 1998). These alterations cumulate in an altered, apoptosis-specific surface pattern representing strong "eat-me" signals for phagocytes.

In addition, phagocytes have to be attracted by so-called "find-me" signals like lyso-phosphatidylcholine (LPC) secreted by apoptotic cells (Lauber et al. 2003). Whether LPC plays a role in vivo is still unclear, since in solid organs removal of apoptotic cells is accomplished by neighbouring cells. Further potential "find me" signals or mediators for phagocytosis released during inflammation are annexin I and its peptide derivatives (Scannell et al. 2007) or lipoxins (Godson et al. 2000; Mitchell et al. 2002).

In contrast to the "eat-me" signals exposed by apoptotic cells, non-apoptotic cells expose on their surface so-called "don´t eat-me" signals that inhibit the binding and activity of phagocytes. Homotypic repulsions between CD31 molecules expressed on the plasma membranes of both the non-apoptotic cell and phagocytes is one such signal (Brown et al. 2002). In addition, interaction of the integrin-associated protein IAP (CD47), which is a ubiquitously expressed cell surface glycoprotein, with its inhibitory receptor signal regulatory protein alpha (SIRPα), present on peripheral blood mononuclear cells (PMNs) and monocyte-derived cells, prevents phagocytosis of viable cells (Oldenborg et al. 2000; Gardai et al. 2005). Apoptotic cells fail to activate SIRPα and are engulfed (Gardai et al. 2005). However, the underlying mechanism, i.e. decrease, inactivation, blocking or redistribution of CD47, is still debated. For circulating red blood cells it was shown that aging and their phagocytic elimination correlates with the decrease in CD47 presence (Khandelwal et al. 2007). In contrast, Tada et al. revealed that CD47 in addition to PS on apoptotic cells is required for engulfment (Tada et al. 2003).

4.1.1.1 Immunological Outcome of Phagocytosis of Apoptotic Cells

Most studies on the immunological consequences of phagocytosis of apoptotic cells by macrophages demonstrate that no inflammatory response is induced or that a pre-existing pro-inflammatory response (secretion of TNF-α, IL-6, IL-8 or IL-12) is down-regulated after phagocytosis (Ucker, Chap. 6; Lacy-Hulbert, Chap. 7, this Vol.; Cocco and Ucker 2001). The anti-inflammatory response of macrophages after ingestion of apoptotic cells is attributed to the up-regulation and secretion of TGF-ß1, platelet activating factor, IL-10, and prostaglandin E2. The anti-inflammatory status of apoptotic cells is also maintained at later stages of cell death, even after entry into secondary necrosis (Patel et al. 2006). However, there are also reports based on in vitro and in vivo experiments suggesting an inflammatory response towards apoptotic cells. Thus, the final outcome of phagocytosis of apoptotic cells might depend on the state of activation of the ingesting macrophage, the type of target cell, the receptors involved in phagocytosis as well as the death-inducing stimulus (for reviews see Cvetanovic and Ucker 2004; Proskuryakov et al. 2005; Krysko et al. 2006a).

In accordance to the majority of findings, which indicate an anti-inflammatory effect of phagocytosis of apoptotic cells by macrophages, many studies reveal that immature dendritic cells (iDCs) do not mature towards T cell stimulating autoantigen-presenting cells after ingestion of apoptotic cells (Lacy-Hulbert, Chap. 7; Divito and Morelli, Chap. 11, this Vol.). Immature DCs are characterized by their high capacity of phagocytosis and their low capacity of stimulating antigen-specific T cell responses. In general, after ingesting antigens, iDCs in the presence of maturation signals or pro-inflammatory cytokines, like bacterial LPS or TNF-α, respectively down-regulate ingestion and up-regulate MHC-peptide complexes. Furthermore, they up-regulate T cell adhesion and T cell co-stimulatory molecules (e.g., ICAM-1, CD80, CD86 and CD40), DC-restricted marker CD83, production of IL-12 and expression of chemokine receptors that guide DC migration into secondary lymphoid organs for priming antigen-specific T cells. These processes are called DC maturation. Ingestion of apoptotic cells by iDCs does not induce full DC maturation (Sauter et al. 2000; Verbovetski et al. 2002; Ip and Lau 2004). Those iDCs that phagocytose apoptotic cells down-regulate the expression of MHC class II, CD40, CD86, and CD83, but acquire the ability to migrate to secondary lymphoid organs by up-regulation of the chemokine receptor CCR7 (Verbovetski et al. 2002; Ip and Lau 2004). These data imply that those DCs induce anergy to autologous antigens in CD4$^+$ T cells (Verbovetski et al. 2002). In contrast, iDCs which have ingested late apoptotic cells, acquire a fully mature DC phenotype with the capability of homing and an enhanced CD4$^+$ T cell stimulatory capacity (Ip and Lau 2004) suggesting that stimulatory signals are released from late apoptotic (secondary necrotic) cells. Such co-stimulatory signals might be also provided by viral infections. Thus, it was shown that the uptake of apoptotic, influenza virus infected cells by iDCs through engagement of their CD36 and αvβ5-integrin receptors leads to cross-presentation of the ingested autoantigens by MHC class I to CD8$^+$ T cells (Albert et al. 1998a; Albert et al. 1998b). In addition, TNF-α functions as a co-stimulatory signal for the maturation of iDCs, which have ingested apoptotic neutrophils. Thus, in the absence of TNF-α iDCs fail to stimulate T cell proliferation due to a reduced expression of the co-stimulatory molecules CD40, CD80, and CD86 (Clayton et al. 2003).

4.1.1.2 "Eat-me" Signals and Their Recognition

The described exterior modifications of apoptotic cells are recognized by specialized receptors or made perceptible by a number of bridging molecules (for reviews see Lauber et al. 2004; Krysko et al. 2006a; Flierman and Daha 2007; Erwig and Henson 2008; Table 4.1). In mammals a number of receptors for apoptotic cells have been identified like the lipopolysaccharide binding receptor CD14, class A scavenger receptor SR-A, class B scavenger receptors CD36 and SR-B1, class E scavenger receptor LOX-1 (lectin-like oxidized low-density lipoprotein receptor 1), integrin receptors like the vitronectin receptor avß3- and avß5 integrin and the complement receptors 3 and 4 (CR3/CR4), the α2-macroglobulin receptor CD91 (LRP; low-density lipoprotein receptor-related protein), the collectin receptor calreticulin (CRT/cC1qR), and the MER receptor tyrosine kinase. The ligands for these receptors can be manifold and their engagement may depend on the activity of bridging molecules (Table 4.1). It has been shown that PS is recognized by a specific receptor (PSR; Fadok et al. 2000). In addition, it has been suggested that a number of bridging molecules (opsonins) support recognition of PS by phagocytes. These bridging molecules include serum protein S, growth arrest-specific gene product (GAS6), milk fat globule protein (MFG-E8), ß2-glycoprotein I (ß2GPI), and annexin I and II. Other bridging molecules like IgM antibodies (Abs), complement components like C1q, collectins, and pentraxins are supposed or proven to bind to further structural alterations on apoptotic cell membranes like PC, PE, lyso-phospholipids, carbohydrates or DNA. Considerable redundancy in receptors for "eat-me" signals and their ligands appears to exist suggesting that a large number of different receptors are engaged in the recognition at the same time. During these multiple ligand-receptor interactions the apoptotic cell and the phagocyte establish a large continuous contact area, for which the term "phagocytic synapse" was coined.

The Phosphatidylserine Receptor (PSR)

Fadok and co-workers identified a putative PSR with the help of the monoclonal antibody mAb217. This antibody inhibited the binding of PS containing liposomes to macrophages. Subsequently a 48 kDa protein was identified as the antigen recognized by mAb217 and cloned. Transfection of negative cells with this clone induced PS-binding and to a lesser degree even the ability to phagocytose apoptotic cells (Fadok et al. 2000). In an elegant study Hoffmann and co-workers dissected the process of engulfment by exposing isolated macrophages to erythrocytes coated with individual "eat-me" signals or antibodies against particular phagocytic receptors. The data demonstrated that the receptors CD14, CD36, and avß3 integrin resulted in attachment (tethering), but not engulfment (Hoffmann et al. 2001). For engulfment to occur the erythrocytes had to additionally expose PS. This switch occurred also in the additional presence of mAB217 as stimulating antibody suggesting engulfment was triggered by PSR ligation. Therefore, the authors suggested that the ingestion of apoptotic cells is a two-step process: an initial tethering step followed by PSR-dependent engulfment (Hoffmann et al. 2001).

Subsequently three groups generated *Psr* knockout mice (Wolf et al. 2007). All mice showed severe malformation of the central nervous system, the lungs or heart pointing to an essential role of this protein during embryogenesis. All mice died perinatally. Böse and co-workers were not able to observe differences in the clearance of apoptotic cells between wild-type (WT) and *Psr* knockout (KO) mice, when they analysed a number of organs during embryonic development like the removal of apoptotic cells of the inter-digital webs during limb development or of other organs. Furthermore, there was no difference in the activity between hepatic macrophages isolated from WT or *Psr* KO mice to engulf apoptotic thymocytes (Bose et al. 2004). These results suggested that the 48 kDa protein is probably not identical to the mAB217 antigen and therefore does not act as a PSR on the plasma membrane of phagocytes. Indeed, in WT mice the 48kDa protein appeared to be localised within the nucleus and was shown to possess 2-oxoglutarate dependent dioxygenase activity (Bose et al. 2004). Thus, the identity of the PSR still remains unsettled.

In a recent study Nagata and co-workers used a similar approach. They screened a library of monoclonal Abs generated against resident mouse peritoneal macrophages and selected one clone (Kat5–18) that inhibited phagocytosis of apoptotic cells by these macrophages. The antigen was identified by expression cloning as TIM4. Murine Tim4 (T cell immunoglobulin- and mucin-domain-containing molecule) is a type I transmembrane protein comprising an extracellular immunoglobulin domain. In the mouse the TIM family has eight, in humans three members (TIM 1, 3 and 4). Both TIM1 and TIM4 were shown to specifically bind PS with high affinity

(Miyanishi et al. 2007). Transfection of TIM4 into NIH3T3 cells conferred the ability to engulf apoptotic thymocytes to these cells previously unable for phagocytosis, which could be inhibited by the monoclonal Kat5–18 and MFG-E8 that is known to mask PS (Hanayama et al. 2002).

The Lipopolysaccharide Binding Receptor CD14

The pattern recognition receptor CD14 interacts with a variety of lipid, carbohydrate and protein ligands especially of microbial origin (Pugin et al. 1994). The analysis of *CD14* KO mice demonstrated that this macrophage receptor plays an essential role for the engulfment of apoptotic cells, since its lack leads to a systemic accumulation of apoptotic cells in multiple tissues like the spleen and thymus of unchallenged mice even though the number of phagocytic cells was not altered in these tissues (Gregory and Pound, Chap. 9, this Vol.; Devitt et al. 2004). These authors also demonstrated that isolated recombinant CD14 binds to apoptotic, but not to normal cells. For leukocytes it was shown that the cell surface adhesion molecule ICAM-3 is modified during apoptosis, leading to their subsequent binding through CD14 by macrophages (Moffatt et al. 1999). These data suggest that modified ICAM-3 is most likely the ligand for CD14 on the surface of apoptotic cells.

Scavenger Receptors (SR-A, CD36, SR-BI, LOX-1)

Scavenger receptors are a group of structurally unrelated molecules of eight subclasses (A-H) which are able to bind modified forms of low-density lipoprotein (LDL) and also apoptotic cells, anionic phospholipids and amyloid. They are believed to be part of the innate immune system in that they are pattern recognition receptors recognizing pathogen-associated molecules, oxidatively modified lipoproteins and apoptotic cells through molecular mimicry (for reviews see Murphy et al. 2005; Moore and Freeman 2006). In contrast to the LDL-receptor, they are able to load macrophages with cholesterol through mechanism not inhibited by cellular cholesterol content (generation of macrophage foam cells); thus, they were demonstrated to be involved in the pathology of atherosclerosis. Especially the ability of oxidized LDL (oxLDL) to interfere with binding of apoptotic cells to phagocytes established the hypothesis that receptors with oxLDL binding capacity are involved in phagocytosis of apoptotic cells and that ligands with similarity to oxLDL might exist on the surface of apoptotic cells.

Expression of class A-scavenger receptor SR-A is mostly restricted to macrophages and facilitates binding of actetylated (acLDL) and oxLDL, β-amyloid peptide, and glycation end-products. In contrast to class B-scavenger receptors, direct binding of SR-A to anionic phospholipids could not be revealed (Lee et al. 1992). Nevertheless, thymic macrophages isolated from *SR-A* KO mice display a reduced clearance of apoptotic thymocytes in vitro, implying that other so far unidentified ligands for SR-A than PS might exist on the surface of apoptotic cells. In contrast to the in vitro experiments, no impairment of apoptotic thymocyte clearance could be observed in vivo neither in unchallenged nor in gamma-irradiated *SR-A* KO in comparison to WT mice (Gregory and Pound, Chap. 9, this Vol.; Platt et al. 2000). Consistent with the finding of SR-A to be of minor importance for the phagocytosis of apoptotic cells are further in vitro experiments demonstrating that murine peritoneal macrophages lacking SR-A show only a minor decrease in binding and uptake of apoptotic thymocytes (reduction of 20–30%; Terpstra et al. 1997). These data underline the redundancy of clearance pathways for apoptotic cells in mammals.

Class B scavenger receptor CD36 is widely expressed on the surface of multiple cell types, like macrophages, adipocytes, platelets, and endothelial cells. SR-Bs possess transmembrane domains with a heavily glycosylated extracellular loop and two short intracellular tails. CD36 does not bind acLDL and extensively oxLDL, but in contrast to SR-A binds native lipoproteins like LDL, HDL and VLDL. In addition, CD36 was shown to bind to anionic phospholipids (Rigotti et al. 1995) and thrombospondin (TSP; Asch et al. 1987) and to be involved in phagocytosis of apoptotic neutrophils, eosinophils and lymphocytes by human monocyte derived macrophages (HMDMs; Savill et al. 1992; Stern et al. 1996). However, phagocytosis was shown to be facilitated by a ternary complex consisting of CD36 and the vitronectin receptor $\alpha v\beta 3$ integrin both bound to TSP (Savill et al. 1992; Ren et al. 1995; Stern et al. 1996). Thrombospondin is a trimeric molecule of ~450 kDa and was first isolated from the platelet granules. It mediates adhesive interactions between activated platelets and other cells and was shown to be secreted by a wide range of cell types including macrophages and neutrophils. In addition to its association with the cell surface of macrophages, it is a transient component of the inflammatory extracellular matrix of healing wounds (reviewed in Savill et al. 1992). The binding sites for TSP on apoptotic cells are unknown and neither *TSP-1*, *TSP-2* nor double-KO mice were reported to display a reduced clearance of apoptotic cells (Agah et al. 2002). Interestingly, *CD36* KO mice show an increase in apoptotic cell numbers in skin wounds, which was attributed to a reduced clearance by macrophages due to a lack of direct interaction between CD36 and oxPS present on the surface of apoptotic cells (Greenberg et al. 2006). Thus, under certain conditions (type of phagocyte, apoptotic target cell) the TSP-CD36-$\alpha v\beta 3$ system might work in separate and different ways with respect to bridging molecules, the apoptotic target structure and the receptor composition. This is also demonstrated for glomerular mesangial cells, which do not express CD36 and phagocytose apoptotic neutrophils via the vitronectin receptor $\alpha v\beta 3$ integrin and TSP as a bridging molecule (Hughes et al. 1997). Immature dendritic cells (Albert et al. 1998a) and retinal pigment epithelial cells (RPE; Finnemann and Rodriguez-Boulan 1999) use $\alpha v\beta 5$ instead of $\alpha v\beta 3$ integrin as a co-receptor for CD36-dependent phagocytosis of apoptotic cells or shed photoreceptor outer segments (POS), respectively, whereas in macrophages the CD36-$\alpha v\beta 3$ integrin receptor system dominates.

Class B scavenger receptor SR-B is a homologue of CD36 expressed as two splice variants (SR-BI and SR-BII) differing in the C-terminal cytoplasmic domain. In addition to macrophages, SR-B is expressed in the liver and steroidogenic tissues such as the adrenal glands, ovaries and testes. It displays the same substrate binding specificities like CD36; however, in addition to the uptake of cholesterol from HDL especially SR-BI mediates cholesterol transfer from cells to HDL making it an important key-player in reverse cholesterol transport. Targeted deletion of the *SR-BI* gene in mice leads to hypercholesterolemia due to a reduced biliary cholesterol secretion. For primary cultures of rat Sertoli cells it was demonstrated that they phagocytose apoptotic spermatogenic cells in a PS-dependent manner. HDL, antibodies against SR-BI and PS-liposomes inhibited phagocytosis, pointing to a role of SR-BI as an apoptotic phagocytic PS-receptor (Shiratsuchi et al. 1999). Injection of a part of the extracellular domain of rat SR-BI fused with human Fc into the seminiferous tubules of living mice increased the numbers of apoptotic spermatogenic cells (Kawasaki et al. 2002). However, a recent study with minks suffering from spontaneous autoimmune orchitis, which causes impaired clearance of apoptotic germ cells, revealed that engulfment of apoptotic cells by Sertoli cells is only partly dependent on SR-BI (Akpovi et al. 2006).

The class E scavenger receptor LOX-1 was identified as an oxLDL receptor in bovine aortic endothelial cells (BAE). Both, cultured BAE and chinese hamster ovary (CHO) cells expressing bovine LOX-1, but not CHO control cells were shown to bind apoptotic cells and aged red blood cells in a LOX-1-PS-dependent manner, which can be inhibited by oxLDL, acLDL, mAbs against LOX-1, recombinant soluble LOX-1 and PS-liposomes (Oka et al. 1998). The authors hypothesized that clearance of apoptotic cells by endothelial cells might be necessary for the prevention of coagulation and thrombosis induced by the pro-coagulant property of PS exposed on the surface of apoptotic cells (Casciola-Rosen et al. 1996).

Complement, Pentraxin and Collectin Receptors

The collectins such as mannose-binding lectin (MBL), bovine conglutinin, surfactant proteins A and D (SP-A and SP-D) are pattern recognition proteins of the innate immune system with the ability to opsonise micro-organisms and to facilitate their removal by phagocytes. They are structurally composed of amino-terminal collagen-like tail regions and carboxy-terminal globular heads containing C-type lectin carbohydrate recognition domains. The complement component C1q possesses collagen-like tails; however, lacks lectin-binding activity. Unlike SP-A and SP-D which are primarily found in the lung, MBL and C1q are serum proteins and able to activate the complement system via the lectin or classical pathway, respectively. In vitro studies revealed that SP-A, SP-D and C1q enhanced phagocytosis of apoptotic cells by human and murine alveolar macrophages (AMs), irrespective of the cell type (Vandivier et al. 2002b), which is consistent with the finding that AMs phagocytose SP-A opsonised apoptotic neutrophils (Schagat et al. 2001). However, in vivo SP-D was found to be mainly important for apoptotic cell clearance in the murine lung (Vandivier et al. 2002b). Collectins and C1q were shown to bind calreticulin (CRT) via their collagenous tails (Malhotra et al. 1990) and therefore, CRT was initially regarded as the collagen C1q receptor (cC1qR) but subsequently on it was termed the collectin receptor (Eggleton et al. 2000). Calreticulin is a chaperone of the rER and implicated in the modulation of cellular calcium signalling (Rauch et al. 2000). Furthermore, it is known to be also present on the surface of viable cells including macrophages (Ogden et al. 2001) and to increase in amount during apoptosis (Gardai et al. 2005). Calreticulin on the surface of macrophages interacts with the endocytic receptor protein CD91 (LRP), which shares homology with CED-1, the main phagocytosis receptor of *Caenorhabiditis elegans* for apoptotic cell clearance (Zhou et al. 2001). Calreticulin-CD91 (LRP)-dependent macropinocytosis of C1q and MBL-opsonised apoptotic cells by HMDMs was first shown by (Ogden et al. 2001). This pathway is also responsible for the phagocytosis of apoptotic cells by human and murine AMs (Vandivier et al. 2002b). In addition to the interaction of CRT-CD91 (LRP) with collectin or C1q opsonised apoptotic cells, it was recently shown that CRT can also be expressed on the surface of apoptotic cells. Then it interacts directly vice versa with CD91 (LRP) on the surface of macrophages and promotes phagocytosis by a concomitant inhibition of the CD47-SIRPα "don't eat-me" signal (Gardai et al. 2005). Mice deficient for C1q, MBL and SP-D proved that all three opsonins, indeed, participate in the removal of apoptotic cells in vivo; however, in a tissue specific manner. Thus, *SP-D* KO mice display a reduced clearance of apoptotic cells in the lung (Vandivier et al. 2002b), whereas *C1q* and *Mbl* KO mice show a reduced clearance of apoptotic cells injected into the peritoneum (Gregory and Pound, Chap. 9, this Vol.; Botto et al. 1998; Stuart et al. 2005). Furthermore, *C1q* KO mice accumulate apoptotic cells within the renal glomeruli (Botto et al. 1998).

The ligands for collectins and C1q on the surface of apoptotic cells are not clarified in detail so far, but seem to be different from PS. It is known that MBL and C1q also bind to viable cells in a diffuse pattern and that patching of the opsonised ligands seems to be necessary for efficient phagocytosis (Ogden et al. 2001; Stuart et al. 2005). It is supposed that carbohydrate moieties, which might be furthermore processed during apoptosis, are the ligands recognized by collectins. Furthermore, it was shown that MBL, SP-A and SP-D bind to pentoses present in DNA-molecules exposed on the surface of apoptotic cells (Palaniyar et al. 2004). Binding of MBL and C1q as well as early complement components to apoptotic cells is a late event, but occurs early during necrosis (Gaipl et al. 2001; Nauta et al. 2003). However, MBL deposition does not lead to the activation of complement via the lectin-pathway, pointing to the fact that collectin opsonised apoptotic cells are primarily phagocytosed by engaging the CRT-CD91 (LRP) receptor on phagocytes (Ogden et al. 2001; Nauta et al. 2003).

In contrast, C1q binding to apoptotic cells activates complement (Mevorach, Chap. 10, this Vol.). C1q either binds directly to apoptotic cells to so far unknown ligands (Korb and Ahearn 1997) and/or acts in a double-bridging way, i.e. binds to IgM Abs (Kim et al. 2002; Peng et al. 2005; Quartier et al. 2005) and pentraxins like serum amyloid protein (SAP; Hicks et al. 1992), C-reactive protein (CRP) (Kaplan and Volanakis 1974), and pentraxin 3 (PTX3) deposited on the cell surface of apoptotic cells (Flierman and Daha 2007). Binding of C1q to apoptotic cells via IgM and pentraxins leads to the activation of the classical complement cascade and deposition of C3 and C4 as well as their degradation products on the apoptotic plasma membrane; however, without lysis of apoptotic cells through the formation of membrane attack complexes (Kaplan and Volanakis 1974; Hicks et al. 1992; Gershov et al. 2000; Kim et al. 2002; Peng et al. 2005; Flierman and Daha 2007). Furthermore, the alternative pathway is also activated at the surface of apoptotic cells, since factor B depleted serum reduced the phagocytosis of apoptotic cells (Mevorach et al. 1998). Subsequent to complement activation, phagocytosis occurs most likely by the interaction of C1q with the CRT-CD91 (LRP) receptor on phagocytes and through binding of C3bi with the monocyte/macrophage complement receptors CR3 (αmβ2 integrin receptor) and CR4 (αxβ2 integrin receptor; Takizawa et al. 1996; Mevorach et al. 1998). In addition, FcγRI, FcγRIIa/b, and FcγRIII were also implicated in the phagocytosis of SAP and CRP opsonised apoptotic cells (Stein et al. 2000; Mold et al. 2002). Although neither IgM nor the pentraxins were described to bind to PS exposed on the surface of apoptotic cells, annexin V, which binds to PS, was demonstrated to significantly reduce C3bi deposition on apoptotic cells (Mevorach et al. 1998). Thus, complement activation might also occur by direct binding of C1q or other complement components to PS, or alternatively annexin V bound to PS sterically hinders C3bi deposition.

The ligands for IgM Abs and the pentraxins on apoptotic cells are only partly described. Natural IgM Abs of normal individuals contain Abs with specificity against LPC exposed on the apoptotic plasma membrane (Kim et al. 2002; Peng et al. 2005). Thus, LPC does not only function as a "find-me"(Peter et al., Chap. 3, this Vol.), but also as an "eat-me" signal. Mice deficient for IgM spontaneously develop IgG anti-DNA autoantibodies and renal immune-complex deposition pointing to the role of IgM Abs in opsonising apoptotic cells and thereby with the removal of potential autoantigens (Ehrenstein et al. 2000). This finding was supported by an enhanced antinuclear autoimmunity found in Lupus-prone C57BL/6 *Fas*[lpr/lpr] mice deficient for secreted but not surface IgM (Boes et al. 2000).

Pentraxins are pentameric acute-phase serum proteins that exhibit calcium-dependent binding to a variety of substrates and are divided into two structural classes, short pentraxins (SAP and CRP) produced by the liver and long pentraxins (PTX3) produced by endothelial cells and phagocytes (van Rossum et al. 2004). Both SAP and CRP are present in the serum of mammals; however, SAP represents a major acute-phase protein in the mouse, whereas in humans CRP fulfils this role. SAP was shown to bind to PE and to facilitate phagocytosis of early and late apoptotic cells by HMDMs (Familian et al. 2001; Bijl et al. 2003; van Rossum et al. 2004). In addition to PS, PE is concomitantly transported from its natural location of the plasma membrane, the inner leaflet, to the outer leaflet during apoptosis (Martin et al. 1995; Familian et al. 2001). Binding of SAP to late apoptotic cells was stronger, pointing to the possibility that SAP is a salvage protein of a rescue pathway for the elimination of apoptotic cells, which were not efficiently phagocytosed by the predominant mechanisms (Familian et al. 2001; Bijl et al. 2003). Indeed, *Sap* KO mice do not display a spontaneous defect in the removal of apoptotic cells but display an enhanced autoimmunity against chromatin (Bickerstaff et al. 1999) consistent with its role as the major DNA and chromatin binding protein in serum (Pepys and Butler 1987; Pepys et al. 1994). Chromatin, which is reported to be exposed on the surface of apoptotic cells (Casciola-Rosen et al. 1994;

Palaniyar et al. 2004), does not seem to function as a further ligand beside PE for SAP on apoptotic cells (Familian et al. 2001).

In humans the acute-phase protein CRP was also demonstrated to bind to apoptotic cells (Gershov et al. 2000). Expression and secretion of CRP is up-regulated in the liver in response to tissue injury and infections after release of pro-inflammatory cytokines like IL-1 and TNF-α. It has been demonstrated that CRP binds in addition to the C-polysaccharide of *Streptococcus pneumonia*, to small nuclear ribonucleoproteins (snRNPs; Pepys et al. 1994), H1-containing chromatin (Robey et al. 1984), and PC (Volanakis and Wirtz 1979; Gershov et al. 2000). For human apoptotic lymphocytes it was shown that they bind CRP, which was presumably facilitated through interaction with PC but not chromatin or snRNPs exposed on the apoptotic plasma membrane (Gershov et al. 2000). Furthermore, CRP-binding occurred later than that of annexin V, demonstrating that CRP does not bind to PS. Interestingly sPLA$_2$, a further acute phase protein, which is known to convert PC to LPC at the surface of apoptotic cells (Atsumi et al. 1997), is described to promote phagocytosis of injured cells and tissue debris in cooperation with CRP (Hack et al. 1997), pointing to the possibility that CRP might not only bind to PC but also to LPC. In addition, CRP-binding resulted in the activation of the classical complement pathway through binding of C1q, although its binding did not induce the formation of membrane attack complexes due to recruitment of factor H, which accelerates the decay of the C3 and C5 convertases (Gershov et al. 2000). Phagocytosis of CRP-opsonised apoptotic cells by macrophages is anti-inflammatory and most likely in analogy to SAP occurs by engagement of the CRT-CD91 (LRP) receptor by bound C1q or by engagement of FcγRI, FcγRIIa/b, and FcγRIII (Stein et al. 2000; Mold et al. 2002).

In contrast to the short pentraxins SAP and CRP, the long pentraxin PTX3 is up-regulated directly at sites of inflammation and secreted by endothelial cells and phagocytes in a milieu of pro-inflammatory signals like LPS, IL-1β and TNF-α (Rovere et al. 2000). Late apoptotic and necrotic Jurkat T cells and neutrophils are specifically opsonised by PTX3; however, PTX3-binding inhibits their phagocytosis by iDCs and macrophages (Rovere et al. 2000; van Rossum et al. 2004). It is supposed that this inhibitory effect might prevent the presentation of autoantigens by maturing DCs through phagocytosis of autologous apoptotic cells at sites of inflammation (Rovere et al. 2000) or to represent a backup mechanism for the clearance of apoptotic cells in situations of high amounts of cell death and a low capacity of clearance (van Rossum et al. 2004). The ligands for PTX3 on the surface of apoptotic cells are unknown so far. The discovery of a factor preventing the phagocytosis of apoptotic cells by phagocytes augment the existing scenario of "don't eat-me", "find-me" and "eat-me" signals by "don't bury-me" signals.

The MER Receptor Tyrosine Kinase

The TAM receptor tyrosine kinase family (RTK) consisting of TYRO3, AXL, and MER are expressed on cells of the reproductive and nervous system as well as on haematopoietic cells like macrophages. They are believed to take part in signal transduction regulating cell growth. They are composed of an extracellular region containing two immunoglobulin-related domains linked to two fibronectin type III repeats, and a cytoplasmic region containing a protein tyrosine kinase domain (reviewed in Hafizi and Dahlback 2006). Especially the family member MER seems to be involved in the clearance of apoptotic cells. This was first demonstrated by the identification of a deletion mutant within the *Mer* gene, which caused the hereditary retinal degeneration found in the Royal College of Surgeons (RCS) rat model (D'Cruz et al. 2000). These rats suffer from an impaired clearance of shed POS by RPE (D'Cruz et al. 2000). It was possible to correct this defect by *Mer* gene transfer (Smith et al. 2003). Knockout mice for the TK domain of MER showed beside a RCS phenotype (Duncan et al. 2003) a reduced clearance of apoptotic cells in the thymus after a challenge with dexamethasone (Gregory and Pound, Chap. 9, this Vol.; Scott et al. 2001). In addition, unchallenged *Mer* KO mice suffer from splenomegaly, accumulation of apoptotic bodies within lymphoid organs, the development of auto-antibodies against chromatin and DNA as well as from a moderate renal pathology (Cohen et al. 2002). Reconstitution experiments transferring bone-marrow of *Mer* KO into irradiated WT and bone-marrow of WT into irradiated *Mer* KO mice revealed that this defect most likely resulted from phagocytosis-incompetent macrophages derived from the bone-marrow of *Mer* KO mice, since this defect could be repaired by transplanting healthy WT-derived bone-marrow (Scott et al. 2001). The clearance defect of macrophages within *Mer* KO mice was not only restricted to the thymus, but was revealed also for peritoneal and splenic macrophages after i.p. or i.v. administration of apoptotic thymocytes or lymphocytes (Scott et al. 2001). Binding of apoptotic cells to macrophages was not impaired by deletion of the cytoplasmic domain of MER, demonstrating that MER is not only involved in tethering, but also in signal transduction leading to phagocytosis (Scott et al. 2001). One ligand of TAM RTKs; especially of MER was demonstrated to be GAS6 which displays growth factor-like properties (Nagata et al. 1996), but in addition was shown to be involved in phagocytosis of POS (Hall et al. 2001). This vitamin-K dependent protein functions as a bridging molecule by its capacity to bind to PS (Nakano et al. 1997). However, *Gas6* KO mice have normal appearing retinas and recent studies revealed that protein S, which is also known to bind to the TAM RTK family members (Stitt et al. 1995), participates as a bridging molecule in the clearance of POS by RPE in a MER-dependent mechanism (Hall et al. 2005). A similar dependence on protein S was shown for the clearance of apoptotic cells by macrophages (Anderson et al. 2003). Protein S is a vitamin-K dependent serum protein, known as a negative regulator of coagulation by binding to anionic phospholipids like PS on the plasma membrane of endothelial cells and activated platelets where it activates protein C, which inactivates the coagulation factors Va and VIIIa (Hafizi and Dahlback 2006). Beside its role as a bridging molecule it appears to be involved in the suppression of coagulation at the surface of apoptotic cells in parallel to annexin V, which is present in plasma only at low levels (Reutelingsperger and van Heerde 1997). Annexin V reduces phagocytosis of apoptotic cells by macrophages (Gaipl et al. 2007), which might be overcome by further PS binding molecules known to act as phagocytosis bridging molecules and to inhibit coagulation like protein S (Hafizi and Dahlback 2006), β2GPI (Nimpf et al. 1986), MFG-E8 or annexins.

Vitronectin Receptors αvβ3 and αvβ5 Integrin

Integrins are a family of heterodimeric transmembrane adhesion receptors each composed of an a- and a b-chain. They are involved in diverse processes like cell adhesion, migration and angiogenesis. In addition, the vitronectin receptors αvβ3 and αvβ5 integrin were shown to be involved in phagocytosis of apoptotic cells by iDCs and macrophages as well as of POS by RPE (Gregory and Pound, Chap. 9, this Vol.). As described above, macrophages use the TSP-CD36-αvβ3 system for phagocytosis, where the class B scavenger receptor CD36 and αvβ3 integrin are linked by the bridging molecule TSP to so far unidentified ligands on the surface of apoptotic cells. In contrast, iDCs phagocytose apoptotic cells through the receptors CD36 and αvβ5 integrin (Albert et al. 1998a). Whether TSP acts as a bridging molecule for iDCs is not clear yet. However, in addition to CD36 and αvβ5 integrin, it was demonstrated that iDCs phagocytose iC3b-opsonised apoptotic cells most likely via CR3 and/or CR4 (Verbovetski et al. 2002).

Phagocytosis of POS by RPE was shown to depend on αvβ5 integrin, although αvβ5 integrin appeared to only participate in tethering, but not internalization (Finnemann et al. 1997). A cooperation of αvβ5 integrin with CD36 seemed to be plausible, since previous in vitro studies revealed the dependence of the internalization process during *POS* phagocytosis on CD36 (Ryeom et al. 1996). However, as described above, in vivo data revealed that MER has a greater impact in this process deduced from the RCS phenotype of RCS rats and *Mer* KO mice (D'Cruz et al. 2000; Duncan et al. 2003; Smith et al. 2003). Internalization of POS by RPE was found to be dependent on MER tyrosine phosphorylation (Feng et al. 2002) and a recent study demonstrated that binding of POS to the αvβ5 integrin on RPE cells results in the formation of a cytoplasmic complex between focal adhesion kinase (FAK) and αvβ5 integrin, a process necessary for POS engulfment (Finnemann 2003). The formation of the FAK-αvβ5 integrin complex correlates with Tyr861 phosphorylation of FAK, which is known to be mediated by SRC family kinases in vascular endothelial growth factor signalling. During VEGF signalling SRC kinases are activated and phosphorylate FAK, which afterwards associates with αvβ5 integrin (Eliceiri et al. 2002). For the αvβ5 integrin-mediated phagocytosis of apoptotic cells it was indeed shown that activation of MER by binding to its ligand GAS6 induces a post-receptor signalling cascade leading to SRC-mediated Tyr861 phosphorylation of FAK and subsequent binding of FAK-Tyr861 to the cytoplasmic tail of αvβ5 integrin (Wu et al. 2005). These events lead to an increased formation of the p130^{CAS}/CRKII/DOCK180 complex and activation of the small GTP-binding protein RAC1 known to be involved in actin filament reorganisation necessary for phagocytosis (Albert et al. 2000; Wu et al. 2005).

The ligand for αvβ5- and αvβ3 integrin on apoptotic cells was shown to be MFG-E8, a 72 kDa glycoprotein secreted from mammary epithelial cells, macrophages, and iDCs (Hanayama et al. 2002; Akakura et al. 2004; Hanayama et al. 2004; Hanayama and Nagata 2005). MFG-E8 binds to oxPS exposed on apoptotic cells (Borisenko et al. 2004) and triggers Dock180-dependent Rac1 activation and phagocytosis of apoptotic cells by iDCs after bridging to αvβ5 integrin (Akakura et al. 2004). In addition, MFG-E8 was shown to be the main bridging molecule for αvβ5 integrin on RPE cells during the circadian rhythm of POS phagocytosis (Nandrot et al. 2007). Although RPE cells of *Mfg-e8* and β5 integrin KO mice display a residual uptake level, they lack the burst of phagocytic activity that follows circadian photoreceptor shedding (Nandrot et al. 2004; Nandrot et al. 2007). This leads to an age-related retinal dysfunction and blindness due to an accumulation of storage bodies containing fluorescent lipofuscin (Nandrot et al. 2004; Nandrot et al. 2007). In addition, *Mfg-e8* KO mice display a reduced ability of tingible body macrophages within the germinal centres of lymph nodes and the spleen to phagocytose apoptotic B cells. The reduced clearance of apoptotic B cells results in splenomegaly and high levels of anti-nuclear autoantibodies, as well as glomerular immune-complex deposition leading to glomerulonephritis and proteinuria (Hanayama et al. 2004). Another phagocytosis defect of *Mfg-e8* KO mice was demonstrated for the involuting mammary gland. Thus, clearance of apoptotic glandular cells and milk fat globules by neighbouring epithelial cells and invading macrophages is impaired leading to mammary duct ectasia with periductal mastitis (Hanayama and Nagata 2005).

Taken together, the receptors MER and the αvβ5 integrin seem to work in concert during phagocytosis of apoptotic cells by binding with their bridging molecules GAS6/protein S and MFG-E8, respectively, to PS or oxPS exposed on the apoptotic cell surface. Their ligation initiates an intracellular signalling cascade leading to the activation of FAK, the recruitment of the p130[CAS]/CRKII/DOCK180 complex, and, activation of the small GTP-binding protein Rac1, which subsequently induces the reorganisation of the actin cytoskeleton necessary for phagocytosis.

In addition to GAS6, protein S, and MFG-E8, the serum glycoprotein β2GPI, which is known to be involved in the regulation of thrombosis and the uptake of lipoproteins by macrophages, was shown to act as a bridging molecule interacting with PS exposed on apoptotic cells in vitro (Balasubramanian et al. 1997; Balasubramanian and Schroit 1998) and in vivo (Balasubramanian et al. 2005) and thereby facilitates the uptake of these cells by macrophages. The macrophage receptors interacting with β2GPI are so far unknown but different from CD36 and CD14 (Balasubramanian and Schroit 1998). Interestingly, it was described that β2GPI binds to annexin II, which is exposed on the surface of activated endothelial cells (Ma et al. 2000). Annexin II is also described to be expressed on the surface of phagocytes and to function as a receptor for the phagocytosis of apoptotic peripheral T lymphocytes (Fan et al. 2004). Thus, it is conceivable that a potential phagocyte receptor for β2GPI is annexin II.

4.1.1.3 Mechanism of Engulfment

Independently of the kind of cell death (apoptotic or necrotic) ingestion of dying or dead cells necessitates a profound rearrangement of the cytoskeleton by the phagocyte in particular its actin containing microfilament system. Phagocytosis is the uptake of particles greater than 0.5 μm. During this process the macrophage normally extends a circular thin, tightly fitting cytoplasmic extension around the particle to be ingested forming the so-called phagocytic cup. The extension of the veil-like cytoplasmic rim around the particle is most probably driven by polarized actin de- and re-polymerisation (actin treadmilling) mechanisms as in lamellipodia (Mannherz et al. 2007). After encircling the particle, the plasma membranes of the opposing sites of the tip of the extension fuse above the particle, which itself becomes embedded in a membrane vesicle—the phagosome that subsequently ripens to the phagolysosome. It has been suggested that phagocytosis of apoptotic cells leads to the formation of a specially structured, more spacious type of phagosomes like those formed during macropinocytosis (Hoffmann et al. 2001; Ogden et al. 2001). In contrast, when the phagocytic activity of macrophages in a cell culture system was analysed by scanning electron microscopy Krysko and co-workers noticed that apoptotic cells were engulfed into tight fitting phagosomes originating from phagocytic cups by the "zipper-like" mechanism described above, whereas necrotic cells or remnants thereof were taken up by macropinocytosis (Krysko et al. 2006b). The discrepancy may have originated from the different morphological procedures employed by these authors.

Very little is known about the signalling processes that must occur between tethering of an apoptotic cell and its engulfment. It appears; however, that the main target of these signalling events is the cytoskeleton. In *Caenorhabditis elegans* a number of genes supposedly involved in the signalling to and activation of the cytoskeleton were identified that form two signalling cascades with partially connected or converging pathways: CED-1, CED-6, CED-7 and CED-2, CED-5, CED-12 (Gronski and Ravichandran, Chap. 5, this Vol.; Kinchen et al. 2005). The transmembrane proteins CED-1 and CED-7 and their mammalian homologues CD91 (LRP) and ABCA1/ABCA7 are presumably involved in recognition or tethering, whereas all other CED-genes most probably code for signalling proteins that are involved in the reorganisation of the actin cytoskeleton by activation of CED-10. The mammalian homologue of CED-10 is the small GTP-binding protein RAC-1, which in haematopoietic cells activates the "Wiskott Aldrich syndrome protein" (WASp) or homologues of it like the N-WASP or SCAR/WAVE complex. These are signalling complexes leading to activation of the heptameric Arp2/3 complex that induces the formation of branched actin filament nets typically present in plasma membrane ruffles and lamellipodia. Indeed, it was shown that macrophages isolated from Wiskott-Aldrich syndrome patients exhibit a reduced phagocytic activity. Similarly, macrophages from WASp-deficient mice showed a clearly reduced ability to engulf apoptotic cells. Concomitant to the formation of the phagocytic cup WASp accumulates together with actin at the site of engulfment of apoptotic cells. Macrophages isolated from WASp-deficient mice are, nevertheless, able to ingest apoptotic cells although with clearly impaired efficiency. Therefore, it was suggested that the related N-WASP/SCAR proteins may compensate the lack of WASp. It has been shown that both CDC42 and Rac-1 activation are able to activate these protein complexes that rely the signal to the Arp2/3 complex, leading to actin polymerisation necessary for phagocytic cup formation and the uptake of apoptotic cells (Leverrier et al. 2001).

4.1.2 Elimination of Necrotic Cells and Cell Debris

The final step during cell death is the digestion of cellular macromolecules to their molecular constituents within the lysosomal compartment of macrophages. During PCD, final digestion is preceded by a coordinated destruction of subcellular structures induced by the activation of specific proteases and nucleases leading to the generation of small "eupeptic" cellular corpses. Digestion of the cellular chromatin for example already starts during the apoptotic suicide program. Thus, the cellular chromatin is cleaved into first high molecular and subsequently into low molecular weight fragments of mono- to oligo-nucleosomal size (DNA-laddering). The main nucleases beside further candidates involved in apoptotic DNA degradation are the caspase activated DNase (CAD, DFF-40) after release from its inhibitor ICAD (DFF-45) due to caspase-3 cleavage (Enari et al. 1998; Liu et al. 1998) and endonuclease G (EndoG) which is released from mitochondria (Li et al. 2001). In addition, DNase γ (DNASE1L3), which is not only secreted, but due to two nuclear localization signals was also detected in the nucleus of certain cells, is supposed to function as an apoptotic nuclease (Shiokawa et al. 1994; Higami et al. 2004). Subsequently final digestion of DNA occurs by the lysosomal *DNase IIα* within the phagolysosomes of macrophages. Interestingly, when heterophagy is disturbed like in the degradation of chromatin within phagocytosed nuclei in DNase IIα KO mice a decreased ability to engulf apoptotic cells results pointing to the fact that PCD, subsequent phagocytosis and heterophagy are interdependent continuous processes (Kawane et al. 2001; Krieser et al. 2002).

However, what is the final doom of primary or secondary necrotic cells? So far, only little information exists about their disposal, nevertheless, two major ways for the elimination of necrotic cell debris are conceivable; (i) like for apoptotic cells phagocytosis of opsonised necrotic cells and released cell debris and/or (ii) digestion of necrotic cell debris by extracellular proteolytic and nucleolytic enzymes followed by resorption of the degraded macromolecules.

4.1.2.1 Phagocytosis of Necrotic Cells

It has been demonstrated that necrotic cells are also removed by phagocytosis (Cocco and Ucker 2001; Krysko et al. 2003; Brouckaert et al. 2004; Bottcher et al. 2006; Krysko et al. 2006b). In contrast to apoptotic cells, necrotic cells undergo membrane disintegration leading after lysis to spreading of cellular contents into the extracellular space that makes it difficult for phagocytes to collect the debris. Thus, phagocytosis of necrotic cells is described to be less efficient and temporarily delayed in comparison to apoptotic cells (Krysko et al. 2003; Brouckaert et al. 2004; Krysko et al. 2006b), although one publication reports a comparable efficiency in the binding and internalization of apoptotic and necrotic cells (Cocco and Ucker 2001). However, in these experiments dead cells were mixed with macrophages at low temperatures (4°C) to discriminate between the efficiency for binding and engulfing. Engulfing only took place at 37°C. Probably, phagocytes bind necrotic cells at 37°C less efficient than apoptotic cells, because the macrophages and their phagocytic receptors respond differently since their altered expression might result in a lower avidity for necrotic cells.

Membrane integrity is not necessary for phagocytosis of necrotic cells and the mechanism of phagocytosis is described to be macropinocytosis. Thus, macrophages develop broad membrane ruffles ending in long fine protrusions which creep around the necrotic material (Krysko et al. 2003; Brouckaert et al. 2004; Krysko et al. 2006b) and facilitate ingestion of the cell debris concomitant with extracellular fluid in macropinosomes (>0.2 μm diameter; Krysko et al. 2003; Brouckaert et al. 2004; Krysko et al. 2006b). As described above, apoptotic cells are ingested by a zipper-like mechanism of phagocytosis, which necessitates the existence of an undamaged plasma membrane. The apparent existence of two different modes of ingestion for apoptotic or necrotic cells implies that at least two different signalling cascades can be activated within macrophages. Indeed, inhibitors of phosphatidylinositol-3'-kinase (PI3K), which is activated during phagocytosis by several receptors (Leverrier and Ridley 2001), reduce the phagocytosis of apoptotic cells whereas the ingestion of necrotic cells remains unaffected (Krysko et al. 2006b). The fact that phagocytosis of apoptotic cells is not completely abolished in the presence of PI3K-inhibitors indicates the existence of a second signalling mechanism for the ingestion of apoptotic cells (Krysko et al. 2006b). Interaction of apoptotic and necrotic cells with macrophages does not only induce divergent signalling events leading to different engulfing mechanisms, but also to differences in the outcome on the immunological status, survival and proliferation of macrophages (Reddy et al. 2002; Patel et al. 2006).

Recognition of necrotic and apoptotic cells by macrophages seems to be partially homologous, i.e. depends on the exposure of PS on the outer layer of the plasma membrane despite other structural differences (Lecoeur et al. 2001; Krysko et al. 2004). Both lead to subsequent phagocytosis (Cocco and Ucker 2001; Hirt and Leist 2003; Brouckaert et al. 2004; Bottcher et al. 2006). Exposure of PS during apoptosis is mainly due to inactivation of the aminophospholipid translocase activity by caspase-3 cleavage. The aminophospholipid translocase activity is oxidant-sensitive, ATP-dependent and probably blocked by high calcium levels. Since all of these events are characteristic also for necrotic cells it is not surprising that necrotic cells expose PS (Fig. 4.1f") and most likely also oxPS. Secondary necrotic cells have more or less gone through the apoptotic suicide process, and therefore, will exhibit structural plasma membrane alterations typical for early apoptotic cells. In vivo the structural composition of the primary necrotic plasma membrane might be altered in a way comparable to those of apoptotic cells by proteolytic and nucleolytic enzymes and the acute phase protein sPLA$_2$ present in serum and the extracellular fluid, by hydrolases released from the dying cells themselves and by enzymes secreted by recruited macrophages and neutrophils. Therefore, it is not surprising that the receptors and bridging molecules involved in phagocytosis of necrotic cells seem to be more or less the same as for apoptotic cells from nematodes to humans (Chung et al. 2000; Bottcher et al. 2006). Thus, it was shown that the TSP-CD36-αvβ3 system, the pattern recognition receptor CD14, and a potential so far unidentified PS-receptor bound by mAb217 facilitate phagocytosis of primary necrotic lymphocytes by HMDMs (Bottcher et al. 2006). An involvement of macrophage receptors directly or indirectly interacting with PS exposed on necrotic cells, like for example the integrin-dependent pathway, was also verified for the macrophage cell line J774A.1 (Cocco and Ucker 2001) demonstrating that PS is not a specific ligand for the recognition of apoptotic cells. Although phagocytosis of both apoptotic and necrotic cells can be effectively inhibited by integrin-binding peptides and PS but not PC vesicles, they do not compete with each other for binding to macrophages (Cocco and Ucker 2001) possibly implying that the avidity of PS binding is higher for apoptotic than necrotic cells. Indeed, many authors suppose that clustering of ligands and their bridging molecules might be necessary for the effective binding and internalization of apoptotic cells and that due to a more homogenous distribution of these ligands on the surface of viable cells engulfment is prevented (Ogden et al. 2001). Probably, due to energy depletion ligand clustering does not or less efficiently occur on necrotic cells resulting in a lower avidity to phagocytes, delayed clearance and a distinct phagocytosis mechanism. Many of the bridging molecules of the collectin, pentraxin and complement family involved in apoptotic cell clearance, were also demonstrated to bind to necrotic cells, especially those which bind preferentially to late apoptotic (secondary necrotic) cells. Thus, complement deposition through activation of the classical pathway by binding C1q is a late event during apoptosis and occurs early during necrosis (Gaipl et al. 2001). Binding of C1q and complement activation thereby occurs predominantly via IgM Abs (Zwart et al. 2004), which are described to bind to LPC (Kim et al. 2002; Peng et al. 2005). LPC on necrotic cells might be generated by sPLA2. Indeed, binding of C1q to necrotic cells was shown to depend on further serum factors (Bottcher et al. 2006) supposedly on IgM Abs or molecules of the pentraxin family like SAP and CRP, which are known to bind preferentially to late apoptotic cells (Bijl et al. 2003; Hart et al. 2005). The long pentraxin PTX-3 binds also to late apoptotic cells, but was shown to rather prevent than to promote phagocytosis (van Rossum et al. 2004). In addition to IgM Abs, complement and the pentraxins, the collectin MBL was shown to bind to both late apoptotic and necrotic cells (Nauta et al. 2003), thus, it might be a further bridging molecule for phagocytosis of necrotic cells via the TSP-CD36-αvβ3 system.

Recently, it was demonstrated that ficolin-2 binds to apoptotic cells and activates complement via the lectin-pathway through interaction with the MBL-associated serine proteases (Kuraya et al. 2005; Jensen et al. 2007). Ficolin-2 is a serum protein produced in the liver and shares structural similarities with the collectins (Kuraya et al. 2005; Jensen et al. 2007). It displays lectin properties with specificity for

N-acetyl-D-glucosamine and binds DNA in a calcium-dependent manner (Kuraya et al. 2005; Jensen et al. 2007). In accordance with the findings for the collectins, ficolin-2 binds to late apoptotic and necrotic cells. This binding is inhibited by DNA demonstrating that its major ligand is DNA supposedly exposed on the plasma membrane of these cells (Jensen et al. 2007).

Histidine-rich glycoprotein (HRG) is a further multidomain protein produced by the liver, which was shown to bind to DNA exposed on apoptotic T cells and facilitating phagocytosis by HMDMs through engagement of the FcγRI (Gorgani et al. 2002). HRG is 75 kDa plasma glycoprotein that binds several ligands like heparan sulphate, plasminogen, TSP, tropomyosin, IgG, C1q, and Fc receptor (for review see Jones et al. 2005a). Recent studies, however, gave contradictory results. Thus, HRG was demonstrated to preferentially bind with high avidity to ligands of the cytoplasm of primary necrotic and late apoptotic Jurkat T cells, whereas it binds only weakly to early apoptotic cells (Jones et al. 2005b). However, the bridging molecules, the ligands within necrotic cells and the phagocytosis receptors involved in the removal of HRG-opsonised necrotic cells are not identified so far; nevertheless, heparan sulphate, tropomyosin and DNA were ruled out as potential ligands (Jones et al. 2005b).

4.1.2.2 Immunological Outcome of Phagocytosis of Necrotic Cells

Correlated with the different mechanisms of dead-cell ingestion are the different immunological outcomes of apoptotic versus necrotic cell ingestion. Whereas apoptotic cell ingestion by macrophages results in anti-inflammatory and abrogation of pro-inflammatory responses, uptake of necrotic cells only enhances, but is not sufficient to induce macrophage activation (Ucker, Chap. 6; Lacy-Hulbert, Chap. 7, this Vol.; Cocco and Ucker 2001). Only lysed neutrophils, but not lysed lymphocytes or apoptotic neutrophils and apoptotic lymphocytes were shown to induce a pro-inflammatory response in murine and human macrophages with the release of macrophage-inflammatory protein 2, IL-8, IL-10 and TNF-α (Fadok et al. 2001). Release of the serine protease elastase from necrotic neutrophils was shown to act as a pro-inflammatory danger signal during the uptake of necrotic neutrophils, whereas cell membranes exposing PS of early and late apoptotic cells and necrotic neutrophils and lymphocytes cells were not able to induce this pro-inflammatory response alone (Fadok et al. 2001). Similar experiments revealed that the uptake of necrotic but not apoptotic neutrophils stimulates antigen presentation by macrophages due to a rapid up-regulation of CD40 (Barker et al. 2002). These data reveal that engulfment of necrotic cells alone is not sufficient to induce a pro-inflammatory response, which apparently necessitates the simultaneous release of danger signals. The nuclear high mobility group box 1 protein (HMGB1) has been shown to act as a further danger signal, since its inflammatory effect is not only due to its role as a chemokine, but also to its ability to induce the secretion of pro-inflammatory cytokines (Peter et al., Chap. 3, this Vol.). However, experiments employing pure recombinant HMGB1 could not verify these results and indicate that HMGB1 functions in cooperation with further danger signals

(Raucci et al. 2007). Indeed, an additive action of danger signals was revealed for heat shock protein Hsp72, which stimulates a more elevated cytokine response by macrophages in cooperation with LPS in comparison to Hsp72 alone (Peter et al., Chap. 3, this Vol.; Campisi et al. 2003).

Although ingestion of lysed lymphocytes by macrophages does not induce a pro-inflammatory response, necrotic murine T cells are able to augment an LPS-induced pro-inflammatory response of murine macrophage cell lines, as measured by the release of IL-6 and TNF-α (Cocco and Ucker 2001). This effect was not attributed to the release of danger signals from necrotic T cells, since further experiments with human Jurkat T cells and murine bone-marrow derived macrophages revealed that apoptotic cells do not become pro-inflammatory upon the loss of membrane integrity, i.e. after undergoing secondary necrosis (Patel et al. 2006).

Several studies investigating the phagocytosis of necrotic in comparison to apoptotic cells by iDCs reveal that ingestion of primary and secondary necrotic cells in contrast to apoptotic cells leads to maturation of iDCs, i.e. expression of chemokine receptors for the migration into secondary lymphoid tissues and efficient presentation of antigens to CD4$^+$ and CD8$^+$ T cells (Divito and Morelli, Chap. 11; Bartunkova and Spisek, Chap. 12, this Vol.; Sauter et al. 2000; Ip and Lau 2004; Kacani et al. 2005). However, one study reported that neither apoptotic nor necrotic neutrophils stimulate DC-dependent T cell proliferation. Thus, although MHC class II complex molecules and CD83 are up-regulated in response to phagocytosis of both cell types, the expression of co-stimulatory molecules like CD40, CD80, and CD86 is down-regulated (Clayton et al. 2003). These findings imply that maturation of iDCs depends on the ingested necrotic cell type or the simultaneous release of danger factors. Indeed, HMGB1 and uric acid are known to be endogenous danger signals, which stimulate iDC maturation (Shi et al. 2003; Rovere-Querini et al. 2004). Thus, necrotic HMGB1 deficient cells have a reduced ability to activate iDCs (Rovere-Querini et al. 2004).

4.1.2.3 Extracellular Mechanisms Involved in the Clearance of Necrotic Cell Debris

The second major way of clearance of necrotic cell debris might be facilitated by extracellular enzymes followed by the resorption of macromolecules by phagocytes. Extracellular clearance of cell debris has so far attracted little attention though the diffusion of extracellular enzymes into necrotic cells is conceivable. The enzymes facilitating heterolysis might include nucleases and proteolytic systems (coagulation and plasminogen system) present in the circulation and extracellular fluid or enzymes secreted by recruited and activated phagocytes. In addition, autolysis by lysosomal enzymes released from the cells themselves might occur under certain circumstances.

In vitro studies have demonstrated that the elimination of necrotic cells is delayed in comparison to apoptotic cells, because after membrane rupture intracellular contents are liberated making it difficult for phagocytes to collect necrotic cell debris.

Thus, it has been described that nuclei of necrotic cells are not ingested by macrophages (Krysko et al. 2003). These findings implicate that the nuclear membrane does not exhibit efficient "eat-me" signals. During erythropoesis the nuclei expelled by reticulocytes are still surrounded by the plasma membrane. These nuclei subsequently expose PS leading to their phagocytosis by surrounding macrophages (Yoshida et al. 2005). These data imply that further mechanisms must exist especially for the elimination of necrotic cell-derived nuclei, which do not possess a remnant plasma membrane.

Indeed, it was demonstrated that extracellular proteins infiltrate necrotic cells and their nuclei. The serum endonuclease DNASE1 and components of the plasminogen system penetrate necrotic cells and after activation of plasminogen at cellular and nuclear structures by its activators tissue-type or urokinase-type plasminogen activator, plasminogen is converted to the active serine protease plasmin. Plasmin degrades structural proteins of the chromatin like histones thereby enhancing chromatin breakdown by DNASE1 (compare Fig. 4.1g', g" with Fig. 4.1h', h"), which has a higher nucleolytic activity against protein-free DNA (Napirei et al. 2004). An additional serum nuclease, with the biochemical properties of recombinant DNase γ (DNASE1L3), was recently discovered by investigating serum from *Dnase1* KO mice (Napirei et al. 2000; Napirei et al. 2005). It is most likely that in addition to DNASE1 this second serum nuclease is also able to penetrate necrotic cells, as shown for recombinant DNASE1L3 in vitro (Napirei et al. 2006b). DNASE1L3 induces inter-nucleosomal chromatin breakdown without proteolytic help and with higher efficiency than DNASE1. In contrast, DNASE1L3 displays a lower affinity for the cleavage of protein-free DNA than DNASE1 (Napirei et al. 2005). Thus, both nucleases might complement each other. C1q, which is known to bind DNA (Jiang et al. 1992), has been shown to also penetrate necrotic cells and to bind to nuclear components (Napirei et al. 2006b). Cooperation between C1q and DNASE1 was recently demonstrated. Thus, C1q augments chromatin degradation by serum DNASE1, probably by altering the chromatin structure, since C1q does not possess a proteolytic or nucleolytic activity. Furthermore, C1q opsonised chromatin enhances the uptake of necrotic cell-derived nuclei and chromatin by macrophages (Gaipl et al. 2004). Similarly, it was shown that intermediate filaments of permeable cells activate the classical complement cascade leading to the binding of monocytes and subsequent phagocytosis (Linder et al. 1983). It is conceivable that further serum proteins like collectins or pentraxins or HRG (Jones et al. 2005b) might also function as bridging molecules for the phagocytosis of components of necrotic cells. SAP is known to bind to chromatin and DNA (Pepys and Butler 1987; Pepys et al. 1994) whereas CRP binds to snRNPs and H1-containing chromatin (Pepys et al. 1994). Furthermore, C1q binds directly to DNA and activates complement (Jiang et al. 1992). Thus, molecules of the innate immune system might not only induce phagocytosis of complete necrotic cells, but furthermore of macro-molecules liberated from these cells.

In vivo necrotic lesions induced for example by ischemia/hypoxia or intoxication, like acetaminophen (paracetamol) induced liver necrosis (Fig. 4.2b and Fig. 4.3b–d), are known to comprise large regions of tissue and it is speculative which of the extracellular mechanisms described for the elimination of necrotic cell debris in vitro do indeed, occur in vivo. Depending on the type of tissue, i.e. rich in lipids and fatty acids (for example brain and adipose tissue), rich in hydrolytic enzymes (digestive glands like the pancreas), or protein-rich (for example muscle and liver) different outcomes of the necrotic lesions are described.

Coagulation necrosis as identified by Carl Weigert and Julius Cohnheim around 1877 is characterized by the transformation of the necrotic tissue into a yellowish white, dry mass and is characteristic for protein-rich tissues (Majno and Joris 1995). Weigert supposed that this transformation resulted from the influx and coagulation of plasma into the necrotic lesion. Today this assumption is contested and it is supposed that due to hypoxia, cells switch to an anaerobic metabolism subsequently leading to a collapse of the cellular energy supply and the uncontrolled inflow of ions. Uncontrolled calcium influx activates the calcium-dependent phospholipase A2 leading to plasma- and organelle-membrane damage and the generation of large amounts of arachidonic acid. Arachidonic acid can be further metabolized for example by cyclooxygenase to prostaglandins and by lipoxygenase to leukotrins, leading to an increased permeability of blood vessels and the recruitment of granulocytes. Due to the acidification of the affected tissue, proteins denature and autolysis by lysosomal enzymes is inhibited due to protein coagulation. Membrane damage induces the breakdown of cell-cell contacts and calcium influx to the closure of gap-junctions in cells of the peri-necrotic region. Furthermore, calcium aggregation leads to calcification of the tissue. The necrotic lesion is surrounded by a peri-focal region of inflammation (hemorrhagic edge zone). Tissue regeneration and removal of cellular debris starts from this area first by recruited granulocytes (early phase) and afterwards by the immigration of monocytes which differentiate to macrophages (histiogenic/late phase). Thus, the necrotic area is most likely degraded by enzymes secreted from these phagocytes and subsequently resorbed (heterolysis). In addition, necrotic cells and their constituents might be opsonised by serum factors diffusing into the necrotic lesion due to an increased permeability of vessels within the inflamed region, followed by phagocytosis by macrophages and granulocytes. During the regeneration process of the tissue, new vessels spread into the lesion and fibroblasts are activated synthesising new extracellular matrix components. This early regenerated tissue is called granulation tissue.

The original supposition of Weigert's plasma coagulation in necrotic tissues does also exist, and today is called fibrinoid necrosis, a special form of the coagulation necrosis. Fibrinoid necrosis is typical for vessel walls, the stratified epithelium of the skin and the epithelium of the gut. These lesions are characterized by proteolytically fragmented collagen, which is embedded in coagulated serum containing fragments of dead cells.

In contrast to coagulation necrosis, there exists a second type of necrotic lesion termed colliquation necrosis. This type of necrosis is characteristic for tissues with a high content of lipids and fatty acids and/or a low content of protein (adipose tissue, mammary gland, brain) and for tissues with a high content of hydrolytic enzymes (pancreas). In principle, the pathological consequences of hypoxia of these tissues are the same as described for the coagulation necrosis. However, these events do not lead to the formation of a firm coagulated protein, but to the fluidization of the tissue. Autolytic processes dominate and hemorrhagic lesions can be found within the necrotic region due to the fluid character of the lesion. In addition, the necrotic area often contains macrophages, which have transformed to foam cells due to the phagocytosis of high amounts of oxidized lipids and cholesterol.

The heterolysis of necrotic tissues by neutrophils is most likely initiated by the secretion of three serine proteases, elastase, proteinase 3 and cathepsin G, which are stored in the primary azurophile granules (Korkmaz et al. 2008). They belong to the non-oxidative antimicrobial pathway and help to kill bacteria in phagolysosomes, but furthermore degrade extracellular matrix components after their secretion leading to the dissolution of tissue lesions. These proteases are also secreted by monocytes. Due to their broad substrate specificity (elastin, fibronectin, laminin, collagens, proteoglycans, plasma proteins) the activity these proteases is controlled by a high amount of natural inhibitors in the plasma like the serpins α1-antitrypsin, antichymotrypsin and protein C inhibitor (Law et al. 2006) and the polyvalent protease inhibitor α2-macroglobulin (Borth 1992). Therefore it is supposed that these proteases are only functional at sites of direct contact between the plasma membrane of phagocytes and their targets. However, it has been reported that during certain diseases like cystic fibrosis neutrophil elastase inhibits the phagocytosis of apoptotic cells by AMs due to the cleavage of the PSR on the phagocyte surface (Vandivier et al. 2002a). Thus, the balance between the level of secreted proteases and the levels of their inhibitors must be tightly controlled in order to achieve tissue repair instead of additional tissue damage.

In addition to proteases, nucleases might also be secreted by phagocytes at sites of inflammation, like the peri-necrotic region. Indeed, it was shown that necrotic cells in vitro do not release soluble DNA-fragments into the extracellular milieu. However, when co-cultured with macrophages they display chromatin breakdown and release of soluble DNA implying that macrophages secrete DNases (Choi et al. 2005) like DNase γ (DNASE1L3), which is known to be secreted by a wide variety of tissue macrophages (Baron et al. 1998). In vivo experiments demonstrate that depletion of macrophages by clodronate treatment of mice followed by an i.p. injection of apoptotic and necrotic cells did not lead to a rise in blood DNA levels, whereas this was the case for untreated mice (Pisetsky and Fairhurst 2007). Thus, macrophages appear to secrete nucleases, which help to clear nuclear debris of necrotic cells. Subsequently, these chromatin fragments might be degraded to completion by serum nucleases like DNASE1 or bind to DNA receptors, present on the surface of leukocytes and monocytes followed by phagocytosis and degradation within their lysosomal compartments (Bennett et al. 1985). In addition, to nucleases secreted by macrophages, it is conceivable that DNASE1 and the second serum nuclease, which displays the characteristics of DNase γ (DNASE1L3), diffuse from the circulation into the necrotic lesions and degrade chromatin within necrotic cell-derived nuclei. Whether serum proteolytic systems like the coagulation system or the plasminogen system are activated within necrotic lesions is speculative. Thus, it has been described that annexin V (Reutelingsperger and van Heerde 1997), protein S and β2GPI inhibit coagulation at PS exposing plasma membranes. However, the occurrence of fibrinoid necrosis reveals that this is not true for all tissues. Whether the plasminogen system is activated within necrotic lesions is not known so far. It is described that plasminogen activator inhibitor 1, which inhibits the tissue- and urokinase-type plasminogen activators, binds in a complex with vitronectin to vimentin-type intermediate filaments of necrotic hepatocytes in a model of hepatic coagulation necrosis. It is supposed that this decoration inhibits plasmin formation at sites of cellular necrosis (Podor et al. 1992). Whether plasminogen is activated within the nuclei of necrotic cells in vivo comparable to the findings in vitro leading to the degradation of necrotic chromatin in a concerted action with serum DNASE1 (Napirei et al. 2004) remains to be analysed.

4.2   Conclusions

Effective elimination of dying cells is essential for the survival of an organism. Ultimately, the dying or dead cell is removed by phagocytosis and subsequent complete intracellular degradation by hydrolytic enzymes. Corpse removal occurs in several steps: recognition, attachment and firm binding to the phagocyte, phagocytic uptake that involves signal induced rearrangement of the phagocytes cytoskeleton, and finally degradation by lysosomal hydrolases. These processes are finely regulated and especially in mammals involve a number of redundant recognition mechanisms including different receptors, bridging molecules, and ligands as well as signalling pathways. It also appears that according to the mode of cell death apoptosis (type I), autophagy (type II) or necrosis (type III) different mechansims exist to guarantee cell corpse removal. Essentially, the dying cells themselves expose "eat-me" signals and release "find-me" or danger molecules that stimulate either professional phagocytes (macrophages and immature dendritic cells) to migrate towards the dying cell or activate cells of the immediate vicinity to become phagocytes. Subsequently, the dying cells have to form intimate and strong contacts to the phagocytes (forming a so-called phagocytic synapse) necessary for signalling the initiation of phagocytosis. Whereas during PCD small eupeptic cell corpses are generated, which can be rapidly and efficiently engulfed, clearance of necrotic cell debris often is delayed due to the lack of cellular shrinkage and the spreading of cellular contents after membrane rupture. Therefore, extracellular mechanisms exist (hydrolytic enzymes, phagocytic opsonins for cellular macromolecules), which supplement classical efferocytosis.

References

Agah A, Kyriakides TR, Lawler J et al (2002) The lack of thrombospondin-1 (TSP1) dictates the course of wound healing in double-TSP1/TSP2-null mice. Am J Pathol 161(3):831–839

Akakura S, Singh S, Spataro M et al (2004) The opsonin MFG-E8 is a ligand for the alphavbeta5 integrin and triggers DOCK180-dependent Rac1 activation for the phagocytosis of apoptotic cells. Exp Cell Res 292(2):403–416

Akpovi CD, Yoon SR, Vitale ML et al (2006) The predominance of one of the SR-BI isoforms is associated with increased esterified cholesterol levels not apoptosis in mink testis. J Lipid Res 47(10):2233–2247

Albert ML, Kim JI, Birge RB (2000) alphavbeta5 integrin recruits the CrkII-Dock180-rac1 complex for phagocytosis of apoptotic cells. Nat Cell Biol 2(12):899–905

Albert ML, Pearce SF, Francisco LM et al (1998a) Immature dendritic cells phagocytose apoptotic cells via alphavbeta5 and CD36, and cross-present antigens to cytotoxic T lymphocytes. J Exp Med 188(7):1359–1368

Albert ML, Sauter B, Bhardwaj N (1998b) Dendritic cells acquire antigen from apoptotic cells and induce class I-restricted CTLs. Nature 392(6671):86–89

Anderson HA, Maylock CA, Williams JA et al (2003) Serum-derived protein S binds to phosphatidylserine and stimulates the phagocytosis of apoptotic cells. Nat Immunol 4(1):87–91

Arur S, Uche UE, Rezaul K et al (2003) Annexin I is an endogenous ligand that mediates apoptotic cell engulfment. Dev Cell 4(4):587–598

Asch AS, Barnwell J, Silverstein RL et al (1987) Isolation of the thrombospondin membrane receptor. J Clin Invest 79(4):1054–1061

Atsumi G, Murakami M, Kojima K et al (2000) Distinct roles of two intracellular phospholipase A2s in fatty acid release in the cell death pathway. Proteolytic fragment of type IVA cytosolic phospholipase A2alpha inhibits stimulus-induced arachidonate release, whereas that of type VI Ca2+-independent phospholipase A2 augments spontaneous fatty acid release. J Biol Chem 275(24):18248–18258

Atsumi G, Murakami M, Tajima M et al (1997) The perturbed membrane of cells undergoing apoptosis is susceptible to type II secretory phospholipase A2 to liberate arachidonic acid. Biochim Biophys Acta 1349(1):43–54

Balasubramanian K, Chandra J, Schroit AJ (1997) Immune clearance of phosphatidylserine-expressing cells by phagocytes. The role of beta2-glycoprotein I in macrophage recognition. J Biol Chem 272(49):31113–31117

Balasubramanian K, Maiti SN, Schroit AJ (2005) Recruitment of beta-2-glycoprotein 1 to cell surfaces in extrinsic and intrinsic apoptosis. Apoptosis 10(2):439–446

Balasubramanian K, Schroit AJ (1998) Characterization of phosphatidylserine-dependent beta2-glycoprotein I macrophage interactions. Implications for apoptotic cell clearance by phagocytes. J Biol Chem 273(44):29272–29277

Barker RN, Erwig LP, Hill KS et al (2002) Antigen presentation by macrophages is enhanced by the uptake of necrotic, but not apoptotic, cells. Clin Exp Immunol 127(2):220–225

Baron WF, Pan CQ, Spencer SA et al (1998) Cloning and characterization of an actin-resistant DNase I-like endonuclease secreted by macrophages. Gene 215(2):291–301

Bennett RM, Gabor GT, Merritt MM (1985) DNA binding to human leukocytes. Evidence for a receptor-mediated association, internalization, and degradation of DNA. J Clin Invest 76(6):2182–2190

Bevers EM, Comfurius P, Dekkers DW et al (1999) Lipid translocation across the plasma membrane of mammalian cells. Biochim Biophys Acta 1439(3):317–330

Bickerstaff MC, Botto M, Hutchinson WL et al (1999) Serum amyloid P component controls chromatin degradation and prevents antinuclear autoimmunity. Nat Med 5(6):694–697

Bijl M, Horst G, Bijzet J et al (2003) Serum amyloid P component binds to late apoptotic cells and mediates their uptake by monocyte-derived macrophages. Arthritis Rheum 48(1):248–254

Bilyy R, Stoika R (2007) Search for novel cell surface markers of apoptotic cells. Autoimmunity 40(4):249–253

Boes M, Schmidt T, Linkemann K et al (2000) Accelerated development of IgG autoantibodies and autoimmune disease in the absence of secreted IgM. Proc Natl Acad Sci U S A 97(3):1184–1189

Borisenko GG, Iverson SL, Ahlberg S et al (2004) Milk fat globule epidermal growth factor 8 (MFG-E8) binds to oxidized phosphatidylserine: implications for macrophage clearance of apoptotic cells. Cell Death Differ 11(8):943–945

Borth W (1992) Alpha 2-macroglobulin, a multifunctional binding protein with targeting characteristics. FASEB J 6(15):3345–3353

Bose J, Gruber AD, Helming L et al (2004) The phosphatidylserine receptor has essential functions during embryogenesis but not in apoptotic cell removal. J Biol 3(4):15

Bottcher A, Gaipl US, Furnrohr BG et al (2006) Involvement of phosphatidylserine, alphavbeta3, CD14, CD36, and complement C1q in the phagocytosis of primary necrotic lymphocytes by macrophages. Arthritis Rheum 54(3):927–938

Botto M, Dell'Agnola C, Bygrave AE et al (1998) Homozygous C1q deficiency causes glomerulonephritis associated with multiple apoptotic bodies. Nat Genet 19(1):56–59

Brouckaert G, Kalai M, Krysko DV et al (2004) Phagocytosis of necrotic cells by macrophages is phosphatidylserine dependent and does not induce inflammatory cytokine production. Mol Biol Cell 15(3):1089–1100

Brown S, Heinisch I, Ross E et al (2002) Apoptosis disables CD31-mediated cell detachment from phagocytes promoting binding and engulfment. Nature 418(6894):200–203

Callahan MK, Williamson P, Schlegel RA (2000) Surface expression of phosphatidylserine on macrophages is required for phagocytosis of apoptotic thymocytes. Cell Death Differ 7(7):645–653

Campisi J, Leem TH, Fleshner M (2003) Stress-induced extracellular Hsp72 is a functionally significant danger signal to the immune system. Cell Stress Chaperones 8(3):272–286

Casciola-Rosen L, Rosen A, Petri M et al (1996) Surface blebs on apoptotic cells are sites of enhanced procoagulant activity: implications for coagulation events and antigenic spread in systemic lupus erythematosus. Proc Natl Acad Sci U S A 93(4):1624–1629

Casciola-Rosen LA, Anhalt G, Rosen A (1994) Autoantigens targeted in systemic lupus erythematosus are clustered in two populations of surface structures on apoptotic keratinocytes. J Exp Med 179(4):1317–1330

Choi JJ, Reich CF, 3rd, Pisetsky DS (2005) The role of macrophages in the in vitro generation of extracellular DNA from apoptotic and necrotic cells. Immunology 115(1):55–62

Chung S, Gumienny TL, Hengartner MO et al (2000) A common set of engulfment genes mediates removal of both apoptotic and necrotic cell corpses in C. elegans. Nat Cell Biol 2(12):931–937

Clayton AR, Prue RL, Harper L et al (2003) Dendritic cell uptake of human apoptotic and necrotic neutrophils inhibits CD40, CD80, and CD86 expression and reduces allogeneic T cell responses: relevance to systemic vasculitis. Arthritis Rheum 48(8):2362–2374

Cocco RE, Ucker DS (2001) Distinct modes of macrophage recognition for apoptotic and necrotic cells are not specified exclusively by phosphatidylserine exposure. Mol Biol Cell 12(4):919–930

Cohen PL, Caricchio R, Abraham V et al (2002) Delayed apoptotic cell clearance and lupus-like autoimmunity in mice lacking the c-mer membrane tyrosine kinase. J Exp Med 196(1):135–140

Cvetanovic M, Ucker DS (2004) Innate immune discrimination of apoptotic cells: repression of proinflammatory macrophage transcription is coupled directly to specific recognition. J Immunol 172(2):880–889

D'Cruz PM, Yasumura D, Weir J et al (2000) Mutation of the receptor tyrosine kinase gene Mertk in the retinal dystrophic RCS rat. Hum Mol Genet 9(4):645–651

deCathelineau AM, Henson PM (2003) The final step in programmed cell death: phagocytes carry apoptotic cells to the grave. Essays Biochem 39:105–117

Devitt A, Parker KG, Ogden CA et al (2004) Persistence of apoptotic cells without autoimmune disease or inflammation in CD14-/- mice. J Cell Biol 167(6):1161–1170

Duncan JL, LaVail MM, Yasumura D et al (2003) An RCS-like retinal dystrophy phenotype in mer knockout mice. Invest Ophthalmol Vis Sci 44(2):826–838

Duvall E, Wyllie AH, Morris RG (1985) Macrophage recognition of cells undergoing programmed cell death (apoptosis). Immunology 56(2):351–358

Eggleton P, Tenner AJ, Reid KB (2000) C1q receptors. Clin Exp Immunol 120(3):406–412

Ehrenstein MR, Cook HT, Neuberger MS (2000) Deficiency in serum immunoglobulin (Ig)M predisposes to development of IgG autoantibodies. J Exp Med 191(7):1253–1258

Eliceiri BP, Puente XS, Hood JD et al (2002) Src-mediated coupling of focal adhesion kinase to integrin alpha(v)beta5 in vascular endothelial growth factor signalling. J Cell Biol 157(1):149–160

Enari M, Sakahira H, Yokoyama H et al (1998) A caspase-activated DNase that degrades DNA during apoptosis, and its inhibitor ICAD. Nature 391(6662):43–50

Erwig LP, Henson PM (2008) Clearance of apoptotic cells by phagocytes. Cell Death Differ 15(2):243–250

Eum HA, Cha YN, Lee SM (2007) Necrosis and apoptosis: sequence of liver damage following reperfusion after 60 min ischemia in rats. Biochem Biophys Res Commun 358(2):500–505

Fadeel B (2004) Plasma membrane alterations during apoptosis: role in corpse clearance. Antioxid Redox Signal 6(2):269–275

Fadok VA, Bratton DL, Guthrie L et al (2001) Differential effects of apoptotic versus lysed cells on macrophage production of cytokines: role of proteases. J Immunol 166(11):6847–6854

Fadok VA, Bratton DL, Rose DM et al (2000) A receptor for phosphatidylserine-specific clearance of apoptotic cells. Nature 405(6782):85–90

Fadok VA, Voelker DR, Campbell PA et al (1992) Exposure of phosphatidylserine on the surface of apoptotic lymphocytes triggers specific recognition and removal by macrophages. J Immunol 148(7):2207–2216

Familian A, Zwart B, Huisman HG et al (2001) Chromatin-independent binding of serum amyloid P component to apoptotic cells. J Immunol 167(2):647–654

Fan X, Krahling S, Smith D et al (2004) Macrophage surface expression of annexins I and II in the phagocytosis of apoptotic lymphocytes. Mol Biol Cell 15(6):2863–2872

Feng W, Yasumura D, Matthes MT et al (2002) Mertk triggers uptake of photoreceptor outer segments during phagocytosis by cultured retinal pigment epithelial cells. J Biol Chem 277(19):17016–17022

Festjens N, Vanden Berghe T, Vandenabeele P (2006) Necrosis, a well-orchestrated form of cell demise: signalling cascades, important mediators and concomitant immune response. Biochim Biophys Acta 1757(9–10):1371–1387

Fiers W, Beyaert R, Declercq W et al (1999) More than one way to die: apoptosis, necrosis and reactive oxygen damage. Oncogene 18(54):7719–1730

Fink SL, Cookson BT (2005) Apoptosis, pyroptosis, and necrosis: mechanistic description of dead and dying eukaryotic cells. Infect Immun 73(4):1907–1916

Finnemann SC (2003) Focal adhesion kinase signalling promotes phagocytosis of integrin-bound photoreceptors. EMBO J 22(16):4143–4154

Finnemann SC, Bonilha VL, Marmorstein A D et al (1997) Phagocytosis of rod outer segments by retinal pigment epithelial cells requires alpha(v)beta5 integrin for binding but not for internalization. Proc Natl Acad Sci U S A 94(24):12932–12937

Finnemann SC, Rodriguez-Boulan E (1999) Macrophage and retinal pigment epithelium phagocytosis: apoptotic cells and photoreceptors compete for alphavbeta3 and alphavbeta5 integrins, and protein kinase C regulates alphavbeta5 binding and cytoskeletal linkage. J Exp Med 190(6):861–874

Flierman R, Daha MR (2007) The clearance of apoptotic cells by complement. Immunobiology 212(4–5):363–370

Gaipl US, Beyer TD, Heyder P et al (2004) Cooperation between C1q and DNase I in the clearance of necrotic cell-derived chromatin. Arthritis Rheum 50(2):640–649

Gaipl US, Kuenkele S, Voll RE et al (2001) Complement binding is an early feature of necrotic and a rather late event during apoptotic cell death. Cell Death Differ 8(4):327–334

Gaipl US, Munoz LE, Rodel F et al (2007) Modulation of the immune system by dying cells and the phosphatidylserine-ligand annexin A5. Autoimmunity 40(4):254–259

Gardai SJ, Bratton DL, Ogden CA et al (2006) Recognition ligands on apoptotic cells: a perspective. J Leukocyte Biol 79(5):896–903

Gardai SJ, McPhillips KA, Frasch SC et al (2005) Cell-surface calreticulin initiates clearance of viable or apoptotic cells through trans-activation of LRP on the phagocyte. Cell 123(2):321–334

Gershov D, Kim S, Brot N et al (2000) C-Reactive protein binds to apoptotic cells, protects the cells from assembly of the terminal complement components, and sustains an antiinflammatory innate immune response: implications for systemic autoimmunity. J Exp Med 192(9):1353–1364

Godson C, Mitchell S, Harvey K et al (2000) Cutting edge: lipoxins rapidly stimulate nonphlogistic phagocytosis of apoptotic neutrophils by monocyte-derived macrophages. J Immunol 164(4):1663–1667

Golstein P, Kroemer G (2007) Cell death by necrosis: towards a molecular definition. Trends Biochem Sci 32(1):37–43

Gorgani NN, Smith BA, Kono DH et al (2002) Histidine-rich glycoprotein binds to DNA and Fc gamma RI and potentiates the ingestion of apoptotic cells by macrophages. J Immunol 169(9):4745–4751

Greenberg ME, Sun M, Zhang R et al (2006) Oxidized phosphatidylserine-CD36 interactions play an essential role in macrophage-dependent phagocytosis of apoptotic cells. J Exp Med 203(12):2613–2625

Gujral JS, Farhood A, Jaeschke H (2003) Oncotic necrosis and caspase-dependent apoptosis during galactosamine-induced liver injury in rats. Toxicol Appl Pharmacol 190(1):37–46

Gujral JS, Knight TR, Farhood A et al (2002) Mode of cell death after acetaminophen overdose in mice: apoptosis or oncotic necrosis? Toxicol Sci 67(2):322–328

Hack CE, Wolbink GJ, Schalkwijk C et al (1997) A role for secretory phospholipase A2 and C-reactive protein in the removal of injured cells. Immunol Today 18(3):111–115

Hafizi S, Dahlback B (2006) Gas6 and protein S. Vitamin K-dependent ligands for the Axl receptor tyrosine kinase subfamily. FEBS J 273(23):5231–5244

Hall MO, Obin MS, Heeb MJ et al (2005) Both protein S and Gas6 stimulate outer segment phagocytosis by cultured rat retinal pigment epithelial cells. Exp Eye Res 81(5):581–591

Hall MO, Prieto AL, Obin MS et al (2001) Outer segment phagocytosis by cultured retinal pigment epithelial cells requires Gas6. Exp Eye Res 73(4):509–520

Hamon Y, Broccardo C, Chambenoit O et al (2000) ABC1 promotes engulfment of apoptotic cells and transbilayer redistribution of phosphatidylserine. Nat Cell Biol 2(7):399–406

Hanayama R, Nagata S (2005) Impaired involution of mammary glands in the absence of milk fat globule EGF factor 8. Proc Natl Acad Sci USA 102(46):16886–16891

Hanayama R, Tanaka M, Miwa K et al (2002) Identification of a factor that links apoptotic cells to phagocytes. Nature 417(6885):182–187

Hanayama R, Tanaka M, Miyasaka K et al (2004) Autoimmune disease and impaired uptake of apoptotic cells in MFG-E8-deficient mice. Science 304(5674):1147–1150

Hart SP, Alexander KM, MacCall SM et al (2005) C-reactive protein does not opsonize early apoptotic human neutrophils, but binds only membrane-permeable late apoptotic cells and has no effect on their phagocytosis by macrophages. J Inflamm (Lond) 2:5.

Hicks PS, Saunero-Nava L, Du Clos TW et al (1992) Serum amyloid P component binds to histones and activates the classical complement pathway. J Immunol 149(11):3689–3694

Higami Y, Tsuchiya T, To K et al (2004) Expression of DNase gamma during Fas-independent apoptotic DNA fragmentation in rodent hepatocytes. Cell Tissue Res 316(3):403–407

Hirt UA, Leist M (2003) Rapid, noninflammatory and PS-dependent phagocytic clearance of necrotic cells. Cell Death Differ 10(10):1156–1164

Hoffmann PR, deCathelineau AM, Ogden CA et al (2001) Phosphatidylserine (PS) induces PS receptor-mediated macropinocytosis and promotes clearance of apoptotic cells. J Cell Biol 155(4):649–659

Hughes J, Liu Y, Van Damme J et al (1997) Human glomerular mesangial cell phagocytosis of apoptotic neutrophils: mediation by a novel CD36-independent vitronectin receptor/thrombospondin recognition mechanism that is uncoupled from chemokine secretion. J Immunol 158(9):4389–4397

Ip WK, Lau YL (2004) Distinct maturation of, but not migration between, human monocyte-derived dendritic cells upon ingestion of apoptotic cells of early or late phases. J Immunol 173(1):189–196

Jacob M, Mannherz HG, Napirei M (2007) Chromatin breakdown by deoxyribonuclease1 promotes acetaminophen-induced liver necrosis: an ultrastructural and histochemical study on male CD-1 mice. Histochem Cell Biol 128(1):19–33

Jaeschke H, Lemasters JJ (2003) Apoptosis versus oncotic necrosis in hepatic ischemia/reperfusion injury. Gastroenterology 125(4):1246–1257

Jehle AW, Gardai SJ, Li S et al (2006) ATP-binding cassette transporter A7 enhances phagocytosis of apoptotic cells and associated ERK signalling in macrophages. J Cell Biol 174(4):547–556

Jensen ML, Honore C, Hummelshoj T et al (2007) Ficolin-2 recognizes DNA and participates in the clearance of dying host cells. Mol Immunol 44(5):856–865

Jiang H, Cooper B, Robey FA et al (1992) DNA binds and activates complement via residues 14-26 of the human C1q A chain. J Biol Chem 267(35):25597–25601

Jones AL, Hulett MD, Parish CR (2005a) Histidine-rich glycoprotein: A novel adaptor protein in plasma that modulates the immune, vascular and coagulation systems. Immunol Cell Biol 83(2):106–118

Jones AL, Poon IK, Hulett MD et al (2005b) Histidine-rich glycoprotein specifically binds to necrotic cells via its amino-terminal domain and facilitates necrotic cell phagocytosis. J Biol Chem 280(42):35733–35741

Kacani L, Wurm M, Schwentner I et al (2005) Maturation of dendritic cells in the presence of living, apoptotic and necrotic tumour cells derived from squamous cell carcinoma of head and neck. Oral Oncol 41(1):17–24

Kagan VE, Borisenko GG, Serinkan BF et al (2003) Appetizing rancidity of apoptotic cells for macrophages: oxidation, externalization, and recognition of phosphatidylserine. Am J Physiol Lung Cell Mol Physiol 285(1):L1–17

Kaplan MH, Volanakis JE (1974) Interaction of C-reactive protein complexes with the complement system. I. Consumption of human complement associated with the reaction of C-reactive protein with pneumococcal C-polysaccharide and with the choline phosphatides, lecithin and sphingomyelin. J Immunol 112(6):2135–2147

Kawane K, Fukuyama H, Kondoh G et al (2001) Requirement of DNase II for definitive erythropoiesis in the mouse fetal liver. Science 292(5521):1546–1549

Kawasaki Y, Nakagawa A, Nagaosa K et al (2002) Phosphatidylserine binding of class B scavenger receptor type I, a phagocytosis receptor of testicular sertoli cells. J Biol Chem 277(30):27559–27566

Kerr JF, Wyllie AH, Currie AR (1972) Apoptosis: a basic biological phenomenon with wide-ranging implications in tissue kinetics. Br J Cancer 26(4):239–257

Khandelwal S, van Rooijen N, Saxena RK (2007) Reduced expression of CD47 during murine red blood cell (RBC) senescence and its role in RBC clearance from the circulation. Transfusion 47(9):1725–1732

Kim SJ, Gershov D, Ma X et al (2002) I-PLA(2) activation during apoptosis promotes the exposure of membrane lysophosphatidylcholine leading to binding by natural immunoglobulin M antibodies and complement activation. J Exp Med 196(5):655–665

Kinchen JM, Cabello J, Klingele D et al (2005) Two pathways converge at CED-10 to mediate actin rearrangement and corpse removal in C. elegans. Nature 434(7029):93–99

Korb LC, Ahearn JM (1997) C1q binds directly and specifically to surface blebs of apoptotic human keratinocytes: complement deficiency and systemic lupus erythematosus revisited. J Immunol 158(10):4525–4528

Korkmaz B, Moreau T, Gauthier F (2008) Neutrophil elastase, proteinase 3 and cathepsin G: Physicochemical properties, activity and physiopathological functions. Biochimie 90(2):227–242

Krieser RJ, MacLea KS, Longnecker DS et al (2002) Deoxyribonuclease IIalpha is required during the phagocytic phase of apoptosis and its loss causes perinatal lethality. Cell Death Differ 9(9):956–962

Krysko DV, Brouckaert G, Kalai M et al (2003) Mechanisms of internalization of apoptotic and necrotic L929 cells by a macrophage cell line studied by electron microscopy. J Morphol 258(3):336–345

Krysko DV, D'Herde K, Vandenabeele P (2006a) Clearance of apoptotic and necrotic cells and its immunological consequences. Apoptosis 11(10):1709–1726

Krysko DV, Denecker G, Festjens N et al (2006b) Macrophages use different internalization mechanisms to clear apoptotic and necrotic cells. Cell Death Differ 13(12):2011–2022

Krysko O, De Ridder L, Cornelissen M (2004) Phosphatidylserine exposure during early primary necrosis (oncosis) in JB6 cells as evidenced by immunogold labeling technique. Apoptosis 9(4):495–500

Kuraya M, Ming Z, Liu X et al (2005) Specific binding of L-ficolin and H-ficolin to apoptotic cells leads to complement activation. Immunobiology 209(9):689–697

Lauber K, Blumenthal SG, Waibel M et al (2004) Clearance of apoptotic cells: getting rid of the corpses. Mol Cell 14(3):277–287

Lauber K, Bohn E, Krober SM et al (2003) Apoptotic cells induce migration of phagocytes via caspase-3-mediated release of a lipid attraction signal. Cell 113(6):717–730

Law RH, Zhang Q, McGowan S et al (2006) An overview of the serpin superfamily. Genome Biol 7(5):216

Lecoeur H, Prevost MC, Gougeon ML (2001) Oncosis is associated with exposure of phosphatidylserine residues on the outside layer of the plasma membrane: a reconsideration of the specificity of the annexin V/propidium iodide assay. Cytometry 44(1):65–72

Lee KD, Pitas RE, Papahadjopoulos D (1992) Evidence that the scavenger receptor is not involved in the uptake of negatively charged liposomes by cells. Biochim Biophys Acta 1111(1):1–6

Leverrier Y, Lorenzi R, Blundell MP et al (2001) Cutting edge: the Wiskott-Aldrich syndrome protein is required for efficient phagocytosis of apoptotic cells. J Immunol 166(8):4831–4834

Leverrier Y, Ridley AJ (2001) Requirement for Rho GTPases and PI 3-kinases during apoptotic cell phagocytosis by macrophages. Curr Biol 11(3):195–199

Li LY, Luo X, Wang X (2001) Endonuclease G is an apoptotic DNase when released from mitochondria. Nature 412(6842):95–99

Linder E, Helin H, Chang CM et al (1983) Complement-mediated binding of monocytes to intermediate filaments in vitro. Am J Pathol 112(3):267–277

Liu X, Li P, Widlak P et al (1998) The 40-kDa subunit of DNA fragmentation factor induces DNA fragmentation and chromatin condensation during apoptosis. Proc Natl Acad Sci U S A 95(15):8461–8466

Luciani MF, Chimini G (1996) The ATP binding cassette transporter ABC1, is required for the engulfment of corpses generated by apoptotic cell death. EMBO J 15(2):226–235

Ma K, Simantov R, Zhang JC et al (2000) High affinity binding of beta 2-glycoprotein I to human endothelial cells is mediated by annexin II. J Biol Chem 275(20):15541–15548

Maiuri MC, Zalckvar E, Kimchi A et al (2007) Self-eating and self-killing: crosstalk between autophagy and apoptosis. Nat Rev Mol Cell Biol 8(9):741–752

Majno G, Joris I (1995) Apoptosis, oncosis, and necrosis. An overview of cell death. Am J Pathol 146(1):3–15

Malhotra R, Thiel S, Reid KB et al (1990) Human leukocyte C1q receptor binds other soluble proteins with collagen domains. J Exp Med 172(3):955–959

Mandal D, Mazumder A, Das P et al (2005) Fas-, caspase 8-, and caspase 3-dependent signalling regulates the activity of the aminophospholipid translocase and phosphatidylserine externalization in human erythrocytes. J Biol Chem 280(47):39460–39467

Mannherz HG, Mach M, Malicka-Blaszkiewicz M et al (2007) Lamellipodial and amoeboid cell locomotion: The role of actin-cycling and bleb formation. Biophysical Reviews and Letters 2(1):5–22

Martin SJ, Reutelingsperger CP, McGahon AJ et al (1995) Early redistribution of plasma membrane phosphatidylserine is a general feature of apoptosis regardless of the initiating stimulus: inhibition by overexpression of Bcl-2 and Abl. J Exp Med 182(5):1545–1556

Mevorach D, Mascarenhas JO, Gershov D et al (1998) Complement-dependent clearance of apoptotic cells by human macrophages. J Exp Med 188(12):2313–2320

Mitchell S, Thomas G, Harvey K et al (2002) Lipoxins, aspirin-triggered epi-lipoxins, lipoxin stable analogues, and the resolution of inflammation: stimulation of macrophage phagocytosis of apoptotic neutrophils in vivo. J Am Soc Nephrol 13(10):2497–2507

Miyanishi M, Tada K, Koike M et al (2007) Identification of Tim4 as a phosphatidylserine receptor. Nature 450(7168):435–439

Moffatt OD, Devitt A, Bell ED et al (1999) Macrophage recognition of ICAM-3 on apoptotic leukocytes. J Immunol 162(11):6800–6810

Mold C, Baca R, Du Clos TW (2002) Serum amyloid P component and C-reactive protein opsonize apoptotic cells for phagocytosis through Fcgamma receptors. J Autoimmun 19(3):147–154

Moore KJ, Freeman MW (2006) Scavenger receptors in atherosclerosis: beyond lipid uptake. Arterioscler Thromb Vasc Biol 26(8):1702–1711

Murphy JE, Tedbury PR, Homer-Vanniasinkam S et al (2005) Biochemistry and cell biology of mammalian scavenger receptors. Atherosclerosis 182(1):1–15

Nagata K, Ohashi K, Nakano T et al (1996) Identification of the product of growth arrest-specific gene 6 as a common ligand for Axl, Sky, and Mer receptor tyrosine kinases. J Biol Chem 271(47):30022–30027

Nakano T, Ishimoto Y, Kishino J et al (1997) Cell adhesion to phosphatidylserine mediated by a product of growth arrest-specific gene 6. J Biol Chem 272(47):29411–29414

Nandrot EF, Anand M, Almeida D et al (2007) Essential role for MFG-E8 as ligand for alphavbeta5 integrin in diurnal retinal phagocytosis. Proc Natl Acad Sci U S A 104(29):12005–12010

Nandrot EF, Kim Y, Brodie SE et al (2004) Loss of synchronized retinal phagocytosis and age-related blindness in mice lacking alphavbeta5 integrin. J Exp Med 200(12):1539–1545

Napirei M, Basnakian AG, Apostolov EO et al (2006a) Deoxyribonuclease 1 aggravates acetaminophen-induced liver necrosis in male CD-1 mice. Hepatology 43(2):297–305

Napirei M, Gultekin A, Kloeckl T et al (2006b) Systemic lupus-erythematosus: Deoxyribonuclease 1 in necrotic chromatin disposal. Int J Biochem Cell Biol 38(3):297–306

Napirei M, Karsunky H, Zevnik B et al (2000) Features of systemic lupus erythematosus in Dnase1-deficient mice. Nat Genet 25(2):177–181

Napirei M, Wulf S, Eulitz D et al (2005) Comparative characterization of rat deoxyribonuclease 1 (Dnase1) and murine deoxyribonuclease 1-like 3 (Dnase1l3). Biochem J 389(Pt 2):355–364

Napirei M, Wulf S, Mannherz HG (2004) Chromatin breakdown during necrosis by serum Dnase1 and the plasminogen system. Arthritis Rheum 50(6):1873–1883

Nauta AJ, Raaschou-Jensen N, Roos A et al (2003) Mannose-binding lectin engagement with late apoptotic and necrotic cells. Eur J Immunol 33(10):2853–2863

Nimpf J, Bevers EM, Bomans P et al (1986) Prothrombinase activity of human platelets is inhibited by beta 2-glycoprotein-I. Biochim Biophys Acta 884(1):142–149

Ogden CA, deCathelineau A, Hoffmann PR et al (2001) C1q and mannose binding lectin engagement of cell surface calreticulin and CD91 initiates macropinocytosis and uptake of apoptotic cells. J Exp Med 194(6):781–795

Oka K, Sawamura T, Kikuta K et al (1998) Lectin-like oxidized low-density lipoprotein receptor 1 mediates phagocytosis of aged/apoptotic cells in endothelial cells. Proc Natl Acad Sci U S A 95(16):9535–9540

Oldenborg PA, Zheleznyak A, Fang YF et al (2000) Role of CD47 as a marker of self on red blood cells. Science 288(5473):2051–2054

Palaniyar N, Nadesalingam J, Clark H et al (2004) Nucleic acid is a novel ligand for innate, immune pattern recognition collectins surfactant proteins A and D and mannose-binding lectin. J Biol Chem 279(31):32728–32736

Patel VA, Longacre A, Hsiao K et al (2006) Apoptotic cells, at all stages of the death process, trigger characteristic signalling events that are divergent from and dominant over those triggered by necrotic cells: Implications for the delayed clearance model of autoimmunity. J Biol Chem 281(8):4663–4670

Peng Y, Kowalewski R, Kim S et al (2005) The role of IgM antibodies in the recognition and clearance of apoptotic cells. Mol Immunol 42(7):781–787

Pepys MB, Booth SE, Tennent GA et al (1994) Binding of pentraxins to different nuclear structures: C-reactive protein binds to small nuclear ribonucleoprotein particles, serum amyloid P component binds to chromatin and nucleoli. Clin Exp Immunol 97(1):152–157

Pepys MB, Butler PJ (1987) Serum amyloid P component is the major calcium-dependent specific DNA binding protein of the serum. Biochem Biophys Res Commun 148(1):308–313

Pisetsky DS, Fairhurst AM (2007) The origin of extracellular DNA during the clearance of dead and dying cells. Autoimmunity 40(4):281–284

Platt N, Suzuki H, Kodama T et al (2000) Apoptotic thymocyte clearance in scavenger receptor class A-deficient mice is apparently normal. J Immunol 164(9):4861–4867

Podor TJ, Joshua P, Butcher M et al (1992) Accumulation of type 1 plasminogen activator inhibitor and vitronectin at sites of cellular necrosis and inflammation. Ann N Y Acad Sci 667:173–177

Proskuryakov SY, Gabai VL, Konoplyannikov AG et al (2005) Immunology of apoptosis and necrosis. Biochemistry (Mosc) 70(12):1310–1320

Pugin J, Heumann ID, Tomasz A et al (1994) CD14 is a pattern recognition receptor. Immunity 1(6):509–516

Quartier P, Potter PK, Ehrenstein MR et al (2005) Predominant role of IgM-dependent activation of the classical pathway in the clearance of dying cells by murine bone marrow-derived macrophages in vitro. Eur J Immunol 35(1):252–260

Raucci A, Palumbo R, Bianchi ME (2007) HMGB1: a signal of necrosis. Autoimmunity 40(4):285–289

Rauch F, Prud'homme J, Arabian A et al (2000) Heart, brain, and body wall defects in mice lacking calreticulin. Exp Cell Res 256(1):105–111

Reddy SM, Hsiao KH, Abernethy VE et al (2002) Phagocytosis of apoptotic cells by macrophages induces novel signalling events leading to cytokine-independent survival and inhibition of proliferation: activation of Akt and inhibition of extracellular signal-regulated kinases 1 and 2. J Immunol 169(2):702–713

Ren Y, Silverstein RL, Allen J et al (1995) CD36 gene transfer confers capacity for phagocytosis of cells undergoing apoptosis. J Exp Med 181(5):1857–1862

Reutelingsperger CP, van Heerde WL (1997) Annexin V, the regulator of phosphatidyl-serine-catalyzed inflammation and coagulation during apoptosis. Cell Mol Life Sci 53(6):527–532

Rigotti A, Acton SL, Krieger M (1995) The class B scavenger receptors SR-BI and CD36 are receptors for anionic phospholipids. J Biol Chem 270(27):16221–16224

Robey FA, Jones KD, Tanaka T et al (1984) Binding of C-reactive protein to chromatin and nucleosome core particles. A possible physiological role of C-reactive protein. J Biol Chem 259(11):7311–7316

Rovere-Querini P, Capobianco A, Scaffidi P et al (2004) HMGB1 is an endogenous immune adjuvant released by necrotic cells. EMBO Rep 5(8):825–830

Rovere P, Peri G, Fazzini F et al (2000) The long pentraxin PTX3 binds to apoptotic cells and regulates their clearance by antigen-presenting dendritic cells. Blood 96(13):4300–4306

Russell L, Waring P, Beaver JP (1998) Increased cell surface exposure of fucose residues is a late event in apoptosis. Biochem Biophys Res Commun 250(2):449–453

Ryeom SW, Sparrow JR, Silverstein RL (1996) CD36 participates in the phagocytosis of rod outer segments by retinal pigment epithelium. J Cell Sci 109 (Pt 2):387–395./bib>

Sauter B, Albert ML, Francisco L et al (2000) Consequences of cell death: exposure to necrotic tumour cells, but not primary tissue cells or apoptotic cells, induces the maturation of immunostimulatory dendritic cells. J Exp Med 191(3):423–434

Savill J, Hogg N, Ren Y et al (1992) Thrombospondin cooperates with CD36 and the vitronectin receptor in macrophage recognition of neutrophils undergoing apoptosis. J Clin Invest 90(4):1513–1522

Scannell M, Flanagan MB, deStefani A et al (2007) Annexin-1 and peptide derivatives are released by apoptotic cells and stimulate phagocytosis of apoptotic neutrophils by macrophages. J Immunol 178(7):4595–4605

Schagat TL, Wofford JA, Wright JR (2001) Surfactant protein A enhances alveolar macrophage phagocytosis of apoptotic neutrophils. J Immunol 166(4):2727–2733

Schroit AJ, Madsen JW, Tanaka Y (1985) In vivo recognition and clearance of red blood cells containing phosphatidylserine in their plasma membranes. J Biol Chem 260(8):5131–5138

Scott RS, McMahon EJ, Pop SM et al (2001) Phagocytosis and clearance of apoptotic cells is mediated by MER. Nature 411(6834):207–211

Chapter 5
Evolutionarily Conserved Pathways Regulating Engulfment of Apoptotic Cells

Matthew A. Gronski and Kodi S. Ravichandran

Abstract: Animal models, such as the nematode *Caenorhabditis elegans* and the fruit fly *Drosophila melanogaster*, combined with the powerful approaches of forward and reverse genetics, have greatly contributed to discovering the mechanisms involved in the clearance of dying cells, and indicated a high degree of evolutionary conservation in many of the signal transduction pathways implicated in the uptake of dying cells. This chapter will discuss the studies that led to the isolation of several genes that regulate recognition, uptake, or processing of apoptotic cells, and provide a review of the mutational analysis approaches that revealed the existence of at least two major genetic pathways for phagocytosis of dying cells in *C. elegans* and mammals. Studies in *D. melanogaster* have helped identify new players, as well as genes that are homologues of those identified in *C. elegans* and mammals. The usage of *D. melanogaster* as one of the model organisms for unraveling the molecular mechanisms of phagocytosis of apoptotic cells will be examined in the light of recently published literature. This chapter will be devoted to a comparative analysis of the biochemical pathways involved in clearance of dead cells, citing studies of phylogenetically diverse multicellular organisms ranging from *C. elegans* and *D. melanogaster*, to mammals.

Keywords: Phagocytosis • Apoptosis • Genetic pathways • Phosphatidylserine • Annexin I

K. S. Ravichandran (✉) · M. A. Gronski
Beirne Carter Immunology Center and the Department of Microbiology
University of Virginia
Charlottesville, VA 22908, USA
Tel.: 434-243-6093
Fax: 434-924-1221
E-mail: ravi@virginia.edu

D.V. Krysko, P. Vandenabeele (eds.), *Phagocytosis of Dying Cells,*
DOI: 10.1007/978-1-4020-9292-3_5, © Springer Science+Business Media B.V. 2009

5.1 Invertebrate Model Systems for Studying Engulfment

5.1.1 C. elegans as a Model System

C. elegans is a small (1 mm long as an adult) and free-living round worm. It normally lives in the soil, where it feeds on bacteria and fungi, but in the lab it is grown on agar plates containing a lawn of *E. coli*. It has been used to address fundamental questions in developmental biology, neurobiology and behavioural biology, because of a number of advantages over other systems.

C. elegans is a eukaryote that shares cellular and molecular structures and control mechanisms with higher organisms. Furthermore, it is a multicellular organism that goes through a complex developmental process, including embryogenesis, morphogenesis and growth into an adult. About 35% of *C. elegans* genes have human homologues. It has been demonstrated that human genes can often replace their *C. elegans* homologues when introduced into the worm. Other useful features of *C. elegans* include its fast and convenient life cycle. Embryogenesis occurs in about 12 hours, development to the adult stage occurs in 2.5 days and the life span is 2–3 weeks, although this is dependent on temperature. Furthermore, the nematode can be frozen, and when thawed is subsequently viable and useful for experimentation.

The *C. elegans* genome is relatively small (97 Megabases) making it easier to analyze compared to the human genome estimated at 3000 Megabases. The entire *C. elegans* genome has been sequenced (1998). Furthermore, because *C. elegans* is small and its anatomy is invariant from one animal to the next, a three-dimensional composite picture of the entire worm has been reconstructed. The development of the worm is known in great detail and because it is transparent, the fate of every one of its 1090 somatic cells during development is known so that any changes from the programmed process can be attributed to a mutation in a gene being studied. 131 of the 1090 cells undergo programmed cell death (Sulston and Horvitz 1977; Kimble and Hirsh 1979; Sulston et al. 1983; Mangahas and Zhou 2005). These apoptotic cells (often called 'cell corpses') can be identified under Nomarski optics and appear as highly refractile button-like objects (Sulston 1976; Sulston and Horvitz 1977; Sulston et al. 1983). In wild-type animals, the engulfment and degradation of dying cells is very quick (about 1 hour) so that very few cell corpses can be seen during late embryonic stages (Ellis et al. 1991), and within an hour of the initial change in refractility the dying cell disappears (Sulston 1976; Sulston and Horvitz 1977).

Using *C. elegans* permits the usage of powerful genetic techniques. For example, it is relatively straightforward to disrupt the function of specific genes using RNA interference (RNAi). The nematode can either be soaked in a solution of double stranded RNA or fed genetically transformed bacteria that express the double stranded RNA of interest.

5.1.2 *Drosophila melanogaster*

Drosophila melanogaster belong to the order of flies and are more commonly referred to as fruit flies. *Drosophila* is probably the most commonly studied organism in genetics and developmental biology research. Ithas many advantages as a model organism. It is small and easy to grow in a lab requiring little lab space. Short generation time and high fecundity (>800 eggs laid per day) allows quick generation and replication. Similarly to *C. elegans, the D. melanogaster* genome has been sequenced (Adams et al. 2000) and 50% of fly protein sequences have mammalian orthologues. Genetic engineering techniques have been available since 1987. Furthermore, *Drosophila* has a number of complex organs not present in *C. elegans,* making it useful for the understanding of development of these organs.

In *Drosophila,* the vast majority of apoptotic cells are cleared by hemocytes, which are a type of blood cell; however, some non-professional phagocytes have also been shown to play a role in engulfment (Wolff and Ready 1991; Sonnenfeld and Jacobs 1995). Moreover, in the nervous system of flies, glial cells have been shown to efficiently engulf corpses. The power of the numerous fly strains with deletions and P-element insertions within the genome, and the ability to generate transgenic flies have greatly aided the identification of specific genes involved in engulfment of cell corpses that appear during fly development.

5.2 Genetics of Cell Corpse Engulfment

Genes that are required for engulfment of apoptotic cells in *C. elegans* were first identified by Hedgecock et al. (1983) in a series of Nomarski screens looking for unengulfed and persistent cell corpses. Later, Ellis et al. (1991) designed a visual screen strategy that allowed the isolation of both maternal-effect and zygotic mutants with persistent cell corpses. The identified seven genes initially identified were placed into two distinct genetic pathways based on additivity of the persistent cell corpse phenotype: loss-of-function alleles in genes of either pathway show only a partial defect in engulfment; double mutants with defects in both pathways show a more severe, persistent cell corpse phenotype, while double mutants affecting genes in the same pathway have a similar defect to single mutants (Ellis et al. 1991; Gumienny et al. 2001). However, because even in double mutants cell corpses eventually disappear, there may be a third, yet undiscovered pathway that inefficiently removes dead cells throughout the life of the worm (Kinchen and Hengartner 2005). The analyses performed place *ced-1*, *ced-6* and *ced-7* genes in one group and *ced-2, ced-5, ced-10* and *ced-12* in the other. None of the triple mutants displayed a stronger phenotype than that seen for the strongest double mutants indicating that none of these seven genes acts in a third pathway (Ellis et al. 1991; Zhou et al. 2001a). Genes that have been identified since then are discussed further below.

5.2.1   CED-1, CED-6 and CED-7 Pathway

The first engulfment group contains CED-1, CED-6 and CED-7, which mediate early events in the activation of a signal transduction cascade (Fig. 5.1). CED-1 encodes a single-pass transmembrane protein, which functions as a receptor that recognizes dying cells. CED-1 has been suggested to be homologous to the SREC (Scavenger Receptor from Endothelial Cells) (Zhou et al. 2001b) and the CD91/LRP1 (Su et al. 2002) scavenger receptors, which have been implicated in the engulfment of apop-

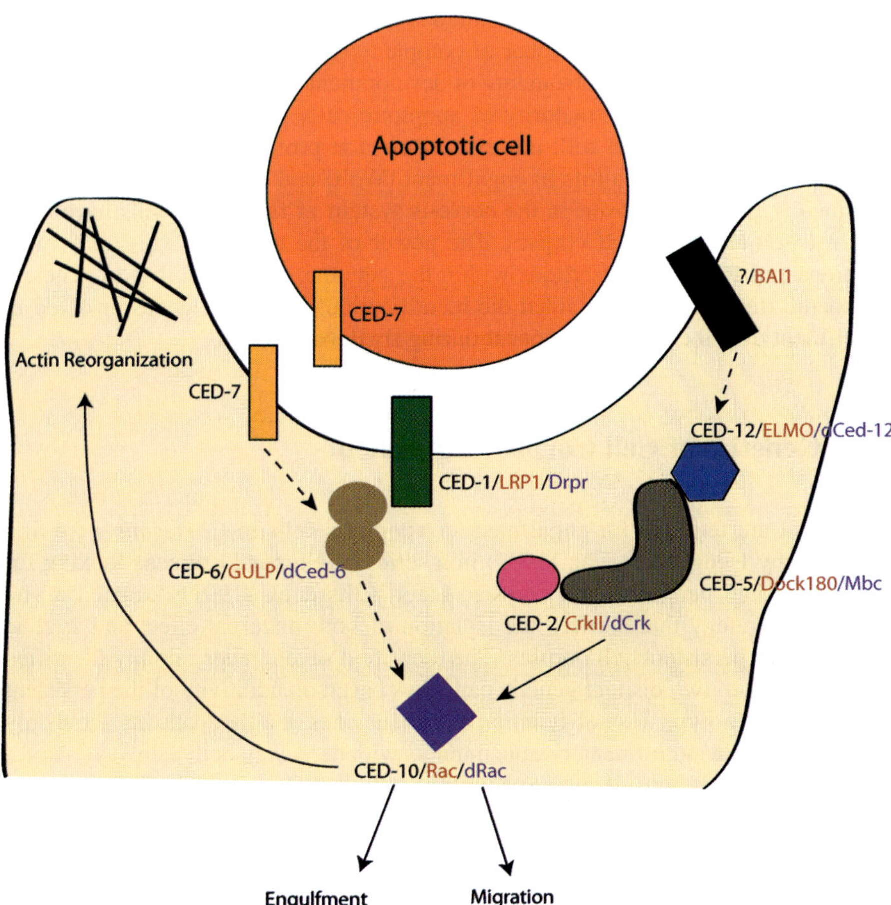

Fig. 5.1 Gene products involved in engulfment in *C. elegans*. The mammalian and Drosophila homologues (if known) are listed in red and purple, respectively. In *C. elegans* seven gene products are grouped into two partially redundant signalling pathways during the removal of apoptotic cells. Loss-of-function mutation in any of these genes results in partial defects in the removal of cell corpses. CED-1/LRP, CED-7 and CED-6/GULP function in one pathway, while CED-2/CrkII, CED-5/Dock-180, CED-12/ELMO function in a different pathway. Signalling from both pathways converges at CED-10/Rac-1. The upstream receptor that activates the second pathway is unknown in *C. elegans*, but BAI-1 is one receptor that has been identified in mammals.

totic cells in mammalian cell culture systems (Ogden et al. 2001). Furthermore, antibody cross-linking of CD91/LRP1 stimulates uptake of bound erythrocytes (Ogden et al. 2001). However, the homology between the intracellular domain of CED-1 and that of SREC (Zhou et al. 2001b; Adachi et al. 1997) and the homology between the extracellular domain of CED-1 and CD91/LRP (Callebaut et al. 2003) is low and the nature of the CED-1 ligand remains elusive. More recently, MEGF10, a protein predicted from human cDNA isolated from the brain was shown to display extensive similarity to CED-1 in both the extracellular and intracellular portions (Nagase et al. 2001). Further studies identified an interaction between MEGF10 and clathrin assembly protein complex 2 medium chain (AP50), a component of clathrin-coated pits supporting a possible role in engulfment (Suzuki and Nakayama 2007). However, MEGF10 expression is primarily restricted to the brain. Moreover, MEGF10 as a transgene fails to rescue the engulfment defect in CED-1-deficient worms, and over-expression of MEGF10 in mammalian cell lines has little effect on engulfment (unpublished observations). Thus, the role of MEGF10 in mammalian engulfment, and whether another CED-1 homologue exists, remains to be determined.

The predicted extracellular domain of CED-1 contains an N-terminal signal peptide and 16 tandem copies of an atypical form of EGF-like repeats, a cysteine-rich motif found in other proteins functioning in adhesive or ligand-receptor interactions (Fig. 5.2). CED-1 also contains a cysteine-rich EMI domain found in many extracellular proteins (Callebaut et al. 2003). The intracellular domain contains NPLY and YASL motifs, which are thought to recruit PTB domain and SH2 domain containing adaptor proteins respectively (Songyang and Cantley 1995).

Evidence indicates that CED-1 recognizes apoptotic cells in vivo. CED-1 is expressed at high levels in cells known to act as engulfing cells in *C. elegans* (Zhou et al. 2001b). It is required in the engulfing cell and not the apoptotic cell for cell corpse clearance. Studies using a CED-1:GFP fusion protein found that CED-1 accumulates at a higher concentration at the phagocytic cup, the region of the engulfing cell that envelops the dying cell (Zhou et al. 2001b). Even in some engulfment mutants such as *ced-6*, the CED-1:GFP fusion protein was observed to accumulate around dying cells, suggesting that CED-1 clustering precedes the initiation of engulfment and is not a consequence of its completion.

In further studies, a truncated CED-1 lacking the intracellular domain could still localize to the plasma membrane of the engulfing cell and cluster around the dying cells, however, its engulfing activity was completely lost (Zhou et al. 2001b). Thus, the intracellular portion of CED-1 is required for downstream signalling necessary for engulfment. Mutational analysis determined that both NPLY and YASL motifs mediate engulfment activity, but are partially redundant (Zhou et al. 2001b).

CED-6 is composed of an N-terminal PTB (phosphotyrosine binding) domain, a leucine zipper and a proline-rich C-terminal region supporting a role as an adapter protein (Fig. 5.2; Liu and Hengartner 1998; Su et al. 2000). Both CED-1 and CD91/LRP have NPXY motifs that interact with CED-6 and its mouse orthologue GULP, suggesting that CED-1 uses CED-6 to transmit a signal into the engulfing cell (Su et al. 2002). CED-6 function is necessary in the engulfing and not the apoptotic cell (Liu and Hengartner 1998). Clustering of the CED-1:GFP fusion protein around

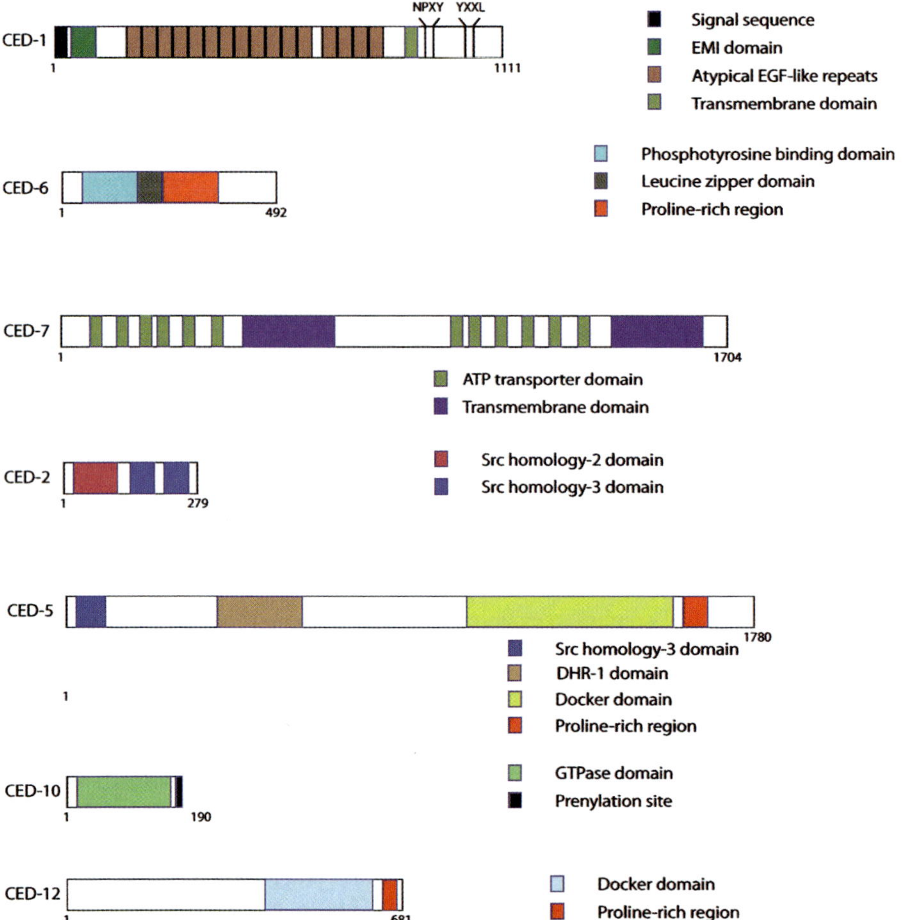

Fig. 5.2 Structures of the seven CED proteins involved in the engulfment of apoptotic cells in *C. elegans* (Zhou et al. 2004).

cell corpses is not affected in *ced-6* mutants, indicating that inactivation of *ced-6* does not block the recognition of 'eat-me' signals, rather it interferes with signalling downstream of CED-1 (Zhou et al. 2001b). Furthermore, over-expression of CED-6 can partially rescue the engulfment defect of *ced-1* mutants (Liu and Hengartner 1998). CED-6 appears to exist basally as a homodimer, with the dimerization being mediated via its leucine zipper motif (Su 2000a). CED-1 and CED-6 are recruited around the apoptotic cell, leading to actin-dependent cytoskeletal reorganization and engulfment of the apoptotic cell (Kinchen et al. 2005).

The importance of *ced-1* and *ced-6* in engulfment was further demonstrated in *Drosophila*. In *Drosophila*, axon pruning is part of the process of neural circuit development and the engulfing action of glial cells is important for this process. Glia specific mutation of *drpr* (*Drosophila ced1* homologue) and knockdown of *ced-6* suppressed engulfment by glial cells resulting in an inhibition of axon pruning

(Awasaki et al. 2006). The authors further showed that *drpr* and *ced-6* interacted genetically in the process of engulfment by glial cells.

CED-7 is also a cell surface protein. However, unlike the other engulfment genes identified in *C. elegans*, *ced-7* function is required in both the engulfing and the dying cell for efficient engulfment (Wu and Horvitz 1998b). In *ced-7* mutant worms, although CED-1 localizes to the cell surface, it does not localize around cell corpses (Zhou et al. 2001b). As mentioned earlier, ABCA1 is the proposed mammalian CED-7 homologue. Although ABCA1 has been proposed to promote PS exposure in certain conditions via an as yet uncharacterized recognition event, these data are controversial and ABCA1 is likely not directly involved in PS flipping. However, ABC transporters actively transport a variety of substances such as sugars, lipids, ions, peptides, proteins and lipoproteins across the plasma membrane (Klein et al. 1999). Which of these possible molecules is important for engulfment as an 'eat-me' signal and is exposed by the actions of CED-7 remains to be determined. Furthermore, what role CED-7 plays in the engulfing cell is unclear. Perhaps it assists CED-1 in recognizing the 'eat-me' signal or functions downstream of CED-1 (Mangahas and Zhou 2005). CED-7 may also expose receptors specific for the 'eat-me' signal on the surface of the engulfing cell (Napirei and Mannherz, Chap. 4; Gregory and Pound, Chap. 9, this Vol.).

The CED group of genes was identified by screening for engulfment defective mutants. This selected genes not crucial for survival, as only viable mutants would be identified. To bypass this limitation, Yu et al. (2006) screened for mutants that both prevented engulfment and arrested during embryonic development. Using this process, they identified *dyn-1*, which is the worm homologue of the mammalian Dynamin. Embryos carrying a *dyn-1* mutation had a major defect in the engulfment of apoptotic cells, however they were still able to concentrate yolk particles in the intestinal cells by endocytosis (Yu et al. 2006). This suggests that *dyn-1* is crucial not for endocytosis, but rather promotes engulfment by other means. In fact, examination of cells by electron microscopy revealed the presence of abnormal vesicles not associated with the plasma membrane in engulfing cells. DYN-1 clustered on the extending phagocytic arms of engulfing cells similarly to CED-1. Co-localization of CED-1 and DYN-1 in the extending arms provided another link between the two proteins. To explain their findings, the authors proposed that DYN-1 protein may actively recruit intracellular vesicles to sites of engulfment. These vesicles, in turn, would promote fusion with the plasma membrane, contribute membrane allowing extension of the arms around the dying cell, while their contents may contribute to the degradation of the dying cell. However, the idea that *dyn-1* promotes vesicle fusion is difficult to reconcile with a large number of studies that have suggested a role for it in vesicle fission (Shaham 2006). A more recent study that also identified DYN-1 as a player involved in corpse clearance in an unbiased screen, showed that *dyn-1* mutants, either in worms or in mammals have no defect in extension of phagocytic arms or internalization of the corpses, rather have a defect that relates to processing of the internalized corpses (Kinchen et al. 2008). Although initially DYN-1 was placed in the CED-1/CED-6 genetic pathway (Yu et al. 2006), in this study, it was observed that DYN-1 functions in both genetic pathways as would be expected if it were involved in processing of internalized corpses.

5.2.2 The CED-2, CED-5, CED-10 and CED-12 Pathway

The second engulfment pathway in the worm includes *ced-2, ced-5, ced-10 and ced-12* (Fig. 5.1). The most upstream member of this family is CED-2. CrkII is its corresponding homologue in mammals. CED-2/CrkII is an adapter molecule composed of one SH2 and two SH3 domains (Fig. 5.2). The N-terminal SH3 domain has been shown to bind CED-5/Dock180 (Matsuda et al. 1996).

CED-5 is a member of the CDM (*ced-5, dock-180,* and **myoblast city**) protein family, which also includes human Dock180 and Dock4 and *Drosophila* Myoblast City (Wu and Horvitz 1998a; Reddien and Horvitz 2000; Yajnik et al. 2003). CED-5 appears to serve as a scaffold on which other proteins assemble and together are recruited to the cell surface to induce 'actin reorganization' around the apoptotic cell (Kinchen and Hengartner 2005). Dock180 interacts with CED-10 (Rac1) via its DOCKER/DHR2 domain (see below) and with CED-12 via an SH3 domain at the N-terminus (Fig. 5.2; Gumienny et al. 2001; Zhou et al. 2001a; Wu et al. 2001).

CED-12/Elmo is the third member of this pathway and again was found by characterization of *C. elegans* mutants with defects in corpse clearance (Gumienny et al. 2001; Zhou et al. 2001a; Wu et al. 2001). CED-12 interacts with CED-5 via a PxxP motif (and a second less defined region adjacent to the PxxP motif). Together with CED-2, CED-5 and CED-10, CED-12 can co-localize to the cell membrane upon induction of membrane ruffling. In addition to playing a role in engulfment, CED-2 / CED-5/CED-12/CED-10 proteins also function during cell migration in *C. elegans* and mammals (Gumienny et al. 2001; Zhou et al. 2001a; Wu et al. 2001; Grimsley et al. 2004). Both *C. elegans* and *Drosophila* have just one CED-12 protein, however humans have three orthologues of ELMO and eleven Dock family members (Cote and Vuori 2002; Meller et al. 2005). Currently, it is unclear if the different Elmo and Dock proteins serve overlapping or distinct functions to those in the nematode.

Initial studies suggested that CED-2 regulated the function of CED-5/CED-12 by their recruitment to the membrane (Gumienny et al. 2001). Recent studies in mammalian systems have shown that both ELMO (CED-12) and Dock180 (CED-5) may use independent membrane-targeting signals and thus may not always require another protein to come to the membrane. However, Dock180 appears to require ELMO to target efficiently to the membrane. CrkII is found as part of an endogenous complex with ELMO and Dock180 in cells. Furthermore, although all three proteins are required for efficient engulfment, an interaction between CrkII and Dock180 is not required for efficient engulfment of apoptotic targets (Tosello-Trampont et al. 2007). Perhaps CrkII brings together Dock180 and a yet unknown protein via its two SH3 domains (Kinchen and Ravichandran 2007).

CED-5/Dock180 functions like other GEFs in that it stabilizes the nucleotide-free transition state as CED-10/Rac cycles from the GDP-bound to the GTP-bound state (Lu et al. 2005). However, unlike conventional GEFs, Dock180 does not contain the tandem Dbl-homology and pleckstrin homology (DH-PH) domains common to most GEFs. Instead, this structure appears to be divided between two proteins. Dock180 contains a Docker domain, which is functionally equivalent, but structurally distinct, to the DH motif (Brugnera et al. 2002; Cote and Vuori 2002). Interestingly, while there is no tandem PH domain immediately following the Docker domain, the PH domain of ELMO1 appears to function in 'trans' to stabilize the Dock180:Rac complex. Normally, Dock180 exists in a closed conformation in which an intramolecular interaction between the N-terminal SH3 domain and the catalytic Docker domain sterically blocks Rac binding (Lu et al. 2005). ELMO binds to the Dock180 SH3 domain relieving its steric inhibition, although the stabilization of nucleotide free Rac happens independently of its binding to the Dock180 SH3 motif (Lu et al. 2004).

A number of findings suggest a variety of means in which the ELMO-Dock180 complex is regulated. Phosphotidyl inositol (3,4,5) triphosphate is enriched at the leading edge during migration and phagocytosis (Cote et al. 2005). Dock180 DHR-1 domain appears capable of binding PIP3 and may play a role in Dock180 recruitment. Interestingly, a large fraction of endogenous Dock180-containing complexes has been found in the nucleus bound to ELMO and this complex has functional Rac-GEF activity (Yin et al. 2004). At this time, however, its role in the nucleus is unclear. Dock180-ELMO may also be regulated via destabilization of the complex. Dock180 is ubiquitinated at the plasma membrane, however, as long as it is in complex with ELMO it appears resistant to degradation (Makino et al. 2006). Over-expression of CrkII increases the ubiquitination of Dock180 resulting in recruitment of an E3 ubiquitin ligase. A recently published work, using an unbiased approach to screen for mutants defective in phagocytosis of apoptotic cells by macrophages in the *Drosophila* system, adds another twist to this notion. Nathalie Franc's group identified Pallbearer, a gene encoding an F box protein, which provides specificity to Skp-Cullin-F-box complexes that serve as E3 ubiquitin ligases promoting ubiquitination of phosphorylated proteins and their degradation by proteasomes (Silva et al. 2007). They suggested a role for ubiquitination and proteasomal degradation for in vivo phagocytosis of apoptotic corpses. However, whether these are the proteins involved in ubiquitination of Dock180 and if a similar mechanism exists in mammals remains to be seen.

BAI1 is a receptor specific for Phosphatydilserine (PS) that has recently been discovered and signals through the CrkII/Dock180/Elmo signalling pathway. BAI1 forms a trimeric complex with Elmo and Dock180 and cooperates with Elmo/Dock180/Rac to promote maximal engulfment of apoptotic cells (Park et al. 2007). However, no clear homologue of BAI1 exists in *C. elegans* suggesting that perhaps another receptor may signal through this complex.

5.2.3   Rho Family GTPases

ced-10 encodes a *C. elegans* homologue of the human Rac1 small GTPase. *Ced-10* mutants suffer from not only engulfment, but also migration defects, including migration of the distal tip cells and axon pathfinding defects (Reddien and Horvitz 2000; Lundquist et al. 2001; Kishore and Sundaram 2002; Soto et al. 2002; Wu et al. 2002; Gitai et al. 2003). Additionally, *ced-10* null mutants display maternal effect lethality, suggesting developmental importance (Lundquist et al. 2001; Soto et al. 2002). Like other GTPases, *ced-10* cycles between a GDP-bound inactive and a GTP-bound active state. GAPs (GTPase activating proteins) promote GTP hydrolysis, while GEFs (guanine nucleotide exchange factors) promote exchange of GDP with GTP. Both engulfment pathways activate CED-10/Rac1 downstream of the engulfment signal (Kinchen et al. 2005).

MIG-2, the nematode orthologue of RhoG plays a key role in cell migration, however, mutation of MIG-2 alone displays only a minor role in corpse removal (Lundquist et al. 2001; deBakker et al. 2004). However, worms carrying mutations of both MIG-2 and a putative partially functional CED-12 have a greater defect in corpse clearance suggesting a genetic/functional link between MIG-2 and CED-12. MIG-2 preferentially interacts with ELMO in its GTP-bound form (Katoh and Negishi 2003). In mammals, however, over-expression of RhoG promotes uptake of apoptotic targets in an ELMO-Dock180-dependent manner (deBakker et al. 2004). RhoG could act as one type of membrane recruiting signal for ELMO, since a version of ELMO with a mutated RhoG binding site was no longer targeted to the membrane ruffles (deBakker et al. 2004).

Work using *ced-10(null)* mutants has shown that a Rac independent pathway may also exist in the nematode, even though this pathway may not be efficient (Kinchen et al. 2005). Previously, experiments using *ced-1; ced-5* double mutants hinted at the existence of this pathway since apoptotic cell corpses are still eventually engulfed. Recent data show that the two engulfment pathways are not required for the removal of the linker cell, which undergoes a caspase-independent death in *C. elegans* (Abraham et al. 2007). However, this third pathway is as of yet completely uncharacterized.

5.3 Phosphatidylserine Recognition and Lessons from Model Organisms

5.3.1 *Phosphatidylserine*

Electron microscopic studies reveal that pseudopods extend from engulfing cells to surround apoptotic cells at a very early stage of cell death, suggesting that dying cells present on their surface some type of signal (now referred to as 'eat-me' signal) that distinguishes them from live cells. A number of possible candidates for this signal have been proposed (Napirei and Mannherz, Chap. 4; Gregory and Pound, Chap. 9, this Vol.).

Cells undergoing apoptosis undergo a variety of cell surface changes leading to a redistribution of phospholipids on the membrane surface (Fadok et al. 1998). One of these signals, Phosphatidylserine (PS) is normally only present on the inner leaflet of the plasma membrane. This status is ensured by an ATP-dependent aminophospholipid translocase activity that removes/translocates any PS on the outer side of the membrane to the inner leaflet (Schlegel et al. 1996). During apoptosis, translocase activity is inhibited and PS is actively transported to the outer leaflet by a transmembrane scramblase (Schlegel et al. 1996; Fadok et al. 2001). PS exposure is a conserved feature of apoptosis in mammals, *Drosophila*, *Xenopus*, and *C. elegans* (Fadok et al. 1992; van den Eijnde et al. 1998). Recent studies have shown that blocking PS exposure in *C. elegans* leads to a reduction in engulfment suggesting that PS plays a role as an 'eat-me' signal (Venegas and Zhou 2007; Zullig et al. 2007). Venegas and Zhou (2007) also identified a phospholipid scramblase PLSC-1 as the scramblase important for exposure of PS on the cell surface of apoptotic germ cells. Interestingly, however, PLSC-1 was not necessary for the removal of apoptotic somatic cells. Somatic cells appear to require CED-7, an ATP-binding cassette (ABC) transporter for PS exposure. This suggests that perhaps different tissues and/or cells at different stages in development differentially regulate PS exposure in the worm. Zullig et al. (2007) found that blocking TAT-1, a flippase/translocase results in disrupted PS exposure on apoptotic cells. This is surprising as flippases are normally responsible for keeping PS on the inner leaflet and hence one would expect the opposite result. How TAT-1 regulates PS is still unclear. Recent work by Ding Xue and colleagues shows an opposite phenotype for TAT-1 and suggests a role for TAT-1 in keeping PS on the inner leaflet (Darland-Ransom et al. 2008).

Further support for PS acting as an 'eat me signal' comes from discoveries of PS recognition receptors on the engulfing cells. Phosphatidylserine receptor (PSR) was the first such molecule (Fadok et al. 2000). PSR was also reported to play a key role in engulfment in mice (Li et al. 2003); however, mutant worms defective in *psr-1*, the *C. elegans* orthologue of PSR, show only weak engulfment defects (Wang et al. 2003) and the original authors of this work have now further limited psr-1 function to only very specific instances during development (Darland-Ransom et al. 2008). Further complicating the interpretation of these studies is the fact that mammalian PSR appears to be predominantly expressed in the nucleus and cannot be visualized on the plasma membrane via either immunostaining or over-expression of GFP tagged constructs (Cui et al. 2004). Perhaps PSR has an indirect role in enhancing engulfment by, for example acting as a transcriptional regulator of another PS receptor (Kinchen and Hengartner 2005). In *Drosophila*, the PSR homologue (dPSR) plays no obvious role in apoptotic cell engulfment, however, it protects cells from apoptosis (Krieser et al. 2007).

Recently, three new possible receptors for PS have been identified in mammalian systems. Brain Angiogenesis Inhibitor-1 (BAI1) is a seven-transmembrane receptor expressed by phagocytes that can directly recognize PS on apoptotic cells and activates downstream engulfment proteins such as Rac (Park et al. 2007). T-cell immunoglobulin domain and mucin domain 4 (TIM-4) and TIM-1 were also identified as PS receptors. They bind to PS and over-expression can enhance engulfment (Miyanishi et al. 2007). However, the signalling pathway activated downstream of the TIMs during engulfment is unknown. Stabilin-2, the third PS receptor identified, recognizes PS on apoptotic cells, enhances phagocytosis and also promotes the release of the anti-inflammatory cytokine TGF-β, which is associated with engulfment of dying cells (Park et al. 2008). Stabilin-2 appears to signal through CED-6/GULP in uptake of aged red blood cells, although this needs to be better defined in the context of other targets and non-over-expressed conditions. There is no obvious homologue for TIM-4 or Stabilin 2 in flies and worms. Although there are a several proteins with TSR-repeats (an essential component of BAI1 in its recognition of PS) in worms and flies, the worm homologues tested to date do not appear to have a defect in engulfment (unpublished observations). Thus, no obvious PS recognition receptors have been identified in either *C. elegans* or *Drosophila* to date and this would be an interesting future challenge.

5.3.2 Annexin I

Annexin V has been utilized for its ability to bind to PS and mark apoptotic cells for a number of years in mammals and more recently in worms. Annexin I has been implicated in promoting efficient engulfment (Arur et al. 2003). Upon induction of apoptosis, Annexin I is cleaved in a caspase-dependent manner to generate a protein that translocates to the outside of the cell and appears to bind PS. Such exposed Annexin I has been linked to efficient engulfment by neighbouring cells. RNA interference (RNAi)-mediated knockdown of the *C. elegans* homologue of annexin I, *nex-1*, results in clearance defects (Arur et al. 2003). However, deletion of the *nex-1* locus did not show a defect in engulfment suggesting that RNAi-mediated knockdown of *nex-1* may have affected other genes besides the intended target. While suggestive of a possible role in engulfment, how Annexin I may contribute to engulfment remains to be determined.

5.3.3 Phagocytosis and Apoptosis—an Inseparable Link?

Both in model organisms and in mammals, very few apoptotic cells are visualized except under conditions when the engulfment pathways are disturbed. The common thinking has been that the engulfment process is rather quick and efficient and that this makes it difficult to see corpses that appear at normal rates. However, even under conditions where a far greater number of cells are induced to undergo apoptosis, uncleared apoptotic cells are rarely seen. Two seminal papers studying apoptosis and corpse clearance in *C. elegans* provided some clues toward this mystery (Hoeppner et al. 2001; Reddien et al. 2001). Although engulfment was thought to function after completion of the apoptotic process, these two papers suggested that the corpses are engulfed while they are still in certain stages of apoptosis, and that the last steps of apoptosis are concluded (or finished off) within the engulfing cell, after the phagocyte has ingested the dying cell. These elegant studies used a combination of worms with a hypomorphic form of the executioner caspase ced-3, and superimposing mutations of engulfment genes (Hoeppner et al. 2001; Reddien et al. 2001). Through the use of elegant 4-D microscopy it was also demonstrated that cells that normally die during *C. elegans* development transiently displayed morphological features of apoptosis, but then recovered (Hoeppner et al. 2001; Reddien et al. 2001). Further careful analysis confirmed that phagocytosis is required to fully complete the developmental apoptotic programme. While similar genetic studies have not been done in mammals, recent studies suggest that recruitment factors or find-me signals secreted by mammalian apoptotic cells during early stages of apoptosis may help recruit monocytes to the site of the dying cell (presumably from another part of the tissue or from circulation) and thereby promote the prompt clearance of the dying cells. This is an important area of investigation that clearly needs to be further explored to fully understand the interplay between the dying cell and the engulfing cell, and how the completion of apoptosis as we generally consider, may be achieved.

5.4 Conclusions

The genetic studies in model organisms and defining the evolutionarily conserved players involved in engulfment have certainly provided a wealth of information and have also suggested further avenues of investigation. However, there are a number of areas where clear homologues of worm or fly genes are not seen in mammals and vice versa. Furthermore, although a number of potential candidate receptors have been identified in the different systems, the effect of loss-of-function mutations on engulfment is often mild (except perhaps CED-1 in worms) and their relative importance for phagocytosis remains to be carefully worked out. Future work, including that in *C. elegans* and *Drosophila,* will hopefully shed more light on this process of apoptotic cell clearance that is important for the normal functioning of all multicellular organisms.

References

Abraham MC, Lu Y, Shaham S (2007) A morphologically conserved nonapoptotic program promotes linker cell death in Caenorhabditis elegans. Dev Cell 12:73–86

Adachi H, Tsujimoto M, Arai H et al (1997) Expression cloning of a novel scavenger receptor from human endothelial cells. J Biol Chem 272:31217–31220

Adams MD, Celniker SE, Holt RA et al (2000) The genome sequence of Drosophila melanogaster. Science 287:2185–2195

Arur S, Uche UE, Rezaul K et al (2003) Annexin I is an endogenous ligand that mediates apoptotic cell engulfment. Dev Cell 4:587–598

Awasaki T, Tatsumi R, Takahashi K et al (2006) Essential role of the apoptotic cell engulfment genes draper and ced-6 in programmed axon pruning during Drosophila metamorphosis. Neuron 50:855–867

Brugnera E, Haney L, Grimsley C et al (2002) Unconventional Rac-GEF activity is mediated through the Dock180-ELMO complex. Nat Cell Biol 4:574–582

Callebaut I, Mignotte V, Souchet M et al (2003) EMI domains are widespread and reveal the probable orthologs of the Caenorhabditis elegans CED-1 protein. Biochem Biophys Res Commun 300:619–623

Cote JF, Motoyama AB, Bush JA et al (2005) A novel and evolutionarily conserved PtdIns(3,4,5)P3-binding domain is necessary for DOCK180 signalling. Nat Cell Biol 7:797–807

Cote JF, Vuori K (2002) Identification of an evolutionarily conserved superfamily of DOCK180-related proteins with guanine nucleotide exchange activity. J Cell Sci 115:4901–4913

Cui P, Qin B, Liu N et al (2004) Nuclear localization of the phosphatidylserine receptor protein via multiple nuclear localization signals. Exp Cell Res 293:154–163

Darland-Ransom M, Wang X, Sun CL et al (2008) Role of C. elegans TAT-1 protein in maintaining plasma membrane phosphatidylserine asymmetry. Science 320:528–531

deBakker CD, Haney LB, Kinchen JM et al (2004) Phagocytosis of apoptotic cells is regulated by a UNC-73/TRIO-MIG-2/RhoG signalling module and armadillo repeats of CED-12/ELMO. Curr Biol 14:2208–2216

Ellis RE, Jacobson DM, Horvitz HR (1991) Genes required for the engulfment of cell corpses during programmed cell death in Caenorhabditis elegans. Genetics 129:79–94

Fadok VA, Bratton DL, Frasch SC et al (1998) The role of phosphatidylserine in recognition of apoptotic cells by phagocytes. Cell Death Differ 5:551–562

Fadok VA, Bratton DL, Rose DM et al (2000) A receptor for phosphatidylserine-specific clearance of apoptotic cells. Nature 405:85–90

Fadok VA, de Cathelineau A, Daleke DL et al (2001) Loss of phospholipid asymmetry and surface exposure of phosphatidylserine is required for phagocytosis of apoptotic cells by macrophages and fibroblasts. J Biol Chem 276:1071–1077

Fadok VA, Voelker DR, Campbell PA et al (1992) Exposure of phosphatidylserine on the surface of apoptotic lymphocytes triggers specific recognition and removal by macrophages. J Immunol 148:2207–2216

Genome sequence of the nematode C. elegans: a platform for investigating biology. (1998) Science 282:2012–2018

Gitai Z, Yu TW, Lundquist EA et al (2003) The netrin receptor UNC-40/DCC stimulates axon attraction and outgrowth through enabled and, in parallel, Rac and UNC-115/AbLIM. Neuron 37:53–65

Grimsley CM, Kinchen JM, Tosello-Trampont AC et al (2004) Dock180 and ELMO1 proteins cooperate to promote evolutionarily conserved Rac-dependent cell migration. J Biol Chem 279:6087–6097

Gumienny TL, Brugnera E, Tosello-Trampont AC et al (2001) CED-12/ELMO, a novel member of the CrkII/Dock180/Rac pathway, is required for phagocytosis and cell migration. Cell 107:27–41

Hedgecock EM, Sulston JE, Thomson JN (1983) Mutations affecting programmed cell deaths in the nematode Caenorhabditis elegans. Science 220:1277–1279

Hoeppner DJ, Hengartner MO, Schnabel R (2001) Engulfment genes cooperate with ced-3 to promote cell death in Caenorhabditis elegans. Nature 412:202–206

Katoh H, Negishi M (2003) RhoG activates Rac1 by direct interaction with the Dock180-binding protein Elmo. Nature 424:461–464

Kimble J, Hirsh D (1979) The postembryonic cell lineages of the hermaphrodite and male gonads in Caenorhabditis elegans. Dev Biol 70:396–417

Kinchen JM, Cabello J, Klingele D et al (2005) Two pathways converge at CED-10 to mediate actin rearrangement and corpse removal in C. elegans. Nature 434:93–99

Kinchen JM, Doukoumetzidis K, Almendinger J et al (2008) A pathway for phagosome maturation during engulfment of apoptotic cells. Nat Cell Biol 10:556–566

Kinchen JM, Hengartner MO (2005) Tales of cannibalism, suicide, and murder: Programmed cell death in C. elegans. Curr Top Dev Biol 65:1–45

Kinchen JM, Ravichandran KS (2007) Journey to the grave: signalling events regulating removal of apoptotic cells. J Cell Sci 120:2143–2149

Kishore RS, Sundaram MV (2002) ced-10 Rac and mig-2 function redundantly and act with unc-73 trio to control the orientation of vulval cell divisions and migrations in Caenorhabditis elegans. Dev Biol 241:339–348

Klein I, Sarkadi B, Varadi A (1999) An inventory of the human ABC proteins. Biochim Biophys Acta 1461:237–262

Krieser RJ, Moore FE, Dresnek D et al (2007) The Drosophila homolog of the putative phosphatidylserine receptor functions to inhibit apoptosis. Development 134:2407–2414

Li MO, Sarkisian MR, Mehal WZ et al (2003) Phosphatidylserine receptor is required for clearance of apoptotic cells. Science 302:1560–1563

Liu QA, Hengartner MO (1998) Candidate adaptor protein CED-6 promotes the engulfment of apoptotic cells in C. elegans. Cell 93:961–972

Lu M, Kinchen JM, Rossman KL et al (2004) PH domain of ELMO functions in trans to regulate Rac activation via Dock180. Nat Struct Mol Biol 11:756–762

Lu M, Kinchen JM, Rossman KL et al (2005) A Steric-inhibition model for regulation of nucleotide exchange via the Dock180 family of GEFs. Curr Biol 15:371–377

Lundquist EA, Reddien PW, Hartwieg E et al (2001) Three C. elegans Rac proteins and several alternative Rac regulators control axon guidance, cell migration and apoptotic cell phagocytosis. Development 128:4475–4488

Makino Y, Tsuda M, Ichihara S et al (2006) Elmo1 inhibits ubiquitylation of Dock180. J Cell Sci 119:923–932

Mangahas PM, Zhou Z (2005) Clearance of apoptotic cells in Caenorhabditis elegans. Semin Cell Dev Biol 16:295–306

Matsuda M, Ota S, Tanimura R et al (1996) Interaction between the amino-terminal SH3 domain of CRK and its natural target proteins. J Biol Chem 271:14468–14472

Meller N, Merlot S, Guda C (2005) CZH proteins: a new family of Rho-GEFs. J Cell Sci 118:4937–4946

Miyanishi M, Tada K, Koike M et al (2007) Identification of Tim4 as a phosphatidylserine receptor. Nature 450:435–439

Nagase T, Nakayama M, Nakajima D et al (2001) Prediction of the coding sequences of unidentified human genes. XX. The complete sequences of 100 new cDNA clones from brain which code for large proteins in vitro. DNA Res 8:85–95

Ogden CA, deCathelineau A, Hoffmann PR et al (2001) C1q and mannose binding lectin engagement of cell surface calreticulin and CD91 initiates macropinocytosis and uptake of apoptotic cells. J Exp Med 194:781–795

Park D, Tosello-Trampont AC, Elliott MR et al (2007) BAI1 is an engulfment receptor for apoptotic cells upstream of the ELMO/Dock180/Rac module. Nature 450:430–434

Park SY, Jung MY, Kim HJ et al (2008) Rapid cell corpse clearance by stabilin-2, a membrane phosphatidylserine receptor. Cell Death Differ 15:192–201

Reddien PW, Cameron S, Horvitz HR (2001) Phagocytosis promotes programmed cell death in C. elegans. Nature 412:198–202

Reddien PW, Horvitz HR (2000) CED-2/CrkII and CED-10/Rac control phagocytosis and cell migration in Caenorhabditis elegans. Nat Cell Biol 2:131–136

Schlegel RA, Callahan M, Krahling S et al (1996) Mechanisms for recognition and phagocytosis of apoptotic lymphocytes by macrophages. Adv Exp Med Biol 406:21–28

Shaham S (2006) The Dynami(n)cs of cell corpse engulfment. Dev Cell 10:690–691

Silva E, Au-Yeung HW, Van Goethem E et al (2007) Requirement for a Drosophila E3-ubiquitin ligase in phagocytosis of apoptotic cells. Immunity 27:585–596

Songyang Z, Cantley LC (1995) Recognition and specificity in protein tyrosine kinase-mediated signalling. Trends Biochem Sci 20:470–475

Sonnenfeld MJ, Jacobs JR (1995) Macrophages and glia participate in the removal of apoptotic neurons from the Drosophila embryonic nervous system. J Comp Neurol 359:644–652

Soto MC, Qadota H, Kasuya K et al (2002) The GEX-2 and GEX-3 proteins are required for tissue morphogenesis and cell migrations in C. elegans. Genes Dev 16:620–632

Su HP, Brugnera E, Van Criekinge W et al (2000) Identification and characterization of a dimerization domain in CED-6, an adapter protein involved in engulfment of apoptotic cells. J Biol Chem 275:9542–9549

Su HP, Nakada-Tsukui K, Tosello-Trampont AC et al (2002) Interaction of CED-6/GULP, an adapter protein involved in engulfment of apoptotic cells with CED-1 and CD91/low density lipoprotein receptor-related protein (LRP). J Biol Chem 277:11772–11779

Sulston JE (1976) Post-embryonic development in the ventral cord of Caenorhabditis elegans. Philos Trans R Soc Lond B Biol Sci 275:287–297

Sulston JE, Horvitz HR (1977) Post-embryonic cell lineages of the nematode, Caenorhabditis elegans. Dev Biol 56:110–156

Sulston JE, Schierenberg E, White JG et al (1983) The embryonic cell lineage of the nematode Caenorhabditis elegans. Dev Biol 100:64–119

Suzuki E, Nakayama M (2007) MEGF10 is a mammalian ortholog of CED-1 that interacts with clathrin assembly protein complex 2 medium chain and induces large vacuole formation. Exp Cell Res 313:3729–3742

Tosello-Trampont AC, Kinchen JM, Brugnera E et al (2007) Identification of two signalling sub-modules within the CrkII/ELMO/Dock180 pathway regulating engulfment of apoptotic cells. Cell Death Differ 14:963–972

Van Den Eijnde SM, Boshart L, Baehrecke EH et al (1998) Cell surface exposure of phosphatidyl-serine during apoptosis is phylogenetically conserved. Apoptosis 3:9–16

Venegas V, Zhou Z (2007) Two alternative mechanisms that regulate the presentation of apoptotic cell engulfment signal in Caenorhabditis elegans. Mol Biol Cell 18:3180–3192

Wang X, Wu YC, Fadok VA et al (2003) Cell corpse engulfment mediated by C. elegans phosphati-dylserine receptor through CED-5 and CED-12. Science 302:1563–1566

Wolff T, Ready DF (1991) Cell death in normal and rough eye mutants of Drosophila. Develop-ment 113:825–839

Wu YC, Cheng TW, Lee MC et al (2002) Distinct rac activation pathways control Caenorhabditis elegans cell migration and axon outgrowth. Dev Biol 250:145–155

Wu YC, Horvitz HR (1998a) C. elegans phagocytosis and cell-migration protein CED-5 is similar to human DOCK180. Nature 392:501–504

Wu YC, Horvitz HR (1998b) The C. elegans cell corpse engulfment gene ced-7 encodes a protein similar to ABC transporters. Cell 93:951–960

Wu YC, Tsai MC, Cheng LC et al (2001) C. elegans CED-12 acts in the conserved crkII/DOCK180/Rac pathway to control cell migration and cell corpse engulfment. Dev Cell 1:491–502

Yajnik V, Paulding C, Sordella R et al (2003) DOCK4, a GTPase activator, is disrupted during tumourigenesis. Cell 112:673–684

Yin J, Haney L, Walk S et al (2004) Nuclear localization of the DOCK180/ELMO complex. Arch Biochem Biophys 429:23–29

Yu X, Odera S, Chuang CH et al (2006) C. elegans Dynamin mediates the signalling of phagocytic receptor CED-1 for the engulfment and degradation of apoptotic cells. Dev Cell 10:743–757

Zhou Z, Caron E, Hartwieg E et al (2001a) The C. elegans PH domain protein CED-12 regulates cytoskeletal reorganization via a Rho/Rac GTPase signalling pathway. Dev Cell 1:477–489

Zhou Z, Hartwieg E, Horvitz HR (2001b) CED-1 is a transmembrane receptor that mediates cell corpse engulfment in C. elegans. Cell 104:43–56

Zhou Z, Mangahas PM, Yu X (2004) The genetics of hiding the corpse: engulfment and degrada-tion of apoptotic cells in C. elegans and D. melanogaster. Curr Top Dev Biol 63:91–143

Zullig S, Neukomm LJ, Jovanovic M et al (2007) Aminophospholipid translocase TAT-1 promotes phosphatidylserine exposure during C. elegans apoptosis. Curr Biol 17:994–999

Chapter 6
Innate Apoptotic Immunity: A Potent Immunosuppressive Response Repertoire Elicited by Specific Apoptotic Cell Recognition

David S. Ucker

Abstract: Phagocytosis, of course, is essential for the clearance of dying cells, and for the degradation of dying cell constituents, including the processing of potential cellular [auto]antigens. Independent of engulfment, the specific recognition of dying cells by phagocytes and other cells elicits a profound repertoire of outcomes. In particular, apoptotic cells are potently suppressive of inflammation and other immunological responses. Immunosuppressive apoptotic effects are elicited in macrophages and dendritic cells, and they are triggered in non-professional phagocytes as well. Among these responses, the immediate-early suppression of specific gene transcription is most evident. The array of genes suppressed upon apoptotic recognition includes pro-inflammatory cytokines and angiogenic factors. The ability of apoptotic cells to elicit these responses depends upon their cell surface expression of specific determinants for recognition. A variety of molecules has been implicated in this process, although the identification of definitive recognition determinants remains incomplete. The suppressive effects exerted by apoptotic cells do not arise as a simple antagonism of stimulatory signals. Selective responsiveness to apoptotic cell recognition occurs independent of known immune receptors and signalling pathways. Remarkably, specific apoptotic recognition, in contrast to all other cases of immune discrimination, exhibits no self-bias. These observations suggest that innate apoptotic immunity represents an unconventional and ubiquitous immune responsiveness—in essence, a second dimension of immunity distinct from the classical self/other axis. The subversion by pathogens (including viruses) of this suppressive responsiveness underscores the physiological significance of innate apoptotic immunity.

D. S. Ucker (✉)
Department of Microbiology and Immunology
University of Illinois College of Medicine
Rm. E803 (MC 790)
835 South Wolcott
Chicago, IL 60612
Tel.: (312) 413 1102
Fax: (312) 996 6415
E-mail: duck@uic.edu

D. V. Krysko, P. Vandenabeele (eds.), *Phagocytosis of Dying Cells*,
DOI: 10.1007/978-1-4020-9292-3_6, © Springer Science+Business Media B.V. 2009

Key words: Apoptosis • Cell death • Inflammation • Innate immunity • Phagocytosis • Signal transduction • Transcriptional repression • Pathogenesis

6.1   Apoptotic Cells Are Affirmatively Anti-inflammatory

The phenomenon of physiological cell death was first observed in the 1880's, in studies of normal metazoan development and metamorphosis (Krysko et al., Chap. 1; Diez-Fraile et al., Chap. 2, this Vol.; reviewed by Clarke and Clarke 1996). Already then, it was recognized that the process of cell death was intimately tied to the elimination of those corpses by phagocytic cells. The physiological significance of what we now know as apoptotic cell death seemed to rest obviously on the simple physical elimination of inappropriate cells.

At about the same time, Metchnikoff (1891) described the phagocytic process, which was accomplished primarily by microphages (neutrophils) and macrophages. Pathogens were the typical targets of phagocytosis; not surprisingly, that process was associated generally with inflammation. In contrast, a most striking aspect of apoptotic cell death is targeted elimination of the corpse *without* inflammation or pathology (Kerr et al. 1972).

Studies with neutrophils brought this seeming conundrum into sharp focus. Neutrophils are inflammatory effectors. When activated, they secrete inflammatory cytokines and chemokines, and release a variety of effector molecules, including matrix metalloproteinases and other proteases, complexes that generate reactive oxygen and nitrogen intermediates, and other anti-microbial proteins (Witko-Sarsat et al. 2000). They also are short-lived cells, and apoptotic neutrophils, like other apoptotic cells, are cleared without inflammation. Indeed, it is estimated that in healthy humans, as many as 10^{11} circulating neutrophils die each day and are cleared without obvious inflammatory or other adverse effect. Morphological studies indicated that the granular contents of neutrophils are not released when they die (Savill et al. 1989), leading to the view that the prevention of cellular leakage is an important aspect of apoptotic cell death and a key to the avoidance of inflammation. Moreover, apoptotic neutrophils do not trigger the release of inflammatory mediators, including chemokines and cytokines such as interleukin (IL)-1β, IL-8, monocyte chemotactic protein 1 (MCP-1), tumour necrosis factor-α (TNFα), and granulocyte/macrophage colony stimulating factor (GM-CSF), by engulfing macrophages (Hughes et al. 1997; Lacy-Hulbert, Chap. 7, this Vol.; Fadok et al. 1998; Meagher et al. 1992).

These observations did not reveal whether the absence of inflammation in apoptotic cell phagocytosis simply reflects the passive avoidance of inflammatory stimulation ("sterile" clearance), or rather is a consequence of an active suppression of inflammation. The answer to this question came with the demonstration that apoptotic neutrophils, as well as other apoptotic cell types, affirmatively exert an anti-inflammatory effect (Voll et al. 1997; Fadok et al. 1998; Cocco and Ucker 2001).

Innate immune receptors that recognize so-called "pathogen-associated molecular patterns" (PAMPs), including the Toll-like receptor (TLR), nucleotide-binding oligomerization domain (NOD)-like receptor, and RIG-I families, initiate inflammatory responses through well-characterized signal transduction cascades (Creagh and O'Neill 2006; Werts et al. 2006). For example, stimulation of macrophages with bacterial lipopolysaccharide (LPS, endotoxin) via the Toll-like receptor 4 (TLR4) signalling complex (Hoshino et al. 1999) triggers the robust secretion of inflammatory cytokines. These secreted cytokines, especially IL-1β and TNFα, act in a paracrine fashion via specific cytokine receptor-mediated signalling to further propagate inflammatory responsiveness (Creagh and O'Neill 2006; Werts et al. 2006). Macrophage interactions with apoptotic cells potently attenuate the response to LPS (Cocco and Ucker 2001; Fadok et al. 1998; Voll et al. 1997) and to other inflammatory agonists (Cvetanovic et al. 2006; Cvetanovic and Ucker 2004).

The inhibitory effect of apoptotic cells is not mediated by soluble factors released from the dying cells. Rather, during the cell death process, apoptotic cells acquire a cell-associated anti-inflammatory signalling activity that overrides pro-inflammatory macrophage responses (Cocco and Ucker 2001). This dominant-acting apoptotic activity appears to be acquired by virtually all cells undergoing apoptotic cell death, regardless of the cell type or the particular death stimulus (Cocco and Ucker 2001; Patel et al. 2006). In a simple model, the anti-inflammatory activity can be viewed, within a continuum of inflammatory signals, as opposing PAMP-triggered and other phlogistic stimuli (Fig. 6.1). In this sense, the apoptotic activity is antagonistic to immunostimulatory (including "danger"; see Matzinger 1994) triggers: in essence, apoptotic cells provide an endogenous "calm" antidote.

The acquired anti-inflammatory activity persists stably in apoptotic cells, even as the cells lose membrane integrity; "early" and "late" apoptotic cells (with intact or compromised plasma membranes, respectively) are equivalent in their anti-inflammatory effect (Cocco and Ucker 2001). By comparison, necrotic cells that die with membrane rupture, although associated with inflammatory pathology in vivo (Henson and Johnston 1987), neither exert anti-inflammatory activity or are substantially pro-inflammatory themselves (Cocco and Ucker 2001; Hirt and Leist 2003; Brouckaert et al. 2004; Cvetanovic et al. 2006; Johann et al. 2006). Manipulations in vitro, including extended post-lethal incubation, extensive cell disruption, and lysate concentration, can enhance the availability of inflammatory molecules, including HMGB1 and uric acid (Sauter et al. 2000; Li et al. 2001; Scaffidi et al.

Fig. 6.1 The anti-inflammatory activity of apoptotic cells antagonizes pro-inflammatory responses A model of inflammatory responsiveness, in which the magnitude of response is determined as a function of agonist (especially pathogen-derived "danger", "+ + +") and antagonist (notably apoptotic "calm", "- - -") signals within a simple continuum.

2002; Shi et al. 2003; Chen et al. 2007). However, the physiological significance of this acquired inflammatory activity is uncertain.

These results refute the notion that the phagocytosis of apoptotic cells must occur prelytically to circumvent the inflammatory release of noxious intracellular contents from ruptured corpses (Savill et al. 1989; Stern et al. 1996), and also counter the associated model of pathological cell death as a source of endogenous "danger" (Matzinger 1994, 2002). Rather, they reinforce the view that the process of apoptotic cell death, beyond assuring the elimination of inapt cells, confers on them a potent immunosuppressive gain of function (Voll et al. 1997; Fadok et al. 1998; Cocco and Ucker 2001; Cvetanovic and Ucker 2004; Cvetanovic et al. 2006).

6.2 The Apoptotic Anti-inflammatory Effect is Exerted on the Level of Transcription in Responder Phagocytes

Pathogens and other innate immune agonists elicit a multitude of inflammatory and other responses, and a principle route of their action is via the cytokines whose expression they trigger in macrophage, dendritic, and other cells. Cytokine expression is regulated primarily on the level of transcription. The relevant transcriptional activators for cytokine gene expression, including Nuclear Factor κB (NF-κB) and additional Rel family members (Collart et al. 1990; Shakhov et al. 1990), are the direct targets of signal transduction pathways initiated by innate immune receptors for PAMPs and other mediators (Medzhitov et al. 1998; Poltorak et al. 1998; Kawai et al. 1999; Akira et al. 2001; Fitzgerald et al. 2001; Horng et al. 2001). Upon LPS treatment, for example, specific adaptor proteins are engaged by a recruitment domain within the cytoplasmic tail of TLR4 (that is shared with all TLRs and the IL-1 Receptor and termed "TIR"). One of these adaptors, MyD88, further recruits IL-1 Receptor associated kinases 1 and 4 (IRAK1/4), as well as kinase-inactive regulators. Autophosphorylation of IRAK1/4 leads to the phosphorylation and activation of TRAF6, which nucleates the kinase complex responsible for the activation of NF-κB and the destruction of NF-κB inhibitors, such as IκBα. In addition, parallel signalling pathways, including MyD88-independent pathways involving distinct TIR-associating adaptor proteins, lead to the activation of members of the mitogen-activated protein kinase (MAPK) family and other transcription factors, including AP1.

In light of the central role of transcriptional regulation in inflammatory responsiveness, it is not surprising that apoptotic repression also is exerted transcriptionally (Cvetanovic and Ucker 2004). Most definitively, quantitative reverse transcriptase—polymerase chain reaction (Q-RT-PCR) analysis of the levels of a number of inflammatory cytokine and chemokine gene transcripts in macrophages confirmed that LPS stimulation results in their rapid up-regulation, and revealed that apoptotic cell interactions trigger an equally rapid blockade or reversal of that induction (Cvetanovic and Ucker 2004). Importantly, the observed inhibition of secretion of these factors, including IL-6, IL-8, TNFα, and macrophage inflammatory protein-1α (MIP-1α), reflects this more rapid transcriptional repression (Cocco and Ucker 2001; Cvetanovic and Ucker 2004; Cvetanovic et al. 2006). These results implicate specific transcriptional regulation as the primary mode by which the apoptotic suppression of inflammatory responsiveness is effected.

Like the induction of cytokine gene expression following the engagement of receptors for PAMPs and other mediators, which ensues in the absence of intervening protein synthesis (Goldfeld et al. 1993; Raabe et al. 1998), the transcriptional repression induced by apoptotic cells is an immediate-early response that occurs in the absence of translation (Cvetanovic and Ucker 2004). This suggests that a distinct and comparably direct signal transduction pathway exists, leading from apoptotic recognition to transcriptional repression.

Cvetanovic et al. (2004, 2006) employed simple transcriptional reporter constructs to dissect the mechanism of apoptotic repression. For example, in cells transfected with a construct wherein expression of the firefly luciferase gene is driven by the transcriptional promoter of the IL-8 gene, luciferase activity was induced by innate agonists and repressed by apoptotic targets. Viable and necrotic cells did not exert such inhibition. More generally, expression of a heterologous reporter gene (not normally regulated by inflammatory stimuli) under the control of a transcriptional promoter or promoter fragment responsive to inflammatory agonists recapitulated patterns of expression of endogenous cytokine genes, both with respect to agonist and apoptotic antagonist responses. These results confirmed that primary apoptotic regulation could not be attributable to post-transcriptional effects (translational control or mRNA stability; see Johann et al. 2008) and that it must be exerted on the level of transcriptional initiation.

These findings encouraged a model of apoptotic anti-inflammatory action in which repression is exerted by a blockade of the transcriptional activators involved normally in cytokine gene expression. Surprisingly, further analysis demonstrated that specific transcriptional activators, including NF-κB, are not the subjects of apoptotic regulation (Cvetanovic and Ucker 2004; Cvetanovic et al. 2006). While a synthetic construct composed of a basal transcriptional promoter fused to an oligomerized NF-κB binding motif is responsive to apoptotic repression, an NF-κB-independent construct responsive to AP1, a distinct transcriptional activator, is equally responsive to apoptotic repression (Cvetanovic and Ucker 2004). Additionally, apoptotic repression of the IL-8 promoter is unaffected by deletion of its cognate NF-κB binding sequence (Cvetanovic et al. 2006). The unimpaired activation of NF-κB in the presence of apoptotic targets, including degradation of its inhibitor IκBα (Cvetanovic and Ucker 2004; Tassiulas et al. 2007), confirms a distinct mode of apoptotic repression. Moreover, these results indicate that proximal steps of TLR-specific signalling are not impaired by apoptotic cell interactions, and that the transcriptional control mediated by apoptotic cells is exerted downstream and independent of transcription factor activation and nuclear mobilization.

Interactions with apoptotic targets do not lead to the global repression of macrophage transcription, or to the loss of macrophage viability (Cvetanovic and Ucker 2004; Stranges et al. 2007). Even some inflammation-related genes are not subject to immediate-early apoptotic repression, including genes for the chemokines MCP-1 (Cvetanovic and Ucker 2004) and CXCL10 (Tassiulas et al. 2007). If apoptotic transcriptional repression neither targets specific activators nor results in the blockade of basal transcription, how is it exerted? Although the details remain to be resolved, repression may be exerted via a common transcriptional co-factor. For example, sequestration of a needed co-activator, recruitment of a co-repressor (Vanden Berghe et al. 1999; Hoberg et al. 2004; Harzenetter et al. 2007), or other transrepressor action (Bailey and Ghosh 2005), could affect susceptible genes selectively. Consistent with such a model, overexpression of the transcriptional co-activator CREB Binding Protein (CBP) or its paralogue p300 (Gerritsen et al. 1997) relieves apoptotic repression (Cvetanovic and Ucker 2004). The peroxisome proliferator-activated receptor-γ (PPARγ), which is activated in macrophages by apoptotic targets (Johann et al. 2006) and has transrepressive activity (Bailey and Ghosh 2005), illustrates a candidate molecule with appropriate regulatory activity, although an essential role for PPARγ itself in apoptotic repression has been excluded (Majái et al. 2007).

6.3 The Anti-inflammatory Response is Triggered by Specific Recognition of Apoptotic Targets, Independent of Engulfment

The supposition that the physical engulfment of apoptotic cells is necessary in order to obviate the potential release of inflammatory contents (Savill et al. 1989; Stern et al. 1996) has directed attention to the process of phagocytosis as a necessary component of their anti-inflammatory regulation. To the contrary, the selective interaction of macrophages with apoptotic cells, leading to suppressive outcomes, occurs on the level of recognition, independent of engulfment (Cocco and Ucker 2001; Cvetanovic and Ucker 2004; Patel et al. 2006).

Apoptotic cells bind to macrophages in a saturable, receptor-mediated process (Cocco and Ucker 2001). Just as anti-inflammatory activity is unimpaired by the loss of membrane integrity (Cvetanovic et al. 2006), so too is recognition unaffected by the state of the plasma membrane: "early" and "late" apoptotic cells compete with each other for macrophage binding. On the other hand, necrotic cell recognition occurs by a distinct and non-competing process (Cocco and Ucker 2001). Experimentally, the recognition of apoptotic and necrotic targets can be dissociated from their engulfment (Cocco and Ucker 2001; Reddy et al. 2002; Cvetanovic and Ucker 2004). For example, treatment of macrophages with cytochalasin D, a pharmacologic agent that inhibits actin polymerization, prevents phagocytosis without affecting binding (Cvetanovic and Ucker 2004). The blockade of engulfment has no effect on macrophage responsiveness to apoptotic targets, as measured by transcriptional repression (Cvetanovic and Ucker 2004).

In parallel with this immediate-early transcriptional response, recognition of apoptotic cells triggers a characteristic set of proximal signalling events in responding macrophages, particularly involving MAPK family members (Reddy et al. 2002; Patel et al. 2006; Patel et al. 2007). Apoptotic cell recognition leads to the inhibition of extracellular signal-regulated kinases 1 and 2 (ERK1/2) as well as to the activation of Jun N-terminal kinases 1 and 2 (JNK1/2) and p38 (Reddy et al. 2002; Patel et al. 2006). This pattern of responses occurs virtually instantaneously upon recognition of apoptotic targets (Cvetanovic et al. 2006; Patel et al. 2006). We have referred to this as the signalling signature of apoptotic recognition (Cvetanovic et al. 2006, Patel et al. 2007). In contrast, necrotic cell recognition leads to the activation ERK1/2 and has no effect on JNK1/2 and p38 activation (Reddy et al. 2002; Patel et al. 2006).

The effects elicited on MAPK modules illustrate the potency of apoptotic cells. Although these responses are not mediated by soluble factors, the full response signature is triggered at less than one apoptotic target per responding macrophage (Patel et al. 2006). These findings are entirely consistent with a requirement for apoptotic cell contact (Patel et al. 2006), and confirm imaging studies that reveal the ability of one target to bind serially to multiple responders (Patel et al. 2007). These results also reinforce the conclusion that the disparate outcomes triggered by apoptotic and necrotic cells are linked to distinct and selective modes of recognition for those targets (Cocco and Ucker 2001), and draw attention to the importance of efforts to identify presumptive receptors for apoptotic recognition determinants.

Apoptotic-target induced signalling, like transcriptional repression, also demonstrates the dominant behavior of apoptotic cells over phlogistic stimuli (Patel et al. 2006). The apoptotic signalling signature contrasts with the expected status of MAPK modules under anti-inflammatory conditions. Activation of JNK1/2 and p38 are not associated typically with an anti-inflammatory milieu; by example, canonical TLR engagement leads to the activation of JNK1/2, p38, and ERK1/2 (Medzhitov et al. 1998; Kawai et al. 1999; Kobayashi et al. 2002; Suzuki et al. 2002; Banerjee et al. 2006). In this context, it is notable that the failure of necrotic cells to trigger JNK1/2 and p38 activation in macrophages accords with their inability to trigger an inflammatory response. The unique apoptotic signalling signature may reveal essential elements of the immunosuppressive pathway engaged upon apoptotic cell recognition.

6.4 Soluble Factors Play only a Secondary Role in Sustaining Apoptotic Suppression

Secondary mechanisms, acting downstream of the initial transcriptional repression, may sustain and enhance apoptotic anti-inflammatory suppression. The triggered release from macrophages of potent anti-inflammatory factors, including IL-10, transforming growth factor-β (TGFβ), and the lipid mediators prostaglandin E2 (PGE2) and platelet activating factor (PAF), has been detected after prolonged interaction with apoptotic targets (Voll et al. 1997; Fadok et al. 1998). These molecules may act in an autocrine fashion, or even in a paracrine manner to disseminate inflammatory suppression. However, the generality and significance of their action in apoptotic suppression is uncertain.

The best evidence for the involvement of a soluble macrophage factor pertains to TGFβ. The immunosuppressive effects of apoptotic targets on macrophages can be mimicked partially by the addition of TGFβ (Fadok et al. 1998; Cvetanovic and Ucker 2004; Johann et al. 2006). In some cases, the addition of TGFβ-specific neutralizing antibody to cultures of macrophages and apoptotic targets was found to restore significantly the secretion of some pro-inflammatory cytokines (Fadok et al. 1998; McDonald et al. 1999). In other cases, however, a role for TGFβ in apoptotic suppression could be excluded (Brouckaert et al. 2004; Cvetanovic and Ucker 2004; Kim et al. 2004; Johann et al. 2006). TGFβ fails to mimic the transient suppression of NADPH oxidase-dependent reactive oxygen species production exerted upon macrophage recognition of apoptotic targets (Johann et al. 2006), for example. Anti-inflammatory TGFβ signalling involves the activation of ERK1/2 (Xiao et al. 2002) and TGFβ-dependent Smad signalling proteins. At early times when apoptotic repression is evident, the inhibition ERK1/2 and the absence of detectable Smad activity, including Smad3-dependent transcription, are inconsistent with TGFβ action (Cvetanovic and Ucker 2004; Cvetanovic et al. 2006; Patel et al. 2006). It remains to be determined whether these disparate observations reflect a dissociation between the recognition-dependent establishment of suppression and its subsequent amplification.

PGE2 and PAF only weakly mimic the immunosuppressive effects of apoptotic targets on macrophages (Fadok et al. 1998), and the pharmacologic blockade of PGE2- or PAF-dependent signalling in macrophages had little effect on the efficacy of apoptotic suppression (McDonald et al. 1999). The release of these lipids is dependent on TGFβ signalling in any case (Freire-de-Lima et al. 2006). IL-10 secretion from macrophages has not been observed generally (Fadok et al. 1998; McDonald et al. 1999; Tassiulas et al. 2007), and the requirement for its de novo synthesis also excludes any potential role in immediate-early apoptotic repression.

Recent data support a role for type I interferon (IFN)-dependent inhibitory effects in later stages of apoptotic suppression. Independent of viral infection (see below), type I IFNs (especially IFNβ) are synthesized and secreted by macrophages via a MyD88-independent signalling pathway engaged upon LPS stimulation (Toshchakov et al. 2002). IFNs, via their related receptors, trigger the activation of Janus protein kinase (Jak) molecules, which phosphorylate and thereby activate members of the "signal transducer and activator of transcription" (STAT) family, a distinct group of IFN-responsive transcriptional activators (Platanias 2005). Type I IFNs lead particularly to the activation of STAT1 and STAT2. The result is the expression of specific target genes involved in inflammatory responses (including inducible nitric oxide synthase [iNOS] and a variety of chemokines) as well as anti-viral reactions (Decker et al. 2002). Another consequence of autocrine IFN stimulation is the production of "suppressor of cytokine synthesis" (SOCS) proteins, which serve as direct feedback inhibitors of the interferon signalling pathway, as well as indirect inhibitors of TLR-mediated signalling (Kinjyo et al. 2002; Nakagawa et al.

2002; Baetz et al. 2004). Interestingly, apoptotic cell interactions suppress interferon responses, including STAT1 activation, while inducing SOCS1/3 expression (Tassiulas et al. 2007). As in the case of TLR responses, the effect of apoptotic cells on interferon responsiveness occurs independent of the route of interferon stimulation (directly by treatment with type I or type II IFN or as consequence of LPS stimulation).

These observations extend the view that apoptotic suppression overrides distinct phlogistic pathways on multiple levels, and does not represent a simple stimulus-specific inhibition (Fig. 6.2). The intriguing observation that apoptotic cell interactions lead to SOCS expression even in the absence of phlogistic stimulation (Tassiulas et al. 2007) raises the possibility of novel routes for SOCS induction, including IL-10/STAT3-dependent and STAT-independent ones. Still, SOCS proteins do not play a role in the establishment of apoptotic suppression, and pre-treatment of responding macrophages with IFNγ does not accelerate or alter the initiation of apoptotic repression (Radke et al. 2008). It will be important to test genetically whether SOCS deficiency affects the duration of apoptotic repression

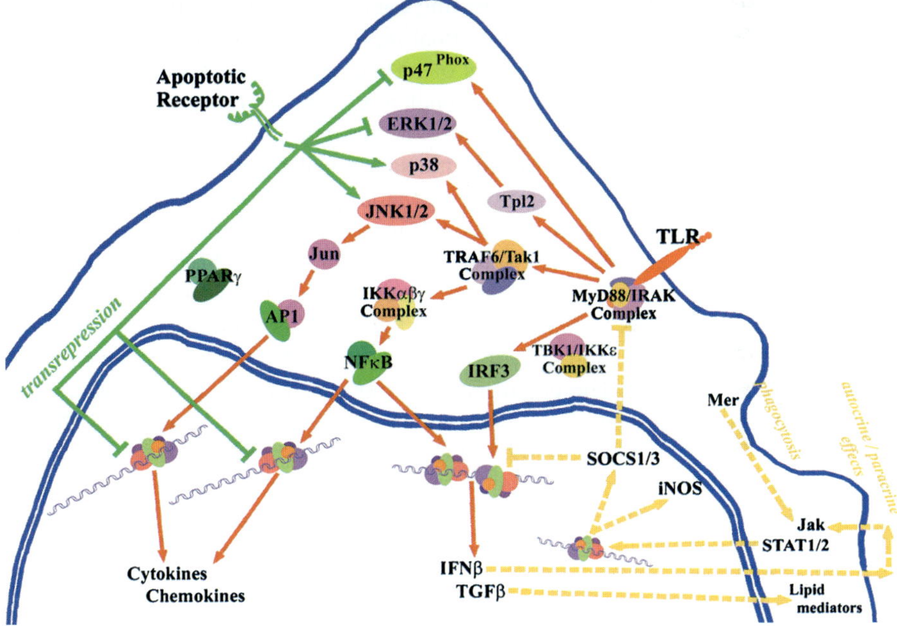

Fig. 6.2 A complex array of signalling events acts to initiate and sustain apoptotic suppression. The profound immunosuppressive outcomes triggered by apoptotic cells involve a complex set of signalling events. Some of the key molecules involved are identified here. Some of these signals are invoked initially by the specific recognition of apoptotic cells. Engagement of the presumptive receptor for conserved apoptotic recognition determinants (referred to as "calm" in the text) affirmatively triggers specific primary signalling events and responses (indicated with green stimulatory and inhibitory arrows). Inflammatory stimuli, such as TLR agonists, initiate distinct signalling events and responses (indicated with red arrows). Secondary responses are engaged subsequently, including upon target engulfment. Some of these (indicated with dashed mustard arrows) can contribute to the persistence and amplification of apoptotic suppression.

This further contrasts apoptotic suppression with other cases of inflammatory hypo-responsiveness. Endotoxin tolerance, for example, represents a self-limiting aspect of TLR signalling engaged by conditioning endotoxin exposure (Medvedev et al. 2006). While multiple levels of suppressive regulation are exerted, including the involvement of SOCS1 (Kinjyo et al. 2002; Nakagawa et al. 2002), endotoxin tolerance is not a state of global unresponsiveness, and it is associated with a very modest degree of cross-tolerance for distinct TLR agonists (Lehner et al. 2001). The canonical TLR pathway is targeted primarily, and endotoxin tolerance depends critically on IRAK-M, the kinase-inactive negative regulator of the MyD88 pathway (Kobayashi et al. 2002; van 't Veer et al. 2007) as well as the specific interference with Rel family members and the activation of NF-κB (Sly et al. 2004; Carmody et al. 2007). Interestingly, although soluble mediators including IL-10 and TGFβ have been proposed to serve as requisite intermediates that elicit endotoxin tolerance, their role in that case also seems not to be substantial (Lehner et al. 2001).

The immunomodulation exerted by neuropeptides, such as calcitonin gene-related peptide (CGRP) provides another interesting contrast. CGRP triggers the induction of ICER/CREM, a pleiotropic transcriptional repressor (Harzenetter et al. 2007). It could be that such secondary repressors of inflammatory gene expression contribute to apoptotic suppression. Transcriptional repression of IL-12 p35, one of the two genes encoding the IL-12 heterodimer, occurs during the apoptotic response, and involves the activation of a specific transcriptional repressor, GC-BP (Kim et al. 2004). The additional possibility of epigenetic silencing (De Santa et al. 2007) in the persistence of the unresponsive state triggered by apoptotic cells has not been explored. An ironic link between apoptotic repression and chromatin-modifying jumonji domain-containing histone demethylases (Chang et al. 2007; De Santa et al. 2007) derives from efforts to identify putative apoptotic recognition receptors.

Finally, phagocytosis itself may provide another set of signals that enhance the suppressive response. The engulfment of apoptotic targets activates the phosphatidylinositol 3-kinase (PI3K)—Akt (protein kinase B) axis (Reddy et al. 2002). Unlike the MAPK signature, Akt signalling in response to apoptotic targets is abrogated by the pharmacologic blockade of phagocytosis (Reddy et al. 2002), although the release of sphingosine-1-phosphate from unengulfed apoptotic cells may augment macrophage Akt stimulation (Weigert et al. 2006). The activation of Akt leads to the phosphorylation of target proteins such as Bad, a death-inducing Bcl-2 family member that is inactivated by phosphorylation, and confers a survival advantage on responding cells (Reddy et al. 2002). Activated Akt also attenuates inflammatory responses (Guha and Mackman 2002). In this regard, phagocyte chemoattractants released from apoptotic cells, such as lysophosphatidylcholine (Lauber et al. 2003), contribute indirectly to the anti-inflammatory milieu, although they do not initiate apoptotic repression (Mitchell and Ucker). Additionally, because the phagocytosis of other particles, including necrotic cells and cell-sized latex beads, leads to Akt activation (Reddy et al. 2002), this aspect of anti-inflammatory regulation is not particular to apoptotic suppression.

6.5 Specific Responsiveness to Apoptotic Targets Is a Ubiquitous Property of Diverse Cell Types

The ability to ingest apoptotic cells is not restricted to professional (migratory) phagocytes, such as macrophages and dendritic cells. The clearance of dying cells in vivo relies in many cases on their engulfment by neighbouring homotypic cells (Wyllie et al. 1980; Parnaik et al. 2000); this likely is of particular importance in the homeostasis of tissues that maintain barrier functions. Mesenchymal and epithelial cells in particular have been shown to engulf apoptotic cells (Saunders 1966; Wood et al. 2000; Lacy-Hulbert, Chap. 7, this Vol.; Monks et al. 2005), and mesenchymal cells have been demonstrated to replace macrophages in the clearance of dead cells during development in mice when the macrophage lineage is ablated (Wood et al. 2000). The role of epithelial-derived cells in the phagocytosis of apoptotic targets is perhaps best characterized in the eye, where the retinal pigment epithelium is responsible for engulfment (Nandrot et al. 2004). Consistent with these observations, a variety of cell types, including epithelial, endothelial, and fibroblastic cells, has been shown to have such phagocytic capacity in culture (Albert et al. 2000; Fadok et al. 2000; Brugnera et al. 2002; Chen et al. 2006; Mitchell et al. 2006).

This clearance by non-professional phagocytes also is linked to specific apoptotic recognition and to anti-inflammatory responsiveness (Monks et al. 2005; Cvetanovic et al. 2006; Mitchell et al. 2006). Although differences exist in the array of cytokines secreted from professional and non-professional phagocytes, the suppression by apoptotic cells of inflammatory cytokine expression is pervasive. For example, the secretion of TNFα by epithelial cells of the mammary gland in response to LPS is suppressed by their engulfment of apoptotic cells (Monks et al. 2005). Apoptotic, but not necrotic, targets also abrogate the synthesis and release of IL-6 by fibroblasts stimulated with IL-1β, and they similarly affect the secretion of IL-8 by stimulated epithelial cells (Cvetanovic et al. 2006).

Apoptotic suppression in non-professional phagocytes is exerted primarily on the level of transcriptional repression, just as it is in macrophages. Again, repression is not an antagonism of particular phlogistic stimuli, does not involve soluble factors, and is independent of proximal steps of characterized innate immune receptor signalling pathways (Cvetanovic et al. 2006). The distinctive signature of early signalling events associated with apoptotic repression in macrophages, especially including the activation of p38 and JNK1/2, and the inhibition of ERK1/2, also is elicited in non-professional phagocytes in response to apoptotic target recognition (Cvetanovic et al. 2006). By these criteria, apoptotic responsiveness—specific apoptotic recognition linked to inflammatory suppression—is an innate property ubiquitous among diverse cell types, even including non-phagocytic cells (Cvetanovic et al. 2006; Patel et al. 2007). That cells derived from all three primary germ layers share this reactivity suggests that apoptotic responsiveness is a fundamental function of metazoan cells, and that its primary elements likely exhibit no tissue- or lineage-restricted expression.

Phagocytosis-dependent signalling arising from activation of receptor tyrosine kinases of the Tyro3/Axl/Mer (TAM) family in dendritic cells has been shown recently to contribute to the persistence of the suppressive state triggered by apoptotic targets (Rothlin et al. 2007). The requirement for de novo gene expression, leading to synthesis of SOCS1/3 (and potentially other) proteins, following TAM engagement identifies this as a secondary mechanism of signalling, which may be operative in macrophages and other cells as well (Fig. 6.2). SOCS expression also is triggered upon dendritic cell activation as a consequence of requisite IFNβ action secondary to TLR engagement (Hoshino et al. 2002). It is not surprising that the pathway of NF-κB activation, including IκBα degradation, is impaired under these conditions (Sen et al. 2007). These observations, rather than suggesting an altered process of initiation of apoptotic suppression in dendritic cells (Sen et al. 2007), are consistent with the existence of a singular and conserved proximal mechanism of apoptotic recognition and signalling leading to the establishment of apoptotic suppression.

Because of their central role in antigen presentation and co-stimulation at the nexus of adaptive immunity, the functional responses of dendritic cells following their interactions with apoptotic targets are of great interest. A number of reports have suggested that, following ingestion by dendritic cells, the processing of apoptotic cells generates antigens that may be presented in a stimulatory form (Albert et al. 1998; Ronchetti et al. 1999; Nouri-Shirazi et al. 2000; Sauter et al. 2000), while others have reported an absence of immunogenicity of apoptotic cell antigens (Liu et al. 2002; Stuart et al. 2002; Blander and Medzhitov 2004).

This contention highlights a central question of immunological tolerance. What is the source of tolerizing autoantigen (especially tolerizing tissue-specific autoantigen in the periphery), and do apoptotic cells and their constituents have a tolerogenic or immunostimulatory role in this context? On the one hand, the homeostatic turnover of differentiated cells in distinct tissues, a process involving apoptotic cell death, leads normally to the clearance of those apoptotic cells. It is generally believed that those apoptotic cells, when engulfed by professional phagocytes, are a source of tolerogenic antigens, especially tissue-specific autoantigens (Škoberne et al. 2005). In contrast, it has become widely accepted that apoptotic cells may be a source of stimulatory autoantigens if, for example, clearance is delayed or reduced (Cohen et al. 2002; Hanayama et al. 2004; but see Devitt et al. 2004). The inherent contradiction of these views rests on the notion that, with delayed clearance, apoptotic cells become necrotic. As discussed above, experimental evidence directly refutes this model (Cocco and Ucker 2001; Cvetanovic et al. 2006; Patel et al. 2006; Patel et al. 2007). Rather, it may be that aberrations in the recognition and signalling responses to apoptotic targets (as opposed to apoptotic cell engulfment per se) contribute to immune pathology (Patel et al. 2007).

Those studies in which the relative efficacies of presentation of antigens derived from apoptotic targets and necrotic or other known immunogenic sources were compared (Ronchetti et al. 1999; Sauter et al. 2000) suggest that apoptotic cells are, at best, only poorly immunogenic. The direct evaluation of the functional status of antigen presenting cells following their interaction with apoptotic targets, as measured by the levels of expression of co-stimulatory molecules (especially CD80 and CD86) and the ability to stimulate antigen-specific T cell responsiveness, indicates that apoptotic cells suppress antigen presenting activity generally (Stuart et al. 2002). These are critical topics, and further analyses, including examination of the extent and generality of the apoptotic suppression of antigen presentation, the potency of apoptotic cells in modulating the effects of immunostimulatory agonists, and the mechanisms involved in this aspect of apoptotic responsiveness, are essential.

Endothelial cells present another interesting intersection of contrasting inflammatory and apoptotic responses. The growth of new vessels is triggered by the production of angiogenic factors, such as vascular endothelial growth factor (VEGF), from a wide variety of cell types in response to inflammatory stimuli and other stressors (Shweiki et al. 1992). Apoptotic cells suppress this angiogenic response (Lacy-Hulbert, Chap. 7, this Vol.). For example, VEGF expression in primary human fibroblasts is induced rapidly and substantially by oxidant stress. The interaction of these non-professional phagocytes with apoptotic, but not with necrotic or viable, targets triggers a rapid repression of VEGF-A gene transcription (Grigera and Ucker 2008). These results parallel the apoptotic repression of pro-inflammatory cytokine gene expression (Cvetanovic and Ucker 2004), and reinforce the view that innate responsiveness to apoptotic targets is common among diverse cell types and may have profound homeostatic effects beyond the immune system (Patel et al. 2007).

Within vessels, inflammatory stimuli trigger the elevated expression of endothelial adhesion molecules, leading to the retention of circulating monocytes and lymphocytes and their subsequent extravasation (Collins 1993). These are hallmarks of the onset and progression of inflammatory disorders such as atherosclerosis. Apoptotic cells appear to suppress this response. A fascinating example has been documented in vascular endothelium (Chen et al. 2006), where homeostatic turnover of trophoblast cells may provide an endogenous apoptotic buffer against endothelial compromise associated with preeclampsia and other stresses and immunological challenges associated with pregnancy. The possibility that the diminished expression of endothelial adhesion molecules, such as the intercellular adhesion molecule 1 (ICAM1, CD54), may be exerted on the level of transcriptional repression has not been explored.

It is interesting to consider that the apoptotic responsiveness of non-professional phagocytes may serve in a tumour suppressive capacity. Apoptotic cells arising within a tumour, either spontaneously or in response to overt stressors (including chemotherapeutic agents), may suppress immune responsiveness to tumour-specific antigens and confer an advantage on the surviving cells. This benefit may be overshadowed, however, by the otherwise inhospitable environment for tumour survival and dissemination that those apoptotic cells simultaneously engender with the suppression of angiogenic and adhesive support. In the balance, apoptotic suppression may even confer a selective disadvantage on transformed cells. It could be that an unusual case of "immunogenic" tumour cell death, in which apoptotic cells lack immunosuppressive activity (Obeid et al. 2007), relates to escape from this tumour suppressive mechanism.

6.6 Apoptotic Responsiveness Represents a Distinct and Unconventional Dimension of Immune Discrimination

Exceptional cases notwithstanding, it is clear that specific recognition determinants that engage an immunosuppressive response repertoire arise in virtually all cases of apoptotic cell death, regardless of cell type or suicidal stimulus. Correspondingly, a similar diversity of cell types innately recognizes and responds specifically to apoptotic cells. This is an unusually ubiquitous immune responsiveness, unrestricted to homotypic (within a tissue, for example) or heterotypic (as in the case of migratory phagocytes) interactions between responders and targets.

Even more remarkable, these innate immune interactions do not manifest species-specific restriction (Voll et al. 1997; McDonald et al. 1999; Cvetanovic and Ucker 2004). Syngeneic, allogeneic, and xenogeneic targets are equally effective in triggering the full repertoire of specific apoptotic immune responses from professional and non-professional phagocyte responders (Cvetanovic and Ucker 2004; Cvetanovic et al. 2006). Beyond its independence from characterized innate immune signalling pathways, the absence of self-restriction marks apoptotic immunity as fundamentally distinct and unconventional. Because the determinants of innate apoptotic recognition are conserved widely throughout metazoan evolution (Mitchell and Ucker), it may be that this is an evolutionarily ancient arm of immunity.

These observations inform our view of apoptotic suppression as more than the simple antagonism of classical immune responsiveness depicted in Fig. 6.1. Both in its innate and adaptive components, the immune system commonly is viewed as a host response to pathogenic or pathological (especially non-self) "danger". In contrast to this essentially unidimensional view of immune provocation, apoptotic "calm" (conserved apoptotic ligand for response modulation) determinants engage a second dimension of immune discrimination devoted to the maintenance of homeostatic moderation and tolerance. Apoptotic "calm" triggers responses in opposition to those elicited by "danger" stimuli, and does so via an independent process of innate immune discrimination. As depicted in Fig. 6.3, immune responsiveness appears to be an integrated function of [at least] these two independent criteria of discrimination (Patel et al. 2007).

6.7 The Enigma of Apoptotic Recognition Determinants

Properties unique to the apoptotic target, particularly cell surface "calm" determinants, allow it to be distinguished innately from viable and other dead cells prior to, and independent of, engulfment. The identity of these "calm" determinants remains elusive. The tractable experimental readout of engulfment, together with the presumed importance of internalization, has focused attention instead on so-called "eat me" determinants (Savill et al. 2002). Numerous molecular species present on the apoptotic cell surface have been identified, and several of these have been

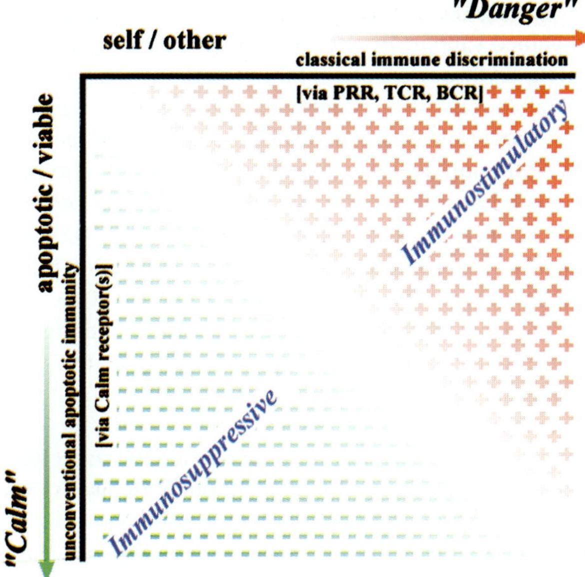

Fig. 6.3 Two dimensions of immune recognition. Immune responsiveness is an integrated function of [at least] two independent criteria of recognition. One dimension is the well-appreciated discrimination along a self vs. other ("danger") axis, involving the antigen-specific receptors of T and B lymphocytes (TCR and BCR) and the innate pattern recognition receptors (PRR). The second reflects the discrimination between viable and dead (or dying) cells. These two distinct criteria of recognition are represented here along orthogonal axes. Whereas non-self or "danger" agonists engage signal transduction pathways linked to immunostimulatory ("+ + +") outcomes, apoptotic ("calm") agonists trigger distinct signalling events that result in immunosuppressive ("- - -") outcomes.

implicated in phagocytosis. None has been linked functionally to specific innate apoptotic recognition, however. Several classes of cell surface molecules illuminate the issues.

Early studies in which carbohydrates (especially amino sugars such as N-acetyl glucosamine) were found to compete with apoptotic targets for binding to macrophages (Duvall et al. 1985; Stern et al. 1996), suggested that recognition was directed to altered glycosyl determinants appearing on the apoptotic cell surface. More recent work has revealed that the exposure of incompletely processed glycosylation intermediates reflects the non-specific loss of membrane integrity, rather than the process of apoptotic cell death, (Franz et al. 2006). These processing intermediates are not involved in specific apoptotic recognition. The comparison of apoptotic targets with other dead cells obviously is critical in these evaluations.

Alterations in lipid composition during the cell death process also have been recognized (Fadok et al. 1992; Schlegel et al. 1993). In particular, phosphatidylserine (PS), an anionic phospholipid normally cloistered in the inner leaflet of the plasma membrane, is externalized (Fadok et al. 1992). The view that externalized PS serves as a ligand for macrophage recognition followed from studies demonstrating that similar changes target aged erythrocytes for clearance (Schroit et al. 1985; McEvoy et al. 1986), and gained support from observations that phospho-L-serine, PS vesicles, and even effete erythrocytes, could inhibit partially the phagocytosis of dying nucleated cells (Fadok et al. 1992; Pradhan et al. 1997). These same observations highlighted that PS externalization is not restricted to apoptotic cells.

A comparison of the interaction of apoptotic versus necrotic cells with macrophages is illuminating. Equivalent externalization of PS occurs during apoptotic and necrotic cell death (Cocco and Ucker 2001; Appelt et al. 2005), and synthetic PS liposomes are equally effective at inhibiting the internalization by macrophages of apoptotic and necrotic targets (Cocco and Ucker 2001). Consistent with these findings, more recent data have demonstrated that PS, and "tethering" molecules from serum that bind externalized PS, are involved in the phagocytosis of already-bound targets (Hoffmann et al. 2001; Hanayama et al. 2004). Apoptotic targets lacking externalized PS remain immunosuppressive, while necrotic targets with externalized PS, like erythrocytes, fail to trigger immunosuppressive effects (Cocco and Ucker 2001; Cvetanovic et al. 2006). Finally, although PS also is exposed on viable macrophages and other cells (Marguet et al. 1999; Callahan et al. 2003), this does not trigger inflammatory suppression or engulfment. Together, these observations demonstrate that PS is neither necessary nor sufficient for specific apoptotic recognition or immunosuppressive response (Cocco and Ucker 2001). PS is not "calm".

It is striking that some of the molecules identified as apoptotic "eat me" determinants engage well-defined pro-inflammatory pathways (Savill et al. 2002). Rather than attribute contrary activities to these molecules, a more parsimonious interpretation is that the distinct, dominant-acting immunosuppressive signalling triggered upon apoptotic recognition overrides any inflammatory signals that might be triggered through the involvement of complement/collectin, Toll-like, or scavenger receptors (Stern et al. 1996; Botto et al. 1998; Mevorach et al. 1998; Devitt et al. 1998; Nauta et al. 2002; Gardai et al. 2005). This again underscores a dissociation between immunosuppressive apoptotic recognition ("calm") and subsequent target engulfment ("eat me").

In this context, the role of calreticulin in an unusual case of "immunogenic" tumour cell death, in which apoptotic cells lack immunosuppressive activity (Obeid et al. 2007), may not be surprising. Cell surface calreticulin is widely expressed and serves as a receptor for complement component C1q; it also is exposed generally on apoptotic cells and then recognized by CD91 on macrophages (Stuart et al. 1997; Gardai et al. 2005). While immunogenicity in this case depends on the phagocytic engulfment and presentation of apoptotic tumour cell antigens, it is likely that the absence of apoptotic suppression reflects a deficiency of "calm" determinants, rather than a gain of a pro-inflammatory "eat me" ligand activity. The clear prediction is that the absence of apoptotic suppression arises selectively in tumour cells (or following long-term cell culture; Kurosaka et al. 1998) as a recessive genetic trait.

Of course, only the identification of "calm" determinants and their receptors (be they singular or multiple) will resolve these issues. Current studies, especially employing proximal and specific readouts of apoptotic recognition (such as transcriptional transrepression; Fig. 6.2), provide some clues. Membrane-associated "calm" determinants emerge on the apoptotic target roughly coincident with, and dependent upon, effector caspase activity during the physiological cell death process, well before the plasma membrane is compromised (Cocco and Ucker 2001; Cvetanovic and Ucker 2004). This appearance shows no species- or tissue-specific restriction (Cvetanovic and Ucker 2004; Cvetanovic et al. 2006). The appearance of "calm" is not dependent on new gene expression (Cvetanovic and Ucker 2004). At the same time, it is not possible to recover active "calm" simply from intracellular organelles or membranes of viable cells (Cvetanovic et al. 2006). "Calm" activity is not dependent on opsonins or other serum factors (including complement), and therefore appears to be intrinsic to the apoptotic target (Cvetanovic et al. 2006). These observations suggest that "calm" is formed by the post-translational modification of evolutionarily conserved and ubiquitously expressed resident molecules in cells dying apoptotically (Cvetanovic et al. 2006). This might even constitute an apoptotic cell-associated molecular pattern, akin to the pattern recognition that typifies classical innate immunity (Savill et al. 2002). An exciting possibility is that "calm" determinants arise from the apoptosis-specific modification of one or more of the numerous normally cytosolic and nuclear proteins that are known to be presented on the apoptotic cell surface (Casciola-Rosen et al. 1994; Cocca et al. 2002).

6.8 Innate Apoptotic Immunity Is Subject to Pathogen Subversion and Is a Target for Therapeutic Intervention

It is not surprising that the potent immunosuppressive pathways engaged by apoptotic cells are targeted by pathogens. This pathogenic subversion takes several forms, two of which are revealed by the closely related intracellular protozoan parasites *Trypanosoma cruzi* and *Leishmania amazonensis*. *T. cruzi* exploits authentic apoptotic suppression. *T. cruzi* growth is enhanced by apoptotic suppression of host macrophages, and the pathogen triggers apoptosis of responding lymphocytes, which promotes this effect (Freire-de-Lima et al. 2000). *L. amazonensis*, in contrast, mimics apoptotic cells in triggering suppression. As viable cells, these protists expose PS on their surface, facilitating their engulfment by macrophages. Simultaneously, the pathogens appear to express "calm"-like immunosuppressive determinants that facilitate a non-inflammatory milieu for their replication and survival (Wanderley et al. 2006).

A third mode of pathogenic manipulation is revealed in viral subversion of apoptotic immunity. Innate host immunity exerts protection against viral infection on several levels, and the array of viral responses, including the blockade of inflammatory mediator production, escape from interferon regulation, and the avoidance of adaptive immunity by down-regulated expression of histocompatibility molecules, are well characterized (Burgert and Kvist 1985; Ray et al. 1992; Foy et al. 2003). The host apoptotic response is another element of this anti-viral reaction. While viruses are known to employ a variety of mechanisms to inhibit apoptosis, enhancing host cell survival transiently and thereby promoting viral replication (Clem and Miller 1993; Debbas and White 1993), this viral interference appears to be a more complex subversion of both apoptosis and apoptotic immunity. In the case of adenoviruses, for example, infection of permissive human cells induces a non-apoptotic cell death response (cytopathic effect) dependent on the viral suppressor of apoptosis, E1B 19 kD (Debbas and White 1993; Chiou et al. 1994). Remarkably, the necrotic cytopathic cells mimic apoptotic targets in repressing inflammatory gene transcription and suppressing cytokine and chemokine secretion from responding macrophages (Radke et al. 2008). Like authentic "calm", the suppressive activity of cytopathic cells is membrane-associated. This apoptotic mimic arises in a caspase-independent manner, however, linked to viral early gene expression (Radke et al. 2008). It is intriguing that this viral subversion, involving concurrent blockade of apoptotic death and induction of apoptotic-like suppression, suggests the existence of alternate routes to "calm".

Innate apoptotic immunity also has been exploited (albeit inadvertently) in a clinical practice of induced immune suppression. "Photopheresis" is a procedure involving the intravenous infusion of syngeneic blood cells after their extracorporeal photochemical treatment (with a psoralen and UV irradiation, leading to apoptotic cell death). It is employed in cases involving pathogenic T cells, including the treatment of cutaneous T cell lymphoma, allograft rejection, and graft vs. host disease (Suchin et al. 1999; Lamioni et al. 2005; Maeda et al. 2005). This practice was established over 30 years ago, and the basis of its effect has remained obscure. Some reports have suggested that the effects of photopheresis are antigen-specific and TGFβ-dependent (Maeda et al. 2005; Kleinclauss et al. 2006). However, the predominant effects of this treatment appear to be antigen non-specific (Suchin et al. 1999; Lamioni et al. 2005). That is, the effects of photopheresis are precisely those of innate apoptotic immunity, in which suppression is evident in the context of a contemporaneous stimulus.

It can be hoped that mechanisms of pathogen subversion will provide insight into the normal process of apoptotic suppression, and that a fuller understanding of innate apoptotic responsiveness will enable interventions that exploit this powerful arm of immunity.

6.9 Prospects in Innate Apoptotic Immunity

It is an astonishing paradox that, as they lose viability and normal function, apoptotic cells gain a potent, new immunosuppressive activity to affirmatively trigger a profound repertoire of responses from cells with which they interact. The obvious purpose of process of physiological cell death is to sculpt and maintain a functionally appropriate cellular network through the targeted elimination of inappropriate cells. Innate apoptotic immunity reveals that the cell death process additionally provides an ongoing immunosuppressive mechanism for the homeostatic maintenance of tolerance and the attenuation of inflammation in support of those surviving cells and the entire organism.

This innate discrimination of apoptotic cells is itself remarkable as a ubiquitous and unconventional dimension of immunity. A number of fundamental issues concerning innate apoptotic immunity remain incompletely explored. Without question, the characterization of "calm" determinants on apoptotic cells and the corresponding receptors on professional phagocytes and other responders will be essential for substantive molecular progress. It also will be important to dissect the signalling leading to transcriptional repression, as well as the mechanism of that repression. The integrated mechanistic understanding of the antagonistic interplay of phlogistic and apoptotic signalling pathways is an essential long-term goal. It also will be interesting to explore the full array of responses targeted by innate apoptotic suppression. Apoptotic effects on reactive oxygen species, for example, likely are important (Johann et al. 2006; Johann et al. 2008), but have not been explored extensively. Finally, it will be valuable to compare innate apoptotic immunity with immune responsiveness to other physiological modes of cell death, such as autophagy. The consequences of autophagic cell death on responder cells currently remain unresolved (Degenhardt et al. 2006; Petrovski et al. 2007).

Given the enormous magnitude of physiological cell death throughout organismal life, the recognition and clearance of apoptotic cells may represent an aspect of innate immunity even more profound for homeostasis than the discrimination of self from other. A deeper understanding of the immunosuppressive functions exerted physiologically by apoptotic cells and of the molecular events involved may reveal new targets for immunological control, with great potential for intervention in cases of pathological inflammation and immunity.

References

Akira S, Takeda K, Kaisho T (2001) Toll-like receptors: critical proteins linking innate and acquired immunity. Nat Immunol 2 (8):675–680

Albert ML, Kim JI, Birge RB (2000) $\alpha v \beta 5$ integrin recruits the CrkII-Dock180-Rac1 complex for phagocytosis of apoptotic cells. Nat Cell Biol 2:899–905

Albert ML, Sauter B, Bhardwaj N (1998) Dendritic cells acquire antigen from apoptotic cells and induce class 1- restricted CTLs. Nature 392:86–89

Appelt U, Sheriff A, Gaipl US et al (2005) Viable, apoptotic and necrotic monocytes expose phosphatidylserine: cooperative binding of the ligand Annexin V to dying but not viable cells and implications for PS-dependent clearance. Cell Death Differ 12 (2):194–196

Baetz A, Frey M, Heeg K et al (2004) Suppressor of cytokine signalling (SOCS) proteins indirectly regulate toll-like receptor signalling in innate immune cells. J Biol Chem 279 (52):54708–54715

Bailey ST, Ghosh S (2005) 'PPAR'ting ways with inflammation. Nat Immunol 6 (10):966–967

Banerjee A, Gugasyan R, McMahon M et al (2006) Diverse Toll-like receptors utilize Tpl2 to activate extracellular signal-regulated kinase (ERK) in hemopoietic cells. Proc Natl Acad Sci U S A 103 (9):3274–3279

Blander JM, Medzhitov R (2004) Regulation of phagosome maturation by signals from toll-like receptors. Science 304:1014–1018

Botto M, Dell'Agnola C, Bygrave AE, et al (1998) Homozygous C1q deficiency causes glomerulonephritis associated with multiple apoptotic bodies. Nat Genet 19:56–59

Brouckaert G, Kalai M, Krysko DV et al (2004) Phagocytosis of necrotic cells by macrophages is phosphatidylserine-dependent and does not induce inflammatory cytokine production. Mol Biol Cell 15:1089–1100

Brugnera E, Haney L, Grimsley C et al (2002) Unconventional Rac-GEF activity is mediated through the Dock180-ELMO complex. Nat Cell Biol 4:574–582

Burgert HG, Kvist S (1985) An adenovirus type 2 glycoprotein blocks cell surface expression of human histocompatibility class I antigens. Cell 41 (3):987–997

Callahan MK, Halleck MS, Krahling S et al (2003) Phosphatidylserine expression and phagocytosis of apoptotic thymocytes during differentiation of monocytic cells. J Leukocyte Biol 74 (5):846–856

Carmody RJ, Ruan Q, Palmer S et al (2007) Negative regulation of toll-like receptor signalling by NF-κB p50 ubiquitination blockade. Science 317 (5838):675–678

Casciola-Rosen LA, Anhalt G, Rosen A (1994) Autoantigens targeted in systemic lupus erythematosus are clustered in two population of surface structures on apoptotic keratinocytes. J Exp Med 179:1317–1339

Chang B, Chen Y, Zhao Y et al (2007) JMJD6 is a histone arginine demethylase. Science 318 (5849):444–447

Chen C-J, Kono H, Golenbock D et al (2007) Identification of a key pathway required for the sterile inflammatory response triggered by dying cells. Nat Med 13 (7):851–856

Chen Q, Stone PR, McCowan LM et al (2006) Phagocytosis of necrotic but not apoptotic trophoblasts induces endothelial cell activation. Hypertension 47 (1):116–121

Chiou S-K, Tseng C-C, Rao L et al (1994) Functional complementation of the adenovirus E1B 19-kilodalton protein with Bcl-2 in the inhibition of apoptosis in infected cells. J Virol 68 (10):6553–6566

Clarke PG, Clarke S (1996) Nineteenth century research on naturally occurring cell death and related phenomena. Anat Embryol 193:81–99

Clem RJ, Miller LK (1993) Apoptosis reduces both the in vitro replication and the in vivo infectivity of a baculovirus. J Virol 67:3730–3738

Cocca BA, Cline AM, Radic MZ (2002) Blebs and apoptotic bodies are B cell autoantigens. J Immunol 169:159–166

Cocco RE, Ucker DS (2001) Distinct modes of macrophage recognition for apoptotic and necrotic cells are not specified exclusively by phosphatidylserine exposure. Mol Biol Cell 12 (4):919–930

Cohen PL, Caricchio R, Abraham V et al (2002) Delayed apoptotic cell clearance and lupus-like autoimmunity in mice lacking the c-mer membrane tyrosine kinase. J Exp Med 196:135–140

Collart MA, Baeuerle P, Vassalli P (1990) Regulation of tumour necrosis factor alpha transcription in macrophages: involvement of four κB-like motifs and of constitutive and inducible forms of NF-κB. Mol Cell Biol 10:1498–1506

Collins T (1993) Endothelial nuclear factor-κB and the initiation of the atherosclerotic lesion. Lab Invest 68 (5):499–508

Creagh EM, O'Neill LA (2006) TLRs, NLRs and RLRs: a trinity of pathogen sensors that co-operate in innate immunity. Trends Immunol 27 (8):352–357

Cvetanovic M, Mitchell JE, Patel V et al (2006) Specific recognition of apoptotic cells reveals a ubiquitous and unconventional innate immunity. J Biol Chem 281:20055–20067

Cvetanovic M, Ucker DS (2004) Innate immune discrimination of apoptotic cells: repression of proinflammatory macrophage transcription is coupled directly to specific recognition. J Immunol 172:880–889

De Santa F, Totaro MG, Prosperini E et al (2007) The histone H3 lysine-27 demethylase Jmjd3 links inflammation to inhibition of polycomb-mediated gene silencing. Cell 130 (6):1083–1094

Debbas M, White E (1993) Wild-type p53 mediates apoptosis by E1A, which is inhibited by E1B. Genes Dev 7:546–554

Decker T, Stockinger S, Karaghiosoff M et al (2002) IFNs and STATs in innate immunity to microorganisms. J Clin Invest 109 (10):1271–1277

Degenhardt K, Mathew R, Beaudoin B et al (2006) Autophagy promotes tumour cell survival and restricts necrosis, inflammation, and tumourigenesis. Cancer Cell 10 (1):51–64

Devitt A, Moffatt OD, Raykundalia C et al (1998) Human CD14 mediates recognition of phagocytosis of apoptotic cells. Nature 392:505–509

Devitt A, Parker KG, Ogden CA et al (2004) Persistence of apoptotic cells without autoimmune disease or inflammation in CD14-/- mice. J Cell Biol 167:1161–1170

Duvall E, Wyllie AH, Morris RG (1985) Macrophage recognition of cells undergoing programmed cell death (apoptosis). Immunology 56:351–358

Fadok VA, Bratton DL, Konowal A et al (1998) Macrophages that have ingested apoptotic cells in vitro inhibit proinflammatory cytokine production through autocrine/paracrine mechanisms involving TGF-β, PGE2, and PAF. J Clin Invest 101:890–898

Fadok VA, Bratton DL, Rose DM et al (2000) A receptor for phosphatidylserine-specific clearance of apoptotic cells. Nature 405:85–90

Fadok VA, Voelker DR, Campbell PA et al (1992) Exposure of phosphatidylserine on the surface of apoptotic lymphocytes triggers specific recognition and removal by macrophages. J Immunol 148:2207–2216

Fitzgerald KA, Palsson-McDermott EM, Bowie AG et al (2001) Mal (MyD88-adapter-like) is required for Toll-like receptor-4 signal transduction. Nature 413:78–83

Foy E, Li K, Wang C et al (2003) Regulation of interferon regulatory factor-3 by the hepatitis C virus serine protease. Science 300 (5622):1145–1148

Franz S, Frey B, Sheriff A et al (2006) Lectins detect changes of the glycosylation status of plasma membrane constituents during late apoptosis. Cytometry A 69 (4):230–239

Freire-de-Lima CG, Nascimento DO, Soares MB et al (2000) Uptake of apoptotic cells drives the growth of a pathogenic trypanosome in macrophages. Nature 403:199–203

Freire-de-Lima CG, Xiao YQ, Gardai SJ et al (2006) Apoptotic cells, through transforming growth factor-β, coordinately induce anti-inflammatory and suppress pro-inflammatory eicosanoid and NO synthesis in murine macrophages. J Biol Chem 281 (50):38376–38384

Gardai SJ, McPhillips KA, Frasch SC et al (2005) Cell-surface calreticulin initiates clearance of viable or apoptotic cells through trans-activation of LRP on the phagocyte. Cell 123:321–334

Gerritsen ME, Williams AJ, Neish AS et al (1997) CREB-binding protein/p300 are transcriptional coactivators of p65. Proc Natl Acad Sci U S A 94:2927–2932

Goldfeld AE, McCaffrey PG, Strominger JL et al (1993) Identification of a novel cyclosporin-sensitive element in the human tumour necrosis factor alpha gene promoter. J Exp Med 178:1365–1379

Grigera F, Ucker DS unpublished data

Guha M, Mackman N (2002) The phosphatidylinositol 3-kinase-Akt pathway limits lipopolysaccharide activation of signalling pathways and expression of inflammatory mediators in human monocytic cells. J Biol Chem 277 (35):32124–32132

Hanayama R, Tanaka M, Miyasaka K et al (2004) Autoimmune disease and impaired uptake of apoptotic cells in MFG-E8-deficient mice. Science 304:1147–1150

Harzenetter MD, Novotny AR, Gais P et al (2007) Negative regulation of TLR responses by the neuropeptide CGRP is mediated by the transcriptional repressor ICER. J Immunol 179 (1):607–615

Henson PM, Johnston RB Jr (1987) Tissue injury in inflammation: oxidants, proteinases, and cationic proteins. J Clin Invest 79:669–674

Hirt UA, Leist M (2003) Rapid, noninflammatory and PS-dependent phagocytic clearance of necrotic cells. Cell Death Differ 10 (10):1156–1164

Hoberg JE, Yeung F, Mayo MW (2004) SMRT derepression by the IκB kinase α: a prerequisite to NF-κB transcription and survival. Mol Cell 16:245–255

Hoffmann PR, deCathelineau AM, Ogden CA et al (2001) Phosphatidylserine (PS) induces PS receptor-mediated macropinocytosis and promotes clearance of apoptotic cells. J Cell Biol 155:649–660

Horng T, Barton GM, Medzhitov R (2001) TIRAP: an adapter molecule in the Toll signalling pathway. Nat Immunol 2:835–841

Hoshino K, Kaisho T, Iwabe T et al (2002) Differential involvement of IFN-β in Toll-like receptor-stimulated dendritic cell activation. Int Immunol 14 (10):1225–1231

Hoshino K, Takeuchi O, Kawai T et al (1999) Toll-like receptor 4 (TLR4)-deficient mice are hyporesponsive to lipopolysaccharide: evidence for TLR4 as the *Lps* gene product. J Immunol 162:3749–3752

Hughes J, Liu Y, Damme JV et al (1997) Human glomerular mesangial cell phagocytosis of apoptotic neutrophils. J Immunol 158:4389–4397

Johann AM, von Knethen A, Lindemann D et al (2006) Recognition of apoptotic cells by macrophages activates the peroxisome proliferator-activated receptor-γ and attenuates the oxidative burst. Cell Death Differ 13 (9):1533–1540

Johann AM, Weigert A, Eberhardt W et al (2008) Apoptotic cell-derived sphingosine-1-phosphate promotes HuR-dependent cyclooxygenase-2 mRNA stabilization and protein expression. J Immunol 180 (2):1239–1248

Kawai T, Adachi O, Ogawa T et al (1999) Unresponsiveness of MyD88-deficient mice to endotoxin. Immunity 11:115–122

Kerr JFR, Wyllie AH, Currie AR (1972) Apoptosis: a basic biological phenomenon with wide-ranging implications in tissue kinetics. Br J Cancer 26:239–256

Kim S, Elkon KB, Ma X (2004) Transcriptional suppression of interleukin-12 gene expression following phagocytosis of apoptotic cells. Immunity 21:643–653

Kinjyo I, Hanada T, Inagaki-Ohara K et al (2002) SOCS1/JAB is a negative regulator of LPS-induced macrophage activation. Immunity 17 (5):583–591

Kleinclauss F, Perruche S, Masson E et al (2006) Intravenous apoptotic spleen cell infusion induces a TGF-β-dependent regulatory T-cell expansion. Cell Death Differ 13:41–52

Kobayashi K, Hernandez LD, Galan JE et al (2002) IRAK-M is a negative regulator of Toll-like receptor signalling. Cell 110 (2):191–202

Kurosaka K, Watanabe N, Kobayashi Y (1998) Production of proinflammatory cytokines by phorbol myristate acetate-treated THP-1 cells and monocyte-derived macrophages after phagocytosis of apoptotic CTLL-2 cells. J Immunol 161:6245–6249

Lamioni A, Parisi F, Isacchi G et al (2005) The immunological effects of extracorporeal photopheresis unraveled: induction of tolerogenic dendritic cells in vitro and regulatory T cells in vivo. Transplantation 79:846–850

Lauber K, Bohn E, Kröber SM et al (2003) Apoptotic cells induce migration of phagocytes via caspase-3-mediated release of a lipid attraction signal. Cell 113:717–730

Lehner MD, Morath S, Michelsen KS et al (2001) Induction of cross-tolerance by lipopolysaccharide and highly purified lipoteichoic acid via different Toll-like receptors independent of paracrine mediators. J Immunol 166 (8):5161–5167

Li M, Carpio DF, Zheng Y et al (2001) An essential role of the NF-κB/Toll-like receptor pathway in induction of inflammatory and tissue-repair gene expression by necrotic cells. J Immunol 166:7128–7135

Liu K, Iyoda T, Saternus M et al (2002) Immune tolerance after delivery of dying cells to dendritic cells in situ. J Exp Med 196:1091–1097

Maeda A, Schwarz A, Kernebeck K et al (2005) Intravenous infusion of syngeneic apoptotic cells by photopheresis induces antigen-specific regulatory T cells. J Immunol 174:5968–5976

Majái G, Sarang Z, Csomós K et al (2007) PPARγ-dependent regulation of human macrophages in phagocytosis of apoptotic cells. Eur J Immunol 37 (5):1343–1354

Marguet D, Luciani MF, Moynault A et al (1999) Engulfment of apoptotic cells involves the redistribution of membrane phosphatidylserine on phagocyte and prey. Nat Cell Biol 1:454–456

Matzinger P (1994) Tolerance, danger, and the extended family. Annu Rev Immunol 12:991–1045

Matzinger P (2002) The danger model: a renewed sense of self. Science 296 (5566):301–305

McDonald PP, Fadok VA, Bratton D et al (1999) Transcriptional and translational regulation of inflammatory mediator production by endogenous TGF-β in macrophages that have ingested apoptotic cells. J Immunol 163 (11):6164–6172

McEvoy L, Williamson P, Schlegel RA (1986) Membrane phospholipid asymmetry as a determinant of erythrocyte recognition by macrophages. Proc Natl Acad Sci U S A 83:3311–3315

Meagher LC, Savill JS, Baker A et al (1992) Phagocytosis of apoptotic neutrophils does not induce macrophage release of thromboxane B2. J Leukocyte Biol 52:269–272

Medvedev AE, Sabroe I, Hasday JD et al (2006) Tolerance to microbial TLR ligands: molecular mechanisms and relevance to disease. J Endotoxin Res 12 (3):133–150

Medzhitov R, Preston-Hurlburt P, Kopp E et al (1998) MyD88 is an adaptor protein in the hToll/IL-1 receptor family signalling pathways. Mol Cell 2:253–258

Metchnikoff E (1891) Leçons sur la pathologie comparée de l'inflammation. Dover Publications, New York, p 224

Mevorach D, Mascarenhas JO, Gershov D et al (1998) Complement-dependent clearance of apoptotic cells by human macrophages. J Exp Med 188:2313–2320

Mitchell JE, Cvetanovic M, Tibrewal N et al (2006) The presumptive phosphatidylserine receptor is dispensable for innate anti- inflammatory recognition and clearance of apoptotic cells. J Biol Chem 281:5718–5725

Mitchell JE, Ucker DS unpublished data

Monks J, Rosner D, Geske FJ et al (2005) Epithelial cells as phagocytes: apoptotic epithelial cells are engulfed by mammary alveolar epithelial cells and repress inflammatory mediator release. Cell Death Differ 12:107–114

Nakagawa R, Naka T, Tsutsui H et al (2002) SOCS-1 participates in negative regulation of LPS responses. Immunity 17 (5):677–687

Nandrot EF, Kim Y, Brodie SE et al (2004) Loss of synchronized retinal phagocytosis and age-related blindness in mice lacking $\alpha v\beta 5$ integrin. J Exp Med 200:1539–1545

Nauta AJ, Trouw LA, Daha MR et al (2002) Direct binding of C1q to apoptotic cells and cell blebs induces complement activation. Eur J Immunol 32:1726–1736

Nouri-Shirazi M, Banchereau J, Bell D et al (2000) Dendritic cells capture killed tumour cells and present their antigens to elicit tumour-specific immune responses. J Immunol 165:3797–3803

Obeid M, Tesniere A, Ghiringhelli F et al (2007) Calreticulin exposure dictates the immunogenicity of cancer cell death. Nat Med 13 (1):54–61

Parnaik R, Raff MC, Scholes J (2000) Differences between the clearance of apoptotic cells by professional and non- professional phagocytes. Curr Biol 10:857–860

Patel VA, Longacre-Antoni A, Cvetanovic M et al (2007) The affirmative response of the innate immune system to apoptotic cells. Autoimmunity 40 (4):274–280

Patel VA, Longacre A, Hsiao K et al (2006) Apoptotic cells, at all stages of the death process, trigger characteristic signalling events that are divergent from and dominant over those triggered by necrotic cells: implications for the delayed clearance model of autoimmunity. J Biol Chem 281 (8):4663–4670

Petrovski G, Zahuczky G, Majái G et al (2007) Phagocytosis of cells dying through autophagy evokes a pro-inflammatory response in macrophages. Autophagy 3 (5):509–511

Platanias LC (2005) Mechanisms of type-I- and type-II-interferon-mediated signalling. Nat Rev: Immunol 5 (5):375–386

Poltorak A, He X, Smirnova I et al (1998) Defective LPS signalling in C3H/HeJ and C57BL/10ScCr mice: mutations in *Tlr4* gene. Science 282 (5396):2085–2088

Pradhan D, Krahling S, Williamson P et al (1997) Multiple systems for recognition of apoptotic lymphocytes by macrophages. Mol Biol Cell 8:767–778

Raabe T, Bukrinsky M, Currie RA (1998) Relative contribution of transcription and translation to the induction of tumour necrosis factor-α by lipopolysaccharide. J Biol Chem 273:974–980

Radke JR, Cook JL, Ucker DS unpublished data

Ray CA, Black RA, Kronheim SR et al (1992) Viral inhibition of inflammation: cowpox virus encodes an inhibitor of the interleukin-1 beta converting enzyme. Cell 69 (4):597–604

Reddy SM, Hsiao KH, Abernethy VE et al (2002) Phagocytosis of apoptotic cells by macrophages induces novel signalling events leading to cytokine-independent survival and inhibition of proliferation: activation of Akt and inhibition of extracellular signal-regulated kinases 1 and 2. J Immunol 169 (2):702–713

Ronchetti A, Rovere P, Iezzi G et al (1999) Immunogenicity of apoptotic cells in vivo: role of antigen load, antigen- presenting cells, and cytokines. J Immunol 163:130–136

Rothlin CV, Ghosh S, Zuniga EI et al (2007) TAM receptors are pleiotropic inhibitors of the innate immune response. Cell 131 (6):1124–1136

Saunders JW Jr (1966) Death in embryonic systems. Science 154:604–612

Sauter B, Albert ML, Francisco L et al (2000) Consequences of cell death: exposure to necrotic tumour cells, but not primary tissue cells or apoptotic cells, induces the maturation of immunostimulatory dendritic cells. J Exp Med 191:423–434

Savill J, Dransfield I, Gregory C et al (2002) A blast from the past: clearance of apoptotic cells regulates immune responses. Nat Rev Immunol 2:965–975

Savill JS, Wyllie AH, Henson JE et al (1989) Macrophage phagocytosis of aging neutrophils in inflammation: programmed cell death in the neutrophil leads to its recognition by macrophages. J Clin Invest 83:865–875

Scaffidi P, Misteli T, Bianchi ME (2002) Release of chromatin protein HMGB1 by necrotic cells triggers inflammation. Nature 418:191–195

Schlegel RA, Stevens M, Lumley-Sapanski K et al (1993) Altered lipid packing identifies apoptotic thymocytes. Immunol Lett 36:283–288

Schroit AJ, Madsen JW, Tanaka Y (1985) In vitro recognition and clearance of red blood cells containing phosphatidylserine in their plasma membranes. J Biol Chem 260:5131–5138

Sen P, Wallet MA, Yi Z, Huang Y et al (2007) Apoptotic cells induce Mer tyrosine kinase-dependent blockade of NF-κB activation in dendritic cells. Blood 109 (2):653–660

Shakhov AN, Collart MA, Vassalli P et al (1990) κB-type enhancers are involved in lipopolysaccharide-mediated transcriptional activation of the tumour necrosis factor α gene in primary macrophages. J Exp Med 171:35–47

Shi Y, Evans JE, Rock KL (2003) Molecular identification of a danger signal that alerts the immune system to dying cells. Nature 425 (6957):516–521

Shweiki D, Itin A, Soffer D et al (1992) Vascular endothelial growth factor induced by hypoxia may mediate hypoxia-initiated angiogenesis. Nature 359 (6398):843–845

Škoberne M, Beignon A-S, Larsson M et al (2005) Apoptotic cells at the crossroads of tolerance and immunity. Curr Top Microbiol Immunol 289:259–292

Sly LM, Rauh MJ, Kalesnikoff J et al (2004) LPS-induced upregulation of SHIP is essential for endotoxin tolerance. Immunity 21 (2):227–239

Stern M, Savill J, Haslett C (1996) Human monocyte-derived macrophage phagocytosis of senescent eosinophils undergoing apoptosis: mediation by αVβ3/CD36/thrombospondin recognition mechanism and lack of phlogistic response. Am J Pathol 149:911–921

Stranges PB, Watson J, Cooper CJ et al (2007) Elimination of antigen-presenting cells and autoreactive T cells by Fas contributes to prevention of autoimmunity. Immunity 26 (5):629–641

Stuart GR, Lynch NJ, Day AJ et al (1997) The C1q and collectin binding site within C1q receptor (cell surface calreticulin). Immunopharmacology 38 (1–2):73–80

Stuart LM, Lucas M, Simpson C et al (2002) Inhibitory effects of apoptotic cell ingestion upon endotoxin-driven myeloid dendritic cell maturation. J Immunol 168 (4):1627–1635

Suchin KR, Cassin M, Washko R et al (1999) Extracorporeal photochemotherapy does not suppress T- or B-cell responses to novel or recall antigens. J Am Acad Dermatol 41:980–986

Suzuki N, Suzuki S, Duncan GS et al (2002) Severe impairment of interleukin-1 and Toll-like receptor signalling in mice lacking IRAK-4. Nature 416:750–756

Tassiulas I, Park-Min K-H, Hu Y et al (2007) Apoptotic cells inhibit LPS-induced cytokine and chemokine production and IFN responses in macrophages. Hum Immunol 68 (3):156–164

Toshchakov V, Jones BW, Perera P-Y et al (2002) TLR4, but not TLR2, mediates IFN-β-induced STAT1α/β-dependent gene expression in macrophages. Nat Immunol 3 (4):392–398

van 't Veer C, Van Den Pangaart PS, van Zoelen MAD et al (2007) Induction of IRAK-M is associated with lipopolysaccharide tolerance in a human endotoxemia model. J Immunol 179 (10):7110–7120

Vanden Berghe W, De Bosscher K, Boone E et al (1999) The nuclear factor-κB engages CBP/p300 and histone acetyltransferase activity for transcriptional activation of the interleukin-6 gene promoter. J Biol Chem 274:32091–32098

Voll RE, Herrmann M, Roth EA et al (1997) Immunosuppressive effects of apoptotic cells. Nature 390:350–351

Wanderley JLM, Moreira MEC, Benjamin A et al (2006) Mimicry of apoptotic cells by exposing phosphatidylserine participates in the establishment of amastigotes of *Leishmania (L) amazonensis* in mammalian hosts. J Immunol 176 (3):1834–1839

Weigert A, Johann AM, von Knethen A et al (2006) Apoptotic cells promote macrophage survival by releasing the antiapoptotic mediator sphingosine-1-phosphate. Blood 108 (5):1635–1642

Werts C, Girardin SE, Philpott DJ (2006) TIR, CARD and PYRIN: three domains for an antimicrobial triad. Cell Death Differ 13 (5):798–815

Witko-Sarsat V, Rieu P, Descamps-Latscha B et al (2000) Neutrophils: molecules, functions and pathophysiological aspects. Lab Invest 80 (5):617–653

Wood W, Turmaine M, Weber R et al (2000) Mesenchymal cells engulf and clear apoptotic footplate cells in macrophageless PU.1 null mouse embryos. Development 127:5245–5252

Wyllie AH, Kerr JFR, Currie AR (1980) Cell death: the significance of apoptosis. Int Rev Cytol 68:251–305

Xiao YQ, Malcolm K, Worthen GS et al (2002) Cross-talk between ERK and p38 MAPK mediates selective suppression of pro- inflammatory cytokines by transforming growth factor-β. J Biol Chem 277 (17):14884–14893

Chapter 7
Comparative Characterization of Non-professional and Professional Phagocyte Responses to Apoptotic Cells

Adam Lacy-Hulbert

Abstract: Cells that die by apoptosis are rapidly removed in vivo, both by neighbouring tissue cells ('non-professional phagocytes') and by the professional phagocytes of the innate immune system, macrophages and dendritic cells. This ongoing process of death and removal of dying cells appears silent at the level of the organism, with cells melting away and replacement cells taking their place. However recent advances in the study of dying cell removal and particularly studies of the consequences of defective apoptotic cell removal reveal that this process is far from silent at the cellular level and contributes critically to tissue homeostasis and immune regulation. This occurs through the stimulation of distinct responses by cells following engagement of apoptotic cells. The nature of these responses and how they appear tailored for the phagocyte and the context of apoptotic cell removal are reviewed and discussed here.

Keywords: Phagocyte • Apoptosis • Response • Tolerance • Regeneration

7.1 Introduction

7.1.1 Apoptosis and Phagocytosis of Apoptotic Cells

In metazoans, cells have the ability to self-destruct, a process termed programmed cell death. Although many forms of such 'cellular suicide' have been described (Krysko et al., Chap. 1; Diez-Fraile et al., Chap. 2, this vol.), the best studied is apoptosis. Apoptosis was originally described as a physiological form of controlled cell death that occurs sporadically in tissues, with dying cells undergoing characteristic

A. Lacy-Hulbert (✉)
Developmental Immunology Program, Department of Pediatrics
Massachusetts General Hospital
Boston, MA 02139, USA
Tel.: (+1) 617 643 5346
Fax: (+1) 617 724 3248
E-mail: adamlh@mit.edu

D. V. Krysko, P. Vandenabeele (eds.), *Phagocytosis of Dying Cells*,
DOI: 10.1007/978-1-4020-9292-3_7, © Springer Science+Business Media B.V. 2009

morphological changes and eventually breaking up into membrane bound vesicles, known as blebs or apoptotic bodies. These apoptotic bodies are swiftly removed by neighbouring phagocytic cells with little or no inflammation or tissue perturbation. Hence apoptosis was described as an active process of cell death and removal, which is able to balance cell proliferation and maintain cell turnover in a multicellular organism.

Perhaps the most dramatic examples of apoptosis occur during development, when coordinated cell proliferation and cell death are required to shape key structures and organs (Jacobson et al. 1997). The subsequent engulfment and removal of apoptotic cells is a key component of this process and the persistence of developmental apoptotic cells is a hallmark of defective cell clearance. In simple metazoans such as *Caenorhabditis elegans*, this property has been exploited to uncover the pathways required for cell clearance; in mammals, disruption of apoptotic cell clearance can lead to a delay of tissue remodeling (Gronski and Ravichandran, Chap. 5, this vol.; Wood et al. 2000). These observations emphasize the important principle that apoptosis and subsequent apoptotic cell removal appear to act in concert and, as discussed in this chapter, may be more closely linked through communication between apoptotic cell and the phagocyte.

Apoptosis also occurs in tissue homeostasis, most notably in the immune system. There is massive turnover of the mammalian immune system, with many millions of unwanted blood cells dying every day by apoptosis and being removed by phagocytes in the bone-marrow and other lymphoid organs. These include developing lymphocytes undergoing deletion in the thymus and bone marrow, and circulating leukocytes that need to be replaced by newly generated cells. During active inflammatory and immune responses, leukocyte production increases dramatically and cells are recruited to tissues and local lymph nodes in enormous numbers. However, once the infection is neutralized or the injury repaired, normal tissue homeostasis is quickly re-established with little or no evidence of inflammation. This is due in large part to local apoptosis of inflammatory cells such as neutrophils and lymphocytes and their removal by macrophages, and it is this process of inflammation resolution that has driven many of the studies of apoptotic cell phagocytosis. Analogous to development, unwanted leukocytes appear to 'melt away' revealing the intact original tissue structure. Disruption of this process would lead to persistence of apoptotic immune cells that would eventually burst, releasing their noxious granule contents, causing tissue damage and perpetuating inflammation. Hence phagocytosis can be considered an essential partner of apoptosis in multicellular organism, ensuring that the dying cell is removed and digested.

7.1.2 Consequences of Apoptotic Cell Uptake

Recent advances, however, show that phagocytosis of apoptotic cells represents much more than waste disposal and that dying cells can have profound effects on the immunological and other responses in the organism (Fig. 7.1). Engagement of

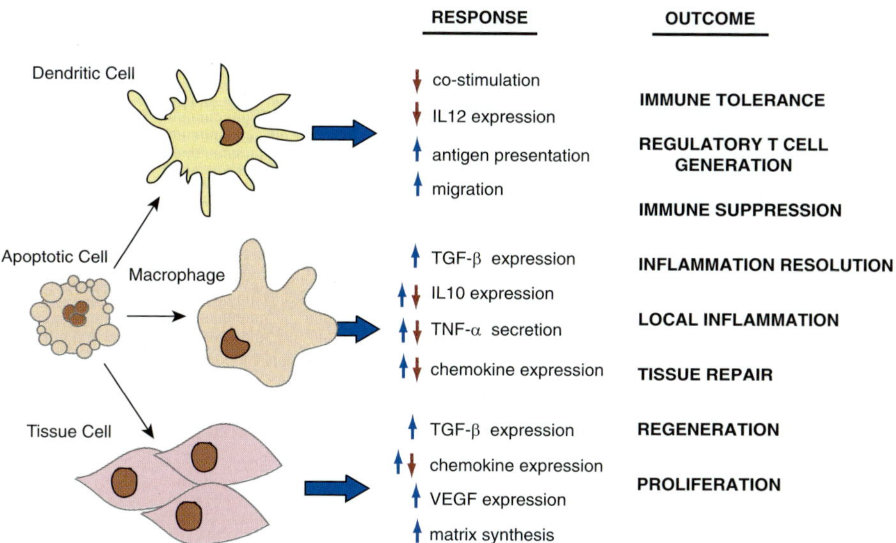

Fig. 7.1   Responses to phagocytosis of apoptotic cells. Recognition and phagocytosis of apoptotic cells stimulates distinctive responses. These lead to changes in cytokine and chemokine expression, cell migration and proliferation. The outcomes of these responses affect a number of processes in the organism, including immune regulation, inflammation resolution and tissue homeostasis. Key responses of professional phagocytes of the immune system (macrophages and DCs) and non-professional phagocytes (tissue cells such as epithelial or muscle cells) are summarized here.

dying cells by the immune phagocytes, macrophages and dendritic cells (DCs), leads to powerful anti-inflammatory and immunosuppressive signalling (Ucker, Chap. 6, this vol.; Fadok et al. 1998; Stuart et al. 2002). Such responses actively promote inflammation resolution and tissue repair (Huynh et al. 2002) and provide an important source of self-antigen for instructing immune tolerance (Liu et al. 2002). These results are underscored by a growing understanding of the consequences of defective removal of dying cells, which in many cases leads to spontaneous inflammation and autoimmunity (Gregory and Pound, Chap. 9, this vol.). Although such phenotypes were initially ascribed to the accumulation of uncleared apoptotic cells leading to pro-inflammatory secondary necrosis and presentation to the immune system (Botto et al. 1998), this seems unlikely given the occasional examples of failed apoptotic cell clearance that lead to persistent dead cells without inflammation (Devitt et al. 2004; Stuart et al. 2005). It is more likely when apoptotic cell clearance is perturbed, anti-inflammatory or tolerogenic responses to dying cells are not generated, and this causes inflammation and autoimmunity. We do not yet have a complete picture of how responses to apoptotic cells are generated or contribute to many cell processes, but we propose that they should be considered a vital third 'partner' in cell death, alongside apoptosis and removal. It is only the concerted action of cell death, removal and response that lead to the efficient and apparently silent removal of cells.

7.1.3   The Legacy of the Dying Cell

These observations have triggered a shift from the study of the process of cell death and phagocytosis, towards the response of living cells to the recently dead, and a realization that this may be the most important legacy of cell death. These responses of course vary depending on the phagocyte, the context and on the type of cell death. The focus of this chapter is on homeostatic responses but cell death can also trigger inflammation and immunity. Understanding the nature of the response of phagocytes to apoptotic cells and how the balance of pro- and anti-inflammatory signals is determined will provide important insights into autoimmunity, chronic inflammation and cancer. Here we will consider how a variety of cell types respond to interaction with apoptotic cells. We will focus particularly on the professional phagocytes of the immune system, macrophages and DCs, but also consider responses of tissue cells such as epithelial and smooth muscle cells to apoptosis of neighbouring cells.

7.2   Apoptotic Cell Uptake by the Immune System: Responses of Professional Phagocytes to Apoptotic Cells

7.2.1   'Professional' Phagocytes

The consequences of apoptotic cell uptake have been studied in most detail in the context of the professional phagocytes of the mammalian immune system, macrophages and DCs, and it is here that we currently have the best understanding of responses to apoptotic cells and the mechanisms by which these occur. Macrophages and DCs are vital components of mammalian immunity and act as sentinels of infection and injury in the body as well as essential effectors of inflammation and immune responses. Both cell types internalize and digest apoptotic cells very efficiently and use this ability to serve two functions: firstly, the removal and recycling of unwanted or damaged cells; and secondly, the acquisition of antigen for instruction of the immune system.

7.2.2   Macrophages: Classical Scavenging Phagocytes of the Immune System

Macrophages are specialized phagocytic cells derived primarily from circulating monocytes. Relatively long-lived populations of macrophages can be found in most tissues and body cavities where they act as a front line of defence against infection and injury, initiating and orchestrating inflammatory responses. These resident macrophages are versatile and plastic cells, which are able to adapt to their environment and context to perform a variety of functions. Often this leads to differentiation into distinct, specialized cell types; these include multinucleated osteoclasts, responsible for resorbtion of bone, Langerhans cell of the skin and Kupffer cells in the liver. Macrophages are also recruited at high numbers to sites of inflammation and infection. These 'inflammatory macrophages' are specialized for phagocytosis and destruction of invading microorganisms, and often secrete potent injurious agents such as reactive oxygen and nitrogen intermediates and powerful immuno-stimulatory cytokines. However, inflammatory macrophages are also recruited to sites of sterile inflammation and wounding where they promote resolution rather than tissue injury.

Although macrophages are generally considered as scavengers, they also possess important synthetic ability, producing local immune effectors (including opsonins, proteases and anti-microbial agents), extracellular matrix, cytokines and chemokines. These functions are tightly controlled and activated on receipt of appropriate environmental cues, such as stimulation by specific cytokines or engagement of microbial components. Hence macrophages act as local 'integrators' of environmental signals and critical orchestrators of immune responses.

7.2.3 Macrophages as Apoptotic Cell Phagocytes

Macrophages are major contributors to removal of apoptotic cells during development, although, as discussed in the section 7.4, local mesenchymal cells can also share the phagocytic burden. Apoptosis also occurs at high levels during lymphocyte development and differentiation as T and B cells assemble and refine their antigen-specific receptor repertoire. Lymphocytes that die during these developmental processes, or surplus lymphocytes generated during immune responses, are removed by a combination of lymph tissue cells (such as thymic epithelial cells) and lymph node-resident macrophages. This is particularly striking in the germinal centers formed in lymph tissues during B cell maturation. These clusters of proliferating B cells contain histological characteristic 'tingible bodies'; these are in fact remnants of apoptotic B cells that have been internalized by macrophages and are undergoing degradation. The fact that engulfed apoptotic cells can be readily seen in lymph tissues but 'free' apoptotic cells are rare, despite the high rate of apoptosis, is testament to the efficient recognition and removal of dead cells by macrophages.

A major driving force in the understanding of apoptotic cell phagocytosis by macrophages came from studies of the removal of neutrophils and other acute inflammatory leukocytes during inflammation resolution. Neutrophils are recruited in massive numbers to sites of injury and infection and must be efficiently removed following elimination of the insult to allow the tissue to return to its pre-perturbed state. It has long been known that this occurs by ingestion of intact neutrophils by macrophages (Metchnikoff 1893; Newman et al. 1982). However, almost 20 years ago it was discovered that this process was triggered by apoptosis of the neutrophil (Savill et al. 1989). It is now known that neutrophils have a short lifespan and are pre-programmed to die by apoptosis within 12 to 48 hours of leaving the bloodstream. Once the apoptotic programme is initiated, neutrophils are rapidly recognized and phagocytosed by macrophages, leading to their safe removal (Savill et al. 1989). The close co-operation between apoptosis and efficient phagocytosis appears to be an essential step in inflammation reduction and resolution and perturbations of these processes are thought to contribute to chronic inflammatory conditions. Hence conditions that inhibit inflammatory cell apoptosis, such as chronic granulomatous disease (CGD), are associated with sterile inflammation (Brown et al. 2003). In contrast, stimulating inflammatory cell apoptosis, such as through the use of cyclin-dependent kinase inhibitors in vivo, promotes resolution of ongoing inflammation (Rossi et al. 2006). Another class of anti-inflammatory drugs, glucocorticoids, may promote resolution through both stimulation of inflammatory cell apoptosis (Meagher et al. 1996) and enhancing subsequent phagocytosis by macrophages (Giles et al. 2001; Gilmour et al. 2006; Liu et al. 1999).

7.2.4 Apoptotic Cell Uptake by Macrophages Regulates Inflammation

Early studies of macrophage encounters with dying cells therefore concentrated on phagocytosis of apoptotic immune cells that would be encountered in inflammatory situations, such as neutrophils and eosinophils, with the view that phagocytosis functioned to facilitate prompt removal of unwanted leukocytes. These studies stressed the 'silent' nature of this removal (Meagher et al. 1992; Savill et al. 1989). Macrophages normally respond to internalization of particles by production of pro-inflammatory mediators. However, macrophages ingesting apoptotic neutrophils do not release pro-inflammatory agents such as thromboxane (Meagher et al. 1992). Importantly, apoptotic cells opsonised with serum, and therefore recognized though complement and/or Fc-receptors, continued to stimulate robust inflammatory responses. These results showed that the lack of inflammatory response to apoptotic cells was due to the mechanism of recognition by the macrophage rather than another inherent property of the apoptotic cell; engagement of specific macrophage receptors for apoptotic cells did not lead to inflammatory responses. This led to the proposal that the receptors for apoptotic cells stimulated phagocytosis but not response, and that apoptotic cell removal occurred without immune consequences.

However, two very important subsequent studies revealed that apoptotic leukocyte clearance was not 'immunologically neutral', but instead actively promoted inflammation resolution. Firstly, Voll and colleagues showed that the cytokine responses of LPS-stimulated monocytes were profoundly affected by co-culture with apoptotic lymphocytes (Ucker, Chap. 6, this vol.; Voll et al. 1997). Whereas LPS normally stimulated production of high levels of TNF-α by monocytes, this was suppressed by apoptotic cell engagement. In contrast, the production of the anti-inflammatory cytokines TGF-β and IL-10 were both increased by the presence of apoptotic cells. Secondly, similar results were reported in human macrophages by Fadok, with apoptotic cell phagocytosis causing suppression of production of IL-1β, GM-CSF, TNF-α and IL-10 (Fadok et al. 1998). In agreement with Voll and colleagues, Fadok showed that apoptotic cell uptake stimulated release of TGF-β by macrophages, as well as other anti-inflammatory mediators. Importantly this study also demonstrated that the anti-inflammatory effects of apoptotic cells were due to TGF-β, prostaglandin E2 and platelet activating factor released from macrophages (Fadok et al. 1998). There is now strong in vivo evidence (Huynh et al. 2002) to support the proposal that macrophage release of TGF-β following apoptotic cell uptake suppresses pro-inflammatory signalling and promotes inflammation resolution. The administration of apoptotic cells to inflamed sites in vivo promotes local TGF-β production, leading to reduced inflammatory cell recruitment and accelerated inflammation resolution (Huynh et al. 2002). The 'active component' of apoptotic cells in these experiments was surface phosphatidylserine (PS), providing some clues to the mechanisms by which apoptotic cells promote these responses.

This phenomenon of active response to apoptotic cells has since been reproduced in primary macrophages/ monocytes and in cell lines from a number of organisms. The suppressive effects of apoptotic cells appear to be widespread, affecting production of many different pro-inflammatory agents following stimulation by most TLR ligands (Lucas et al. 2003). From such studies a model for the effects of apoptotic cell uptake has emerged, with engagement of key receptors on the monocyte or macrophage surface leading to production of immunomodulatory cytokines, particularly TGF-β, which regulate cytokine production (Fig. 7.2). From this model, two predictions can be made regarding apoptotic cell responses: (1) that many of the immuno-suppressive effects of apoptotic cell uptake would only follow release and action of TGF-β; and (2) that the effects of apoptotic cell uptake would spread to neighbouring cells due to paracrine signalling by TGF-β. In studies of cytokine release by LPS stimulated macrophages, both of these results are seen. Production of TNF-α and MCP-1 are only reduced 6-12 hours following apoptotic cell engagement, corresponding to the time course of release of TGF-β (Lucas et al. 2003). Furthermore, the effects of apoptotic cell uptake are seen in all cells in culture, when only 20–40% of macrophages have ingested apoptotic cells (Lucas et al. 2006). Further analysis of this mouse macrophage system has shown that this 'paracrine' effect requires initial contact with apoptotic cells to license macrophages for later response to TGF-β (Lucas et al. 2006; Fig. 7.2), representing a further control of the immunosuppressive effects of apoptotic cells. The possible mechanisms for this effect and the receptors required for response and initiation of anti-inflammatory signalling are discussed in more detail later in this chapter.

Macrophage responses to apoptotic cells are not universally anti-inflammatory. More detailed analysis of cytokine production following apoptotic cell uptake revealed that apoptotic neutrophils cause an initial increase in production of TNF-α shortly (2-4 hours) after LPS stimulation although TNF-α production was significantly inhibited relative to LPS treatment at 24 hours (Lucas et al. 2003). The early production of chemokines MCP-1 and MIP-1α were also increased by apoptotic cells (Lucas et al. 2003). These early effects of apoptotic cells are also only revealed when macrophages are stimulated with microbial components such as LPS,

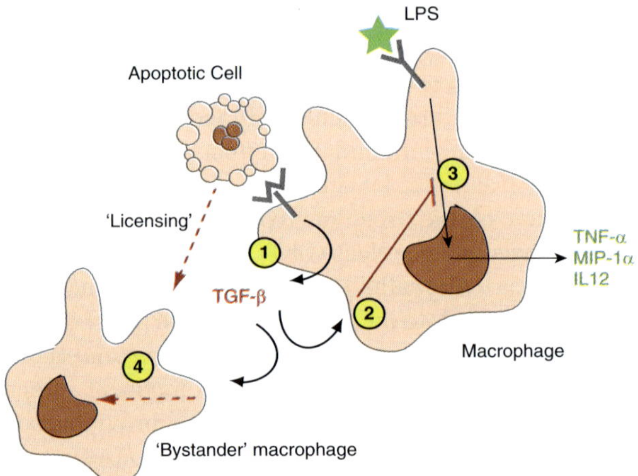

Fig. 7.2 TGF-β-mediated model of apoptotic cell effects. In a model based on the original studies of Fadok and colleagues (Fadok et al. 1998), apoptotic cell engagement stimulates release of TGF-β and other anti-inflammatory mediators (1). These then bind TGF-β receptors on the phagocytosing cell (2) and regulate responses to pro-inflammatory stimulation by LPS or other TLR ligands, inhibiting production of TNF-α and other cytokines (3). Secreted TGF-β can also act on neighbouring cells (4), but these may need to first encounter apoptotic cells to 'license' them to respond fully (Lucas et al. 2006).

suggesting that phagocytosis may prime or enhance TLR signalling in macrophages. Similar transient pro-inflammatory effects of apoptotic cell uptake on macrophages have been reported by others. Macrophage cell lines and primary mouse macrophages produced increased amounts of MIP-2 following apoptotic cell uptake (Kurosaka et al. 2001) and this stimulates neutrophil recruitment in vivo following ectopic administration of apoptotic cells. The production of pro-inflammatory cytokines is dependent on the stage of apoptosis of the phagocytosed cell, with cells at very early stages of apoptosis failing to stimulate pro-inflammatory cytokine production (Kurosaka et al. 2003). The addition of serum or bystander monocytic cells suppressed IL-8/ MIP-2 production, partly through the production of anti-inflammatory cytokines IL-10 and TGF-β (Kurosaka et al. 2002; Takahashi et al. 2004), more consistent with the anti-inflammatory effects of apoptotic cell uptake described above. Chemokine production by macrophages is also suppressed by additional factors, such as Nitric Oxide (NO) and adiponectin, which are produced following phagocytosis of early and late apoptotic cells respectively (Saijo et al. 2005; Shibata et al. 2002).

These apparently paradoxical pro- and anti-inflammatory effects of macrophages appear to reflect the complex requirements for macrophage responses to apoptotic cells. We have proposed that this 're-programming' of macrophage phenotype stimulates tissue repair and inflammation resolution by a limited recruitment of additional innate immune cells coupled with suppression of adaptive immune responses. Hence significant amounts of apoptosis would stimulate local inflammation but also promote resolution and immune tolerance. In support of this, treatments that induce widespread cell death, such as X-irradiation, lead to recruitment of inflammatory cells including neutrophils to sites of apoptosis. Furthermore, rather than stimulating further inflammation, recruited neutrophils actually accelerate apoptotic cell phagocytosis and digestion by macrophages (Iyoda et al. 2005).

Phagocytosis of apoptotic cells may also act as a brake on immune cell generation. The phagocytosis of neutrophils by macrophages in peripheral tissues reduces expression of the cytokine IL-23. IL-23 normally stimulates the expression of IL-17 by subpopulations of lymph node and spleen T cells, which, in turn, promotes production of G-CSF and granulopoiesis (Stark et al. 2005). Hence, the presence of apoptotic neutrophils in tissue ultimately downregulates granulopoiesis. These results are interesting as they suggest that the control point for neutrophil homeostasis is at the removal of unwanted apoptotic cells, rather than monitoring of newly generated or circulating cells. In addition, this control appears to originate from the intestine, a major source of both IL-23 and IL-17, and a site of constant challenge from microbial stimuli.

7.2.5   DCs: Specialized Antigen Presenting Cells

DCs are a highly specialized population of immune phagocytes, which serve to instruct antigen-dependent responses by the adaptive immune system. DCs begin their life as immature cells in the peripheral tissues, specialized for antigen capture. Through a combination of macropinocytosis, endocytosis and phagocytosis they efficiently scavenge components of their surroundings. They also bear a range of pattern recognition and cytokine receptors to sense local infection and injury. When stimulated through these receptors, DCs become activated and mature, migrating to local lymph nodes and differentiating into antigen presenting cells (APCs). Here they present their acquired antigen to naïve T cells and initiate adaptive immune responses.

7.2.6   DCs Phagocytose Apoptotic Cells and Cross Present Apoptotic Antigens

The observation that DCs could ingest apoptotic cells and present antigens derived from dying cells (Albert et al. 1998a; Albert et al. 1998b; Rubartelli et al. 1997) provided a route for antigen from virally infected or tumor cells to enter DCs and be presented to the immune system. In seminal studies by Albert and colleagues (Albert et al. 1998a; Albert et al. 1998b), human monocytes that were infected with influenza virus and underwent apoptosis were efficiently phagocytosed by DCs, which were then able to induce virus-specific CD8[+] cytotoxic T cells (CTLs) (Albert et al. 1998b). Furthermore, the pathway of uptake of apoptotic cells was similar to that used by macrophages, involving αv integrins and the scavenger receptor CD36 (Albert et al. 1998a). These studies were particularly important as they described an efficient pathway for cellular antigens to access the MHC class I pathway in DCs and initiate CD8 CTL responses, a process termed cross priming or cross-presentation.

CD8[+] CTLs mediate immunity to viral infection and tumors by recognizing antigens on the surface of infected cells. Classically these antigens are expressed in the cytoplasm of the target cell, processed and presented on MHC class I. This pathway of antigen presentation is normally restricted to endogenous cellular antigens; in contrast, antigen internalized from the surroundings enters the MHC class II pathway and is recognized by CD4[+] T cells. The ability of exogenous cellular antigens to be acquired and presented to CTLs had been previously described, for example following transplantation, but the description of apoptotic cell uptake and antigen capture by DCs provided a universal pathway by which this could occur. Furthermore, apoptotic cells provided a much more efficient route than other antigen delivery methods for entering the cross-presentation pathway. The details of how phagocytosed antigen is able to access the Class I pathway and why this is particularly favored for apoptotic cells remain unclear. However, it is clear that this represents a specialization of DCs over macrophages and non-professional phagocytes and is probably in part due to the mechanism of phagocytosis employed (Houde et al. 2003). It has also recently been shown that apoptosis can 'pre-process' antigens in internalized cells, raising the interesting possibility that the programme of apoptosis may favor presentation of certain cellular antigens by DCs (Blachere et al. 2005).

7.2.7 *Apoptotic Cell Uptake by DCs Regulates Immune Responses*

These interesting results nevertheless raised the important question of how the immune system avoids constant presentation of self-antigen from normal dying cells, which would be likely to lead to autoimmunity. The answer to this lies in the specific response of DCs to apoptotic cell uptake. Through mechanisms similar to those reported in macrophages, apoptotic cells alter the maturation programme of DCs to suppress immune responses to apoptotic cells. Cultured bone marrow-derived DCs that have internalized apoptotic cells and are then stimulated with the TLR-ligand LPS express lower levels of cytokines (IL-12) and co-stimulatory molecules (CD40, CD80 and CD86) than DCs stimulated with LPS alone (Stuart et al. 2002). These DCs consequently have a reduced ability to stimulate T cell proliferation (Stuart et al. 2002) and programme T cells for subsequent apoptosis and deletion (Stuart and Lacy-Hulbert, *unpublished data*). The induction of these 'tolerogenic' apoptotic cell-containing DCs occurred in apparently uniform cultures of DCs and only affected DCs that internalized apoptotic cells, with neighbouring DCs unaffected (Stuart et al. 2002). These results demonstrate that apoptotic cell uptake directly affects DC phenotype, rather than marking a subpopulation of DCs capable of internalizing apoptotic cells. We believe that this phagocytosis-dependent mechanism restricts the immunosuppressive effects of apoptotic cells to apoptotic cell-derived antigen. This level of control would be important to maintain protective immunity to pathogens during infections associated with apoptotic cell death. Furthermore, the skewing of tolerogenic responses by apoptotic cells can be overcome by strong stimulation with TLR ligands such as LPS (Stuart et al. 2002) or other DC stimuli such as anti-CD40 antibody (Liu et al. 2002), demonstrating that apoptotic cell-containing DCs are capable of differentiating into immuno-stimulatory DCs. Hence we would propose that induction of immune tolerance is a specific response of DCs to apoptotic cell uptake, rather than being tolerance and phagocytosis being two shared properties of a DC subtype. Similar effects of apoptotic cells on DCs have subsequently been reported for primary human and mouse DCs, demonstrating that this effect is widespread and not limited to specific DC subpopulations (Ip and Lau 2004; Stuart et al. 2002; Verbovetski et al. 2002). It is notable that certain aspects of DC maturation are unaffected, or even stimulated by apoptotic cell uptake, including upregulation of MHC class II, antigen presentation, loss of phagocytic ability and migration to Lymph Nodes (Stuart et al. 2002; Verbovetski et al. 2002). Hence apoptotic cells induce a distinct maturation programme in DCs that does not prevent presentation of apoptotic cell derived (self) antigen but instead skews the outcome of the immune response to that antigen towards immune tolerance.

7.2.8 Apoptotic Cell Uptake by DCs Induces Immune Tolerance

This programme of apoptotic cell responses is used in the body to establish and maintain peripheral tolerance in the steady state. Peripheral tissue-resident DCs constitutively internalize apoptotic cells and traffic to lymph nodes, where they induce tolerance to apoptotic cell derived antigens (Savill et al. 2002; Steinman et al. 2000). This has been well characterized in the intestine, a site of constant immunological challenge where strong immune regulation is essential. The intestinal epithelial layer is constantly regenerated by local proliferation of stem cells and apoptosis of aging epithelial cells. Although many of these dying cells are shed into the lumen of the intestine, a significant number are phagocytosed by underlying macrophages and DCs and it has been possible to identify DCs bearing these apoptotic epithelial cells in the draining lymph and mesenteric lymph nodes (Huang et al. 2000; Liu et al. 1998). These DCs express low levels of costimulatory molecules, are poor stimulators of T cell proliferation and are likely to be involved in mediating peripheral tolerance (Huang et al. 2000; Jang et al. 2006). These may represent a distinct population of DCs that are specialized for phagocytosis of apoptotic cells and are hard-wired to induce immune tolerance. However, their close phenotypic and morphologic similarity to cultured DCs that have internalized apoptotic cells suggest that the recognition and response to apoptotic cells is an essential determinant of their tolerogenic capacity.

Conversely, defective clearance of apoptotic cells has been linked to development of autoimmunity in human diseases and genetically altered mouse models (Botto et al. 1998; Cohen et al. 2002; Hanayama et al. 2004b; Savill et al. 2002). Historically this has been attributed to the accumulation of uncleared apoptotic cell debris that stimulates inflammation and is then internalized and presented by DCs. However, our growing understanding of the mechanisms by which apoptotic cells play an active part in immune tolerance suggests that instead autoimmunity arises from failed recognition or clearance of apoptotic cells by DCs, blocking a normal pathway of immunological regulation. Apoptotic cells have been used in a number of systems to 'immunize' mice against apoptotic or apoptotic cell-associated antigens. In many of these cases, high numbers of apoptotic cells are used or adjuvants are delivered with the apoptotic cell to induce immunity. However, administration of apoptotic cells in the absence of adjuvants or inflammation has been found to lead to immune tolerance, in agreement with in vitro effects of apoptotic cells on DCs. Injected circulating apoptotic cells are rapidly internalized by splenic CD8α DCs which induce antigen-specific T cell proliferation, subsequent deletion and tolerance (Liu et al. 2002). This tolerance is associated with down-regulation of cytokine production by splenic DCs, as seen for in vitro culture of DCs and macrophages with apoptotic cells (Morelli et al. 2003). Building on these results, Morelli and colleagues have been able to induce tolerance to transplanted organs by co-administration of donor apoptotic cells, demonstrating that these responses to apoptotic cells can be manipulated therapeutically (Divito and Morelli, Chap. 11, this vol.; Morelli 2006; Wang et al. 2006).

7.2.9 Apoptotic Cells and Regulatory T cells

Immune tolerance in response to apoptotic cells is probably mediated in part by the generation of regulatory T cells (Treg cells). Treg cells are CD4[+] T cells that suppress T cell responses and act as critical mediators of immune tolerance. It has been proposed that two principal populations of Treg cells exist. 'Innate' Treg cells are generated in the thymus and regulate immune responses to classic thymic self-antigens. In contrast, 'adaptive' Treg cells are generated in the periphery from naïve or activated T cells. Adaptive Treg cells mediate tolerance to innocuous or 'self-associated' antigens that are not expressed in the thymus, such as peripheral tissue-specific proteins or antigens from commensal bacteria and loss of these adaptive cells is associated with colitis in experimental mouse models (Coombes et al. 2007; Lacy-Hulbert et al. 2007). The induction of tolerance by antigen-laden apoptotic cells is associated with expansion of CD4[+] CD25[+] FoxP3[+] adaptive Treg cells (Kleinclauss et al. 2006; Saas et al. 2007). Generation of Treg cells and subsequent tolerance in this system requires both uptake of apoptotic cells by DCs, and production of TGF-β by macrophages. This fits well with what is known of adaptive Treg cells, which can be generated in vitro by sub-optimal stimulation (such as by immature DCs) in the presence of TGF-β. Hence the maturation programme initiated by apoptotic cells in DCs and macrophages generates ideal conditions for Treg generation.

7.2.10 Pro-immunogenic Responses to Apoptotic Cells

Many of the early reports of apoptotic cell uptake by DCs reported efficient antigen presentation and cross presentation and robust T cell responses, in contrast to the apparent anti-inflammatory responses described above. This probably reflects the type of apoptotic cell challenge and the maturation stimulus used experimentally. In most cases, immune activation by apoptotic cells requires a strong 'danger' signal which is able to overcome the natural anti-inflammatory response to the apoptotic cell (Gallucci et al. 1999). These danger signals include components of necrotic tumor cells and 'stressed' apoptotic cells, which are probably exposed heat shock proteins and type 1 cytokines such as IFN-γ, released during certain types of pathological cell death (Feng et al. 2002; Feng et al. 2003; Sauter et al. 2000). This is in agreement with our observations of macrophage responses to apoptotic cells, in which cytokine production by IFN-γ-stimulated macrophages was not affected by apoptotic cells, unlike responses to a range of TLR ligands (Lucas et al. 2003). However, TLR signals may be sufficient to induce immunity to apoptotic cells, such as in the context of infection-induced apoptosis. Viral infection often leads to apoptotic cell death and under certain circumstances DCs recognize virus-infected apoptotic cells and respond by stimulating cross-presentation and immunity rather than tolerance. In this case, productive immunity is stimulated by recognition of nucleic acids in the cytosol of infected cells by TLR3 in the DC (Schulz et al. 2005). It is interesting that many pathogen-associated ligands, recognized by TLRs, are regulated by apoptotic cells but other signals, particularly endogenous 'danger' signals, are not; this may represent a natural 'hierarchy' of DC responses to immunogenic and tolerogenic antigens. Such a model would help explain how apoptotic cells can participate in both tolerance and immunity and emphasizes the importance of the phagocytes response to the dying cell in determining the final outcome of cell death.

7.3 Phagocytosis of Apoptotic Cells During Development and Homeostasis: Responses of Non-classical Phagocytes to Dying Cells

Phagocytosis is a basic cellular process that is not limited to the professional phagocytes of the immune system. Almost all mesenchymal cells are capable of phagocytosis of relatively large particles, including apoptotic cells and apoptotic debris. Removal by close neighbours is therefore the most likely route of disposal of sporadic apoptosis in tissue. Mesenchymal cells are also capable of removal of larger numbers of apoptotic cells present during development or tissue remodeling. This process is normally carried out by macrophages, as is apparent, for example, in remodeling of the developing mammalian footplate to generate interdigital spaces (Hopkinson-Woolley et al. 1994). Here macrophages rapidly internalize dying cells,

phagocytes are present and available. Furthermore, phagocytosis by tissue cells provides more than a low level or 'back-up' mechanism for apoptotic cell clearance; apoptotic cells stimulate powerful responses in neighbouring phagocytosing tissue cells, similar to the immuno-modulatory effects of apoptotic cells on phagocytes of the immune system. Recognition of apoptotic cells by tissue cells promotes cell proliferation, angiogenesis and matrix synthesis whilst stimulating local inflammation, all key elements of tissue repair and regeneration. Hence, the process of cell death in damaged or wounded tissue, itself a contributor to tissue damage in the short term, provides the necessary signals to promote subsequent repair.

7.4 Mechanisms of Response to Apoptotic Cells

The exact mechanism of apoptotic cell responses remains poorly understood. Although it is clear that local production of TGF-β plays an important role in the downstream effects of apoptotic cell uptake, the receptors responsible for recognition of apoptotic cells and the early signalling events are not fully defined. There is consensus that contact with apoptotic cells, but not necessarily phagocytosis, is required for apoptotic cell responses, that this leads to intracellular signalling that has broad effects on gene transcription, and that a major consequence is suppression of expression of pro-inflammatory cytokines (Ucker, Chap. 6, this vol.; Cvetanovic et al. 2006). However, the extent to which suppression of cytokine production is primarily due to apoptotic cell engagement or secondary to release of key anti-inflammatory mediators is debated. In many studies apoptotic cell uptake triggers release of TGF-β, IL-10 or other factors which then mediate anti-inflammatory responses (as first reported by Fadok et al. 1998). In other systems, the effects of apoptotic cells on cytokine production appears to be a direct results of apoptotic cell engagement (Ucker, Chap. 6, this vol.; Cvetanovic et al. 2006). Such controversies probably arise from the many different systems used to study these effects, complicated by the large number of potential apoptotic cell receptors and widespread and varied reports of apoptotic cell effects. One potential mediator of apoptotic cell immunosuppressive effects, IL-10, has been reported to be both stimulated and repressed by apoptotic cell uptake (Fadok et al. 1998; Voll et al. 1997). Likewise, our own studies show that secretion of the pro-inflammatory cytokines TNF-α and MCP-1 are both stimulated and inhibited by apoptotic cell uptake depending on the time at which the cytokines are measured (Lucas et al. 2003). Although this diversity of cell systems can complicate study of the details of apoptotic cell phagocytosis and response, the convergence of these studies of many different cell types suggest that the mechanisms of response to apoptotic cells are broadly conserved. In this section, the current understanding of the mechanisms of apoptotic cell response will be summarized, focusing on the events that occur at different stages of apoptotic cell response: engagement by the phagocyte, receptor signalling and transcriptional activation/ suppression.

7.4.1   Cell Surface Recognition of Apoptotic Cells

In almost all studies of phagocyte response to dying cells, there is an absolute requirement for contact with the apoptotic cell and others (Cvetanovic and Ucker 2004; Fadok et al. 1998; Lucas et al. 2003; Lucas et al. 2006). Hence, in experiments where apoptotic cells are co-cultured but physically separated from responding cells, there is little suppression of pro-inflammatory cytokine production (Cvetanovic and Ucker 2004; Lucas et al. 2006). Likewise, conditioned culture medium from intact apoptotic cells generally does not induce anti-inflammatory effects in macrophages. As the apoptosis programme progresses, fragments of apoptotic cells (blebs or apoptotic bodies) are released, and these particles can mimic apoptotic cell responses (Peter et al., Chap. 3, this vol.) Such microparticles may be responsible for some reports of 'contact-independent' apoptotic cell effects. However, in general, responses to apoptotic cells are mediated by specific recognition of components of the dying cell surface (Napirei and Mannherz, Chap. 4; Gregory and Pound, Chap. 9, this vol.).

These recognized components represent changes in the surface composition of a cell as it dies, as it is clear that interaction with live cells does not induce the same anti-inflammatory signalling. There is debate over whether cells that die a non-programmed cell death, or necrosis, induce pro- or anti-inflammatory effects. There are many reports of pro-inflammatory stimulation by preparations of cells that have undergone a variety of physical insults (such as freezing, boiling or squashing) or that have simply fallen apart late in the apoptosis programme (secondary necrosis). However, other studies show that necrotic and apoptotic cells induce indistinguishable anti-inflammatory effects on phagocytes. The most likely explanation is that necrotic cells express the same surface components as apoptotic cells, but often also present strong pro-inflammatory signals (such as heat shock proteins, granule contents or nuclear components) that can overcome the anti-inflammatory effects of the apoptotic cell surface.

Cells at all stages of apoptotic cell death have been shown to stimulate anti-inflammatory signalling, suggesting that the key surface changes either occur very early in apoptosis, or that multiple cell surface changes can initiate signalling (Patel et al. 2006). One likely component is PS, which is exposed on the outer leaflet of dying cells early in the apoptosis programme. Lipid vesicles comprised of PS can mimic many of the effects of apoptotic cells, including release of TGF-β and suppression of anti-inflammatory signalling, both in vitro and in vivo (Huynh et al. 2002). However, apoptotic cells that do not express PS can also trigger anti-inflammatory effects (Cocco and Ucker 2001; Cvetanovic et al. 2006), suggesting that additional apoptotic surface components can initiate these signals.

7.4.2   Apoptotic Cell Receptors and Response

The requirement for contact with dying cells to initiate responses implicates the involvement of specific receptors in signalling the presence of apoptotic cells.

Apoptotic cells can be recognized and phagocytosed by a multitude of different receptors and opsonins, binding a variety of ligands on the apoptotic cell surface (Peter et al., Chap. 3; Napirei and Mannherz, Chap. 4; Gregory and Pound, Chap. 9; Dini and Vergallo, Chap. 15; this vol.). Several of these receptors have also been implicated in responses to apoptotic cells. Engagement of the scavenger receptor CD36 or integrin $\alpha v \beta 3$ using antibodies has been shown to mimic anti-inflammatory effects of apoptotic cells and stimulate production of TGF-β (Freire-de-Lima et al. 2000; Voll et al. 1997). However, macrophages from mice lacking either CD36 or $\beta 3$ integrins show normal anti-inflammatory signalling following apoptotic cell engagement, although phagocytosis is compromised (Gregory and Pound, Chap. 9; this vol.; Lucas et al. 2006). Therefore, although these receptors are involved in apoptotic cell phagocytosis and may be capable of initiating anti-inflammatory signalling, they are not essential for response to apoptotic cells. Similar results have been reported for many of the apoptotic cell receptors and opsonins, which make a major contribution to apoptotic cell phagocytosis but do not seem to be required for response. These include CD14 (Devitt et al. 2004), $\beta 3$, $\beta 5$, SRA and CD36 (Lucas et al. 2006). It is striking that a number of molecules implicated in apoptotic cell uptake and/or responses (CD36, CD14, SRA, C1q, MBL) are also involved in phagocytosis and response to TLR-bearing particles. In pathogen-associated molecular pattern recognition signalling, these molecules are thought to act by presenting potential ligands to TLRs and perhaps even in regulating assembly of functional TLR signalling modules. It is possible that they have similar roles in apoptotic cell recognition, presenting apoptotic cell surface molecules to recognition receptors and coordinating assembly of anti-inflammatory signalling modules. The presence of several different 'apoptotic cell-associated molecular patterns' would allow signalling to proceed in the absence of one receptor, as seen in knockout mice. However, there is currently no firm evidence for such a 'phagocytic synapse' in apoptotic cell recognition.

Several molecules mediate recognition of apoptotic cells through binding to surface PS. These include soluble proteins that act as opsonins (MFGE8, Gas6, Del1, Protein S; Anderson et al. 2003; Hanayama et al. 2004a; Hanayama et al. 2002; Ishimoto et al. 2000) and direct 'PS receptors' which include TIM-1, TIM-4, BAI-1 and stabilin 2 (Kobayashi et al. 2007; Miyanishi et al. 2007; Park et al. 2007; Park et al. 2008). The ability of PS to trigger anti-inflammatory signalling similar to that induced by apoptotic cells strongly implicates at least one of these receptors in apoptotic cell responses. The PS opsonins bind to two main receptors, Mer tyrosine kinase (MerTK; binds Protein S and Gas6) and αv integrins (bind to RGD sequences in MFGE8 and Del1). Mice in which either MerTK or αv are deleted have defective apoptotic cell clearance, inflammation and autoimmunity, suggesting on first glance that these may be the key receptors for apoptotic cell response but it is not yet possible to draw a firm conclusion as both genes play a number of roles in immune function making interpretation of the knockout phenotypes complex (Cohen et al. 2002; Lacy-Hulbert et al. 2007; Scott et al. 2001). Nonetheless, there is good evidence that both Mer and αv integrins are important components of the anti-inflammatory responses to apoptotic cells.

MerTK, along with Tyro and Axl, make up the family of TAM receptor tyrosine kinases, and have been implicated in homeostatic regulation of the immune system. Mice lacking one or more of the TAM receptors have hyperactivated macrophages and DCs, and develop spontaneous inflammation and autoimmunity (Lu and Lemke 2001). These receptors have now been shown to function as pleitrophic inhibitors of TLR signalling (Rothlin et al. 2007), being stimulated during TLR responses of macrophages and DCs and suppressing inflammatory responses through expression of SOCS1 and SOCS3. TAM receptors also bind apoptotic cells through the opsonins Gas6 and Protein S, which is therefore likely to stimulate similar suppressive responses; indeed it has recently been shown that deletion of MerTK or loss of Gas6 prevents apoptotic cell inhibition of DC maturation (Wallet et al. 2008).

MerTK mediates apoptotic cell uptake in concert with αv integrins, particularly αvβ5 (Wu et al. 2005). αv knockout mice have a lethal phenotype, dying at birth with developmental defects. However, using a conditional knockout strategy, mice lacking αv integrins only in the immune system have been generated (Lacy-Hulbert et al. 2007). These mice have impaired apoptotic cell clearance with reduced phagocytosis by both macrophages and DCs and persistence of apoptotic cells in vivo. αv-conditional knockout mice develop spontaneous inflammation (colitis) and autoimmunity, and this phenotype is due to specific loss of αv from macrophages and DCs. However the disruption of functions of αv integrins additional to apoptotic cell recognition are also likely to contribute to the development of colitis, particularly the essential role for αv integrins in activating TGF-β (Munger et al. 1999; Yang et al. 2007). The local activation of TGF-β by DCs is required for the generation of gut-homing Treg cells, themselves essential components of gut immune homeostasis, and deletion of αv integrins causes a reduction in the numbers of tolerogenic DCs and subsequent loss of gut Treg cells. Similar phenotypes are seen when β8 integrins are deleted from DCs, or when TGF-β signalling is disrupted in T cells, further implicating loss of αvβ8-mediated TGF-β as a contributor to colitis in αv-knockout mice (Li et al. 2007; Travis et al. 2007). However, considering the association between apoptotic cell uptake and TGF-β production, it is possible that the engagement of αv integrins during phagocytosis allows local TGF-β activation and signalling (Lucas et al. 2006). This may be particularly important in the cases where apoptotic cells secrete or carry surface TGF-β (Chen et al. 2001).

7.4.3 *Intracellular Signalling*

The intracellular pathways of apoptotic cell response signalling are now beginning to emerge. Phagocytosis of apoptotic cells by macrophages stimulates both activation of Akt and inhibition of extracellular signal-regulated kinase (ERK)1 and ERK2 (Ucker, Chap. 6, this vol.). This specific combination of signalling events was linked with the promotion of macrophage survival and may be the basis of the growth promoting effects of apoptotic cell uptake (Cui et al. 2007; Reddy et al. 2002). Although these studies did not investigate the apoptotic cell receptor responsible for activating these signalling pathways, subsequent studies have shown that MerTK can stimulate phosphatidylinositol 3-kinase (PI3K)/ Akt signalling. Following apoptotic cell engagement, MerTK associates with the catalytic subunit of PI3K through an YDIM motif in the MerTK cytoplasmic tail. This leads to PI3K activation and phosphorylation and activation of Akt (Sen et al. 2007). Importantly, pharmacological blockade of PI3K/ Akt leads to a reversal of apoptotic cell-induced suppression of TNF-α production (Sen et al. 2007). PI3K and Akt act to inhibit cytokine production by negative regulation of the transcription factor NF-κB, an essential component of pro-inflammatory cytokine transcription. TLR signalling normally activates the IκB-kinase, IKK, causing NF-κB to dissociate from its inhibitor IκB and move to the nucleus. PI3K/ Akt blocks the activation of IKK, preventing generation of nuclear NF-κB and inhibiting cytokine transcription (Sen et al. 2007). This study therefore links engagement of a cell surface apoptotic cell receptor with suppression of an important transcription factor. Targeting NF-κB certainly provides a plausible mechanism for the down-regulation of expression of multiple pro-inflammatory cytokines and inhibition of NF-κB-mediated transcription has been reported in many different cell types following apoptotic cell uptake. However, it should be noted there is disagreement as to whether this effect is at the level of NF-κB activation or transcriptional activity (Cvetanovic et al. 2006; Sen et al. 2007).

7.4.4 Transcriptional Regulation

The exact effects of apoptotic cell uptake on transcription have now been dissected at the level of individual promoters to begin to identify the key transcription factors involved (Ucker, Chap. 6, this vol.). There is evidence emerging that apoptotic cells regulate cytokine production through both transcriptional stimulation and repression. In a mouse model system (J774 or RAW264.7 macrophage cell lines) the mechanism of apoptotic cell suppression of IL-12 production has been carefully investigated (Kim et al. 2004). Co-engagement of apoptotic cells has been widely reported to block LPS-induced up-regulation of IL-12 production. In this study, Kim and colleagues show that this occurs independently of IL-10 or TGF-β production by the macrophage and is mediated through inhibition of transcription of the IL-12p35 gene. Using promoter-reporter constructs, the authors were able to map this suppression to a specific site in the IL-12 promoter and identify a zinc finger-containing protein, GC-BP that binds this site in response to apoptotic cells and represses IL-12p35 transcription. This GC-BP repressive activity is normally inactivated by tyrosine phosphorylation when macrophages respond to LPS. When macrophages see apoptotic cells, LPS-induced tyrosine phosphorylation of GC-BP does not occur and GC-BP suppresses IL-12p35 transcription. The upstream events in this pathway are not yet identified, although there is evidence that the signals originate from binding of apoptotic cell PS (Kim et al. 2004).

Analysis of the IL-10 promoter has recently revealed a potential pathway for activation of gene transcription by apoptotic cells (Chung et al. 2007). Under certain

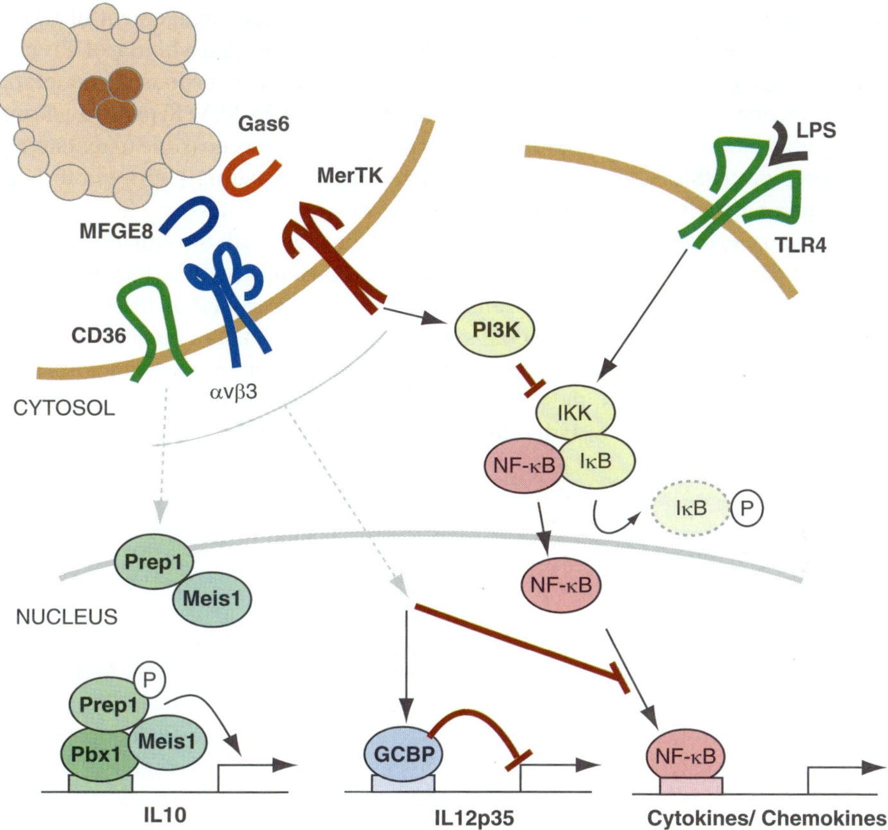

Fig. 7.3 Potential mechanisms for apoptotic cell response signalling. Apoptotic cells bind surface receptors on macrophages, including CD36, integrin αvβ3 and MerTK as well as additional receptors not depicted. Signals generated by apoptotic cells block pro-inflammatory signalling from TLR binding microbial components such as LPS. This may be through blockade of NF-κB activation (such as reported for PI3K activation, shown) or through alterations of NF-κB activity (signals may originate from multiple receptors). Apoptotic cells are known to directly regulate transcription of IL-10 and IL-12p35 through indicated transcription factors. For full details and references, refer to text (sect. 7.5).

conditions, apoptotic cells stimulate IL-10 production by macrophages. An apoptotic-cell response element (ACRE) was identified in the IL-10 promoter and specifically mapped to a pre-B cell leukemia transcription factor-1 (Pbx1) binding site (Chung et al. 2007). Macrophages in which Pbx1 was deleted did not produce IL-10 in response to apoptotic cells, showing that Pbx1 was the mediator of these effects. Furthermore, this was specific to IL-10 expression as apoptotic cell-induced production of TGF-β and LPS-induced cytokine production were unaffected. However, apoptotic cell uptake does not act directly on Pbx-1 production or phosphorylation but regulates the activity of Pbx-1-containing transcription factor complexes. Pbx proteins act as cofactors for Hox-family homeodomain-containing transcription factors and two Pbx-1 dimerising partners, Meis-1 and Prep-1, were shown to bind the ACRE following apoptotic cell encounter. This required engagement of CD36 on the macrophage and was associated with tyrosine phosphorylation of Prep-1. Therefore the production of immuno-modulatory cytokines in response to apoptotic cells is mediated in part by CD36, signalling through a p38 MAPK pathway. From these initial studies of two apoptotic cell-regulated genes, it is possible to tentatively link receptors (PSR/ CD36) through tyrosine kinase signalling pathways to both transcriptional activation and repression.

In summary, the studies summarized here demonstrate both the apparent conservation of mechanisms of apoptotic cell effects (mediated by cell surface receptors, convergence on NF-κB-mediated transcription; Fig. 7.3) as well as the differences and discrepancies between individual systems. There are many components of the signalling pathways that are yet to be identified and, as seen in the characterization of apoptotic cell responses, these are likely to reflect both common shared mechanisms and specializations of individual cell types and systems.

7.5   Conclusions

Phagocytosis of apoptotic cells is an evolutionarily conserved process, leading to the silent removal of dying cells and the induction of anti-inflammatory and reparative responses. This can be seen in many different types of cells, including professional and non-professional phagocytes, and is a response to specific recognition of the dying cell. Analysis of the basic mechanisms of apoptotic cell responses suggest that these are shared between many cell types and this has been proposed as a universal 'innate immune' recognition pathway (Cvetanovic et al. 2006).

The recognition of apoptotic cells can lead to widespread modulation of transcriptional activity. Through these effects, apoptotic cells act on the ongoing 'programme' of the phagocyte; hence the response to apoptotic cells is determined in large part by the normal function of the phagocyte. For many non-professional cells, this means promoting cell survival and local regeneration, as well as the induction of a local, limited inflammatory response. The response of macrophages, the near-ubiquitous professional phagocyte, appears more complex. This reflects both the multifunctional nature of macrophages, which participate in many different processes in resting and inflamed tissue, and the much greater number of studies of macrophage responses to apoptotic cells. When considered more broadly, macrophages generally respond by down-regulating pro-inflammatory cytokine production and stimulating tissue repair. Both professional and non-professional phagocytes secrete TGF-β in response to apoptotic cell uptake providing further evidence of a shared pathway of apoptotic cell response. DCs are highly specialized for interaction with adaptive immune cells and apoptotic cell uptake therefore regulates this process, reducing cytokine production and co-stimulatory molecule expression, but inducing migration and antigen processing in common with other phagocytosed particles.

All of these responses to apoptotic cells have roles in normal development and function, and perturbation of apoptosis, apoptotic cell clearance or apoptotic cell response can have profound effects. Hence apoptotic cell phagocytosis and the subsequent response to apoptotic cells should be considered an integral part of the apoptotic programme, applicable to all organisms that undergo programmed cell death.

References

Albert ML, Pearce SF, Francisco LM et al (1998a) Immature dendritic cells phagocytose apoptotic cells via alphavbeta5 and cd36, and cross-present antigens to cytotoxic t lymphocytes. J Exp Med 188:1359–1368

Albert ML, Sauter B, Bhardwaj N (1998b) Dendritic cells acquire antigen from apoptotic cells and induce class i- restricted ctls. Nature 392:86–89

Anderson HA, Maylock CA, Williams JA et al (2003) Serum-derived protein s binds to phosphatidylserine and stimulates the phagocytosis of apoptotic cells. Nat Immunol 4:87–91

Blachere NE, Darnell RB, Albert ML (2005) Apoptotic cells deliver processed antigen to dendritic cells for cross-presentation. PLoS Biol 3:E185

Botto M, Dell'Agnola C, Bygrave AE et al (1998) Homozygous c1q deficiency causes glomerulonephritis associated with multiple apoptotic bodies. Nat Genet 19:56–59

Brown JR, Goldblatt D, Buddle J et al (2003) Diminished production of anti-inflammatory mediators during neutrophil apoptosis and macrophage phagocytosis in chronic granulomatous disease (cgd). J Leukocyte Biol 73:591–599

Chen W, Frank ME, Jin W et al (2001) Tgf-beta released by apoptotic t cells contributes to an immunosuppressive milieu. Immunity 14:715–725

Chung EY, Liu J, Homma Y et al (2007) Interleukin-10 expression in macrophages during phagocytosis of apoptotic cells is mediated by homeodomain proteins pbx1 and prep-1. Immunity 27:952–964

Cocco RE, Ucker DS (2001) Distinct modes of macrophage recognition for apoptotic and necrotic cells are not specified exclusively by phosphatidylserine exposure. Mol Biol Cell 12:919–930

Cohen PL, Caricchio R, Abraham V et al (2002 Delayed apoptotic cell clearance and lupus-like autoimmunity in mice lacking the c-mer membrane tyrosine kinase. J Exp Med 196:135–140

Coombes JL, Siddiqui KR, Arancibia-Carcamo CV et al (2007) A functionally specialized population of mucosal cd103+ dcs induces foxp3+ regulatory t cells via a tgf-beta and retinoic acid-dependent mechanism. J Exp Med 204:1757–1764

Cui D, Thorp E, Li Y et al (2007) Pivotal advance: Macrophages become resistant to cholesterol-induced death after phagocytosis of apoptotic cells. J Leukocyte Biol 82:1040–1050

Cvetanovic M, Mitchell JE, Patel V et al (2006) Specific recognition of apoptotic cells reveals a ubiquitous and unconventional innate immunity. J Biol Chem 281:20055–20067

Cvetanovic M, Ucker DS (2004) Innate immune discrimination of apoptotic cells: Repression of proinflammatory macrophage transcription is coupled directly to specific recognition. J Immunol 172:880–889

Devitt A, Parker KG, Ogden CA et al (2004) Persistence of apoptotic cells without autoimmune disease or inflammation in cd14-/- mice. J Cell Biol 167:1161–1170

Fadok VA, Bratton DL, Konowal A et al (1998) Macrophages that have ingested apoptotic cells in vitro inhibit proinflammatory cytokine production through autocrine/paracrine mechanisms involving tgf-beta, pge2, and paf. J Clin Invest 101:890–898

Feng H, Zeng Y, Graner MW et al (2002) Stressed apoptotic tumor cells stimulate dendritic cells and induce specific cytotoxic t cells. Blood 100:4108–4115

Feng H, Zeng Y, Graner MW et al (2003) Exogenous stress proteins enhance the immunogenicity of apoptotic tumor cells and stimulate antitumor immunity. Blood 101:245–252

Freire-de-Lima CG, Nascimento DO, Soares MB et al (2000) Uptake of apoptotic cells drives the growth of a pathogenic trypanosome in macrophages. Nature 403:199–203

Fries DM, Lightfoot R, Koval M et al (2005) Autologous apoptotic cell engulfment stimulates chemokine secretion by vascular smooth muscle cells. Am J Pathol 167:345–353

Gallucci S, Lolkema M, Matzinger P (1999) Natural adjuvants: Endogenous activators of dendritic cells. Nat Med 5:1249–1255

Giles KM, Ross K, Rossi AG et al (2001) Glucocorticoid augmentation of macrophage capacity for phagocytosis of apoptotic cells is associated with reduced p130cas expression, loss of paxillin/pyk2 phosphorylation, and high levels of active rac. J Immunol 167:976–986

Gilmour JS, Coutinho AE, Cailhier JF et al (2006) Local amplification of glucocorticoids by 11 beta-hydroxysteroid dehydrogenase type 1 promotes macrophage phagocytosis of apoptotic leukocytes. J Immunol 176:7605–7611

Golpon HA, Fadok VA, Taraseviciene-Stewart L et al (2004) Life after corpse engulfment: Phagocytosis of apoptotic cells leads to vegf secretion and cell growth. FASEB J 18:1716–1718

Hamon Y, Broccardo C, Chambenoit O et al (2000) Abc1 promotes engulfment of apoptotic cells and transbilayer redistribution of phosphatidylserine. Nat Cell Biol 2:399–406

Hanayama R, Tanaka M, Miwa K et al (2004a) Expression of developmental endothelial locus-1 in a subset of macrophages for engulfment of apoptotic cells. J Immunol 172:3876–3882

Hanayama R, Tanaka M, Miwa K et al (2002) Identification of a factor that links apoptotic cells to phagocytes. Nature 417:182–187

Hanayama R, Tanaka M, Miyasaka K et al (2004b) Autoimmune disease and impaired uptake of apoptotic cells in mfg-e8-deficient mice. Science 304:1147–1150

Hopkinson-Woolley J, Hughes D, Gordon S et al (1994) Macrophage recruitment during limb development and wound healing in the embryonic and foetal mouse. J Cell Sci 107 (Pt 5): 1159–1167

Houde M, Bertholet S, Gagnon E et al (2003) Phagosomes are competent organelles for antigen cross-presentation. Nature 425:402–406

Huang FP, Platt N, Wykes M et al (2000) A discrete subpopulation of dendritic cells transports apoptotic intestinal epithelial cells to t cell areas of mesenteric lymph nodes. J Exp Med 191:435–444

Huh JR, Guo M, Hay BA (2004) Compensatory proliferation induced by cell death in the drosophila wing disc requires activity of the apical cell death caspase dronc in a nonapoptotic role. Curr Biol 14:1262–1266

Huynh ML, Fadok VA, Henson PM (2002) Phosphatidylserine-dependent ingestion of apoptotic cells promotes tgf-beta1 secretion and the resolution of inflammation. J Clin Invest 109:41–50

Ip WK, Lau YL (2004) Distinct maturation of, but not migration between, human monocyte-derived dendritic cells upon ingestion of apoptotic cells of early or late phases. J Immunol 173:189–196

Ishimoto Y, Ohashi K, Mizuno K et al (2000) Promotion of the uptake of ps liposomes and apoptotic cells by a product of growth arrest-specific gene, gas6. J Biochem 127:411–417

Iyoda T, Nagata K, Akashi M et al (2005) Neutrophils accelerate macrophage-mediated digestion of apoptotic cells in vivo as well as in vitro. J Immunol 175:3475–3483

Jacobson MD, Weil M, Raff MC (1997) Programmed cell death in animal development. Cell 88:347-354

Jang MH, Sougawa N, Tanaka T et al (2006) Ccr7 is critically important for migration of dendritic cells in intestinal lamina propria to mesenteric lymph nodes. J Immunol 176:803–810

Kim S, Elkon KB, Ma X (2004) Transcriptional suppression of interleukin-12 gene expression following phagocytosis of apoptotic cells. Immunity 21:643–653

Kirsch T, Woywodt A, Beese M et al (2007) Engulfment of apoptotic cells by microvascular endothelial cells induces proinflammatory responses. Blood 109:2854–2862

Kleinclauss F, Perruche S, Masson E et al (2006) Intravenous apoptotic spleen cell infusion induces a tgf-beta-dependent regulatory t-cell expansion. Cell Death Differ 13:41–52

Kobayashi N, Karisola P, Pena-Cruz V et al (2007) Tim-1 and tim-4 glycoproteins bind phosphatidylserine and mediate uptake of apoptotic cells. Immunity 27:927–940

Kurosaka K, Takahashi M, Watanabe N et al (2003) Silent cleanup of very early apoptotic cells by macrophages. J Immunol 171:4672–4679.

Kurosaka K, Watanabe N, Kobayashi Y (2001) Production of proinflammatory cytokines by resident tissue macrophages after phagocytosis of apoptotic cells. Cell Immunol 211:1–7

Kurosaka K, Watanabe N, Kobayashi Y (2002) Potentiation by human serum of anti-inflammatory cytokine production by human macrophages in response to apoptotic cells. J Leukocyte Biol 71:950–956

Lacy-Hulbert A, Smith AM, Tissire H et al (2007) Ulcerative colitis and autoimmunity induced by loss of myeloid alphav integrins. Proc Natl Acad Sci U S A 104:15823–15828

Li MO, Wan YY, Flavell RA (2007) T cell-produced transforming growth factor-beta1 controls t cell tolerance and regulates th1- and th17-cell differentiation. Immunity 26:579–591

Liu K, Iyoda T, Saternus M et al (2002) Immune tolerance after delivery of dying cells to dendritic cells in situ. J Exp Med 196:1091–1097

Liu L, Zhang M, Jenkins C et al (1998) Dendritic cell heterogeneity in vivo: Two functionally different dendritic cell populations in rat intestinal lymph can be distinguished by cd4 expression. J Immunol 161:1146–1155

Liu Y, Cousin JM, Hughes J et al (1999) Glucocorticoids promote nonphlogistic phagocytosis of apoptotic leukocytes. J Immunol 162:3639–3646

Lu Q, Lemke G (2001). Homeostatic regulation of the immune system by receptor tyrosine kinases of the tyro 3 family. Science 293:306–311

Lucas M, Stuart LM, Savill J et al (2003) Apoptotic cells and innate immune stimuli combine to regulate macrophage cytokine secretion. J Immunol 171:2610–2615

Lucas M, Stuart LM, Zhang A et al (2006) Requirements for apoptotic cell contact in regulation of macrophage responses. J Immunol 177:4047–4054

Meagher LC, Cousin JM, Seckl JR et al (1996) Opposing effects of glucocorticoids on the rate of apoptosis in neutrophilic and eosinophilic granulocytes. J Immunol 156:4422–4428

Meagher LC, Savill JS, Baker A et al (1992) Phagocytosis of apoptotic neutrophils does not induce macrophage release of thromboxane b2. J Leukocyte Biol 52:269–273

Metchnikoff E (1893) Lectures on the comparative pathology of inflammation (London: Kegan, Paul, Trench and Trubner).

Milan M, Campuzano S, Garcia-Bellido A (1997) Developmental parameters of cell death in the wing disc of drosophila. Proc Natl Acad Sci U S A 94:5691–5696

Miyanishi M, Tada K, Koike M et al (2007) Identification of tim4 as a phosphatidylserine receptor. Nature 450:435–439

Morelli AE (2006) The immune regulatory effect of apoptotic cells and exosomes on dendritic cells: Its impact on transplantation. Am J Transplant 6:254–261

Morelli AE, Larregina AT, Shufesky WJ et al (2003) Internalization of circulating apoptotic cells by splenic marginal zone dendritic cells: Dependence on complement receptors and effect on cytokine production. Blood 101:611–620

Munger JS, Huang X, Kawakatsu H et al (1999) The integrin alpha v beta 6 binds and activates latent tgf beta 1: A mechanism for regulating pulmonary inflammation and fibrosis. Cell 96:319–328

Newman SL, Henson JE, Henson PM (1982) Phagocytosis of senescent neutrophils by human monocyte-derived macrophages and rabbit inflammatory macrophages. J Exp Med 156:430–442

Park D, Tosello-Trampont AC, Elliott MR et al (2007) Bai1 is an engulfment receptor for apoptotic cells upstream of the elmo/dock180/rac module. Nature 450:430–434

Park SY, Jung MY, Kim HJ et al (2008) Rapid cell corpse clearance by stabilin-2, a membrane phosphatidylserine receptor. Cell Death Differ 15:192–201

Patel VA, Longacre A, Hsiao K et al (2006) Apoptotic cells, at all stages of the death process, trigger characteristic signalling events that are divergent from and dominant over those triggered by necrotic cells: Implications for the delayed clearance model of autoimmunity. J Biol Chem 281:4663–4670

Perez-Garijo A, Martin FA, Morata G (2004) Caspase inhibition during apoptosis causes abnormal signalling and developmental aberrations in drosophila. Development 131:5591–5598

Perez-Garijo A, Martin FA, Struhl G et al (2005) Dpp signalling and the induction of neoplastic tumors by caspase-inhibited apoptotic cells in drosophila. Proc Natl Acad Sci U S A 102:17664–17669

Reddy SM, Hsiao KH, Abernethy VE et al (2002) Phagocytosis of apoptotic cells by macrophages induces novel signalling events leading to cytokine-independent survival and inhibition of proliferation: Activation of akt and inhibition of extracellular signal-regulated kinases 1 and 2. J Immunol 169:702–713

Rossi AG, Sawatzky DA, Walker A et al (2006) Cyclin-dependent kinase inhibitors enhance the resolution of inflammation by promoting inflammatory cell apoptosis. Nat Med 12:1056–1064

Rothlin CV, Ghosh S, Zuniga EI et al (2007) Tam receptors are pleiotropic inhibitors of the innate immune response. Cell 131:1124–1136

Rubartelli A, Poggi A, Zocchi MR (1997) The selective engulfment of apoptotic bodies by dendritic cells is mediated by the alpha(v)beta3 integrin and requires intracellular and extracellular calcium. Eur J Immunol 27:1893–1900

Ryoo HD, Gorenc T, Steller H (2004) Apoptotic cells can induce compensatory cell proliferation through the jnk and the wingless signalling pathways. Dev Cell 7:491–501

Saas P, Bonnefoy F, Kury-Paulin S et al (2007) Mediators involved in the immunomodulatory effects of apoptotic cells. Transplantation 84:S31–S34

Saijo S, Nagata K, Nakano Y et al (2005) Inhibition by adiponectin of il-8 production by human macrophages upon coculturing with late apoptotic cells. Biochem Biophys Res Commun 334:1180–1183

Sauter B, Albert ML, Francisco L et al (2000) Consequences of cell death: Exposure to necrotic tumor cells, but not primary tissue cells or apoptotic cells, induces the maturation of immunostimulatory dendritic cells. J Exp Med 191:423–434

Savill J, Dransfield I, Gregory C et al (2002) A blast from the past: Clearance of apoptotic cells regulates immune responses. Nat Rev Immunol 2:965–975

Savill JS, Wyllie AH, Henson JE et al (1989) Macrophage phagocytosis of aging neutrophils in inflammation. Programmed cell death in the neutrophil leads to its recognition by macrophages. J Clin Invest 83:865–875

Schulz O, Diebold SS, Chen M et al (2005) Toll-like receptor 3 promotes cross-priming to virus-infected cells. Nature 433:887–892

Scott RS, McMahon EJ, Pop SM et al (2001) Phagocytosis and clearance of apoptotic cells is mediated by mer. Nature 411:207–211

Sen P, Wallet MA, Yi Z et al (2007) Apoptotic cells induce mer tyrosine kinase-dependent blockade of nf-kappab activation in dendritic cells. Blood 109:653–660

Shibata Y, Hishikawa Y, Izumi S et al (2002) Involvement of fas/fas ligand in the induction of apoptosis in chronic sialadenitis of minor salivary glands including sjogren's syndrome. Hum Cell 15:52–60

Stark MA, Huo Y, Burcin TL et al (2005) Phagocytosis of apoptotic neutrophils regulates granulopoiesis via il-23 and il-17. Immunity 22:285–294

Steinman RM, Turley S, Mellman I et al (2000) The induction of tolerance by dendritic cells that have captured apoptotic cells. J Exp Med 191:411–416

Stuart LM, Lucas M, Simpson C et al (2002) Inhibitory effects of apoptotic cell ingestion upon endotoxin-driven myeloid dendritic cell maturation. J Immunol 168:1627–1635

Stuart LM, Takahashi K, Shi L et al (2005) Mannose-binding lectin-deficient mice display defective apoptotic cell clearance but no autoimmune phenotype. J Immunol 174:3220–3226

Takahashi M, Kurosaka K, Kobayashi Y (2004) Immature dendritic cells reduce proinflammatory cytokine production by a coculture of macrophages and apoptotic cells in a cell-to-cell contact-dependent manner. J Leukocyte Biol 75:865–873

Travis MA, Reizis B, Melton AC et al (2007) Loss of integrin alpha(v)beta(8) on dendritic cells causes autoimmunity and colitis in mice. Nature

Verbovetski I, Bychkov H, Trahtemberg U et al (2002) Opsonization of apoptotic cells by autologous ic3b facilitates clearance by immature dendritic cells, down-regulates dr and cd86, and up-regulates cc chemokine receptor 7. J Exp Med 196:1553–1561

Voll RE, Herrmann M, Roth EA et al (1997) Immunosuppressive effects of apoptotic cells. Nature 390:350–351

Wallet MA, Sen P, Flores RR et al (2008) Mertk is required for apoptotic cell-induced t cell tolerance. J Exp Med 205:219–232

Wang Z, Larregina AT, Shufesky WJ et al (2006) Use of the inhibitory effect of apoptotic cells on dendritic cells for graft survival via t-cell deletion and regulatory t cells. Am J Transplant 6:1297–1311

Wood W, Turmaine M, Weber R et al (2000) Mesenchymal cells engulf and clear apoptotic footplate cells in macrophageless pu.1 null mouse embryos. Development 127:5245–5252

Wu Y, Singh S, Georgescu MM et al (2005) A role for mer tyrosine kinase in alphavbeta5 integrin-mediated phagocytosis of apoptotic cells. J Cell Sci 118:539–553

Yang Z, Mu Z, Dabovic B et al (2007) Absence of integrin-mediated tgf-beta 1 activation in vivo recapitulates the phenotype of tgf-beta 1-null mice. J Cell Biol 176:787–793

Chapter 8
Methods Used to Study Apoptotic Cell Clearance

Uriel Trahtemberg and Dror Mevorach

Abstract: The phagocytosis of dying cells is a complex process involving interactions between many molecules on the dying cell, on the phagocyte, and in the micro-environment. Although much is known on the subject, many questions and unknown variables remain under investigation. Knowledge of appropriate methods for studying phagocytosis of dying cells is essential for acquisition of a deeper understanding of programmed cell death, phagocytosis, dying cell clearance, and homeostasis. In this chapter we will define a consensual vocabulary and discuss general and specific considerations regarding the different methods and models used to study the clearance of dying cells. In the second part of this chapter we will describe in further detail selected techniques and protocols.

Keywords: Apoptosis • Necrosis • Cell death • Clearance • Phagocytosis • Methods

8.1   Introduction

Clearance of dying cells has become an important field of research, both to increase our understanding of the mechanisms for uptake and also because cell clearance is a basic mechanism in tissue homeostasis, cancer, resolution of inflammation, induction of tolerance, and autoimmunity. Research in invertebrates such as *C. elegans* and *D. melanogaster* is generally based on genetic mutations and morphological observations and will not be reviewed in this chapter. In mammals, during the early days of research on phagocytosis, uptake was evaluated using the phagocytosis index. A few

D. Mevorach (✉) · U. Trahtemberg
The Laboratory for Cellular and Molecular Immunology, Rheumatology Unit
Department of Medicine
Hadassah-Hebrew University Medical Center
Jerusalem, 91120, Israel
Tel.: 972-2-677-7317
Fax: 972-2-646-8801
E-mail: mevorachd@hadassah.org.il

D.V. Krysko, P. Vandenabeele (eds.), *Phagocytosis of Dying Cells,*
DOI: 10.1007/978-1-4020-9292-3_8, © Springer Science+Business Media B.V. 2009

hundred apoptotic cells within or near phagocytes were counted using light microscopy. The phagocytosis index led to remarkable discoveries, but nowadays it is not considered sufficient on its own for evaluation of professional phagocytosis.

Before we discuss other methods and their applications, we should agree on an accepted vocabulary and definitions. Investigators are using similar names for a variety of physiological conditions, which can lead to misleading interpretations of the results of other researchers. Thus, our first section will provide a core vocabulary, terminology, and brief definitions of processes such as programmed cell death (PCD), apoptosis, necrosis, phagocytosis. The subsequent section will deal with the actual study of phagocytosis of dying cells, and in the last section we will present selected experimental methods and protocols.

8.2 Vocabulary

8.2.1 Programmed Cell Death

Programmed cell death (PCD) refers to the process by which a cell, once the decision to die is taken, undergoes a series of defined molecular transformations that bring about an ordered demise (Chiarugi 2005; Kroemer and Martin 2005; Zong and Thompson 2006). The terms apoptosis and PCD were once used interchangeably. However, today the term PCD can be subdivided into at least three death pathways: apoptosis, necrosis, and autophagy (Krysko et al., Chap. 1; Diez-Fraile et al., Chap. 2, this Vol.; Chiarugi 2005; Kroemer and Martin 2005; Zong and Thompson 2006; Levine and Deretic 2007; Maiuri et al. 2007). Issues relevant to PCD include the trigger leading to the decision to die, the death pattern involved, the duration of the process, and the other programs that are activated. In some circumstances, cells may switch from apoptosis to necrosis or autophagy, and vice versa.

8.2.2 Apoptosis

Apoptosis is a type of cell death that is defined, first and foremost, by a set of morphological changes, not all of which are always present: nuclear shrinkage and fragmentation, chromatin condensation, membrane blebbing, and the release of apoptotic bodies (Krysko et al., Chap. 1; Diez-Fraile et al., Chap. 2, this Vol.; Darzynkiewicz et al. 1997). Another important aspect of apoptosis is that it is an active, ATP-dependent cell death, which led to its designation as 'cellular suicide'. It requires energy for completion and involves a large set of events (Zong and Thompson 2006). Interestingly, some events during apoptosis are of functional importance for the organism, and not directly related to its demise (Krispin et al. 2006). The main molecular events during apoptosis are now known, although many details still require elucidation. Apoptosis is not an entirely stereotypic process; a variety of molecular triggers can lead to apoptosis through different pathways (Jaattela and Tschopp 2003; Kroemer and Martin 2005). Thus, several different methods are used to determine whether a cell is undergoing apoptosis. It is important to consider the advantages and disadvantages of each method ('Dying cells: sources and identification', Sect. 8.4.1). Different modes of apoptosis may involve varying patterns of death, death kinetics, and phagocytosis kinetics, as well as different responses of the phagocyte and the organism.

8.2.3 Primary Necrosis

Necrosis has a different morphological appearance from apoptosis. It is characterized by cellular swelling, chromatin digestion, organelle vacuolation, and plasma membrane disruption (Darzynkiewicz et al. 1997). Primary necrosis was for a long time considered as an 'accidental cell death', in contrast to 'programmed cell death.' However, there is now extensive evidence that necrosis can be due to both accidental and programmed cell death (Chiarugi 2005; Kroemer and Martin 2005; Zong and Thompson 2006). In general, death is considered to be necrotic when the cell cannot maintain its membrane integrity. In primary necrosis, this is due to abrupt and rapid cellular insults that overwhelm the cell's processes (Zong and Thompson 2006). Thus, necrosis proceeds down energetic and thermodynamic gradients that usually proceed to completion without the need for generation of energy (Zong and Thompson 2006). Several methods for inducing primary necrosis have been reported, such as freezing and thawing, exposure to H_2O_2, heating, and others. These may lead to different cell morphologies and even to immediate lysis. All cells undergoing necrosis will, by definition, eventually end in lysis and spillage of cellular contents, but the timing of these events differs depending on the mode of necrosis. For example, following a few cycles of freezing and thawing, most cells will lyse, and any evidence of cell morphology will disappear. Thus, when talking about clearance of necrotic cells, one must show the cell's morphology, for example, using light scatter changes in flow cytometry, or by direct microscopic observation, as well as give details about the type and kinetics of the induction of necrosis ('Dying cells: sources and identification', Sect. 8.3.1).

8.2.4 Secondary Necrosis

Cells undergoing apoptosis are normally rapidly recognized at early stages by phagocytes or by sister cells. However, if they are not cleared by phagocytes, they will eventually be unable to maintain integrity of their membranes, and they will undergo secondary necrosis (Chiarugi 2005). This happens mainly in three situations: the first is in vitro studies where the normal phagocytic system is not present; second, in sites of inflammation and injury, where the phagocytic system is not recruited, is unable to cope with the large quantity of dying cells, or is hampered by the processes causing injury and inflammation (Torre et al. 2002; Kirkham 2007; Bianchi et al. 2008); and, finally, in humans or animal models with specific or general defects in the apoptotic cell clearance mechanisms (Scott et al. 2001; Potter et al. 2003; Devitt et al. 2004). It is important to mention that although secondary necrotic cells may also eventually lyse, the consequences of secondary necrosis may be different from primary necrosis (Mevorach 2000). This is because the signals that are emitted by the apoptotic cell and that are present on its membrane before its necrotic death, as well as the contents that are spilled, are substantially different from those of primary necrotic cells (Sauter et al. 2000; Krysko et al. 2006a; Patel et al. 2006).

8.2.5 *Cell Lysate*

This term refers to laboratory procedures in which cells are intentionally and rapidly disrupted to obtain a preparation containing the cells' contents in a state that has not been modified by PCD. Lysates may or may not include free nucleic acids, depending on the method used. Cells are usually lysed by cyclic freezing and thawing or by using detergents. When used in experiments, cell lysates may yield substantially different results from preparations in which necrosis is induced by other methods (Mevorach 2000).

8.2.6 *Autophagy*

Autophagy is an intracellular degradation system that delivers cytoplasmic constituents to the lysosome. This process is either a response triggered by conditions of extreme duress, or part of the normal turnover of cellular contents. Autophagy allows the cell to consume parts of itself for energy to increase its chance of survival during a stressful period. Despite the apparent simplicity of the concept, recent progress has demonstrated that autophagy plays a wide variety of physiological and pathophysiological roles that are sometimes complex (Levine and Deretic 2007; Maiuri et al. 2007). Autophagy progresses through several ordered steps: sequestration of the cellular contents, lysosomal fusion, degradation of the contents, and finally extraction and use of the degradation products (Klionsky 2007; Mizushima 2007). As a natural part of cellular homeostasis, autophagy usually takes place unnoticed by neighbouring cells. However, if cells undergoing autophagy become stressed to the point of resembling dying cells, or if they switch to PCD, they may be cleared by phagocytes (Krysko et al., Chap. 1; Diez-Fraile et al., Chap. 2, this Vol.)

8.2.7 *Cell Viability*

In the context of this chapter, a cell is considered viable when it is not undergoing any form of death process. This means that it does not show any features of apoptosis, autophagy, or necrosis. Electron microscopy (EM) is still considered the gold standard for this determination. However, because doing EM studies routinely is laborious and expensive for most laboratories, simpler methods are generally used. These simple methods usually involve demonstration that the cells exclude a DNA-specific, membrane-impermeable dye, such as propidium iodide (PI) or trypan blue, or that they do not express phosphatidylserine on the outer leaf of the plasma membrane, as shown by an Annexin V binding assay. Another common but less popular method consists of demonstrating a persistent mitochondrial membrane potential. Nonetheless, many researchers routinely examine morphology either by light microscopy alone, or by PI or trypan blue exclusion alone. These techniques may be insensitive to early PCD, and they are also considerably less effective at distinguishing between the different modes of PCD. Important exceptions exist, such as cells that transiently or constantly expose phosphatidylserine but do not undergo cell death (Krysko et al. 2006a), or cells that allow limited entry of DNA-specific, membrane exclusion dyes without being on a death process (Frey 1995, 1997).

8.2.8   Cell Type

When studying dying cells and phagocytes, it is very important to specify the experimental model being used. Primary cells are cells that have been taken from a living organism and are used directly or following differentiation for experiments in vitro or ex-vivo. They more accurately represent cells living in the organism as well as their developmental state. Cell lines, by contrast, are immortalized cells derived from a specific progenitor. Although readily available and easier to use, they are not considered identical surrogates for primary cells. Rather, they are alternative, though similar, cellular models. The advantages of cell lines are their homogeneity and their invariance, allowing different laboratories to work on the same model with high reproducibility. Still, it is important to remember that some cell lines have a propensity to accumulate mutations with time, and they should be assessed for such changes (Yabuta et al. 2004; Inaba et al. 2005; Liu et al. 2007).

8.2.9   Professional and Non-professional Phagocytes

The term professional phagocyte refers mainly to macrophages and dendritic cells. However, there are different populations of macrophages and dendritic cells, both in humans and mice, and they have different capacities and/or patterns of clearance ('Phagocytes', Sect. 8.3.2; Ucker, Chap. 6, Lacy-Hulbert, Chap. 7, Gregory and Pound, Chap. 9, this Vol.). Moreover, even cells localized at a single site may perform differently under different conditions. For example, murine peritoneal macrophages interact differently with dying cells, depending on whether they are resident peritoneal macrophages or thioglycollate induced macrophages (inflammatory macrophages; Gregory and Devitt 2004; Xu et al. 2006a). Thus, attention should be given to the type of phagocyte being used. Non-professional phagocytes are usually localized near the dying cell (neighbouring cells) and each one of them can engulf one or two dying cells (Parnaik et al. 2000; Wood et al. 2000; Gregory and Devitt 2004).

8.2.10 Clearance of Dying Cells

This refers to the phagocytic engulfment or uptake of apoptotic or necrotic cells by professional and/or neighbouring phagocytes (Figs. 8.1 to 8.6).

8.2.11 Efferocytosis

Efferocytosis is a term used instead of engulfment, uptake, or clearance to refer to the phagocytosis of apoptotic cells in particular and to distinguish it from the general term phagocytosis. This term, suggested by the Henson group, is taken from the Latin root 'effero', meaning 'to take to the grave' or 'bury' (Vandivier et al. 2006; Ravichandran and Lorenz 2007).

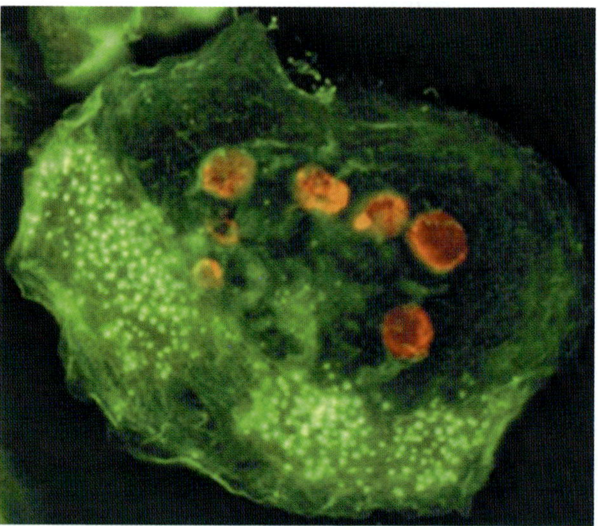

Fig. 8.1 Phagocytosis of apoptotic Jurkat cells by a human monocyte-derived macrophage.This is a high quality confocal microscopy reconstruction of a human monocyte-derived macrophage (actin filaments stained green) that has taken up the remnants of apoptotic Jurkat cells (stained red). Reprinted with permission from Macmillan Publishers Ltd: Nature 407:784, copyright 2000.

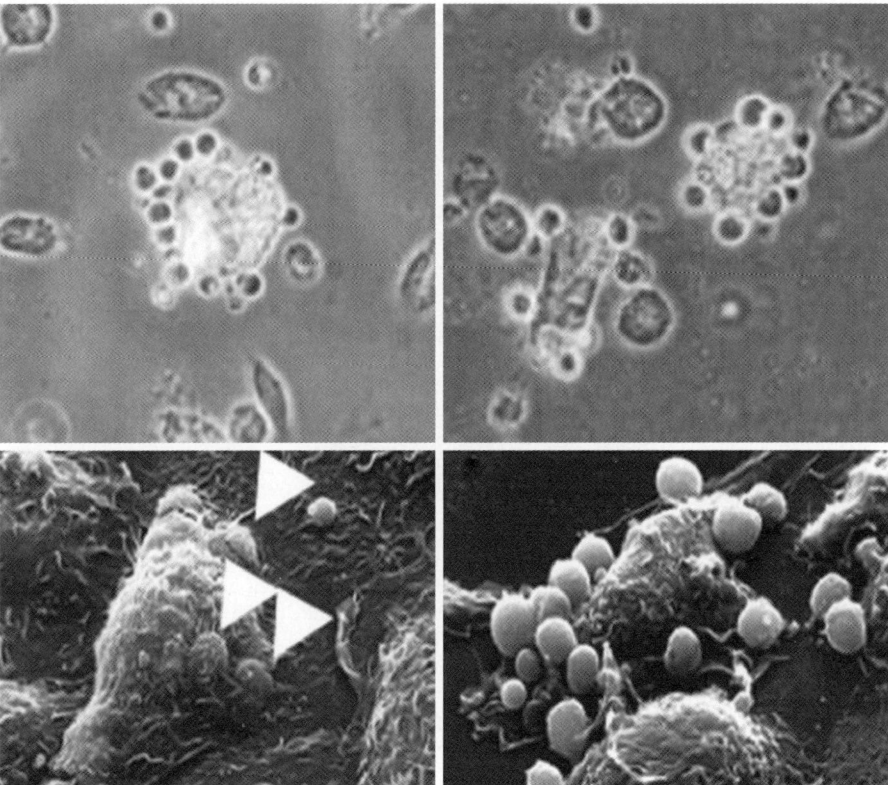

Fig. 8.2 Light and electron microscopy of wild-type and mutant murine macrophages after interaction with apoptotic thymocytes. Murine apoptotic thymocytes were offered to murine wild-type macrophages (left column) and mutant macrophages with a defective MER protein (right column). In the top row we can see light contrast microscopy of their interaction, and in the bottom row the corresponding scanning electron microscopy (SEM) images. In the lower left panel, apoptotic thymocyte profiles during different stages of internalization (arrow-heads) by a wild-type macrophage can be seen in the SEM images. In the lower right panel, in comparison, apoptotic thymocytes are only adhering to the mutant macrophage but are not being internalized. SEM images clearly distinguish between adherence and internalization, but it is very difficult to make such a distinction with light microscopy. This is an example why phagocytosis index should be called the interaction index, of the prime importance of using EM for morphologic and interaction determinations, and of the pitfalls of light microscopy if used alone. It also underscores the high quality images that SEM can provide. Reprinted with permission from Macmillan Publishers Ltd: "Phagocytosis and clearance of apoptotic cells is mediated by MER" Scott et al. Nature 411:207, copyright 2001.

8.2.12 Tethering and Adhesion

It has been proposed that binding and ingestion are separable events, each mediated by a variety of receptors. The requirement for both tethering and engulfment receptors is important for the ability of professional phagocytes to rapidly and efficiently

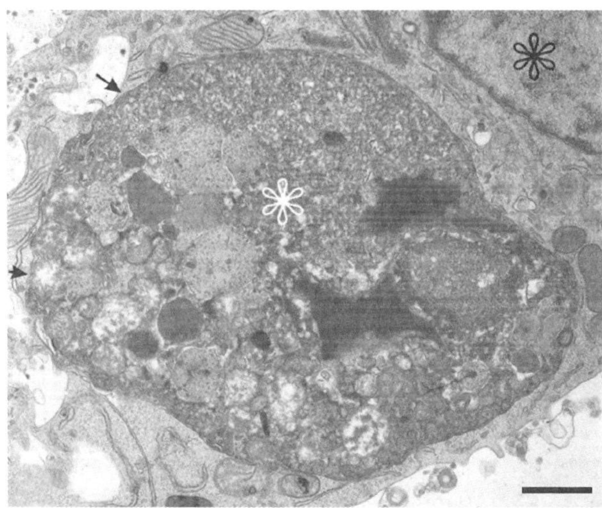

Fig. 8.3 Electron microscopy image of an apoptotic body inside a macrophage. This transmission electron microscopy image shows in great detail an apoptotic body (white asterisk) that has been completely internalized by a macrophage. The classic appearance of an apoptotic cell remnant can be seen (white asterisk), in contrast to the normal architecture of the surrounding macrophage (black asterisk, inside the macrophage nucleus). The arrows point to the double membrane structure, indicating that this is a tightly fitting phagosome. Scale bar: 1 μm. Reprinted with permission from Wiley-Liss, Inc., a subsidiary of John Wiley & Sons, Inc: "Mechanisms of internalization of apoptotic and necrotic L929 cells by a macrophage cell line studied by electron microscopy" Krysko et al. J. Morphol 258:336, copyright 2003.

clear apoptotic cells. In contrast, non-professional phagocytes may not require both types of receptors. Recognition of apoptotic ligands or opsonins on apoptotic cells by tethering receptors on the phagocyte is associated with initiation of a series of signalling pathways in the phagocyte, regardless of whether or not the cell will be ingested thereafter (Gronsi and Ravichandran, Chap. 5, this Vol.; Lucas et al. 2006; Figs. 8.2 and 8.7).

8.2.13   Phagocytosis Index

The phagocytosis index is defined as the number of apoptotic cells that are attached to 100 phagocytes. Our group prefers the term 'interaction index' (Shoshan et al. 2001), which describes the association between apoptotic cells and phagocytes, including both tethering/adhesion and internalization without making a distinction between these aspects. We believe that unless additional proof of ingestion vs. adhesion of the dying cells is given, there can be no claim to actual phagocytosis, certainly not under the usual conditions that are used for the determination of the phagocytosis index, namely, light microscopy alone. Some authors use trypsin with

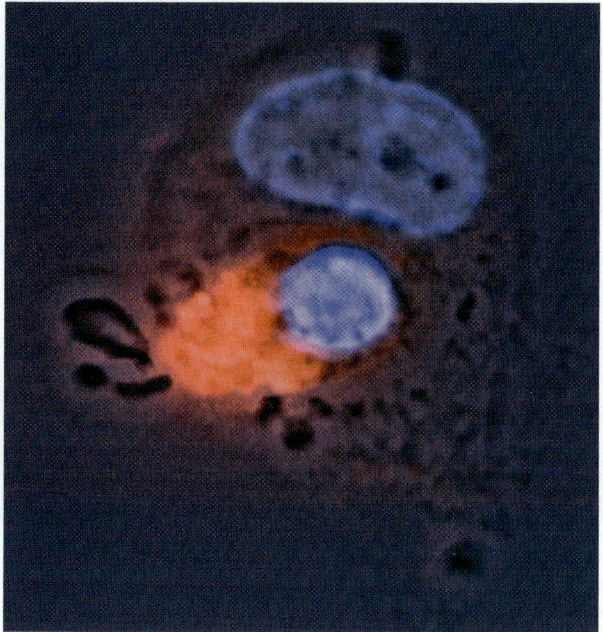

Fig. 8.4 Human monocyte-derived dendritic cell after engulfing an apoptotic monocyte. Wide-field fluorescence microscopy image of a human monocyte-derived dendritic cell that has engulfed a DiI-stained apoptotic monocyte (red). DAPI (blue) was used to demonstrate the nuclei, with an overlay of the phase contrast profile. Note the difference between the lightly fluorescent dendritic cell nucleus and the condensed and highly fluorescent chromatin of the apoptotic monocyte nucleus. Original magnification, 1000x.

EDTA to separate the adherent cells and thus distinguish between binding or tethering and phagocytosis (Savill et al. 1990; Savill et al. 1992). However, the accuracy of this method is not clear (Figs. 8.2, 8.6, and 8.8).

8.3   Methods of Study

To study the phagocytosis of dying cells effectively, one must take into consideration several important factors when deciding on the type and form of experiment. Counting the number of dying cells poses a few difficulties. Dying cells are susceptible to mechanical disruption, lysis (primary or secondary), and fragmentation, precluding their accurate counting. The same logic applies when trying to quantify phagocytosis based on the amount of material acquired by the phagocytes: acquisition of fragments and debris can easily skew the measurements (Leers et al. 2002). For this reason, it is better to use cells that are as stable as possible. Early apoptotic cells are the least likely to present problems in that regard (Fig. 8.9). Kinetic studies that

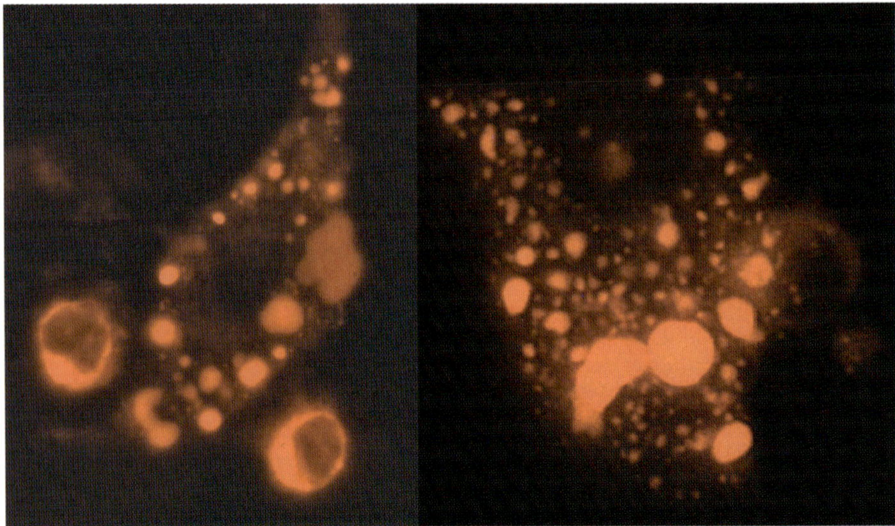

Fig. 8.5 Human monocytes-derived dendritic cells after interaction with apoptotic, DiI-labeled human monocytes. Widefield fluorescence microscopy images of human monocyte-derived dendritic cells after interaction in vitro with DiI-labeled apoptotic human monocytes (red). The dendritic cells have acquired significant amounts of apoptotic debris and are also in contact with intact apoptotic cells. Left panel: Reprinted with permission from Blood Journal: "Apoptotic cell thrombospondin-1 and heparin-binding domain lead to dendritic-cell phagocytic and tolerizing states" Krispin et al. Blood 2006; 108:3580 © the American Society of Hematology.

determine the rate of spontaneous 'disappearance' of dying cells are important as a control. Other important measures include viable phagocytes that have not been left standing on ice for too long. Our experience shows that dendritic cells down-regulate their uptake capacity after incubation on ice, although the uptake capacity of macrophages is less susceptible to this effect. In all cases, it is of prime importance to keep the number of target cells and phagocytes, as well as the conditions of their interactions, constant. For example, comparison of the interactions between phagocytes and apoptotic cells at respective ratios of 1:2 and 1:8 cells should be done in the same volume of material. The medium used is important as well. The presence of serum, even fetal calf serum, may influence uptake due to the presence of natural opsonins, such as protein S (Anderson et al. 2003), complement (Mevorach et al. 1998a), autoantibodies (Manfredi et al. 1998), and others. There may even be differences between frozen and fresh sera. For example, frozen murine serum looses many complement factors upon thawing, and if complement is to be included in the evaluation, it should be prepared fresh. Another important general consideration is that macrophages tend to adhere to plastic. If macrophages are needed in suspension, trypsin with EDTA is commonly used to detach them, but that treatment may alter the cells and change the results. An alternative approach is to use a type of plastic to which macrophages do not adhere, such as polypropylene.

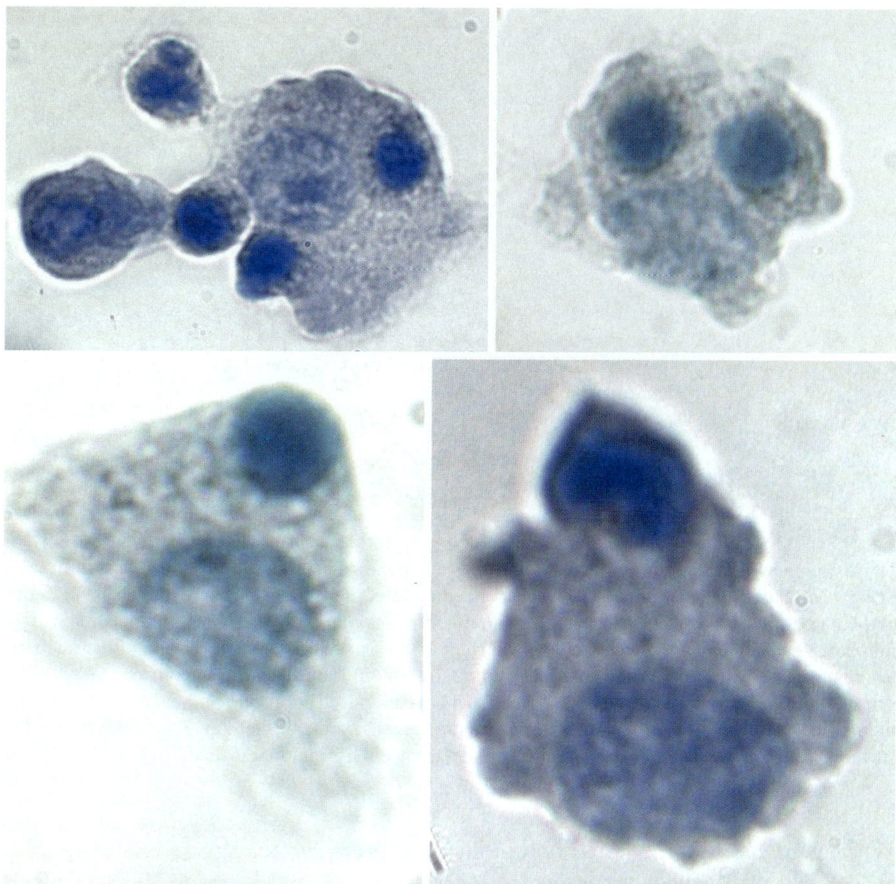

Fig. 8.6 Human monocytes-derived dendritic cells after interaction with apoptotic monocytes. Light microscopy of hematoxylin-stained, human monocyte-derived dendritic cells, after their interaction with human apoptotic monocytes. These images are examples of the level of detail that can be obtained with light microscopy. Note the differential staining of the nuclei of dendritic cells and apoptotic monocytes. Top left panel, 1000x original magnification; all other panels, 600x original magnification.

8.3.1 *Dying Cells: Sources and Identification*

When studying phagocytosis of dying cells, one should indicate which death process the cells are undergoing. There are many protocols and methods for the induction of programmed cell death using both extrinsic and intrinsic pathways (Krysko et al., Chap. 1, this Vol.), and caspase-dependent and caspase-independent patterns. These patterns may differ in their effects on the plasma membrane, death kinetics, and other aspects of the death process. Before a particular method is used to study the clearance of dying cells, it must be carefully studied to ensure that the

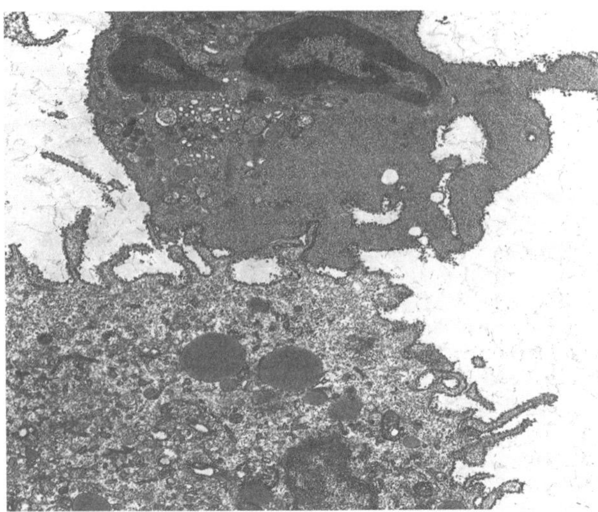

Fig. 8.7 Transmission electron micrograph of an apoptotic neutrophil adhering to a macrophage. At the top is an apoptotic human neutrophil. It has typical early apoptotic morphology, with condensed and clumped chromatin, and small concentrated contents (suggested by the electron-dense cytosol). At the bottom, a portion of a human macrophage can be seen. The macrophage contains numerous organelles and vesicles, and at the right is what might be a small portion of the nucleus. Some of the vesicles contain electron-dense material of similar consistency as the apoptotic neutrophil, indicating that these are phagosomes that have engulfed apoptotic bodies or remnants. The membrane of the macrophage shows numerous protrusions, some of them in close contact with the surface of the apoptotic neutrophil, indicating adherence. Original magnification, 6300x.

experimental design includes appropriate conditions. Some triggers induce apoptosis in one type of cells but necrosis in another (Jaattela and Tschopp 2003; Kroemer and Martin 2005). Similarly, one concentration (e.g. of H_2O_2) can induce apoptosis while a higher concentration induces necrosis. In addition, as time passes, cells that undergo apoptosis may become secondarily necrotic unless they are cleared. Contradictions between reports on the immunologic effects of apoptotic and necrotic cells are evident (Ronchetti et al. 1999; Sauter et al. 2000; Savill and Fadok 2000; Verbovetski et al. 2002; Goldszmid et al. 2003; Hirt and Leist 2003; Brouckaert et al. 2004; Casares et al. 2005; Kacani et al. 2005). These discrepancies are often due to the use of different populations of dying cells and the adoption of definitions that are not clear.

Frequently it is almost impossible to obtain pure populations of apoptotic or necrotic cells (Fig. 8.9). It is therefore advisable to have control sets to verify the state of the cells being used in any given experiment, although it is sometimes impossible to obtain viable cells uncontaminated with apoptotic cells. As mentioned above, careful preparatory studies of the cells of interest, as well as meticulous consistency across experiments, are the best ways to obtain specific and reproducible results.

Programmed cell death can be initiated by several triggers, and the timing of the apoptotic program execution can vary among cells depending on the circum-

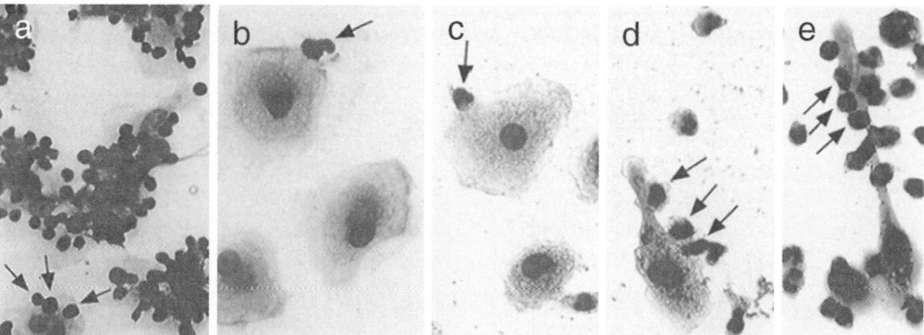

Fig. 8.8 The interaction index: phagocytosis of human and murine apoptotic cells by human macrophages. In panels a and b, human macrophages were exposed to apoptotic murine thymocytes, and in panels c to e, to autologous apoptotic neutrophils. In panels a, d, and e, 15% human serum was added during the interaction, and in panels b and c it was not added. This is an example of a phagocytosis index experiment, where the number of dying cells interacting with the phagocytes is determined and counted using light microscopy (Wright stain). The arrows point to examples of apoptotic cells interacting with the phagocytes. The difficulty of confidently determining the status of such interactions is readily apparent; in this image it is not possible to discern whether the apoptotic cells are adhering, internalized, or superimposed on the phagocyte (see also Fig. 8.2). Moreover, in cases of strong interaction, such as in panel a, even counting the events of interaction can be very difficult. In this experiment the opsonization of apoptotic cells by complement was studied. The comparison of panels b and c, where there was no human serum added to the interaction and therefore no complement was available to opsonize the dying cells, with panels a, and d, and e, show that in these conditions, opsonization by complement greatly increases the capacity for interaction between apoptotic cells and macrophages. Original magnification, 400x. Reprinted with permission from The Journal of Experimental Medicine. Copyright 1998, The Rockefeller University Press. "Complement-dependent clearance of apoptotic cells by human macrophages" Mevorach et al. 1998 188: 2313–2320.

stances. Nonetheless, it is useful to generalize by stating that the early events are dissipation of the mitochondrial potential and exposure of phosphatidylserine, that early-to-intermediate events are the readily measurable caspase activation and the compromise of plasma membrane integrity, and that the intermediate-to-late event is DNA degradation. Here, we present some of the methods most commonly used to detect the processes just mentioned. Many manipulations and refinements of these methods are available, and other methods also exist.

8.3.1.1 Morphological Appearance

The classic changes associated with apoptosis include membrane blebbing, shrinkage, nuclear fragmentation, chromatin condensation and the release of apoptotic bodies (Diez-Fraile, Chap. 2, this Vol.; Darzynkiewicz et al. 1997; Fig. 8.10). These can be seen to different extents with all types of microscopic methods; EM is the gold standard. When using flow cytometry, rough apoptotic changes can be observed as a reduction in forward scatter, and depending on the cell type, an increase or decrease

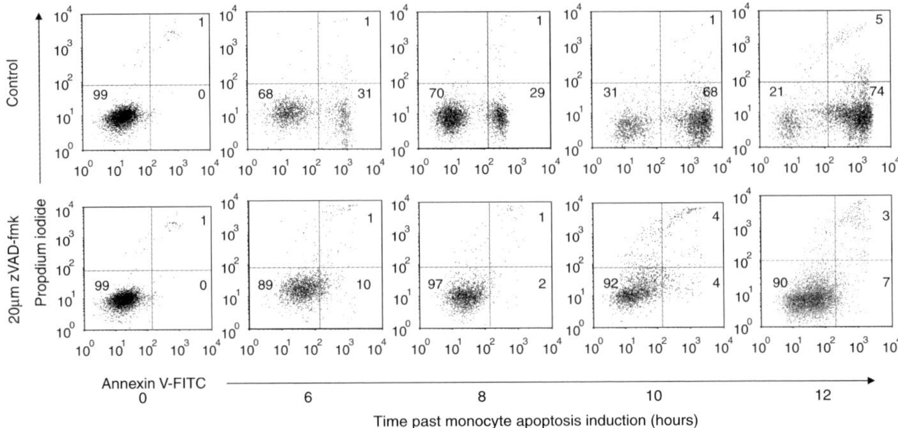

Fig. 8.9 Kinetics of serum-withdrawal apoptosis in human monocytes. Human blood monocytes were incubated under serum withdrawal conditions, and their viability was assessed using Annexin V and propidium iodide (PI). In every plot, the lower left quadrant shows viable cells that are negative for Annexin V and PI, the lower right quadrant, early apoptotic cells that are Annexin V positive, and the upper right quadrant, late apoptotic and secondarily necrotic cells that are positive for Annexin V and PI. In the top row, it can be seen that monocytes undergo an ordered progression of cell death, with slow conversion from viable to early apoptosis. Late apoptosis with secondary necrosis is almost not seen. This allows establishment of a population that is 75% enriched for pure apoptotic cells after 12 hours. When the pan-caspase inhibitor zVAD-fmk was added, the progression of apoptosis was halted but, importantly, a switch towards a necrotic-type of cell death was not observed. Reprinted with permission from Blood Journal: "Apoptotic cell thrombospondin-1 and heparin-binding domain lead to dendritic-cell phagocytic and tolerizing states" Krispin et al. Blood 2006 108:3580 © the American Society of Hematology.

in side scatter (Darzynkiewic et al. 1997; Rasola and Geuna 2001). In contrast to apoptosis, necrosis is associated with cellular swelling and chromatin digestion, followed by DNA hydrolysis, organelle vacuolation, and plasma membrane disruption, and ending in lysis and spillage of the cellular contents (Darzynkiewicz et al. 1997). As with apoptosis, these changes can be seen with microscopy, but they are harder to observe because necrotic cells are very susceptible to damage during the staining procedure and they lyse easily. In this way necrotic cells leave no evidence of their presence other than scattered debris. With flow cytometry, necrotic cells usually show a marked increase in both side and forward scatter (Darzynkiewicz et al. 1997; Rasola and Geuna 2001).

8.3.1.2 Membrane Integrity Probes

Although plasma membrane integrity is maintained during the early stages of apoptosis, it can become compromised during cell death, allowing leakage of cellular contents as well as entry of extracellular molecules. Several available dyes are excluded by viable cells but can enter through the compromised membranes of

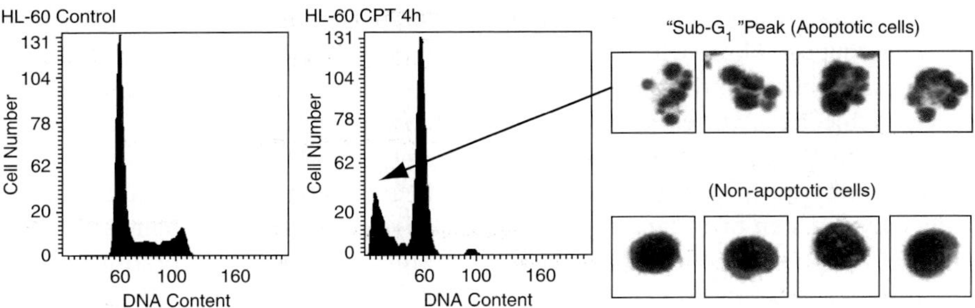

Fig. 8.10 Concomitant analysis of DNA content and cellular morphology. HL-60 cells were induced to undergo apoptosis by incubation for 4 hours with campothecin. They were then fixed on slides with 70% ethanol and stained with propidium iodide (PI). To avoid RNA staining by PI, they were fixed in the presence of RNAse. Cells were then analysed using a laser scanning cytometer. This machine uses a laser to scan slides with adhering cells, thus resembling the capability of a confocal microscope, while collecting the scattered light and fluorescence using photomultipliers, a task typically performed using a flow cytometer. This dual capability enables the machine to provide images of the cells together with accurate light scatter and fluorescent measures. Apoptotic cells undergo DNA fragmentation, and small DNA fragments are washed following ethanol fixation, as was done in this example. As a result, apoptotic cells characteristically present less than diploid DNA, and are called the "sub-G1" population. The advantage of the laser scanning cytometer is that it can show both accurate cell fluorescence measurements, as seen in the DNA content plots at the left, as well as excellent images of exactly the same cells, as seen in images at the right, in both the top and bottom rows.Reprinted with permission from Wiley-Liss, Inc., a subsidiary of John Wiley & Sons, Inc: "Analysis of apoptosis by laser scanning cytometry", Bedner et al. Cytometry 35:181 copyright 1999.

dying cells. Some examples include nucleic acid-specific dyes, such as propidium iodide (PI), 7AAD, and the Sytox family of dyes, all of which are fluorescent, or trypan blue which is not fluorescent but stains cells blue and can be detected by light microscopy. The fluorescent dyes bind DNA (and sometimes RNA) with high specificity, and significantly increase their fluorescence, effectively marking the compromised cells (Frey 1995, 1997; Gaforio et al. 2002). Necrotic cells, per definition, are permeable to such membrane-exclusion dyes (Fig. 8.11). As a general rule, apoptotic cells are not labelled by such dyes in the earliest stages of programmed cell death, but they do become labeled at later stages. Nevertheless, some types of apoptotic cells can become permeable to DNA-specific, membrane-excluded dyes of small molecular weight relatively early during the cell death process (Frey 1995, 1997; Zamai et al. 1996; Zamai et al. 2001). The cells become fluorescent, but with less intensity than late apoptotic or frankly necrotic cells. Some authors use the term 'PI dim' staining to indicate the low PI penetration that characterizes apoptotic cells that are not yet necrotic. An alternative method based on the same principle uses fluorogenic esterase substrates (Frey 1997). These are ester derivatives of fluorophores that are non-fluorescent and electrically neutral, and thus can freely diffuse into cells. Once inside, they are cleaved by ubiquitous esterases, becoming fluorescent and negatively charged. Viable cells with impermeable membranes can sequester the fluorophores inside, whereas cells with compromised plasma membrane integrity leak the compounds. Thus, whereas classic membrane integrity probes, such as PI, label cells with compromised membranes, this alternative method fluorescently labels viable cells. Different types of these fluorophores are available, the classic being fluorescein diacetate (FDA). As with the necrotic exclusion dyes, different fluorophores diffuse in and out of the cells to different extents, and apoptotic cells may differ in their capacity to retain the dyes depending on their origin and stage of apoptosis. As mentioned before, careful titration and calibration studies are always mandatory to validate these assays for any given model of study.

8.3.1.3   Annexin V

One of the earliest processes occurring during apoptosis is the exposure of phosphatidylserine on the outer leaflet of the plasma membrane, whereas viable cells restrict phosphatidylserine to the inner leaflet (Fadok et al. 1992; Fadok et al. 1998a; Holthuis and Levine 2005). In the presence of calcium, Annexin V binds phosphatidylserine with high affinity and specificity, allowing the detection of early apoptosis (Trahtemberg et al. 2007; Figs. 8.9 and 8.11). It is important to note that even relatively low concentrations of calcium are enough to facilitate effective Annexin V-phosphatidylserine binding. Use of media with high calcium concentrations (such as the media commonly provided with Annexin V kits) can affect the death process and kill cells, especially primary cells (Trahtemberg et al. 2007). Of note, some cells fail to expose phosphatidylserine upon apoptosis (Frey 1997), and some do so upon activation or even constitutively (Krysko et al. 2006a).

8.3.1.4   Mitochondrial Membrane Potential

Upon apoptosis, the mitochondrial membrane loses its potential before or concomitantly with phosphatidylserine exposure (Castedo et al. 1996; Trahtemberg et al. 2007). Several available dyes can be used to monitor the mitochondrial membrane potential, and two of the most useful are $DiOC_6(3)$ (Ozgen et al. 2000) and TMRM (Rasola and Geuna 2001). These dyes accumulate inside the negatively charged mitochondrion, but they cannot accumulate if the mitochondrion's membrane potential is dissipated. Therefore, viable cells fluoresce more intensely if their mitochondria maintain their activity and membrane potential. (Fig. 8.11). When conducting such experiments, the dyes should be used at very low concentrations, on the order of a few to a few tens of nanomoles. At higher concentrations the dyes' fluorescence loses its dependence on mitochondrial potential (Shapiro 2003). Since significant variations of the staining patterns can be observed in different cells, careful titrations should be performed with the specific model being used and positive controls should be included, such as the ionopore CCCP, which artificially dissipates the mitochondrial membrane potential.

8.3.1.5   Caspase Activation

One of the most frequently studied phenomena relating to apoptotic programmed cell death is caspase activation. Caspases are a family of proteases of broad specificity that are inactive under normal circumstances (Earnshaw et al. 1999). Caspases cleave many important targets inside cells as part of the death program, preparing the cells for their demise and future ingestion, and activating other proteins that are involved in intra and extracellular signalling (Earnshaw et al. 1999; Kagan et al. 2002). Of note, it has been found that some of the caspases are not part of the apoptotic program, but instead participate in the cleavage of proteins, such as cytokines, as part of normal cellular functions (Taylor et al. 2008). Fluorescent caspase inhibitors, which enter cells, react with the active caspases and bind them, are commonly used in the detection of caspase activity. An alternative but similar approach uses fluorogenic caspase substrates. Some caspase substrates and inhibitors have a high specificity for individual caspases whereas others have a broad range (Kagan et al. 2002; Pozarowski et al. 2003). Added caspase substrates and inhibitors might compete with the endogenous substrates and affect the cell death process (Smolewski et al. 2002; Pozarowski et al. 2003). False-positives can also occur, apparently due to retention of substrates by factors other than caspases (Pozarowski et al. 2003). Another option is the use of an anti-activated caspase 3 antibody (Gardai et al. 2005), which is used for immunohistochemistry and Western blots. Many other proteolytic enzymes involved in apoptotic, necrotic and other cell death processes have been discovered (Jaattela and Tschopp 2003; Kroemer and Martin 2005). As our knowledge advances, new techniques to test enzyme activity and new caspase assays are being developed.

8.3.1.6   DNA Degradation

One of the best-known and most-studied processes occurring during apoptotic cell death is the cleavage of genomic DNA at regular intervals, a process that produces a characteristic "laddering" when the DNA is run on a gel. The most common method used to test for this type of DNA degradation is the terminal uridine deoxynucleotidyl transferase dUTP nick end labelling (TUNEL) assay (Gavrieli et al. 1992). This method takes advantage of the fact that the DNA degradation process proceeds through the creation of nicks in the DNA strands: the enzyme terminal uridine deoxynucleotidyl transferase (TdT) can insert fluorescent or labeled dUTP nucleotides on those nicks, thus labelling the dying cells (Gavrieli et al. 1992; Darzynkiewicz et al. 1997). It has been suggested that this method should not be used as the sole means of discriminating between apoptosis and necrosis, because a large amount of DNA degradation can also occur during necrotic cell death. Although DNA degradation is not as orderly during necrosis as during apoptosis, it still produces false positives (Darzynkiewicz et al. 1997; Maciorowski et al. 2001). During fixation, DNase can enter viable cells and degrade their DNa, also producing false positives. Another way to asses DNA degradation is to fix the apoptotic cells with ethanol and

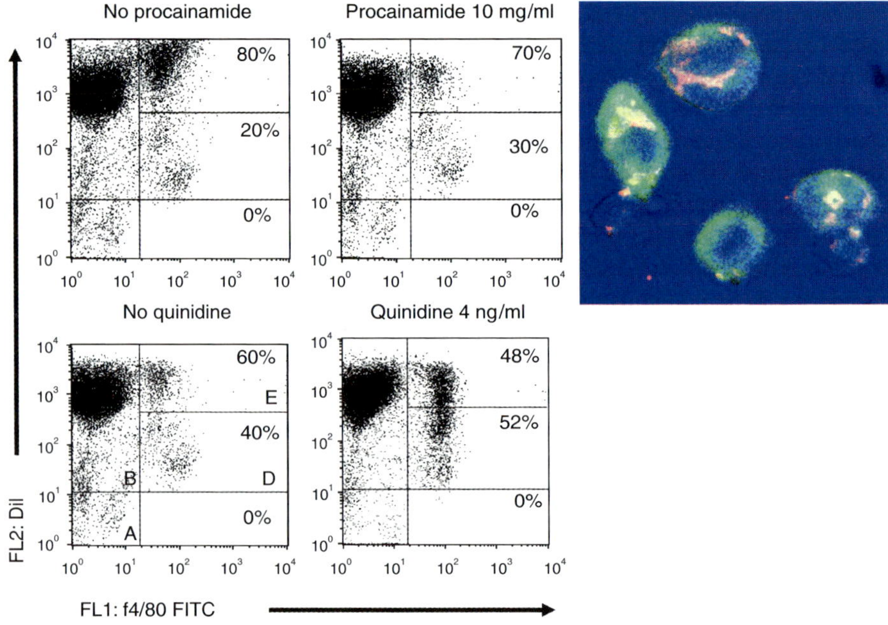

Fig. 8.12 Phagocytosis of necrotic murine thymocytes by murine peritoneal macrophages. In this study, murine thymocytes were labeled with DiI and induced to undergo heat necrosis by incubation at 56°C for 10 minutes. The thymocytes were then offered to murine peritoneal macrophages in vitro. Analysis was performed by flow cytometry and confocal microscopy. Flow cytometry plots on the left show macrophages after staining with the macrophage marker f4/80, horizontal axis, with DiI fluorescence on the vertical axis. The far left section of the plots (represented by "A" and "B" in the lower left plot) corresponds to f4/80-negative thymocytes that were not taken up by macrophages. The far right section ("C" to "E") corresponds to f4/80-positive macrophages, and their DiI fluorescence corresponds to the level of phagocytosis of necrotic thymocytes. In the lower left plot, section "D" represents a low level of uptake, whereas section "E" represents a high level of uptake. In the two experiments represented here, the effects of procainamide and quinidine on the phagocytosis of necrotic thymocytes was explored. Both procainamide and quinidine significantly impaired macrophage phagocytosis, as can be seen from the different percentages of macrophages with high uptake levels in section "E" vs. macrophages with low uptake level in section "D". On the right, a confocal microscopy image of a sample from the same experiment is shown, stained in the same manner. Green-labeled macrophages have internalized remnants of red-labeled thymocytes. This is an example of confirmation of results by complementary techniques.Reprinted with permission from Springer Science and Business Media. "Quinidine and procainamide inhibit murine macrophage uptake of apoptotic and necrotic cells: a novel contributing mechanism of drug-induced-lupus" Ablin et al. Apoptosis 10:1009 2005.

8.3.3.2   Transfection

The transfection of receptors into cells lines that are not naturally phagocytic is a powerful technique for studying specific receptors in isolation (Ren et al. 1995;

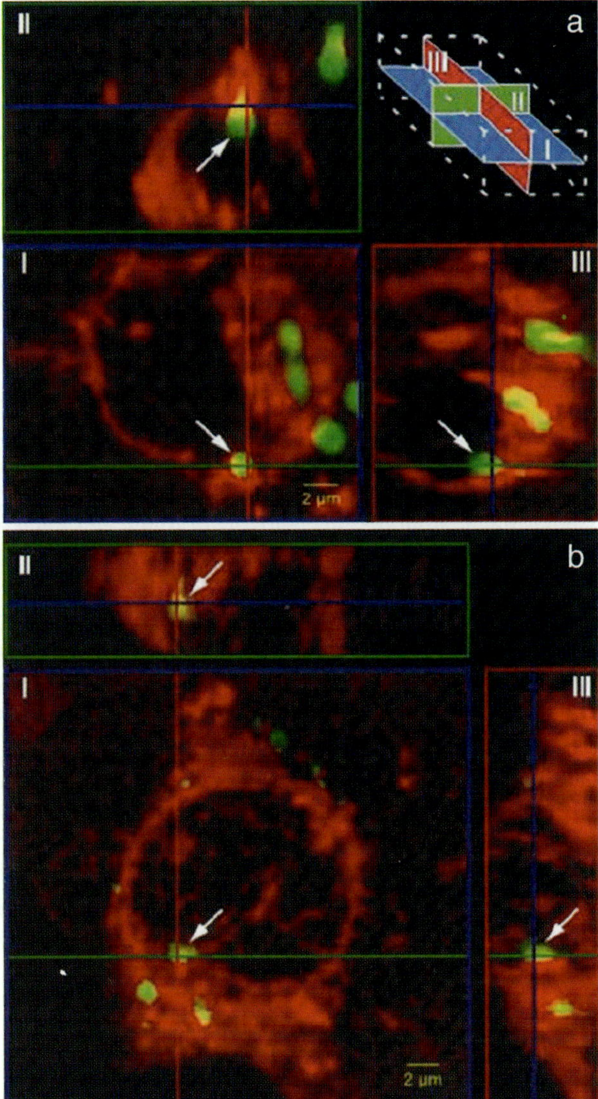

Fig. 8.13 Confocal microscopy of macrophage phagocytosis. Confocal laser scanning microscopy image of a) J774 macrophage, labeled in red, following phagocytosis of green fluorescent beads, and b) apoptotic U937 cells, labelled with green fluorescence, after 1 hour of coincubation. This is an example of a high quality, high resolution confocal image, which unambiguously shows that the particles marked with arrows are inside the macrophage. This is done by making a 3D reconstruction and determining that the particle is surrounded by cytoplasm in each of tree perpendicular planes. Reprinted with permission from The Journal of Immunological Methods, 287:101, Schrijvers et al. "Flow cytometric evaluation of a model for phagocytosis of cells undergoing apoptosis" Copyright 2004.

Mevorach et al. 1998a; Fig. 8.14). The main disadvantage of this technique is that many receptors do not work alone, and transfection of a single receptor may produce results that do not accurately define the receptor's role in vivo. Alternatively, the transfected cell line may lack some of the molecular machinery needed for the normal function of the transfected receptor.

8.3.3.3 Opsonins

The use of opsonins involves the study of the effects of opsonization on the phagocytosis of dying cells (Verbovetski et al. 2002; Anderson et al. 2003; Gardai et al. 2005; Fig. 8.8).

8.3.3.4 Consequences of Interaction with Apoptotic Cells

In addition to studies elucidating the mechanisms of dying cell uptake, experiments are also performed to evaluate the consequences of such uptake (Ucker, Chap. 6; Lacy-Hulbert, Chap. 7; Gregory-Pound, Chap. 9, this Vol.). These experiments commonly look for effects at the level of signal transduction (Tassiulas et al. 2007), cytokine or chemokine secretion (Fadok et al. 1998c; Sauter et al. 2000; Morelli et al. 2003; Kim et al. 2004), or changes in the transcriptome, including the use of RNA microarrays (Grolleau et al. 2003; Brouckaert et al. 2004; Kim et al. 2004). There are also functional assays for studying the consequences of phagocytosis of dying cells. A common method used in immunology consists of offering dying cells to phagocytes, and then using these phagocytes for a second interaction round with other cells of the immune system, such as B, T or NK cells. The outcomes of such interactions can be studied by various methods (Sauter et al. 2000; Goldszmid et al. 2003; Miyake et al. 2007). One example of these studies is the mixed lymphocyte reaction (MLR), in which dendritic cells are incubated with dying cells and then mixed with allogeneic lymphocytes that have been previously marked with a tracker label such as CFSE. As the lymphocytes divide, they split their contents, including the CFSE label. The reduction in CFSE labelling over 5 to 6 days is used as a measure of the proliferation induced by the dendritic cells (Hoeffel et al. 2007; Johansson et al. 2007).

8.3.4 *In vivo Interactions*

Two common, non-mammalian model organisms exist for the in vivo study of phagocytosis of dying cells, *C. elegans* and *D. melanogaster*. The body of work on *C. elegans* is extensive, to the point that it is known exactly how many cells comprise the adult organism, exactly how many cells die in the process of maturation, and when (Krysko et al. 2006a; Lettre and Hengartner 2006; Fig. 8.15). Many of the most important genes and processes involved in the phagocytosis of dying cells

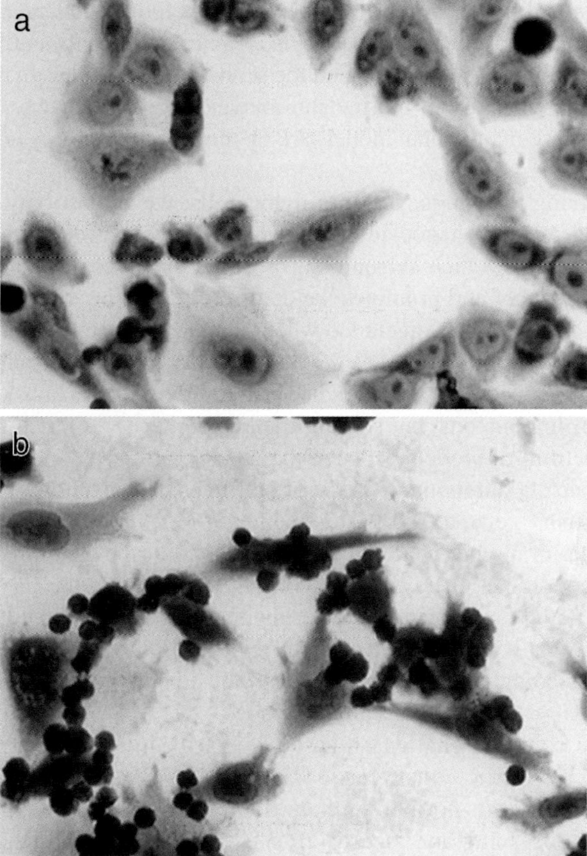

Fig. 8.14 Phagocytosis of murine apoptotic thymocytes by CHO cells transfected with human CR3, with or without exposure to 15% human serum. In this experiment, the relatively non-phagocytic CHO cell line was stably transfected with the human CR3 receptor. Apoptotic murine thymocytes were offered to the cells, either in the absence (top panel, a) or presence (bottom panel, b) of human serum. The cells were then fixed and labeled with Wright stain. As can be seen, addition of human serum to CR3 transfected cells conferred upon them the capacity to interact with and take up the apoptotic cells. Non-transfected CHO cells did not take up apoptotic cells, even in the presence of human serum (not shown). Original magnification, 400x. Reprinted with permission from The Journal of Experimental Medicine "Complement-dependent clearance of apoptotic cells by human macrophages" Mevorach et al. 1998, 188: 2313–2320. Copyright 1998, The Rockefeller University Press.

have been discovered in *C. elegans* (Gronski and Ravichandran, Chap. 5, this Vol.; Chung et al. 2000; Lettre and Hengartner 2006); one of the most striking discoveries is the degree of conservation of the molecular machinery (Krysko et al. 2006a; Lettre and Hengartner 2006). Studies in *D. melanogaster*, due to its higher complexity and phylogenetic closeness to humans, have steadily increased its appeal as a model organism (Krysko et al. 2006a; Stuart and Ezekowitz 2008). Of special importance are the professional phagocytic cells, the hemocytes found in *D. melanogaster*, the equivalent of which do not exist in *C. elegans* (Franc et al. 1999). Further discussion of these two model organisms is beyond the scope of this chapter. In this section we will concentrate on the use of murine and human models of phagocytosis of dying cells, because these mammalian models are by the most commonly used in this area of study.

When studying the phagocytosis of dying cells, we are not only studying programmed cell death and phagocytes, but the interactions between these cells as well as their interaction with their surroundings. This is underscored by the finding that phagocytosis is part of and promotes the demise of the dying cell, to the point that some types of cell death are precluded when phagocytic mechanisms are disrupted (Hoeppner et al. 2001; Reddien et al. 2001). While these effects are more easily dissected in vitro, that is clearly an artificial setting. In vivo experiments allow for the full richness of interactions, but solving the puzzle becomes much more difficult. There is a spectrum of models between the in vitro and in vivo extremes, ranging from the completely autonomous study of cell lines, to the study of ex vivo tissue samples or primary cells, to studies that combine in vivo with ex vivo stages. The main advantages of in vitro work are the complete control of the experimental environment and conditions, the availability of the samples, and the far greater ability to manipulate both the cells and the experimental model. The main disadvantage of in vitro work is that it uses isolated systems to study limited interactions, which is a far cry from the complexity of living tissues. The advantages and disadvantages of in vivo and in vitro work are basically opposites.

Beyond the common challenges associated with in vivo experiments in animals, there is a specific constraint to the study of phagocytosis of dying cells in vivo, namely, that in normal animals this process is very fast and efficient. Consequently, observing and measuring it is difficult, even in tissues with very high rates of cell death and turnover, such as the young thymus or bone-marrow (Scott et al. 2001; Ravichandran and Lorenz 2007). For this reason, many in vivo studies include active interventions. A common, simple approach to in vivo studies is tissue sample staining (Scott et al. 2001; Leers et al. 2002; Bose et al. 2004; Hanayama et al. 2004), using histochemical stains, either alone or augmented by reagents that detect cell death such as the TUNEL assay (Leers et al. 2002; Bose et al. 2004; Hanayama et al. 2004), or phagocytes, including specific antibodies (Leers et al. 2002; Bose et al. 2004; Hanayama et al. 2004). Another common model is based on ex vivo cell labelling, followed by induction of cell death and injection into the original or another animal, either intravenously (Scott et al. 2001; Cohen et al. 2002; Morelli et al. 2003), intraperitoneally (Ronchetti et al. 1999; Hu et al. 2000; Scott et al. 2001), subcutaneously (Goldszmid et al. 2003), or even intratracheally (Hu et al. 2000). After a specific time lapse, usually a few hours to one day, the animal is sacrificed and tissues are scanned microscopically for fluorescence, or by flow cytometry following tissue disruption and extraction of the phagocytes (Hu et al. 2000; Cohen et al. 2002; Morelli et al. 2003). In many cases cell labelling is complemented by labelling of specific markers on the phagocytes (Goldszmid et al. 2003; Morelli

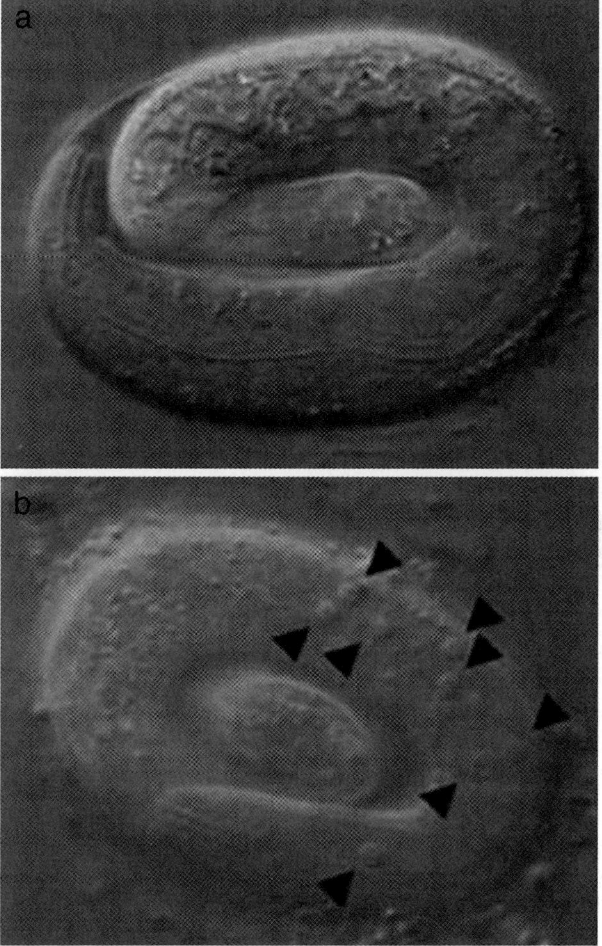

Fig. 8.15 Late C. elegans embryos. a) wild-type C. elegans embryo at the three-fold stage, and b) a mutant strain. In this example of in vivo imaging, there is a failure in the clearance of dying cells (arrow-heads) in the mutant specimen, whereas in the wild-type embryo all dead and dying cells have been completely removed. Reprinted with permission from Cell 107:27, Gumienny et al. "CED-12/ELMO, a novel member of the CrkII/Dock180/Rac pathway, is required for phagocytosis and cell migration" Copyright 2001.

et al. 2003). Another option is induction of extensive cell death by irradiation, ischemia, or toxins, followed by immunohistochemical tissue studies, fluorescence microscopy, or flow cytometry, as mentioned above (Haruta et al. 2001; Lorimore et al. 2001; Scott et al. 2001). Some of the most important in vivo studies are performed using mutant organisms, such as knockout mice, to study the effects of specific defects of the phagocytic machinery or cell death molecular machinery (Gumienny et al. 2001; Haruta et al. 2001; Scott et al. 2001; Cohen et al. 2002). Many times these knockout animals are used also as sources of mutant cells for in vitro experiments (Fig. 8.2). Although these studies are considered state of the art, they may be misleading, because the richness of interactions and complexity of the organism do not always lend themselves to interpretations that assume a one-to-one co-relation between cause and effect. Moreover, these studies are complex undertakings, and technical errors can lead to mistakes or flawed interpretations (Fadok et al. 2000; Bose et al. 2004; Bygrave et al. 2004; Kunisaki et al. 2004; Mitchell et al. 2006). It is also worth remembering that there are still many unknown variables and genes that interact to produce observed phenotypes, and some of the effects attributed to missing genes in knockout mice may in fact be related to other factors (Bygrave et al. 2004). These methods have been further refined in elegant studies that create conditional knockout or conditional expression, where the molecule of interest is selectively repressed or expressed in a specific tissue type or after the injection of a specific ligand, thus avoiding many of the pitfalls that can plague classic transgenic or knockout studies (Glaser et al. 2005; Chen and Sheppard 2007; Mishina and Sakimura 2007).

One of the central reasons for doing in vivo studies is determining the consequences to the whole organism of phagocytosis of a specific set of dying cells, or phagocytosis of dying cells under specific conditions. A common method used in these studies is to inject marked or unmarked dying cells into animals and then to obtain tissue or phagocyte samples for further study (Ronchetti et al. 1999; Hu et al. 2000; Scott et al. 2001; Cohen et al. 2002; Morelli et al. 2003; Bondanza et al. 2004; Kunisaki et al. 2004). Another common method is to determine the clinical outcome or phenotype of the whole organism (Mevorach et al. 1998b; Goldszmid et al. 2003; Bondanza et al. 2004; Casares et al. 2005; Miyake et al. 2007). When using this approach, it is recommended that cells from the same species be used, because xenogeneic experiments are complicated by the large number of interspecies interactions that can confound and compromise the results (Taylor et al. 2000).

When using murine models, certain organs and conditions can simulate human diseases in which phagocytosis of dying cells has special importance. Examples include knockout mice that develop (or fail to develop) autoimmune diseases due to defects in cell death or phagocytosis (Scott et al. 2001; Cohen et al. 2002; Devitt et al. 2004; Hanayama et al. 2004), ingestion of apoptotic cells by lung alveolar macrophages, usually following acute lung infections or injury (Hu et al. 2000; Henson and Tuder 2008), and toxic liver injury (Haruta et al. 2001).

As mentioned in the previous section, many studies combine in vitro and in vivo approaches. For example, dying cells may be offered to dendritic cells in vitro, and the dendritic cells are then injected into the mouse or patient (Goldszmid et al. 2003). This is the most commonly used method for intense scientific investigation of the effects of dying cells on the induction of tolerance to allogeneic transplants or in autoimmune diseases (Liu et al. 2006; Tzeng et al. 2006; Miyake et al. 2007), or the induction of immunity in HIV and cancer models (Ronchetti et al. 1999; Goldszmid et al. 2003; Bondanza et al. 2004; Hoeffel et al. 2007).

8.3.5 Advanced Methods

8.3.5.1 The Study of Live Cells

Cells must often be fixed when performing imaging studies. Fixation allows manipulations that are impossible to perform on living cells. Fixation allows the study of cells in a specific state, and preserves samples for future analysis. The main disadvantage of fixing cells is the introduction of morphological and molecular artifacts, including modification of proteins or the leakage of contents. There are many approaches for imaging live cells in vitro. The simplest example is the study of adherent cells, such as macrophages (Cocco and Ucker 2001), but this approach is somewhat limited because cells must be constantly covered by an aqueous phase, and special inverted microscopes or water immersion lenses are required. Moreover, even adherent cells are not necessarily completely immobile, but this is sometimes an advantage, for example, in time-lapse microscopy. Also, many cells are not adherent. To study living cells that are usually non-adherent or only loosely adherent, various adhesive surfaces comprised of amino acid or polypeptide coatings have been developed, one of the most common being poly-L-lysine (Albert et al. 2000). An attractive alternative is using a mounting material that is compatible with living cells (Biostatus' CyGel;[tm] Upton et al. 2007), which makes it possible to image cells on standard slides and coverslips, without the need for fixation and its associated artifacts. Alternatively, matrix gels or connective tissue scaffolds that enable cell embedding during the experiment can be used for live cell imaging (Tasaki et al. 2004). This approach provides a physiologic scaffold for the cells, but limits their manipulation. Temperature control is essential during live cell imaging (Cocco and Ucker 2001; Bondanza et al. 2004). Cells kept at 37°C in a viable metabolic environment continue their metabolic and physiologic processes, including the phagocytosis of the dying cells. On the other hand, if cells are maintained at 4°C, their metabolism slows down. This provides more time to study them in a stable state and permits the metabolic requirements of the medium to be relaxed. Nonetheless, after some hours at that temperature many cells will begin to degrade. However, most microscopes do not have heated or cooled stages.

8.3.5.2 Quenching of Fluorescent Dyes, pH Dependency and Soluble Dyes

One of the less considered properties of fluorescent probes is their quenching. This can take several forms. One of the most significant and overlooked forms is fluorescence intensity reduction with pH changes, notably at low pH. The classic example is fluorescein, whose fluorescence emission drops significantly when the pH falls below 7 (Ge and Chen 2007). Thus, after probes labeled with fluorescein are internalized, their fluorescence decreases as the phagosomes mature. This phenomenon can affect the results, but it can also be exploited by using fluorescence reduction as evidence of phagocytosis (Haggie and Verkman 2007). The pH dependence of fluorescence is observed for many fluorescent probes and no probe should be assumed to be stable at varying pH levels unless this has been determined by the experimenter or manufacturer. This problem can be surmounted by using fluorescent probes that exhibit stable fluorescent emission across a broad range of pH. The availability of pH-insensitive probes and pH measuring labels provide great versatility for studies of internal phagosome processes. Some of these probes undergo a specific change in their fluorescence intensity or spectrum as a function of pH (Frey 1997). pH-sensitive and pH-insensitive dyes can be added to the extracellular medium (Hoffmann et al. 2001), or they can be used to label dying cells (Ogden et al. 2001; Gardai et al. 2005), allowing the researcher to track the phagosome and to assess its maturation (Dunn et al. 1994; Hackam et al. 1997; Haggie and Verkman 2007). In a related application, pH-insensitive dyes are added to the medium together with dying cells that are offered to the phagocytes. Following their interaction, the medium is washed and the cells are examined by microscopy. The presence or absence of the dyes together with dying cells inside the phagosome (their colocalization), can serve as an indicator of whether internalization was by membrane ruffling and macropynocytosis, or by zipper-like tight ingestion of the target cell (Ogden et al. 2001; Gardai et al. 2005; Krysko et al. 2006b). It is important to note that a given phagocyte can indulge in both phagocytosis of target cells and macropinocytosis of extracellular medium. Therefore it is essential to examine the co-localization of the free dye and the ingested cell, and not only for the presence of both labels in a given phagocyte. Chemical quenching occurs when the addition of a quenching molecule reduces the emission of the fluorophore. A classic example of this application is the addition of trypan blue to the medium after interaction between fluorescently-labeled dying cells and phagocytes. Trypan blue quenches the fluorescence of both free and adherent cells but does not affect the fluorescence of cells that have been taken up completely because they are not in contact with the dye (Bjerknes and Bassoe 1984; Wan et al. 1993; Finnemann et al. 1997)

8.3.5.3 The ImageStream® Platform

Amnis corporation has developed a new analytical instrument that combines microscopy and flow cytometry; references relevant to the study of phagocytosis include (George et al. 2004; Tibrewal et al. 2007; Boettner et al. 2008). Instead of using photomultiplier detectors as in flow cytometers, the ImageStream uses charge-coupled devices of the type used in digital cameras. The device tracks cells as they flow in front of the detectors, integrating and combining their signals to provide a total value for the fluorescence of the whole cell as in flow cytometry, a wide field fluorescent image of every cell, four fluorescence channels, and both wide field and dark field images. The combination of imaging and analysis of large numbers of cells facilitates studies that had been impossible or very difficult to perform. Some of the disadvantages of the ImageStream include the limited number of fluorescent channels, the relatively limited speed of acquisition, limited ability to determine colocalizations with certainty due to the wide-field image format, and the high cost of the machine. Nevertheless, this platform has enormous potential for future applications in the study of the phagocytosis of dying cells.

8.3.5.4 Lipomics

These techniques study changes in lipids that constitute the membranes of dying cells and the effects of these changes on subsequent phagocytosis (Ritov et al. 1996; Kagan et al. 2002; Kagan et al. 2004; Matsura et al. 2004). Important insights have been gained through such techniques, such as determining that it is oxidized phosphatidylserine that interacts with CD36 during the phagocytosis of dying cells (Greenberg et al. 2006). Some of the methods used in these studies include the following: insertion of specific lipids into cells, which are then induced to die and offered to phagocytes; the creation of lipid vesicles that may or may not be oxidized, which are then offered to phagocytes so that they would compete with dying cells for uptake; and the study of the lipid composition of cells by HPLC and mass spectrometry; and others.

8.3.5.5 Proteomics

This term refers to the large scale study of proteins and the use of advanced methods for their identification (Cravatt et al. 2007). For example, we recently collected supernatants of apoptotic monocytes, separated the secreted proteins by gel electrophoresis, and analysed them with mass spectrometry (Krispin et al. 2006). Another example is the elucidation of the role of Annexin I in phagocytosis of dying cells, which was also accomplished using a proteomic approach (Arur et al. 2003). The proteomic paradigm takes a broader approach to the study of proteins; instead of looking for specific proteins that are either secreted, membrane-bound or intracellular, a whole set of proteins is analysed in large scale, followed by a more in-depth study of specific targets (Stuart et al. 2007). Advances in proteomics are revealing great versatility in protein modifications, beyond their amino acid sequences, and diverse glycosylations that can affect the structure and function of the proteins (Siuti and Kelleher 2007; Witze et al. 2007). However, proteomics methods are complex, they require extensive knowledge of the chemistry and molecular biology involved in the processes under study, and technical errors may compromise the results. As with all large-scale approaches that look for patterns of expression, inadvertent biases can be introduced in the preparation or analysis of such large sets. Some examples of possible pitfalls include protein contamination during the separation of the compartment to be studied, for example by unwanted organelles during the separation of subcellular compartments, or by inclusion of proteins derived from different cells in the culture or tissue sample. Moreover, proteins may become

8.4.1.3 DNA Content Determination

Selected references: (Nicoletti et al. 1991; Gong et al. 1994; Bedner et al. 1999).

- The cells of interest should be centrifuged, the supernatant removed, and the pellets disrupted. Then, slowly add 70–80% ethanol dropwise while stirring the tube, and allow fixation to continue for about 30 minutes. Then, wash the cells and resuspend them in PBS. After fixation, the cells can be kept at 4°C for several weeks. Use PBS for all subsequent steps.
- An optional step is the incubation of the cells for 5–10 minutes in a high molarity phosphate-citrate buffer, which makes small DNA fragments easier to extract. The buffer is prepared by mixing 0.2 M sodium phosphate and 0.1 M citric acid at a 24:1 ratio. After this treatment, the cells should be washed again.
- Add 200 mg of PI and 2 mg of DNAse-free RNAse to 10 ml cell suspension in PBS, and leave to stain for 30 minutes. The RNAse treatment is needed because PI binds RNA as well as DNA. Finally, the cells should be washed and analysed at the flow cytometer.
- The cells with a reduced DNA content are the apoptotic cells, also called hypodiploid or sub-G1 cells.

8.4.1.4 Detection of Apoptosis using Fluorescent Caspase Inhibitors

Selected reference: (Pozarowski et al. 2003).

- This method applies equally well to cells in suspension and to adherent cells. Substrates specific for individual caspases are available, as well as substrates with broader caspase specificity.
- During the last stages of experiments, or after their completion, fluorescent caspase inhibitors are added to the tissue culture and incubated for 30 minutes to 1 hour. Reagent suppliers usually recommend suitable concentrations of about 10 μM. After this incubation, the cells are washed and resuspended in medium without fluorescent caspase inhibitors.
- After that cells should be analysed as soon as possible. Adherent cells can be fixed and mounted to allow time for their analysis, whereas cells in suspension should be analysed promptly without fixation. Additional dyes or fluorescent markers can be combined as needed.
- The control treatment consists of the addition of unlabelled caspase inhibitors 30 minutes to one hour before adding the fluorescent inhibitors. The unlabelled inhibitors should be added at a higher concentration than the fluorescent inhibitors. This control is critical for assessment of the specificity of the assay for caspase activation.

8.4.2 *Phagocytes*

8.4.2.1 Human Monocyte Derived Macrophages

Selected references: (Mevorach et al. 1998a; Flora and Gregory 1994).

- PBMCs are isolated from peripheral blood or standard donor buffy coats using a Ficoll density gradient. After washing, the cells are plated at 10×10^6 cells per mL, in RPMI in 3.5 cm Petri dishes and incubated for 1–2 hours at 37°C. This step allows monocytes to adhere, while other cells stay in suspension or adhere only lightly.
- The PBMCs are vigorously washed at least three times with 2–3 ml of RPMI each time, using a transfer pipette. The remaining cells are cultured in 3 ml of RPMI or Iscove's medium with antibiotics and 10% human serum (either pooled AB serum or autologous serum).
- Three days after incubation, the medium is exchanged with fresh medium and serum.
- On day six, the adhering cells will have differentiated into macrophages and are ready for further experimentation.

8.4.2.2 Other Human Macrophages

Selected reference: (Xu et al. 2006b).

- Human macrophages can be polarized in vitro into pro-inflammatory and anti-inflammatory cells by GM-CSF and M-CSF (CSF-1), respectively. M-CSF derived macrophages were shown to favour the binding and ingestion of early apoptotic cells.

8.4.2.3 Isolation of Murine Resident and Thioglycollate-stimulated Peritoneal Macrophages

Selected references: (Fadok et al. 1992; Licht et al. 1999).

- After sacrificing or anaesthetizing the mouse, peritoneal lavage is performed using 10ml of ice cold PBS with 10% FCS, and the solution is then aspirated. The cells obtained are washed once and resuspended in the same medium.
- After counting, the cells are seeded on small Petri dishes or flat-bottom plates, usually at no more than 10^6 cells per well or dish. They are incubated for 1.5–2 hours at 37°C to allow macrophage to adhere.
- Non-adherent cells are then removed by vigorous washing with a transfer pipette and fresh medium.

- At this point, resident macrophages will have adhered to the plastic surface and will be ready for further experimentation, either immediately or after up to 24 hours of culture. Approximately 10^6 macrophages can be obtained from each mouse.
- To obtain thioglycollate-treated, inflammatory macrophages, 0.5–1 ml of 4% Brewer's thioglycollate is injected intraperitoneally, and 2–4 days later the macrophages are obtained as detailed above. Approximately 2 x 10^6 macrophages can be obtained from each mouse.

8.4.2.4 Generation of Human Monocyte-Derived Dendritic Cells

Selected references: (Romani et al. 1994; Verbovetski et al. 2002).

- PBMCs are isolated from peripheral blood or standard donor buffy coats using a Ficoll density gradient. Monocytes are then separated by using CD14-conjugated magnetic beads or by adherence to plastic and seeded in six-well plates at 2.5 x 10^6 cells in 3 mL of RPMI supplemented with L-glutamine, antibiotics, and either 10% FCS or 1% human serum, preferentially autologous. 1000 U/ml of GM-CSF and 500 U/mL IL-4 are added as well.
- On days 2 and 4 of the culture, 300 µl of medium is extracted carefully without disturbing the cells, which coalesce in the middle of the bottom of the wells. 0.5 ml of fresh medium containing serum and cytokines is then added.
- On day six, cells will have differentiated into immature dendritic cells, and are ready for experimentation. Pipette the well up and down 3–5 times with 1 ml medium to detach the cells before removing them.
- If mature dendritic cells are required, the medium should be refreshed again on day six, followed by the addition of cytokines (e.g. TNF) or inflammatory molecules (e.g. LPS), as required by the experimental protocol. Mature dendritic cells can be harvested 1–3 days later.

8.4.2.5 Transfection of a Cell Line with a Specific Uptake Receptor to Obtain Receptor-Specific Phagocytes

Selected references: (Savill et al. 1992; Mevorach et al. 1998a).

- These experiments require the generation of cell lines, such as COS or CHO, that have been transfected with a specific uptake receptor, either in a transient or stable way. For many receptors, such cell lines are already available from repositories such as ATCC, or through contributions from fellow researchers.
- Once cells are obtained, they can be used to study the role of a specific receptor in the phagocytosis of dying cells.
- It is important to bear in mind that two controls are necessary for a full understanding of the role of the receptor being studied: non-transfected cells of the same origin, and professional phagocytes that are known to express the same receptor.

8.4.3 In Vitro Methods for Evaluating the Interactions of Phagocytes with Dying Cells

8.4.3.1 Phagocytosis (Interaction) Index

Selected references: (Hoffmann et al. 2001; Shoshan et al. 2001).

- This simple, classical method is still in widespread use, but it should not be used alone to study the phagocytosis of dying cells. Many modifications to the simple method described below have been reported; these modifications are aimed at increasing the specificity or sensitivity of the phagocytosis index for the object of interest ('In vitro interactions,' Sect. 8.3.3).
- It is very difficult and sometimes impossible to distinguish, by light microscopy alone, between adhesion and internalization of dying cells by phagocytes. Therefore, we have proposed the term 'interaction index', which describes the actual observations more accurately. Unless adhesion can be reliably distinguished from internalization, all contacts between the dying cells and the phagocytes should be counted and considered interactions.
- The cells used are almost invariably phagocytes that are adhering to plastic or glass surfaces. After their interaction with the dying cells, the slide is washed with cold media (in order to wash out non-adhered or phagocytosed cells), fixed with ethanol or paraformaldehyde, and then usually treated with histological stains, such as Giemsa, Wright or hematoxylin-eosin.
- Dying cells can be identified in most instances not only because they are smaller than phagocytes, but also because of apoptosis-related shrinkage and the condensation of nuclei, which are characteristics of apoptosis.
- Alternatively, a stain that is specific for phagocytes or dying cells can be used to discriminate between them.
- The slide is examined under a microscope (or computer assisted microscope), and 100 to 400 phagocytes are counted. For every phagocyte counted, the number of dying cells interacting with it is recorded and a final tally is prepared. The interaction index is then calculated as the average number of interacting dying cells per phagocyte, or per 100 phagocytes.

8.4.3.2 Flow Cytometric Determination of the Uptake of Labelled Dying Cells

Selected references: (Hess et al. 1997; Ablin et al. 2005).

- This technique is based on two basic procedures: 1) labelling cells with a tracker fluorescent dye and then inducing them to die, and 2) specific staining of the phagocytes after their interaction with the labelled dying cells.
- The cells that are to be induced to die can be labeled with a variety of tracker dyes, three of the most popular being CFSE, PKH, and DiI (and related dyes,

such as DiD and DiO). CFSE is a derivative of fluorescein that labels intracellular proteins, while both PKH and DiI are lipophilic molecules that insert themselves into the membranes of the labelled cells.

- After labelling, the cells are induced to die according to the experimental design. Two important technical considerations to be taken into account are: 1) a substantial number of cells may be lost, and 2) labelling may change the response to the death stimulus in comparison with the unlabelled controls.
- After inducing cell death in the labelled cells, they are usually offered to phagocytes and allowed to interact at 37°C.
- The next step is detaching the phagocytes and staining them with a specific marker so that they can be identified accurately. A typical example is the marker f4/80 for murine macrophages.
- Finally, cells are analysed by flow cytometry. Phagocytes are gated according to their specific marker, and the level of phagocytosis is determined based on the color that was used to stain the dying cells. The analysis can include both the number of phagocytes that have taken up dying cells (percentage of uptake) and by the intensity of their fluorescence (level of uptake).
- One of the pitfalls of this technique is the implicit and mistaken assumption that phagocytes take up only complete cells. On the contrary, both apoptotic and necrotic cells release fragments, and phagocytes are quite adept at their uptake.
- Another issue is distinguishing between internalization and adhesion. This is one reason why the gating of phagocytes should be performed with a specific fluorescent marker and not by light scatter alone: dying cells adhered to phagocytes can change their light scatter characteristics and thus be excluded from the analysis. Moreover, and of greater concern, dying cells that adhere to each other or that swell during the death process can enter the light scatter gate of the phagocytes and contaminate the results. When specific marking of the phagocytes is not desired or not possible, these difficulties may be overcome by using the area vs. the width profiles to discriminate between single cells and doublets. Another option is the addition of an external quencher, such as trypan blue, but then the specific marker of the phagocytes is also lost unless they have been previously marked with an internal probe. Finally, another option for discrimination between adhering and internalized dying cells is to add a third stain that is specific for the dying cells. Thus, only dying cells that have not been internalized will be marked, whereas completely internalized cells will not be available for staining. The problem with this approach is that in many instances the phagocytes take up several cells or cellular fragments, and if they still have a single adhering cell they could be excluded from the analysis.

8.4.3.3 Fluorescence Microscopy of the Phagocytosis of Dying Cells

Selected references: (Hess et al. 1997; Schrijvers et al. 2004; Jones et al. 2005).

- This technique is essentially similar to flow cytometry, except that cells are analysed on a slide rather than in suspension. The properties of phagocytes determine how they are mounted on slides. Adherent phagocytes can be grown directly and stained on slides, with or without fixation; phagocytes grown in suspension can be cytospinned onto slides and fixed either before or after the cytospin.

- Monoclonal antibodies can be used as specific markers for phagocytes, but phagocytes can also be stained before interaction with dying cells using a tracking dye colored to contrast with the dye used to stain dying cells. Phagocytes can be distinguished from dying cells by microscopy.

- Another important difference from flow cytometry is that microscopy allows for better qualitative determinations, but the number of cells that can be studied and the capacity to quantify results are limited. On the other hand, flow cytometry enables analysis of large cell populations as well as accurate determination of fluorescence intensity. Thus, these two technologies are excellent complements of each other.

- Microscopy also makes it possible to differentiate between the phagocytosis of fragments and the uptake of entire cells. It may be assumed that distinguishing between adherence and internalization is also resolved by microscopy, but this is not necessarily true. To confirm internalization, confocal microscopy must be performed to show that the object is surrounded in three dimensions by the phagocyte's membrane

8.4.3.4   Electron Microscopy

Selected references: (Scott et al. 2001; Hisatomi et al. 2003; Krysko et al. 2003).

- Electron microscopy is divided into transmission (TEM) and scanning (SEM) microscopy. For TEM, cells are fixed and embedded according to special protocols, sliced with a microtome, and thin sections are mounted on carriers. TEM produces subcellular images at extremely high magnification. Samples for SEM are prepared in a different way, and the cells are covered with a thin layer of metal and scanned to produce detailed 3D surface images at very high magnifications. It is outside the scope of this chapter to delve into the complex protocols for TEM and SEM. These experiments usually require advice and assistance from specialized technicians unless the research group has extensive experience with EM and access to all necessary facilities.

8.4.3.5   Studies of the Effects of Proteins and Cytokines on Phagocytosis of Dying Cells

Selected reference: (Krispin et al. 2006).

- The experimental system in these studies is set up according to the objects of study. Specific molecules are added to the phagocytes before or together with the dying cells, and the molecules' effects on phagocytosis are studied by a variety of methods.

- For example, addition of thrombosponding-1 (TSP-1) to a mixture of human monocyte-derived dendritic cells and human apoptotic monocytes greatly increased uptake as assessed by flow cytometry (Krispin, et al. 2006)

8.4.3.6　The Use of Monoclonal Antibodies to Block Specific Receptors

Selected references: (Savill et al. 1990; Ren et al. 1995; Brown et al. 2002; Gardai et al. 2005).

- This technique is similar to the one described in the previous section, but instead of activating a cellular mechanism, the monoclonal antibodies disrupt it. By adding a monoclonal antibody that blocks the action of a specific target molecule, cellular processes can be studied under conditions in which that molecule's function is abrogated.
- Three of the selected references deal with molecules ($\alpha_v\beta_3$, CD36, and CD31) that, when blocked, cause a reduction in the phagocytosis of apoptotic cells. It is of interest that, although not specific, $\alpha_v\beta_3$ (the vitronectin receptor) can also be blocked by RGD sequences, which compete for the binding site for apoptotic cells on the surface of phagocytes. This is an example of the use of competing molecules to disrupt the interactions between cells, as an alternative approach to the use of monoclonal antibodies. Gardai et al. (2005) studied an interesting phenomenon, CD47 is discovered to be a repulsive surface molecule. Once this receptor was blocked, phagocytosis of viable cells increased.

8.4.4　In Vivo Methods for Evaluating Interaction of Phagocytes with Dying Cells

8.4.4.1　Tissue Stains and Histology

Selected references: (Franc et al. 1999; Gumienny et al. 2001; Scott et al. 2001; Leers et al. 2002; Bose et al. 2004; Kuhn et al. 2006).

- These methods involve the fixation of tissue samples, usually tissue sections, followed by staining with a variety of formulations, including common histological stains such as hematoxylin-eosin, immunohistochemical stains such as TUNEL, antibody labels such as anti-activated caspase 3, fluorescent dyes such as 7AAD, or even by fluorescence of intracellularly expressed GFP constructs.
- Tissues can be obtained from untreated animals that have a high level of cellular turnover, such as murine embryos, from treated animals, such as thymus samples after dexamethazone treatment, or from human samples, such as irradiated epidermis from SLE patients. Two of the selected references show the use of invertebrates, *D. melanogaster* embryos and *C. elegans*, which can be stained and imaged without sectioning or disruption.

8.4.4.2 Intraperitoneal Infusion of Dying Cells into Mice Followed by Removal of Peritoneal Macrophages

Selected references: (Ronchetti et al. 1999; Potter et al. 2003).

- Cells are labeled with fluorescent trackers (so that they can be counted easily later), cell death is induced, and then they are injected into the peritoneum of mice.
- After 30 minutes to a few hours of interaction with peritoneal macrophages, peritoneal cells are extracted as described in Sect. 8.4.2.4. The results of the interaction between the dying cells in vivo and the macrophages can be analysed by flow cytometry or by cell cytospinning and microscopy.

8.4.4.3 Intravenous Injection of Dying Cells or Phagocytes That Have Taken Up Dying Cells into Mice

Selected references: (Divito and Morelli, Chap. 11; Bartunkova and Spisek, Chap. 12, this Vol.; Mevorach et al. 1998b; Ronchetti et al. 1999; Iyoda et al. 2002; Goldszmid et al. 2003; Morelli et al. 2003).

- This is a very general method that lends itself to many types of studies.
- Cell death is induced, usually after labelling the cells with a tracker dye such as DiI, PKH, or CFSE, and the cells are then injected into mice. Interaction time can be as short as a few hours or as long as several weeks, depending on the target endpoint.
- The targets studied after the injection of dying cells depends on the experiment's objective. Illustrative examples are production of autoantibodies, analysis of plasma cytokine levels, generation of reactive T and B cells, or the study of tissue sections, such as spleen or lymph nodes, either by fluorescent microscopy or flow cytometry.
- An alternative protocol for eliciting systemic responses to dying cells is generation of syngeneic phagocytes, in vitro incubation of the phagocytes with dying cells, and then injection of the phagoctes into the animal. With this technique, interaction between phagocytes and dying cells occurs in vitro instead of in vivo, and the interaction between phagocytes and the whole animal is left intact.

8.4.4.4 Knock-out Mouse Studies

Selected references: (Gregory and Pound, Chap. 9, this Vol.; Fadok et al. 2000; Scott et al. 2001; Bose et al. 2004; Bygrave et al. 2004; Glaser, et al. 2005; Mitchell et al. 2006).

- The generation of knockout mice is a complex undertaking, and even more so the generation of conditional knockouts. Detailed descriptions of these processes are

Cocco REUcker DS (2001) Distinct modes of macrophage recognition for apoptotic and necrotic cells are not specified exclusively by phosphatidylserine exposure. Mol Biol Cell 12(12):919–930

Cohen PL, Caricchio R, Abraham V, et al (2002) Delayed apoptotic cell clearance and lupus-like autoimmunity in mice lacking the c-mer membrane tyrosine kinase. J Exp Med 196(196):135–140

Cravatt BF, Simon GMYates JR, 3rd (2007) The biological impact of mass-spectrometry-based proteomics. Nature 450(450):991–1000

Dalgaard J, Beckstrom KJ, Jahnsen FL, et al (2005) Differential capability for phagocytosis of apoptotic and necrotic leukemia cells by human peripheral blood dendritic cell subsets. J Leukoc Biol 77(77):689–698

Darzynkiewicz Z, Juan G, Li X, et al (1997) Cytometry in cell necrobiology: Analysis of apoptosis and accidental cell death (necrosis). Cytometry 27(27):1–20

den Haan JM, Lehar SMBevan MJ (2000) Cd8(+) but not cd8(-) dendritic cells cross-prime cytotoxic t cells in vivo. J Exp Med 192(192):1685–1696

Devitt A, Moffatt OD, Raykundalia C, et al (1998) Human cd14 mediates recognition and phagocytosis of apoptotic cells. Nature 392(392):505–509

Devitt A, Parker KG, Ogden CA, et al (2004) Persistence of apoptotic cells without autoimmune disease or inflammation in cd14-/- mice. J Cell Biol 167(167):1161–1170

Dogusan Z, Montecino-Rodriguez EDorshkind K (2004) Macrophages and stromal cells phagocytose apoptotic bone marrow-derived b lineage cells. J Immunol 172(172):4717–4723

Dunn KW, Mayor S, Myers JN, et al (1994) Applications of ratio fluorescence microscopy in the study of cell physiology. Faseb J 8(8):573–582

Earnshaw WC, Martins LMKaufmann SH (1999) Mammalian caspases: Structure, activation, substrates, and functions during apoptosis. Annu Rev Biochem 68(68):383–424

Evers JF, Schmitt S, Sibila M, et al (2005) Progress in functional neuroanatomy: Precise automatic geometric reconstruction of neuronal morphology from confocal image stacks. J Neurophysiol 93(93):2331–2342

Fadok VA, Bratton DL, Frasch SC, et al (1998a) The role of phosphatidylserine in recognition of apoptotic cells by phagocytes. Cell Death Differ 5(5):551–562

Fadok VA, Bratton DL, Konowal A, et al (1998c) Macrophages that have ingested apoptotic cells in vitro inhibit proinflammatory cytokine production through autocrine/paracrine mechanisms involving tgf-beta, pge2, and paf. J Clin Invest 101(101):890–898

Fadok VA, Bratton DL, Rose DM, et al (2000) A receptor for phosphatidylserine-specific clearance of apoptotic cells. Nature 405(405):85–90

Fadok VA, de Cathelineau A, Daleke DL, et al (2001) Loss of phospholipid asymmetry and surface exposure of phosphatidylserine is required for phagocytosis of apoptotic cells by macrophages and fibroblasts. J Biol Chem 276(276):1071–1077

Fadok VA, Voelker DR, Campbell PA, et al (1992) Exposure of phosphatidylserine on the surface of apoptotic lymphocytes triggers specific recognition and removal by macrophages. J Immunol 148(148):2207–2216

Fadok VA, Warner ML, Bratton DL, et al (1998b) Cd36 is required for phagocytosis of apoptotic cells by human macrophages that use either a phosphatidylserine receptor or the vitronectin receptor (alpha v beta 3). J Immunol 161(161):6250–6257

Finnemann SC, Bonilha VL, Marmorstein AD, et al (1997) Phagocytosis of rod outer segments by retinal pigment epithelial cells requires alpha(v)beta5 integrin for binding but not for internalization. Proc Natl Acad Sci U S A 94(94):12932–12937

Flora PKGregory CD (1994) Recognition of apoptotic cells by human macrophages: Inhibition by a monocyte/macrophage-specific monoclonal antibody. Eur J Immunol 24(24):2625–2632

Franc NC, Heitzler P, Ezekowitz RA, et al (1999) Requirement for croquemort in phagocytosis of apoptotic cells in drosophila. Science 284(284):1991–1994

Frey T (1995) Nucleic acid dyes for detection of apoptosis in live cells. Cytometry 21(21):265–274

Frey T (1997) Correlated flow cytometric analysis of terminal events in apoptosis reveals the absence of some changes in some model systems. Cytometry 28(28):253–263

Gaforio JJ, Serrano MJ, Algarra I, et al (2002) Phagocytosis of apoptotic cells assessed by flow cytometry using 7-aminoactinomycin d. Cytometry 49(49):8–11

Gardai SJ, McPhillips KA, Frasch SC, et al (2005) Cell-surface calreticulin initiates clearance of viable or apoptotic cells through trans-activation of lrp on the phagocyte. Cell 123(123):321–334

Gavrieli Y, Sherman YBen-Sasson SA (1992) Identification of programmed cell death in situ via specific labeling of nuclear DNA fragmentation. J Cell Biol 119(119):493–501

Ge FYChen LG (2007) Ph fluorescent probes: Chlorinated fluoresceins. J Fluoresc

Geissmann F, Prost C, Monnet JP, et al (1998) Transforming growth factor beta1, in the presence of granulocyte/macrophage colony-stimulating factor and interleukin 4, induces differentiation of human peripheral blood monocytes into dendritic langerhans cells. J Exp Med 187(187):961–966

George TC, Basiji DA, Hall BE, et al (2004) Distinguishing modes of cell death using the imagestream multispectral imaging flow cytometer. Cytometry A 59(59):237–245

Glaser S, Anastassiadis KStewart AF (2005) Current issues in mouse genome engineering. Nat Genet 37(37):1187–1193

Goldszmid RS, Idoyaga J, Bravo AI, et al (2003) Dendritic cells charged with apoptotic tumor cells induce long-lived protective cd4+ and cd8+ t cell immunity against b16 melanoma. J Immunol 171(171):5940–5947

Gong J, Traganos FDarzynkiewicz Z (1994) A selective procedure for DNA extraction from apoptotic cells applicable for gel electrophoresis and flow cytometry. Anal Biochem 218(218):314–319

Gordon S (2007) The macrophage: Past, present and future. Eur J Immunol 37 Suppl 1(37 Suppl 1):S9–17

Gordon STaylor PR (2005) Monocyte and macrophage heterogeneity. Nat Rev Immunol 5(5):953–964

Greenberg ME, Sun M, Zhang R, et al (2006) Oxidized phosphatidylserine-cd36 interactions play an essential role in macrophage-dependent phagocytosis of apoptotic cells. J Exp Med 203(203):2613–2625

Gregory CDDevitt A (2004) The macrophage and the apoptotic cell: An innate immune interaction viewed simplistically? Immunology 113(113):1–14

Grolleau A, Misek DE, Kuick R, et al (2003) Inducible expression of macrophage receptor marco by dendritic cells following phagocytic uptake of dead cells uncovered by oligonucleotide arrays. J Immunol 171(171):2879–2888

Gumienny TL, Brugnera E, Tosello-Trampont AC, et al (2001) Ced-12/elmo, a novel member of the crkii/dock180/rac pathway, is required for phagocytosis and cell migration. Cell 107(107):27–41

Hackam DJ, Rotstein OD, Zhang WJ, et al (1997) Regulation of phagosomal acidification. Differential targeting of na+/h+ exchangers, na+/k+-atpases, and vacuolar-type h+-atpases. J Biol Chem 272(272):29810–29820

Haggie PMVerkman AS (2007) Cystic fibrosis transmembrane conductance regulator-independent phagosomal acidification in macrophages. J Biol Chem 282(282):31422–31428

Hanayama R, Tanaka M, Miyasaka K, et al (2004) Autoimmune disease and impaired uptake of apoptotic cells in mfg-e8-deficient mice. Science 304(304):1147–1150

Haruta I, Kato Y, Hashimoto E, et al (2001) Association of aim, a novel apoptosis inhibitory factor, with hepatitis via supporting macrophage survival and enhancing phagocytotic function of macrophages. J Biol Chem 276(276):22910–22914

Henson PMTuder RM (2008) Apoptosis in the lung: Induction, clearance and detection. Am J Physiol Lung Cell Mol Physiol

Hess KL, Babcock GF, Askew DS, et al (1997) A novel flow cytometric method for quantifying phagocytosis of apoptotic cells. Cytometry 27(27):145–152

Hirt UALeist M (2003) Rapid, noninflammatory and ps-dependent phagocytic clearance of necrotic cells. Cell Death Differ 10(10):1156–1164

Hisatomi T, Sakamoto T, Sonoda KH, et al (2003) Clearance of apoptotic photoreceptors: Elimination of apoptotic debris into the subretinal space and macrophage-mediated phagocytosis via phosphatidylserine receptor and integrin alphavbeta3. Am J Pathol 162(162):1869–1879

Hoeffel G, Ripoche AC, Matheoud D, et al (2007) Antigen crosspresentation by human plasmacytoid dendritic cells. Immunity 27(27):481–492

Hoeppner DJ, Hengartner MOSchnabel R (2001) Engulfment genes cooperate with ced-3 to promote cell death in caenorhabditis elegans. Nature 412(412):202–206

Hoffmann PR, deCathelineau AM, Ogden CA, et al (2001) Phosphatidylserine (ps) induces ps receptor-mediated macropinocytosis and promotes clearance of apoptotic cells. J Cell Biol 155(155):649–659

Holthuis JCLevine TP (2005) Lipid traffic: Floppy drives and a superhighway. Nat Rev Mol Cell Biol 6(6):209–220

Hu B, Sonstein J, Christensen PJ, et al (2000) Deficient in vitro and in vivo phagocytosis of apoptotic t cells by resident murine alveolar macrophages. J Immunol 165(165):2124–2133

Inaba H, Shimada K, Zhou YW, et al (2005) Acquisition of fas resistance by fas receptor mutation in a childhood b-precursor acute lymphoblastic leukemia cell line, mml-1. Int J Oncol 27(27):573–579

Iwasaki A (2007) Mucosal dendritic cells. Annu Rev Immunol 25(25):381–418

Iyoda T, Shimoyama S, Liu K, et al (2002) The cd8+ dendritic cell subset selectively endocytoses dying cells in culture and in vivo. J Exp Med 195(195):1289–1302

Jaattela MTschopp J (2003) Caspase-independent cell death in t lymphocytes. Nat Immunol 4(4):416–423

Jares-Erijman EAJovin TM (2003) Fret imaging. Nat Biotechnol 21(21):1387–1395

Jehle AW, Gardai SJ, Li S, et al (2006) Atp-binding cassette transporter a7 enhances phagocytosis of apoptotic cells and associated erk signaling in macrophages. J Cell Biol 174(174):547–556

Johansson U, Walther-Jallow L, Smed-Sorensen A, et al (2007) Triggering of dendritic cell responses after exposure to activated, but not resting, apoptotic pbmcs. J Immunol 179(179):1711–1720

Jones AL, Poon IK, Hulett MD, et al (2005) Histidine-rich glycoprotein specifically binds to necrotic cells via its amino-terminal domain and facilitates necrotic cell phagocytosis. J Biol Chem 280(280):35733–35741

Kacani L, Wurm M, Schwentner I, et al (2005) Maturation of dendritic cells in the presence of living, apoptotic and necrotic tumour cells derived from squamous cell carcinoma of head and neck. Oral Oncol 41(41):17–24

Kagan VE, Borisenko GG, Tyurina YY, et al (2004) Oxidative lipidomics of apoptosis: Redox catalytic interactions of cytochrome c with cardiolipin and phosphatidylserine. Free Radic Biol Med 37(37):1963–1985

Kagan VE, Gleiss B, Tyurina YY, et al (2002) A role for oxidative stress in apoptosis: Oxidation and externalization of phosphatidylserine is required for macrophage clearance of cells undergoing fas-mediated apoptosis. J Immunol 169(169):487–499

Kerr JF, Wyllie AHCurrie AR (1972) Apoptosis: A basic biological phenomenon with wide-ranging implications in tissue kinetics. Br J Cancer 26(26):239–257

Kim S, Elkon KBMa X (2004) Transcriptional suppression of interleukin-12 gene expression following phagocytosis of apoptotic cells. Immunity 21(21):643–653

Kirkham P (2007) Oxidative stress and macrophage function: A failure to resolve the inflammatory response. Biochem Soc Trans 35(35):284–287

Klionsky DJ (2007) Autophagy: From phenomenology to molecular understanding in less than a decade. Nat Rev Mol Cell Biol 8(8):931–937

Krispin A, Bledi Y, Atallah M, et al (2006) Apoptotic cell thrombospondin-1 and heparin-binding domain lead to dendritic-cell phagocytic and tolerizing states. Blood 108(108):3580–3589

Kroemer GMartin SJ (2005) Caspase-independent cell death. Nat Med 11(11):725–730

Krysko DV, Brouckaert G, Kalai M, et al (2003) Mechanisms of internalization of apoptotic and necrotic l929 cells by a macrophage cell line studied by electron microscopy. J Morphol 258(258):336–345

Krysko DV, Denecker G, Festjens N, et al (2006b) Macrophages use different internalization mechanisms to clear apoptotic and necrotic cells. Cell Death Differ 13(13):2011–2022

Krysko DV, D'Herde KVandenabeele P (2006a) Clearance of apoptotic and necrotic cells and its immunological consequences. Apoptosis 11(11):1709–1726

Kuhn A, Herrmann M, Kleber S, et al (2006) Accumulation of apoptotic cells in the epidermis of patients with cutaneous lupus erythematosus after ultraviolet irradiation. Arthritis Rheum 54(54):939–950

Kumar SJack R (2006) Origin of monocytes and their differentiation to macrophages and dendritic cells. J Endotoxin Res 12(12):278–284

Kunisaki Y, Masuko S, Noda M, et al (2004) Defective fetal liver erythropoiesis and t lymphopoiesis in mice lacking the phosphatidylserine receptor. Blood 103(103):3362–3364

Laskin DL, Weinberger BLaskin JD (2001) Functional heterogeneity in liver and lung macrophages. J Leukoc Biol 70(70):163–170

Leers MP, Bjorklund V, Bjorklund B, et al (2002) An immunohistochemical study of the clearance of apoptotic cellular fragments. Cell Mol Life Sci 59(59):1358–1365

Lettre GHengartner MO (2006) Developmental apoptosis in c. Elegans: A complex cednario. Nat Rev Mol Cell Biol 7(7):97–108

Levine BDeretic V (2007) Unveiling the roles of autophagy in innate and adaptive immunity. Nat Rev Immunol 7(7):767–777

Licht R, Jacobs CW, Tax WJ, et al (1999) An assay for the quantitative measurement of in vitro phagocytosis of early apoptotic thymocytes by murine resident peritoneal macrophages. J Immunol Methods 223(223):237–248

Liu G, Wu C, Wu Y, et al (2006) Phagocytosis of apoptotic cells and immune regulation. Scand J Immunol 64(64):1–9

Liu YJ (2005) Ipc: Professional type 1 interferon-producing cells and plasmacytoid dendritic cell precursors. Annu Rev Immunol 23(23):275–306

Liu Z, Wan G, Heaphy C, et al (2007) A novel loss-of-function mutation in tp53 in an endometrial cancer cell line and uterine papillary serous carcinoma model. Mol Cell Biochem 297(297):179–187

Lorimore SA, Coates PJ, Scobie GE, et al (2001) Inflammatory-type responses after exposure to ionizing radiation in vivo: A mechanism for radiation-induced bystander effects? Oncogene 20(20):7085–7095

Lucas M, Stuart LM, Zhang A, et al (2006) Requirements for apoptotic cell contact in regulation of macrophage responses. J Immunol 177(177):4047–4054

Maciorowski Z, Klijanienko J, Padoy E, et al (2001) Comparative image and flow cytometric tunel analysis of fine needle samples of breast carcinoma. Cytometry 46(46):150–156

Maderna P, Yona S, Perretti M, et al (2005) Modulation of phagocytosis of apoptotic neutrophils by supernatant from dexamethasone-treated macrophages and annexin-derived peptide ac(2-26). J Immunol 174(174):3727–3733

Maiuri MC, Zalckvar E, Kimchi A, et al (2007) Self-eating and self-killing: Crosstalk between autophagy and apoptosis. Nat Rev Mol Cell Biol 8(8):741–752

Manfredi AA, Rovere P, Galati G, et al (1998) Apoptotic cell clearance in systemic lupus erythematosus. I. Opsonization by antiphospholipid antibodies. Arthritis Rheum 41(41):205–214

Matsura T, Kai M, Jiang J, et al (2004) Endogenously generated hydrogen peroxide is required for execution of melphalan-induced apoptosis as well as oxidation and externalization of phosphatidylserine. Chem Res Toxicol 17(17):685–696

Mevorach D (2000) Opsonization of apoptotic cells. Implications for uptake and autoimmunity. Ann N Y Acad Sci 926(926):226-235

Mevorach D, Mascarenhas JO, Gershov D, et al (1998a) Complement-dependent clearance of apoptotic cells by human macrophages. J Exp Med 188(188):2313–2320

Mevorach D, Zhou JL, Song X, et al (1998b) Systemic exposure to irradiated apoptotic cells induces autoantibody production. J Exp Med 188(188):387–392

Mishina MSakimura K (2007) Conditional gene targeting on the pure c57bl/6 genetic background. Neurosci Res 58(58):105–112

Mitchell JE, Cvetanovic M, Tibrewal N, et al (2006) The presumptive phosphatidylserine receptor is dispensable for innate anti-inflammatory recognition and clearance of apoptotic cells. J Biol Chem 281(281):5718–5725

Miyake Y, Asano K, Kaise H, et al (2007) Critical role of macrophages in the marginal zone in the suppression of immune responses to apoptotic cell-associated antigens. J Clin Invest 117(117):2268–2278

Miyanishi M, Tada K, Koike M, et al (2007) Identification of tim4 as a phosphatidylserine receptor. Nature 450(450):435–439

Mizushima N (2007) Autophagy: Process and function. Genes Dev 21(21):2861–2873

Morelli AE, Larregina AT, Shufesky WJ, et al (2003) Internalization of circulating apoptotic cells by splenic marginal zone dendritic cells: Dependence on complement receptors and effect on cytokine production. Blood 101(101):611–620

Morelli AEThomson AW (2007) Tolerogenic dendritic cells and the quest for transplant tolerance. Nat Rev Immunol 7(7):610–621

Nicoletti I, Migliorati G, Pagliacci MC, et al (1991) A rapid and simple method for measuring thymocyte apoptosis by propidium iodide staining and flow cytometry. J Immunol Methods 139(139):271–279

Ogden CA, deCathelineau A, Hoffmann PR, et al (2001) C1q and mannose binding lectin engagement of cell surface calreticulin and cd91 initiates macropinocytosis and uptake of apoptotic cells. J Exp Med 194(194):781–795

Ozgen U, Savasan S, Buck S, et al (2000) Comparison of dioc(6)(3) uptake and annexin v labeling for quantification of apoptosis in leukemia cells and non-malignant t lymphocytes from children. Cytometry 42(42):74–78

Park SY, Jung MY, Kim HJ, et al (2008) Rapid cell corpse clearance by stabilin-2, a membrane phosphatidylserine receptor. Cell Death Differ 15(15):192–201

Parnaik R, Raff MCScholes J (2000) Differences between the clearance of apoptotic cells by professional and non-professional phagocytes. Curr Biol 10(10):857–860

Patel VA, Longacre A, Hsiao K, et al (2006) Apoptotic cells, at all stages of the death process, trigger characteristic signaling events that are divergent from and dominant over those triggered by necrotic cells: Implications for the delayed clearance model of autoimmunity. J Biol Chem 281(281):4663–4670

Potter PK, Cortes-Hernandez J, Quartier P, et al (2003) Lupus-prone mice have an abnormal response to thioglycolate and an impaired clearance of apoptotic cells. J Immunol 170(170):3223–3232

Pozarowski P, Huang X, Halicka DH, et al (2003) Interactions of fluorochrome-labeled caspase inhibitors with apoptotic cells: A caution in data interpretation. Cytometry A 55(55):50–60

Rasola AGeuna M (2001) A flow cytometry assay simultaneously detects independent apoptotic parameters. Cytometry 45(45):151–157

Ravichandran KSLorenz U (2007) Engulfment of apoptotic cells: Signals for a good meal. Nat Rev Immunol 7(7):964–974

Reddien PW, Cameron SHorvitz HR (2001) Phagocytosis promotes programmed cell death in c. Elegans. Nature 412(412):198–202

Ren Y, Silverstein RL, Allen J, et al (1995) Cd36 gene transfer confers capacity for phagocytosis of cells undergoing apoptosis. J Exp Med 181(181):1857–1862

Ritov VB, Banni S, Yalowich JC, et al (1996) Non-random peroxidation of different classes of membrane phospholipids in live cells detected by metabolically integrated cis-parinaric acid. Biochim Biophys Acta 1283(1283):127–140

Romani N, Gruner S, Brang D, et al (1994) Proliferating dendritic cell progenitors in human blood. J Exp Med 180(180):83–93

Ronchetti A, Rovere P, Iezzi G, et al (1999) Immunogenicity of apoptotic cells in vivo: Role of antigen load, antigen-presenting cells, and cytokines. J Immunol 163(163):130–136

Sallusto FLanzavecchia A (1994) Efficient presentation of soluble antigen by cultured human dendritic cells is maintained by granulocyte/macrophage colony-stimulating factor plus interleukin 4 and downregulated by tumor necrosis factor alpha. J Exp Med 179(179):1109–1118

Sauter B, Albert ML, Francisco L, et al (2000) Consequences of cell death: Exposure to necrotic tumor cells, but not primary tissue cells or apoptotic cells, induces the maturation of immunostimulatory dendritic cells. J Exp Med 191(191):423–434

Savill J, Dransfield I, Hogg N, et al (1990) Vitronectin receptor-mediated phagocytosis of cells undergoing apoptosis. Nature 343(343):170–173

Savill J, Hogg N, Ren Y, et al (1992) Thrombospondin cooperates with cd36 and the vitronectin receptor in macrophage recognition of neutrophils undergoing apoptosis. J Clin Invest 90(90):1513–1522

Savill JFadok V (2000) Corpse clearance defines the meaning of cell death. Nature 407(407):784–788

Schrijvers DM, Martinet W, De Meyer GR, et al (2004) Flow cytometric evaluation of a model for phagocytosis of cells undergoing apoptosis. J Immunol Methods 287(287):101–108

Schulz OReis e Sousa C (2002) Cross-presentation of cell-associated antigens by cd8alpha+ dendritic cells is attributable to their ability to internalize dead cells. Immunology 107(107):183–189

Scott RS, McMahon EJ, Pop SM, et al (2001) Phagocytosis and clearance of apoptotic cells is mediated by mer. Nature 411(411):207–211

Shapiro HM (2003) Practical flow cytometry. Wiley-Liss.

Shortman KLiu YJ (2002) Mouse and human dendritic cell subtypes. Nat Rev Immunol 2(2):151–161

Shortman KNaik SH (2007) Steady-state and inflammatory dendritic-cell development. Nat Rev Immunol 7(7):19–30

Shoshan Y, Shapira I, Toubi E, et al (2001) Accelerated fas-mediated apoptosis of monocytes and maturing macrophages from patients with systemic lupus erythematosus: Relevance to in vitro impairment of interaction with ic3b-opsonized apoptotic cells. J Immunol 167(167):5963–5969

Siuti NKelleher NL (2007) Decoding protein modifications using top-down mass spectrometry. Nat Methods 4(4):817–821

Smolewski P, Grabarek J, Lee BW, et al (2002) Kinetics of hl-60 cell entry to apoptosis during treatment with tnf-alpha or camptothecin assayed by the stathmo-apoptosis method. Cytometry 47(47):143–149

Steinman RM (2007) Dendritic cells: Understanding immunogenicity. Eur J Immunol 37 Suppl 1(37 Suppl 1):S53–60

Stout RDSuttles J (2004) Functional plasticity of macrophages: Reversible adaptation to changing microenvironments. J Leukoc Biol 76(76):509–513

Stuart LM, Boulais J, Charriere GM, et al (2007) A systems biology analysis of the drosophila phagosome. Nature 445(445):95–101

Stuart LMEzekowitz RA (2008) Phagocytosis and comparative innate immunity: Learning on the fly. Nat Rev Immunol 8(8):131–141

Tasaki A, Yamanaka N, Kubo M, et al (2004) Three-dimensional two-layer collagen matrix gel culture model for evaluating complex biological functions of monocyte-derived dendritic cells. J Immunol Methods 287(287):79–90

Tassiulas I, Park-Min KH, Hu Y, et al (2007) Apoptotic cells inhibit lps-induced cytokine and chemokine production and ifn responses in macrophages. Hum Immunol 68(68):156–164

Taylor PR, Carugati A, Fadok VA, et al (2000) A hierarchical role for classical pathway complement proteins in the clearance of apoptotic cells in vivo. J Exp Med 192(192):359–366

Taylor RC, Cullen SPMartin SJ (2008) Apoptosis: Controlled demolition at the cellular level. Nat Rev Mol Cell Biol 9(9):231–241

Tibrewal N, Wu Y, D'Mello V, et al (2007) Autophosphorylation docking site tyr867 in mertk allows for dissociation of multiple signaling pathways for phagocytosis of apoptotic cells and down-modulation of lps-inducible nf-kappa b transcriptional activation. J Biol Chem

Torre D, Gennero L, Baccino FM, et al (2002) Impaired macrophage phagocytosis of apoptotic neutrophils in patients with human immunodeficiency virus type 1 infection. Clin Diagn Lab Immunol 9(9):983–986

Trahtemberg U, Atallah M, Krispin A, et al (2007) Calcium, leukocyte cell death and the use of annexin v: Fatal encounters. Apoptosis 12(12):1769–1780

Tzeng TC, Suen JLChiang BL (2006) Dendritic cells pulsed with apoptotic cells activate self-reactive t-cells of lupus mice both in vitro and in vivo. Rheumatology (Oxford) 45(45):1230–1237

Ueno H, Klechevsky E, Morita R, et al (2007) Dendritic cell subsets in health and disease. Immunol Rev 219(219):118–142

Upton JP, Valentijn AJ, Zhang L, et al (2007) The n-terminal conformation of bax regulates cell commitment to apoptosis. Cell Death Differ 14(14):932–942

Vandivier RW, Henson PMDouglas IS (2006) Burying the dead: The impact of failed apoptotic cell removal (efferocytosis) on chronic inflammatory lung disease. Chest 129(129):1673–1682

Verbovetski I, Bychkov H, Trahtemberg U, et al (2002) Opsonization of apoptotic cells by autologous ic3b facilitates clearance by immature dendritic cells, down-regulates dr and cd86, and up-regulates cc chemokine receptor 7. J Exp Med 196(196):1553–1561

Verreck FA, de Boer T, Langenberg DM, et al (2004) Human il-23-producing type 1 macrophages promote but il-10-producing type 2 macrophages subvert immunity to (myco)bacteria. Proc Natl Acad Sci U S A 101(101):4560–4565

Villadangos JASchnorrer P (2007) Intrinsic and cooperative antigen-presenting functions of dendritic-cell subsets in vivo. Nat Rev Immunol 7(7):543–555

Wan CP, Park CSLau BH (1993) A rapid and simple microfluorometric phagocytosis assay. J Immunol Methods 162(162):1–7

Wan HDupasquier M (2005) Dendritic cells in vivo and in vitro. Cell Mol Immunol 2(2):28–35

Witze ES, Old WM, Resing KA, et al (2007) Mapping protein post-translational modifications with mass spectrometry. Nat Methods 4(4):798–806

Wood W, Turmaine M, Weber R, et al (2000) Mesenchymal cells engulf and clear apoptotic footplate cells in macrophageless pu.1 null mouse embryos. Development 127(127):5245–5252

Xu W, Roos A, Daha MR, et al (2006a) Dendritic cell and macrophage subsets in the handling of dying cells. Immunobiology 211(211):567–575

Xu W, Roos A, Schlagwein N, et al (2006b) Il-10-producing macrophages preferentially clear early apoptotic cells. Blood 107(107):4930–4937

Yabuta T, Shinmura K, Yamane A, et al (2004) Effect of exogenous msh6 and pold1 expression on the mutation rate of the hprt locus in a human colon cancer cell line with mutator phenotype, dld-1. Int J Oncol 24(24):697–702

Yates RMRussell DG (2005) Phagosome maturation proceeds independently of stimulation of toll-like receptors 2 and 4. Immunity 23(23):409–417

Zamai L, Canonico B, Luchetti F, et al (2001) Supravital exposure to propidium iodide identifies apoptosis on adherent cells. Cytometry 44(44):57–64

Zamai L, Falcieri E, Marhefka G, et al (1996) Supravital exposure to propidium iodide identifies apoptotic cells in the absence of nucleosomal DNA fragmentation. Cytometry 23(23):303–311

Zhou LJTedder TF (1996) Cd14+ blood monocytes can differentiate into functionally mature cd83+ dendritic cells. Proc Natl Acad Sci U S A 93(93):2588–2592

Zong WXThompson CB (2006) Necrotic death as a cell fate. Genes Dev 20(20):1–15

Part II
Impairment of Phagocytosis of Dying Cells and its Role in the Development of Diseases

Chapter 9
Results of Defective Clearance of Apoptotic Cells: Lessons from Knock-out Mouse Models

Christopher D. Gregory and John D. Pound

Abstract: The current stupefying array of molecules associated with the celearance of cells undergoing apoptosis not only demonstrates complexity in the process but also suggests redundancy. The molecular complexity is rooted in multiple cellular events, including (1) multiple stages in the apoptosis programme, (2) multiple steps in the processes that culminate in engulfment and (3) multiple responses of different phagocytes. While there is almost relentless discovery of molecules that are implicated in these events, studies of knock-out mice are beginning to reveal their biological significance. Here we review these investigations, which demonstrate that, in certain circumstances, defective apoptotic-cell clearance can be associated with the pathogenesis of autoimmune disease. However, these studies also show that persistence of apoptotic cells as a consequence of defective clearance is not, of necessity, pro-inflammatory and immunostimulatory. Indeed, it appears that, under appropriate circumstances, persistent apoptotic cells may provide prolonged anti-inflammatory signals.

Keywords: Autoimmunity • Clearance defect • Inflammation • Knock-out • Phagocyte response

9.1   Introduction

From an historical and simplistic standpoint, cells dying purposefully by programmed cell death or apoptosis are held as being phagocytosed not only by mechanisms that fail to activate inflammatory or immune reactions but also in some cases

C. D. Gregory (✉) · J. D. Pound
MRC Centre for Inflammation Research
University of Edinburgh, Queen's Medical Research Institute
47 Little France Crescent, EH16 4TJ
Edinburgh, UK
Tel.: +44 131 242 9170
Fax: +44 131 242 9171
E-mail: chris.gregory@ed.ac.uk

D.V. Krysko, P. Vandenabeele (eds.), *Phagocytosis of Dying Cells,*
DOI: 10.1007/978-1-4020-9292-3_2, © Springer Science+Business Media B.V. 2009

by mechanisms that are actively anti-inflammatory and tolerogenic. Drawing on the classical descriptions of the pro-inflammatory properties of necrotic cells, it is widely assumed that the inefficient clearance of apoptotic cells leads to their acquisition of necrotic characteristics and ultimately pathological consequences. While there is no doubt that the failed clearance of apoptotic cells can have immunostimulatory sequelae that can lead to the development of features of autoimmune disease, this is by no means a straightforward default, but rather is dependent upon multiple factors, not least the nature of the clearance failure. Given the apparent molecular complexity of the clearance process, it is perhaps not surprising that defects in different facets of the process have different outcomes. In this Chapter, we will review present perspectives on the molecular mechanisms of apoptotic-cell clearance and, with particular reference to those knock-out mouse models that have been studied to date, we will illustrate the importance of efficient clearance for homeostasis and begin to rationalise the complexity of the current molecular picture.

9.2　Molecular Complexity in Apoptotic-Cell Clearance Mechanisms

The current, astonishingly complex array of molecules that has been implicated in the clearance of apoptotic cells is summarized in Fig. 9.1 (Napirei and Mannherz, Chap. 4; Dini and Vergallo, Chap. 15, this Vol.). Much of the detail in this molecular picture is based on in vitro studies, but as we shall discuss, the significance of individual elements is beginning to be clarified in mammalian in vivo models.

In order to be engulfed selectively, an apoptotic cell must obviously at a critical point become exquisitely distinguishable from its viable counterparts. The mobile, professional scavengers of apoptotic cells, the macrophages, use specific chemotactic mechanisms to sense apoptotic cells and navigate towards them (Peter et al., Chap. 3, this Vol.). Neighbouring, juxtaposed phagocytes—and the importance of these cells in apoptotic-cell clearance should not be under estimated—seem likely to engage with dying cells more directly via processes that are not preceded by such chemotactic mechanisms, although motility of certain of non-professional phagocytes towards apoptotic cells would be predicted to occur via chemotactic mechanisms. The most renowned feature of the apoptotic cell surface is the altered plasma membrane lipid topology, in particular the loss of phospholipid asymmetry that results in the redistribution of the anionic phospholipid, phosphatidylserine (PS) from the inner plasma membrane leaflet to the outer (Fadok et al. 1992; Martin et al. 1995; Bratton et al. 1997; Gardai et al. 2006). Exposed PS can engage with a number of glycoproteins which act as bridging molecules (e.g. MFG-E8 or Gas6) to phagocyte surface structures (e.g. integrins $\alpha_v\beta_{3/5}$ or Mer; Scott et al. 2001; Hanayama et al. 2002) or directly with integral PS receptors of the phagocyte membrane such as TIM-1/4 (Kobayashi et al. 2007; Miyanishi et al. 2007; Santiago et al. 2007), BAI1 (Park et al. 2007) and Stabilin-2 (Park et al. 2008; Fig. 9.1). Note that the earlier-reported PSR (Fadok et al.

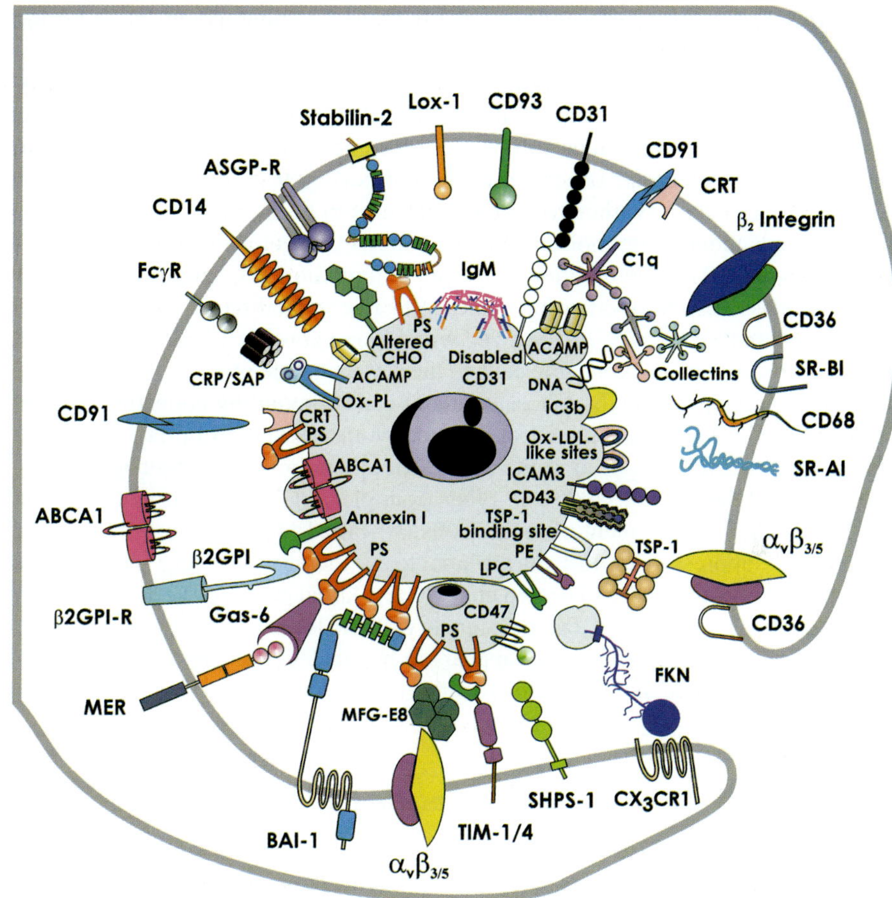

Fig. 9.1 Schematic representation of molecules involved in the interaction mechanisms between phagocytes and apoptotic cells that underlie the clearance process. Consensus apoptotic cell and phagocyte are represented and molecules used by non-macrophages as well as macrophages are included. $\alpha_v\beta_3$, $\alpha_v\beta_5$ vitronectin receptor integrins; ABCA1, ATP-binding cassette transporter A1; ACAMPs, apoptotic cell-associated molecular patterns; ASGP-R, asialoglycoprotein receptor; β2GPI, β2 glycoprotein I; β2GPI-R, β2GPI-receptor; β2 integrins include CR3 and CR4; BAI1, brain-specific angiogenesis inhibitor 1; C1q, first component of complement; CHO, carbohydrate; CRP, C-reactive protein; CRT, calreticulin; CX3CR1, fractalkine receptor; Del-1, developmental endothelial locus-1; FKN, fractalkine; Gas-6, growth arrest specific-6; iC3b, inactivated complement fragment C3b; ICAM-3 (CD50), intercellular adhesion molecule-3; Lox-1, oxidized low density lipoprotein receptor 1; LPC, lysophosphatidylcholine; MER, myeloid epithelial reproductive tyrosine kinase; MFG-E8, milk fat globule epidermal growth factor-8; Ox-PL oxidized phospholipids; PE, phosphatidylethanolamine; PS, phosphatidylserine; SAP, serum amyloid protein, SHPS-1, Src homology 2 domain-bearing protein tyrosine phosphatase substrate-1; SR-AI, scavenger receptor AI; SR-BI, scavenger receptor BI; TIM-1/4, T-cell immunoglobulin-and mucin-domain-containing molecule; TSP-1, thrombospondin-1. Adapted from Gregory and Devitt (2004).

2000) proved not to be a surface PS-receptor but rather a Jumonji domain-containing nuclear protein (Bratton and Henson 2008).

While further, detailed information on the molecular architecture of the apoptotic-cell surface is eagerly awaited, it is evident that, in addition to phospholipids, alterations in plasma membrane components that encompass carbohydrate, protein and nucleic acid categories occur during apoptosis (Fig. 9.1). An emerging principle is that, rather than synthesizing new molecules to be used as 'eat-me' signals, apoptotic cells redistribute molecules they already express to signal their readiness for phagocytosis. Thus, in addition to the redistribution of PS, apoptotic cells alter the location/association of other molecules to enable or facilitate clearance: calreticulin, for example, becomes co-localized with PS and annexin 1 becomes relocated from its intracellular sites in viable cells to the surface of apoptotic cells. DNA, too, is available at apoptotic-cell surfaces, where it can be bound by collectins (Palaniyar et al. 2004). The large subunit of the eukaryotic translation initiation factor 3 (eIF3a) also becomes exposed at the surface of apoptotic cells for recognition by macrophages (Nakai et al. 2005). In other examples, moieties such as CD31 and CD47 that inhibit engulfment of viable cells (so-called 'don't eat me' signals) are functionally suppressed or lost during apoptosis, thereby directing the selective engulfment of apoptotic cells (Brown et al. 2002; Gardai et al. 2005). Therefore, by combining the appearance of 'eat me' signals with the loss of 'don't eat me' signals, the appropriate molecular topology appears to be created at the apoptotic-cell surface to ensure efficient and selective engulfment of dying, but not viable, cells.

9.2.1 *Pattern Recognition in Apoptotic-Cell Clearance*

As well as the integrins and PS-receptors already mentioned, additional molecules implicated in the surface interactions between phagocytes and apoptotic cells include lectins, scavenger receptors, complement receptors and receptors such as CD14, complement components and collectins that are better known for their role in innate responses to microbes. More details of these molecules can be found in Fig. 9.1 (Napirei and Mannherz, Chap. 4; Dini and Vergallo, Chap. 15, this Vol.). Of particular note are the findings that interactions between phagocytes and apoptotic cells involve many of the molecules that had first gained renown as elements of the innate immune system involved in interaction with microbial structures. Proponents of 'pattern recognition' classify these elements as PRRs—pattern recognition receptors—that have innate capacity to interact with conserved microbial structures—PAMPs: pathogen-associated molecular patterns, such as LPS—leading to host responses that stem pathogen invasion and expansion (Medzhitov and Janeway 1997; Janeway 1989). Discovery of the involvement of PRRs, including the prototypic PRR, CD14 in the clearance of apoptotic cells has led to the proposal that apoptotic cells present ACAMPs to the immune system: apoptotic-cell-associated molecular patterns (Franc et al. 1999; Gregory 2000) that may share structural homology with PAMPs. We have previously speculated that CD14, for example, a prototypic PRR that is linked closely with host responses to lipopolysaccharides (LPS), which are prototypic PAMPs, has an affinity for ACAMPs displaying LPS-like patterns in three-dimensional space. This notion predicts that certain molecules that bind LPS would also have demonstrable capacity to bind apoptotic cells—via ACAMPs. In this context it is interesting to note recent studies indicating that, in addition to soluble CD14 that is known to bind apoptotic cells, other LPS-binding proteins, including anti-LPS antibodies, share this ability, most likely through the translocation of intracellular protein molecules to the cell surface during apoptosis, allowing them to contribute to ACAMP formation in the altered environment of the plasma membrane of the apoptotic cell (Tennant et al. 2008).

Translocation of molecules from intracellular to extracellular sites, as we have discussed briefly above, is an attractive possible mechanism of ACAMP generation that permits discrimination of viable and dying cells by phagocytes. Loss of inhibitory molecules and subtle changes in molecular architecture of plasma membrane domains may be critically important. Exposure of PS (Fadok et al. 2001b) along with its subsequent oxidation, at least in certain cases (Kagan et al. 2002), appears to be a general feature of the apoptotic-cell surface and is probably a necessary pre-requisite for clearance. Exposed PS may prove to be a key component of putative ACAMPs, but PS exposure is not sufficient to ensure cell clearance (Anderson et al. 2002; Devitt et al. 2003), indicating that further structural changes on the apoptotic-cell surface are required, in cooperation with the altered phospholipid. In recent years it has become clear that PS-rich domains on apoptotic cells contain additional components from various sources, both intracellular and extracellular. These include, as mentioned above, calreticulin, which is present on the membranes of viable cells, but colocalizes with PS in the absence of functional CD47 during apoptosis (Gardai et al. 2005), and annexin I (Arur et al. 2003), which is recruited from its cytosolic location following caspase activation to colocalize with PS in the plasma membrane of apoptotic cells. Other molecules, including CD43, and less well defined entities such as the moieties bound by the globular heads of C1q and the collectin mannose-binding lectin (MBL) as well as other molecules are also likely to contribute to ACAMP formation because of their ability to localise in the plasma membrane in discrete patterns (Fujii et al. 2001; Navratil et al. 2001; Ogden et al. 2001; Eda et al. 2004; Tennant et al. 2008).

Whatever the molecular architecture of ACAMPs, the parallels that have been drawn between these apparently conserved apoptotic-cell structures and the PAMPs of the microbial world must terminate at the PRR recognition step, because phagocyte responses to PAMPs and ACAMPs are divergent: the former activate pro-inflammatory responses whereas the latter are, by contrast, non-phlogistic. Toll-like receptors (TLRs) that are known to cooperate with PRR in providing coupled downstream signalling pathways in response to PAMPs could provide a molecular basis for the differential PRR-dependent responses to PAMPs vs. ACAMPs. In simplest terms, PAMP responsiveness is TLR-dependent in contrast to ACAMP responsiveness which appears to be TLR-independent. Supportive evidence has been provided recently: not only is it clear that TLR4 plays no role in the engulfment of apoptotic cells (Shiratsuchi et al. 2004) but also, phagocytosis of bacteria—but not of apoptotic cells—is impaired in the absence of TLR signals (Blander and Medzhitov 2004).

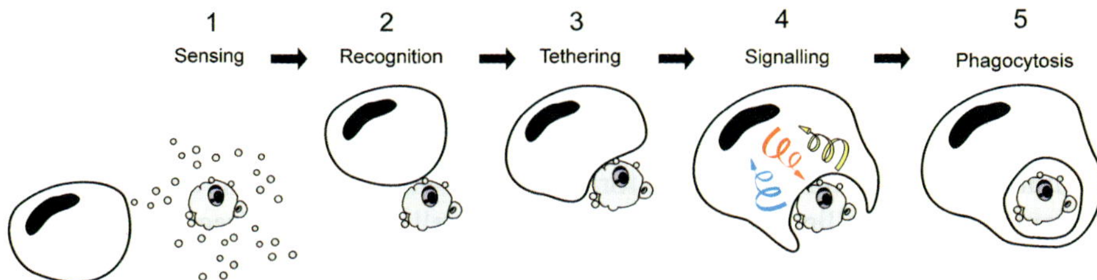

Fig. 9.3 Phases in the interactions between apoptotic cells and phagocytes. Chronologically, these can be divided into (1) sensing that includes chemotactic navigational mechanisms, (2) recognition leading to (3) binding events that result in strong intercellular adhesion and the formation of a molecular complex or "phagocytic synapse" that is functional in (4) signalling: initiation of engulfment and together with additional responses of the phagocyte such as production of anti-inflammatory mediators like TGF-β1. Phagocytic signalling culminates in (5) phagocytosis which results in confinement of apoptotic cells and bodies, degradation, antigen processing and presentation. Adapted from Gregory and Devitt (2004).

Chap. 6; Lacy-Hulbert, Chap. 7, this Vol.). These include dendritic cells (DC), fibroblasts, epithelial cells, endothelial cells, kidney mesangial cells, smooth muscle cells, hepatocytes, Sertoli cells of the testis, thecal cells of the ovary and stromal cells of the bone marrow (Bursch et al. 1985; Bennett et al. 1995; Albert et al. 1998; Svensson et al. 1999; Cao et al. 2004; Dogusan et al. 2004; reviewed in Gregory and Devitt 2004). Macrophages have been the focus of most attention because of their obvious association with apoptotic cellsin vivo. Thus, the routine histological picture of apoptosis—which is rarely seen except when apoptosis occurs at high rate such as during tissue remodelling or immune selection—is that of the remnants of apoptotic cells in association with macrophages (Kerr et al. 1972), examples of renown being the macrophage clearance of cells during the embryonic development of the interdigital spaces and the tingible body macrophages of the germinal centers of lymphoid follicles (see Gregory and Devitt 2004 for illustrations). It is likely that the full extent of cell death occurring by apoptosis and other forms of programmed cell death in tissues has yet to be realized because cell deletion is difficult to measure in the face of rapid clearance. In support of this argument, engulfment in vivoof cells showing normal gross morphology but membrane changes indicative of apoptosis has been demonstrated (Van den Eijnde et al. 1997) and kinetics of tumour growth point to 'invisible' cell deletion (Wyllie 1992).

Given the available, albeit limited, evidence, it seems likely that the list of 'amateur' phagocytes of apoptotic cells will prove to be extensive. It may be predicted that these cells would frequently engulf their juxtaposed neighbours by common mechanisms and this may occur through combinations of loss of 'don't eat me' and acquisition of eat me signals as described above—but it seems probable that, in different cell or tissue contexts, different molecular mechanisms could be involved, especially with respect to effector mechanisms triggered by the clearance events, including induction of tissue repair, immunological tolerance, and resolution of inflammation (see below). In this respect, even macrophages at different sites have been suggested to use different PS-interacting molecules to engage with apoptotic cells. For example, MFG-E8 is used by elicited peritoneal macrophages and tingible body macrophages whereas fetal liver and thymic macrophages deploy the related protein, Del-1 (Hanayama et al. 2004a); resting peritoneal macrophages use a further PS-receptor Tim-4 (Miyanishi et al. 2007). Other clearance molecules have importance in different tissues or at different stages in the apoptosis programme. C1q, for example, appears to be particularly important for the clearance of apoptotic cells in the kidney (Pickering et al. 2001). As we shall describe below, the relationships between cellular and molecular complexity have important consequences in vivo.

9.2.2.2 Phases in the Apoptosis Programme and Clearance Process

In addition to the multiplicity of phagocytes mediating clearance of apoptotic cells, distinct or overlapping mechanisms may support the clearance of cells at different stages during the apoptosis programme. Evidence has been presented, for example, that CD43 is involved in clearance of early-stage apoptotic cells whereas the pentraxins and collectins have been implicated in the clearance of cells further advanced in the programme (Bijl et al. 2003; Nauta et al. 2003). Additional molecules also appear to be preferentially involved with clearance of cells at specific stages in the apoptosis programme (see for example Fujii et al. 2001; Nakai et al. 2005).

Further molecular complexity lies in the multiple phases that are involved in the physical interactions between phagocytes and apoptotic cells. Some of these steps have been previously encompassed in the term 'tether' and 'tickle' (Hoffmann et al. 2001) to indicate the initial binding phase and subsequent signalling events. Here, we divide the processes chronologically into (1) sensing—including chemotactic mechanisms that direct phagocytes to apoptotic cells (Peter et al., Chap. 3, this Vol.; Lauber et al. 2003; Truman et al. 2008), (2) recognition (the beginnings of the intercellular interaction such as phagocyte CD31 interacting with disabled CD31 on the apoptotic cell (Napirei and Mannherz, Chap. 4, this Vol.; Brown et al. 2002), which leads on to (3) binding—tethering events that provide in strong intercellular adhesion and result in the formation of a molecular complex that is responsible for (4) signalling firstly the initiation of engulfment and secondly, additional responses of the phagocyte including production of anti-inflammatory mediators like TGF-β1 (Gronski and Ravichandran, Chap. 5; Ucker, Chap. 6; Lacy-Hulbert, Chap. 7, this Vol.). Initiation of phagocytic signalling that is mediated by Rho family GTPases, especially Rac, culminates in (4) phagocytosis which results in confinement of the apoptotic cell or body within the phagocyte, ultimately degradation and antigen processing and presentation (Divito and Morelli, Chap. 11, this Vol.; Fig. 9.3). While the specific roles of individual molecular players in the binding and signalling phases are not yet fully understood, some interactions, for example CD14 on the phagocyte interacting with an undefined entity on the apoptotic cell primarily mediate tethering (Devitt et al. 2004), while others, such as Mer/Gas6/PS (Scott et al. 2001) and $\alpha_v\beta_3$/MFG-E8/PS (Hanayama et al. 2002) appear to have a primary role in mediating engulfment signalling and possibly signalling for other events too. It has been suggested that, in a manner similar to the immunological synapse formed between T cells and antigen presenting cells, the signalling complex required for engulfment of apoptotic cells may involve multiple, low affinity receptor-ligand interactions that come together to form a high-avidity signalling complex or "phagocytic synapse" (Fadok et al. 2001a; Savill et al. 2002; Grimsley and Ravichandran 2003). Certain categories of molecules within such a complex would be predicted to provide both adhesive and signal transduction functions.

### 9.2.2.3	Responses of Phagocytes to Apoptotic Cells

Earlier "accepted wisdom" about cell death by apoptosis being non-phlogistic and, by contrast, necrotic cell death being necessarily pro-inflammatory, has been challenged in recent years as tenuous and overly simplistic (Ucker, Chap. 6; Lacy-Hulbert, Chap. 7, this Vol.; Gregory and Devitt 2004; Henson and Hume 2006). Furthermore, of particular importance to the issue of failed clearance of apoptotic cells in vivo, the assumption that "post-apoptotic" or secondarily necrotic cells have similar pro-inflammatory properties to primary necrotic lesions is incorrect (Gregory and Devitt 2004). As a consequence of engaging with apoptotic cells, phagocytes are capable of mounting neutral (Meagher et al. 1992; Stern et al. 1996; Kurosaka et al. 2003b), pro- or anti-inflammatory responses.

Current knowledge of the anti-inflammatory effects of apoptotic cells is largely based on in vitro responses of mononuclear phagocytes (Huynh et al. 2002), although the inflammatory and immunological consequences of apoptotic cells in vivo is beginning to be understood (see below). Mononuclear phagocytes responding to apoptotic cells release anti-inflammatory cytokines, including IL-10 and TGF-β1, which may act as key local autocrine or paracrine anti-inflammatory factors and immunosuppressants (Voll et al. 1997; Fadok et al. 1998; McDonald et al. 1999; Ogden et al. 2005). Furthermore, IL-10 secretion may be stimulated in mononuclear phagocytes via annexin 1-dependent mechanisms (Parente and Solito 2004). PS receptors are likely to be key mediators of anti-inflammatory mediator release. Stabilin-2, for example, signals TGF-β1 production (Park et al. 2008).

In addition to production of soluble anti-inflammatory mediators from phagocytes responding to apoptotic cells, direct suppression of pro-inflammatory responses in phagocytes may follow apoptotic-cell interaction. For example, the kinase domain of Mer has the potential to generate direct anti-inflammatory macrophage responses to apoptotic cells following binding of PS on apoptotic cells via the bridging molecule Gas6 (Savill et al. 2002; Lemke and Lu 2003).

Further interactive mechanisms between apoptotic cells and mononuclear phagocytes appear to create a micro-environment that is not only immunosuppressive and anti-inflammatory, but also promotes apoptotic cell clearance. Apoptotic cells can themselves produce anti-inflammatory mediators such as IL-10 (Gao et al. 1998) and TGFβ-1 (Chen et al. 2001), each of which is capable of up-regulating the capacity of macrophages to clear apoptotic cells (Szondy et al. 2003; Ogden et al. 2005). In addition, mechanisms dependent on mere contact—rather than phagocytosis—between apoptotic cells and phagocytes promote local anti-inflammatory effects. For example, contact between apoptotic neutrophils and activated monocytes causes switching of the latter from a pro-inflammatory to an anti-inflammatory state (Byrne and Reen 2002). Furthermore, contact between apoptotic cells and phagocytes has been shown to activate a "kiss and tell" response by the phagocyte: a contact-dependent licensing of macrophages to the anti-inflammatory effects of TGF-β1 that may serve to limit the extent of the anti-inflammatory responses to the local milieu of the dying cell (Lucas et al. 2006).

While the arguments surrounding the differential responses of phagocytes to apoptotic vs. necrotic cells are well rehearsed, with apoptotic cells inhibiting pro-inflammatory responses and necrotic cells not necessarily capable of eliciting inflammation, the notion that apoptotic cells can be pro-inflammatory in nature is less well accepted. Autoantibodies, including those against PS or nuclear autoantigens displayed in blebs at the apoptotic-cell surface can activate pro-inflammatory responses in macrophages that engulf the immunoglobulin-opsonised apoptotic cells through Fc receptor pathways (reviewed in Savill et al. 2002; Hart et al. 2004b). Furthermore, IgG immune complexes that can bridge apoptotic neutrophils to macrophages via Fcγ receptors, trigger pro-inflammatory responses in the phagocytes (Hart et al. 2004a). Many examples of apoptotic cells stimulating pro-inflammatory responses can be found in the literature (Muhl et al. 1999; Uchimura et al. 2000; Kawagishi et al. 2001; Kurosaka et al. 2003a; Lorimore et al. 2003; Iyoda et al. 2005) and, while in some cases this may be related to post-apoptotic, cytolytic events with pro-inflammatory characteristics, in other cases it appears to be a feature of the kinetics of the interaction between the apoptotic cell and phagocyte, with early pro-inflammatory cytokine release presaging that of anti-inflammatory mediators. Thus, one study reported that macrophages responding to apoptotic cells in combination with TLR ligands exhibited an early secretion of pro-inflammatory cytokines including TNF-α that was followed by TGF-β1 release (Lucas et al. 2003).

Responses of phagocytes to apoptotic cells are not limited to those that regulate inflammatory and immune responses. Ample evidence indicates that exposure to apoptotic cells elicits repair responses including growth factor and angiogenic factor production (Morimoto et al. 2001; Golpon et al. 2004; Hristov et al. 2004). Indeed it appears that merely the presence of apoptotic cells stimulates responses by neighbours, indicating that apoptotic cells condition the tissue micro-environment independently of phagocytic events. In this respect, a fascinating study revealed that apoptotic cells are capable of initiating directional endothelial cell sprouting—towards the dying cells—by an electrostatic mechanism resulting from their increased surface negative charge consequent to PS exposure (Weihua et al. 2005).

9.3 Mouse Knock-out Models Relevant to Apoptotic-Cell Clearance: General Principles

Given the complexities, reviewed above, of the interactions between apoptotic cells and phagocytes at both the cellular and the molecular levels, together with the range of responses that can be elicited following engagement of an apoptotic cell with a phagocyte or non-phagocytic neighbour, it is obvious that interpretation of mouse clearance-knock-out models will be far from straightforward. As we have discussed above, complex variables in the clearance process include a multiplicity of cells and of sensing, recognition, binding and response mechanisms. Added to these is the important issue of pleiotropy: many of the molecules implicated in the clearance process have multiple functions, as we have already observed with PRRs: TSP, $\alpha_v\beta_3$ and BAI1 for example, are known to play independent roles in regulating angiogenesis.

Genetic studies in the worm *C. elegans* have identified a number of evolutionarily conserved genes required for the efficient phagocytosis of cells that are deleted normally and reproducibly during worm development (Gronski and Ravichandran, Chap. 5, this Vol.; Ellis et al. 1991; Hengartner 2001). These genes function in the engulfment of neighbours by non-specialized phagocytes since *C. elegans* does not contain macrophages. Evolutionary conservation has also been noted in *Drosophila*, which does possess an ancient macrophage, the haemocyte (Franc et al. 1999; Gumienny and Hengartner 2001). The functional activities of these conserved molecules encompass surface receptor-ligand interactions (e.g. CED-1, -7) and intracellular signalling processes (CED-6, -10, -2, -5 and 12), notably in relation to the extension of cell surfaces that occurs in migration as well as phagocytosis (Wu and Horvitz 1998; Tosello-Trampont et al. 2001; Su et al. 2002; Grimsley and Ravichandran 2003; Grimsley et al. 2004). These highly conserved molecules provide a minimalist backdrop for the mammalian clearance process with mammalian orthologues identified: LRP/CD91 is the mammalian orthologue of CED-1, CED-7 is represented in mammals by ABCA-1, CED-6 by the engulfment adapter protein, GULP, CED-10 by Rac and CED-2/-5/-12 by the signalling complex CrkII/Dock180/ELMO (Gronski and Ravichandran, Chap. 5, this Vol.). MEGF10 has also been proposed as a functional orthologue in mammals of CED-1 (Hamon et al. 2006).

Fundamental problems are inherent with mouse knock-out models, including genetic background, lethality and compensatory mechanisms, although at least some of these issues can be addressed using conditional knock-out approaches. While most of the mouse knock-out models relevant to apoptotic-cell clearance studied to date are single-gene defects affecting all somatic cells, elegant work on conditional knock-outs has recently begun to provide detailed insight into the pleiotropy of molecules involved in the clearance process (Lacy-Hulbert et al. 2007).

One example of a possible case of compensation is the PU.1[−/−] mouse. Lack of the transcription factor, PU.1, which is necessary for mononuclear phagocyte development, results in altered apoptotic-cell clearance during development, exemplified by a small inhibition in clearance in the interdigital webs. At these sites, neighbouring mesenchymal cells were observed to function as 'reserve' phagocytes, albeit with reduced efficiency in comparison with macrophages, and by the time of birth, normal digits have been sculpted (Wood et al. 2000). Although the absolute absence of macrophages at these sites in PU.1[−/−] mice has been challenged (Lichanska et al. 1999) the observation that neighbouring mesenchymal cells in these embryos have the power to engulf apoptotic cells when required demonstrates that so-called non-professional phagocytes can play a significant role in apoptotic-cell clearance. Indeed, as we have argued previously (Gregory and Devitt 2004), although the full extent of the role of non-professional phagocytes in the clearance of cells undergoing programmed death is not understood, 'invisible' clearance of deleted cells by neighbours is likely to play a substantial role in tissue homeostasis.

As discussed below, thymic nurse epithelial cells may play the principal part in clearance of apoptotic thymocytes in situ, while stromal cells play a key role in clearance of apoptotic B cells in the bone-marrow. Observation of tissues of mice defective in clearance genes will not only provide information on the role of specific genes in the clearance process in specific tissues but, by making the deleted cells visible, will also supply much-needed data that addresses the extent of cell death in specific tissues.

9.3.1 Clearance-Defect Phenotypes

Intuitively, in view of the complexity and possible redundancy in molecules contributing to the clearance process, the phenotypic consequences of a single-gene defect would be predictably small, if apparent at all. Theoretically, the consequences of defective apoptotic cell clearance would be (1) neutral, (2) pro-inflammatory and immunostimulatory, coupling the defect to the development of autoimmunity, (3) anti-inflammatory and (4) pro-resolution (reparatory). Perhaps surprisingly, it is clear from Table 9.1 that single-gene defects in clearance molecules—such as C1q or MFG-E8—can indeed produce prominent phenotypes in mice that can lead to SLE-like autoimmune disease (Botto et al. 1998; Hanayama et al. 2004b). By contrast, in other cases like CD14$^{-/-}$ and MBL$^{-/-}$ mice, clearance defects resulting in apoptotic-cell persistence cause no overt pathology (Devitt et al. 2004; Stuart et al. 2005). These observations provide two important lessons about apoptotic-cell clearance in vivo: firstly, clearance defects can be caused by loss of a single class of receptor or bridging molecule, despite the fact that in vitro studies implicate multiple such molecules in the process, suggesting conceptually that blockade of multiple receptors would be required in order to achieve a phenotype. This argues against redundancy in the function of these molecules, although it is important to note that in all cases studied, inhibition of clearance is incomplete: the process is slowed rather than blocked. Second, persistence of apoptotic cells need not have pro-inflammatory consequences. Therefore, persisting apoptotic cells, which are likely to progress to secondary necrosis are not, of necessity, pro-inflammatory or immunostimulatory. This indicates that the key to the consequences of the clearance defect lie not in the failed clearance per se but in the role of the defective molecule in the clearance process. For example, knock-out of a molecule that subserves only recognition and binding of apoptotic cells is likely to have less of an impact on the animal than one which orchestrates responses to apoptotic cells.

The appearance of a clearance defect may not be apparent during normal homeostasis but may become apparent following apoptosis challenge, or may be apparent in some tissues and not others. Such differential phenotypes allow conclusions to be drawn as to the role of specific molecules in different tissue contexts or under pro-apoptotic physiological or pathological conditions. For example, in unchallenged *mer*^{kd} mice, which lack Mer kinase activity, clearance of apoptotic cells in

Table 9.1 Genetically modified mice with altered phenotypes of apoptotic-cell clearance.

Modification	Phenotype		Tissue/Cell Affected	Inflammation /Autoimmunity	Refs
	in vivo	in vitro			
C1qa[-/-]	Yes	Yes	Kidney, peritoneal macrophage	Increased anti-histone, ANA antibodies, glomerulone-phritis	(Botto et al. 1998; Taylor et al. 2000)
C4[-/-]	Yes	nt	Elicited peritoneal macrophage	nt	(Taylor et al. 2000)
CD93[-/-]	Yes	No	Elicited peritoneal macrophage	Normal	(Norsworthy et al. 2004)
SR-A[-/-]	No	Yes	nd	nt	(Platt et al. 2000)
ABCA1[-/-]	Yes	Yes	Embryonic footplate, elicited peritoneal macrophage	Glomerulone-phritis	(Christiansen-Weber et al. 2000; Hamon al. 2000)
PU.1[-/-]	Yes	nt	Embryonic footplate, macrophage	nt	(Wood et al. 2000)
merkd	Yes		Thymus (induced apoptosis), elic-ited peritoneal macrophage	Increased anti-chromatin, anti-DNA, anti-IgG antibodies, mild renal pathology	(Cohen et al. 2002; Schie-mann et al. 2001)
TGase2[-/-]	Yes	Yes	Thymus (induced apoptosis), liver (induced apoptosis), peritoneal macro-phage, Kupffer cell	Increased autoantibodies, glomerulone-phritis	(Szondy et al. 2003; Fesus and Szondy 2005)
Gal3[-/-]	Yes	Yes	Elicited perito-neal macrophage. Phagocytic defect not restricted to apoptotic cells	nt	(Sano et al. 2003)
MFG-E8[-/-]	Yes	Yes	Lymphoid, tingible body macrophage, elicited peritoneal macrophage	Late-onset increased anti-dsDNA, anti-nuclear antibodies, glomerulone-phritis	(Hanayama et al. 2004b; Hanayama and Nagata 2005)

Table 9.1 (continued)

Modification	Phenotype		Tissue/Cell Affected	Inflammation /Autoimmunity	Refs
	in vivo	in vitro			
CD14$^{-/-}$	Yes	Yes	Primary and secondary lymphoid, lung, liver, gut, peritoneal macrophage, bone marrow macrophage	None (normal)	(Devitt et al. 2004; Truman et al. 2008)
MBL$^{-/-}$	Yes	nt	Elicited peritoneal macrophage	None (but increased B1 B	(Stuart et al. 2005)
ApoE$^{-/-}$	Yes	Yes	Liver, lung, brain	Mild inflammation	(Grainger et al. 2004
α_v $^{-/-}$ (α_v-tie2; α_v-LysM conditional)	Yes	Yes	Colonic mucosa, peritoneal macrophages	Ulcerative colitis; inflammation of peritoneum, liver and respiratory tract	(Lacy-Hulbert et al. 2007)

α_v-tie2, conditional knock-out of α_v integrin in haemopoietic and endothelial cells; α_v-LysM, conditional knock-out of α_v integrin in myeloid lineage; ApoE, Apolipoprotein E; Gal-3, galectin-3; merkd, kinase defective mutant of mer; nd, none detected; nt, not tested; SR-A, scavenger receptor-A; TGase2, tissue transglutaminase. See Fig. 9.1 for other abbreviations.

the thymus appears to be normal (Scott et al. 2001), suggesting that Mer-independent mechanisms of apoptotic-cell clearance normally operate in this tissue. However, Mer-dependent mechanisms are brought into play in the thymus as revealed by defective clearance in *mer*^{kd} thymi following induction of synchronous thymocyte apoptosis by dexamethasone (Scott et al. 2001). More work is required to investigate these issues further as, to date, studies of relevant KO mice are limited in the tissues observed in depth or in the apoptosis stimuli used.

Further clues as to the identification of the important clearance molecules amongst the myriad illustrated in Fig. 9.1 can be drawn from the comparison between in vitroand in vivomodels. Such comparisons not only identify the relative importance to the clearance process of a specific molecule expressed by a particular phagocyte but also provide valuable information about the importance of that phagocyte in a specific tissue in vivo. The class A scavenger receptor, SR-A, provides a case in point. This receptor is expressed by thymic macrophages in situ and this expression is retained following the isolation of these phagocytes. Isolated thymic macrophages engulf apoptotic thymocytes in vitro and the dependence of this clearance process on SR-A can be demonstrated using blocking antibody against SR-A and using macrophages isolated from SR-A knock-out mice (Platt et al. 1996; Platt et al. 2000). However, SR-A deficiency failed to generate a clearance defect in vivo, even when γ-irradiation was used to synchronize and increase the frequency of apoptosis and to stimulate "back-up" clearance mechanisms. While these results could indicate that there are compensatory mechanisms for the SR-A defect in vivo, it seems more likely—unless clearance mechanisms are fundamentally different in vitro and in vivo—that thymic macrophages are not the principal phagocytes of apoptotic thymocytes in vivo, even though they are capable of acting in this capacity when directly fed apoptotic cells in vitro. In this context, thymic nurse cell structures comprising epithelial cells and macrophages have been reported to be the principal sites of apoptosis and clearance of thymocytes in situ (Hiramine et al. 1996), although earlier studies suggested that apoptotic thymocytes are cleared throughout the cortex by F4/80[+] macrophages (Surh and Sprent 1994). Whatever the case, this study serves to illustrate that clearance defects demonstrable artificially in vitro are not necessarily recapitulated in vivo.

9.4 Mouse Knock-out Models of Clearance in Relation to Inflammation and Disease Pathogenesis

As we have described above, the key to the generation of a phenotype caused by a particular clearance gene defect is determined not simply by the persistence of apoptotic cells but by the functional role of the affected gene in the clearance process. A molecule simply contributing towards tethering might be expected to produce less of an impact than a molecule involved in signalling pro-inflammatory responses or immune tolerization. Mouse mutants of particular relevance are summarized in Table 9.1.

9.4.1 Apoptotic-Cell Clearance and Autoimmune Disease

9.4.1.1 C1q$^{-/-}$ Mice

The first in vivo evidence supporting the long-standing notion that apoptotic cells provide an important source of autoantigens and that inefficient clearance of apoptotic cells could lead to autoimmune disease was provided a decade ago (Mevorach, Chap. 10, this Vol.). C1qa$^{-/-}$ mice were generated to investigate mechanistically the link between complement deficiency and the development of SLE. Monitored over a period of 8 months, these mice exhibited an increased mortality rate and produced higher titers of autoantibodies than wild-types. 25% of the C1qa$^{-/-}$ animals developed glomerulonephritis with increased numbers of apoptotic cells in glomeruli, providing evidence for the first time that the phenotype of autoimmunity was linked to defective clearance of apoptotic cells (Botto et al. 1998). The phenotype is dependent on a mixed genetic background (129 x C57BL/6 or MRL/Mp; Mitchell et al. 2002) and is tissue specific indicating, not only that additional genetic factors contribute to the development of autoimmunity but also that the role of C1q in apoptotic cell clearance in vivo may be of particular importance in the kidney. As discussed above, the full contributions of individual molecules to apoptotic-cell clearance in different tissues under basal and challenging conditions has not been exhaustively documented and C1q is no exception, although its functional capabilities in clearing apoptotic cells are not limited to the kidney. Thus, for example, in the first clear demonstration of inhibition of clearance of apoptotic cells by macrophages in vivo, C1q was found to be required for efficient engulfment by inflammatory macrophages of apoptotic cells instilled into the peritoneal cavity (Taylor et al. 2000). Other evidence indicates that complement components, including C3 which is deposited on apoptotic cells in concert with IgM (Kim et al. 2003; Quartier et al. 2005) are likely to have multiple roles in regulating apoptotic cell recognition, engulfment and response (Mevorach et al. 1998; Gaipl et al. 2001; Nauta et al. 2002; Hart et al. 2004b).

9.4.1.2 *mer*[kd] Mice

Additional mutant mice exhibit autoimmune features associated with apoptotic-cell clearance defects. Animals that are functionally deficient in the Mer tyrosine kinase (*mer*[kd] mice) produce increased titres of lupus-like autoantibodies (Scott et al. 2001; Cohen et al. 2002), while MFG-E8-knock-out animals similarly develop high titers of autoantibodies that lead to glomerulonephritis (Hanayama et al. 2004b). Each of these mutations allows the tethering stage to proceed unhindered but the engulfment phase is inhibited. In *mer*[kd] mice, because the signalling events associated with the observed endogenous apoptotic-cell clearance defect were subtle, only being apparent—at least with respect to the thymus, the only tissue documented—upon challenge of the animals with a massive apoptosis stimulus, dexamethasone treatment, it is logical to conclude that the engulfment process normally proceeded in these animals at a rate that was close to that of wild-types and that only by overloading the clearance system could the engulfment-signalling role of Mer be revealed. Given the autoimmune phenotype in the apparent absence of overt clearance defects in untreated *mer*[kd] animals, this suggests that the anti-inflammatory signalling role of Mer in apoptotic-cell clearance may be a more important primary function of this molecule in apoptotic-cell clearance than its engulfment-signalling role.

9.4.1.3 MFG-E8[-/-] Mice

The MFG-E8[-/-] mouse provides an example of the significance of tissue context in apoptotic-cell clearance and its association with SLE-like autoimmune disease. In these animals, the absence of MFG-E8 compromises the capacity of tingible body macrophages—the macrophages of germinal centers of lymphoid follicles—to engulf apoptotic cells, presumably mostly B cells undergoing selection at these sites. As with the *mer*[kd] mutation, the defect does not inhibit the tethering phase. MFG-E8[-/-] mice display marked splenomegaly at 40 weeks of age paralleled by massively increased titers of autoantibodies and glomerular immune complex deposition (Hanayama et al. 2004b). Treatment of normal mice with a recombinant RGD point-mutant form of MFG-E8 that retains the ability to bind PS but which is unable to bridge to integrins produces similar autoimmune features (Asano et al. 2004). Intriguingly, it has been reported that tingible body macrophages of patients with SLE also have reduced capacity for apoptotic-cell engulfment (Baumann et al. 2002). These results also indicate that tingible body macrophages play more than a mere disposal route for deleted B cells in germinal centers. More work is required before the functional attributes of these cells are fully understood, but it seems likely that these macrophages, through interacting with and responding to apoptotic B cells provide important tolerizing signals that militate against the generation of autoreactive B cells in germinal centers. Recent work shows that recruitment of macrophages to germinal centers is dependent on the chemokine receptor CX3CR1 (fractalkine receptor). As we have discussed above, fractalkine (CX3CL1) is released by apoptotic lymphocytes as a bleb-associated chemoattractant of macrophages and this may form the basis of the recruitment of macrophages to germinal centers (Truman et al. 2008). Interestingly, it has been shown that fractalkine can up-regulate MFG-E8 expression in phagocytes (Leonardi-Essmann et al. 2005) and enhance apoptotic-cell clearance (Miksa et al. 2007). Therefore fractalkine provides an important link between the sensing and engulfment phases of apoptotic-cell clearance by tingible body macrophages. Detailed analysis of the phenotype of CX3CR1-deficient mice in relation to apoptotic-cell clearance is required to determine the full extent of this relationship in different tissue or cell contexts. In this respect it is important to note that MFG-E8 is involved not only in the clearance of apoptotic cells by macrophages but also immature DC and mammary epithelial cells (Miyasaka et al. 2004; Hanayama and Nagata 2005). Failure in apoptotic-cell engulfment and response by DC is likely to contribute significantly to the development of autoimmunity. Involution of mammary glands is impaired in MFG-E8$^{-/-}$ mice and leads to ductal ectasia and periductal mastitis. This is associated with impaired ability of mammary gland macrophages and mammary epithelial cells to engulf apoptotic cells. Furthermore, the deficiency in involution caused impairment of mammary gland re-development (Hanayama and Nagata 2005).

9.4.1.4 Conditional Knock-outs of the α_v Integrin

Recent elegant studies in an epic paper by Lacy-Hulbert and colleagues have used conditional knock out mice to investigate further links between apoptotic cell clearance and regulation of autoimmunity. Expression of the αv integrin chain—a component of the $\alpha_v \beta_3$ and $\alpha_v \beta_5$ integrins implicated in apoptotic cell clearance—was manipulated in haemopoietic and endothelial cells by generating conditional knock out mice through targeting via tie2 or LysM: α_v-tie2 mice were deficient in α_v in both haemopoietic and endothelial cells while α_v-LysM mice lacked α_v on macrophages, neutrophils and DCs, (Lacy-Hulbert et al. 2007). α_v-tie2 mice were found initially to develop normally but, from about 12 weeks developed chronic and progressive inflammation of the colon and cecum that led to premature death and development of adenocarcinoma in older animals. The colitis was characterized by infiltration of monocytes, plasma cells and lymphocytes, the latter being in the majority and mainly T cells. The mesenteric lymph nodes and Peyer's patches of α_v-tie2 mice contained higher than normal levels of activated CD4$^+$ T cells (CD62L^{low}CD44^{high}). Cytokines IL-4, -5, -6, TNFα and IFNγ were increased but IL-12 and IL-23 were not. Intriguingly, enlargement of mesenteric lymph nodes, increased T-cell activation, IL-4 and IFNγ production in these animals were demonstrable at 3 weeks of age, long before overt signs of colonic inflammation, pointing to an early defect in T-cell regulation resulting from the lack of α_v. Accordingly, the authors found that the number of adaptive regulatory T-cells (Tregs) in the colons of α_v-tie2 mice were reduced to around half that of control animals. Adoptive transfer experiments indicated that the cellular origins of the phenotype of the α_v-tie2 mice lay with haemopoietic cells and not endothelial cells. α_v-LysM mice also developed colitis, narrowing the cellular defect to macrophages, neutrophils and/or DCs. Further studies indicated that mucosal DCs from α_v-tie2 mice are deficient in their ability to generate adaptive Tregs and in vitro investigations of the capacity of DCs and macrophages from α_v-tie2 mice to clear apoptotic cells was impaired providing a tantalizing possible link between the observed inflammatory disease development and impaired αv-dependent apoptotic-cell clearance by innate immune cells. In vitro clearance defects were borne out in vivoboth using a peritoneal macrophage clearance assay of ip-injected apoptotic cells and in histological counts demonstrating persistent apoptotic cells in the colons of αv-deficient animals. The affected animals developed high titers of autoantibodies including those directed at dead/dying cell components, including anti-PS, anti-dsDNA and anti-nuclear antibodies. These results suggest that αv on myeloid cells plays a critical role in regulating immune responses in the colon, perhaps through impairment of mucosal Treg production that is dependent upon effective apoptotic-cell clearance mechanisms mediated by αv integrins.

There is extensive evidence that $\alpha_v\beta_3$ and $\alpha_v\beta_5$ integrins are involved in apoptotic-cell clearance in vitro. However, since $\beta 3$ and $\beta 5$-deficient mice do not develop autoimmune disease (Lacy-Hulbert et al. 2007), it follows that αv-dependent apoptotic-cell clearance defects are not sufficient to cause the inflammatory colitis generated in α_v-deficient mice. Other evidence implicating $\beta 8$ in generation of colitis indicates that $\alpha_v\beta_8$ could be a crucial integrin in the pathogenesis of this disease (Travis et al. 2007). Interestingly, α_v-tie2 mice also show inflammatory disease at other sites, including the respiratory tract and liver, (Lacy-Hulbert et al. 2007). This model serves particularly well to illustrate the complex level of interpretation that is required to unravel the specific roles of individual molecules in disease processes resulting from defects in apoptotic-cell clearance functions. Many important questions remain, including the nature of the pro-inflammatory stimuli that are regulated by αv, the relative importance of myeloid cell types (e.g. DC vs. neutrophils vs. macrophages) in controlling the inflammation, the link with apoptotic-cell clearance and the possible local TGF-β1 production, particularly since adaptive Treg cells require TGF-β1 and TGF-β1-deficient mice succumb to similar inflammatory colitis as αv-deficient animals (Li et al. 2006). In this context it is of interest to note that TGase2-deficient animals produce high titers of autoantibodies and develop immune complex glomerulonephritis. This may be due in part to defective macrophage clearance of apoptotic cells displayed by these animals and in part by the absence of active TGF-β1 production (Szondy et al. 2003). Furthermore, Apolipoprotein E-deficient mice show defective clearance of apoptotic cells in multiple tissues and a mild pro-inflammatory phenotype that could be linked to the decreased levels of TGF-β1 production in these animals (Grainger et al. 2004).

9.4.2   Apoptotic-Cell Clearance and Anti-inflammatory Effects

In marked contrast to the models outlined above, defects in other genes implicated in the apoptotic-cell clearance process that have been studied to date either produce no phenotype or produce a phenotypic clearance defect with no pathology. Animals lacking ABCA1, for example, display delayed apoptotic-cell clearance during footplate development but by birth fail to show any developmental abnormalities. ABCA1-null mice suffer, however, from fundamental problems with membrane lipid turnover and many animals die after birth from perivisceral hemorrhage that is unlikely to be related to apoptotic-cell clearance defects since no clearance-defective phenotype is apparent at full term (Hamon et al. 2000). CD36[-/-] mice have no reported clearance defect in situ and SR-A[-/-] CD36[-/-] are similarly normal with respect to apoptotic-cell clearance, at least as far as is documented. However, given the inflammatory mechanisms that are at play in such animals, especially in the cardiovascular system (Kuchibhotla et al. 2008), it seems likely that the relationships between these molecules and apoptotic cells are mainly directed towards the regulation of the response to the apoptotic cell rather than of engulfment itself. Similar arguments can be made for TSP-1, since TSP-1[-/-] mice have not been reported as displaying apoptotic-cell clearance defects but do succumb to lung inflammation (Lawler et al. 1998).

CD14[-/-] animals, by contrast, have a readily discernible clearance defect in many tissues including the thymus, spleen, gut and lung, apoptotic cells persisting in these tissues even under basal conditions of normal cell turnover (Devitt et al. 2004).

Although not studied in depth throughout multiple tissues, synchronization of apoptosis in the thymus or increased apoptosis in germinal centers of spleen also revealed clearance defects in challenged animals (Devitt et al. 2004; Truman et al. 2008). Despite the persistence of apoptotic cells at many sites including secondary lymphoid tissues, CD14[-/-] mice are normal, healthy and fertile and are not predisposed to development of autoimmunity. Detailed analysis of the role of CD14 has indicated that its function in the clearance process is mainly as a tethering receptor. Significantly, apoptotic cells can still effectively induce production of TGF-β1 in CD14[-/-] macrophages and can inhibit their pro-inflammatory responses, just as in wild-type macrophages (Devitt et al. 2004). The apparently simple tethering role— at least from a 'minimalist' perspective—of CD14 is in marked contrast to that of MFG-E8, for example, absence of which, as we have already observed above, also leads to persistence of apoptotic cells in germinal centers but, unlike absence of CD14, does lead to SLE-like autoimmunity. This comparison suggests that tingible body macrophages, when diminished in apoptotic-cell clearance capacity, affect autoimmunity in different ways: when tethering is inhibited, there appear to be no detrimental consequences, but when engulfment and/or response signalling events are inhibited, the consequences are pathological. The combined effects of inhibiting both tethering and engulfment/response are as yet unknown.

The MBL$^{-/-}$ mouse is also known to produce a phenotype of apoptotic-cell persistence in the absence of increased autoimmunity, providing further evidence to uncouple the causative association between failed (or more accurately reduced) clearance and development of autoimmune disease. As we have already discussed here and elsewhere, secondarily necrotic cells generated through failed clearance (and indeed primary-necrotic cells), are not necessarily pro-inflammatory or sufficiently immuno-stimulatory to break tolerance. Because of the anti-inflammatory constitution and potential anti-inflammatory triggering properties of apoptotic cells, their accumulation under appropriate circumstances could, by contrast, prolong or enhance anti-inflammatory signalling, rather than having pro-inflammatory and autoimmune consequence. In this respect it is noteworthy that systemic administration of apoptotic cells causes TGFβ-1-dependent up-regulation of Treg numbers (Kleinclauss et al. 2006). Furthermore, infused apoptotic cells have been shown to induce regulatory B-cell and IL-10 production and to ameliorate autoimmune chronic arthritic inflammation (Gray et al. 2007). Therefore failure in appropriate responses to apoptotic cells of B-cell as well as T-cell arms of the adaptive immune system are likely to be key to the autoimmune effects of apoptotic cells.

9.5 Conclusions

The nature of the association between apoptotic cell clearance defects and development of autoimmunity and inflammatory disorders is still far from clear, arguments for causation vs. exacerbation vs. coincidental association having been made in detail previously (Gregory and Devitt 2004; Stuart et al. 2005) and elsewhere in this book. Further studies of knock-out mice will continue to define the complexity, redundancy, cell biological and immunological pathways involved in the clearance process. While such studies might be expected ultimately to help in simplifying the overall mechanistic picture, that picture remains complex for now, not least as a result of the pleiotropy amongst the molecular players in the process together with the contributions of genetic background and inflammatory mileu. Indeed, the importance of the inflammatory context and micro-environment of the dying cell should not be underestimated (Albert 2004): it is noteworthy that the inflammatory micro-environment, specifically TNF-α, can render macrophages deficient in their ability to undertake apoptotic-cell clearance (McPhillips et al. 2007).

Mouse knock-out studies with current and new models are undoubtedly continuing in this field and these investigations are likely to provide further insight into the mechanisms underlying apoptotic-cell clearance and autoimmunity. For example, newly discovered PS-receptors show promise as molecules that contribute to the control of autoimmunity. Notably, administration of anti-TIM-4 antibody to mice led to increased titers of autoantibodies, suggesting that TIM-4$^{-/-}$ animals might be

Grainger DJ, Reckless J, McKilligin E (2004) Apolipoprotein E modulates clearance of apoptotic bodies in vitro and in vivo, resulting in a systemic pro-inflammatory state in apolipoprotein E-deficient mice. J Immunol 173:6366–6375

Gray M, Miles K, Salter D et al (2007) Apoptotic cells protect mice from autoimmune inflammation by the induction of regulatory B cells. Proc Natl Acad Sci U S A 104:14080–14085

Gregory CD, Devitt A (1999) CD14 and apoptosis. Apoptosis 4:11–20

Gregory CD (2000) CD14-dependent clearance of apoptotic cells: relevance to the immune system. Curr Opin Immunol 12:27–34

Gregory CD, Devitt A (2004) The macrophage and the apoptotic cell: an innate immune interaction viewed simplistically? Immunology 113:1–14

Grimsley C, Ravichandran KS (2003) Cues for apoptotic cell engulfment: eat-me, don't eat-me and come-get-me signals. Trends Cell Biol 13:648–656

Grimsley CM, Kinchen JM, Tosello-Trampont AC et al (2004) Dock180 andELMO1 proteins cooperate to promote evolutionarily conserved Rac-dependent cell migration. J Biol Chem 279:6087–6097

Gumienny TL, Hengartner MO (2001) How the worm removes corpses: the nematode *C. elegans* as a model system to study engulfment. Cell Death Differ 8:564–568

Hamon Y, Broccardo C, Chambenoit O et al (2000) ABC1 promotes engulfment of apoptotic cells and transbilayer redistribution of phosphatidylserine. Nat Cell Biol 2:399–406

Hamon Y, Trompier D, Ma Z et al (2006) Cooperation between engulfment receptors: the case of ABCA1 and MEGF10. PLoS ONE 1: E120

Hanayama R, Tanaka M, Miwa K et al (2002) Identification of a factor that links apoptotic cells to phagocytes. Nature 417:182–187

Hanayama R, Tanaka M, Miwa K et al (2004a) Expression of developmental endothelial locus-1 in a subset of macrophages for engulfment of apoptotic cells. J Immunol 172:3876–3882

Hanayama R, Tanaka M, Miyasaka K et al (2004b) Autoimmune disease and impaired uptake of apoptotic cells in MFG-E8-deficient mice. Science 304:1147–1150

Hanayama R, and Nagata S (2005) Impaired involution of mammary glands in the absence of milk fat globule EGF factor 8. Proc Natl Acad Sci U S A 102:16886–16891

Hart SP, Alexander KM, Dransfield I (2004a) Immune complexes bind preferentially to Fc gamma RIIA (CD32) on apoptotic neutrophils, leading to augmented phagocytosis by macrophages and release of pro-inflammatory cytokines. J Immunol 172:1882–1887

Hart SP, Smith JR, Dransfield I (2004b) Phagocytosis of opsonized apoptotic cells: roles for 'old-fashioned' receptors for antibody and complement. Clin Exp Immunol 135:181–185

Hengartner MO (2001) Apoptosis: Corralling the corpses. Cell 104:325–328

Henson PM, Hume DA (2006) Apoptotic cell removal in development and tissue homeostasis. Trends Immunol 27:244–250

Hiramine C, Nakagawa T, Miyauchi A et al (1996) Thymic nurse cells as the site of thymocyte apoptosis and apoptotic cell clearance in the thymus of cyclophosphamide-treated mice. Lab Invest 75:185–201

Hoffmann PR, de Cathelineau AM, Ogden CA et al (2001) Phosphatidylserine (PS) induces PS receptor-mediated macropinocytosis and promotes clearance of apoptotic cells. J Cell Biol 155:649–659

Hristov M, Erl W, Linder S et al (2004) Apoptotic bodies from endothelial cells enhance the number and initiate the differentiation of human endothelial progenitor cells in vitro. Blood 104:2761–2766

Huynh ML, Fadok VA, Henson, PM (2002) Phosphatidylserine-dependent ingestion of apoptotic cells promotes TGF-beta1 secretion and the resolution of inflammation. J Clin Invest 109:41–50

Iyoda T, Nagata K, Akashi M et al (2005) Neutrophils accelerate macrophage-mediated digestion of apoptotic cells in vivo as well as in vitro. J Immunol 175:3475–3483

Janeway CA (1989) Approaching the Asymptote -Evolution and Revolution in Immunology. Cold Spring Harb Symp Quant Biol 54:1–13

Kagan VE, Gleiss B, Tyurina YY et al (2002) A role for oxidative stress in apoptosis: Oxidation and externalization of phosphatidylserine is required for macrophage clearance of cells undergoing Fas-mediated apoptosis. J Immunol 169:487–499

Kawagishi C, Kurosaka K, Watanabe N et al (2001) Cytokine production by macrophages in association with phagocytosis of etoposide-treated P388 cells in vitro and in vivo. Biochim Biophys Acta-Mol Cell Res 1541:221–230

Kerr JFR, Wyllie AH, Currie AR (1972) Apoptosis: a basic biological phenomenon with wide-ranging implications in tissue kinetics. Br J Cancer 26:239–257

Kim SJ, Gershov D, Ma X et al (2003) Opsonization of apoptotic cells and its effect on macrophage and T cell immune responses. Ann N Y Acad Sci 987:68–78

Kleinclauss F, Perruche S, Masson E et al (2006) Intravenous apoptotic spleen cell infusion induces a TGF-beta-dependent regulatory T-cell expansion. Cell Death Differ 13:41–52

Kobayashi N, Karisola P, Pena-Cruz V et al (2007) TIM-1 and TIM-4 glycoproteins bind phosphatidylserine and mediate uptake of apoptotic cells. Immunity 27:927–940

Kuchibhotla S, Vanegas D, Kennedy DJ et al (2008) Absence of CD36 protects against atherosclerosis in ApoE knock-out mice with no additional protection provided by absence of scavenger receptor A I/II. Cardiovasc Res 78:185–196

Kurosaka K, Takahashi M, and Kobayashi Y (2003a) Activation of extracellular signal-regulated kinase 1/2 is involved in production of CXC-chemokine by macrophages during phagocytosis of late apoptotic cells. Biochem Biophys Res Commun 306:1070–1074

Kurosaka K, Takahashi M, Watanabe N et al (2003b) Silent cleanup of very early apoptotic cells by macrophages. J Immunol 171:4672–4679

Lacy-Hulbert A, Smith AM, Tissire H et al (2007) Ulcerative colitis and autoimmunity induced by loss of myeloid alpha v integrins. Proc Natl Acad Sci U S A 104:15823–15828

Lauber K, Bohn E, Krober SM et al (2003) Apoptotic cells induce migration of phagocytes via caspase-3-mediated release of a lipid attraction signal. Cell 113:717–730

Lawler J, Sunday M, Thibert V et al (1998) Thrombospondin-1 is required for normal murine pulmonary homeostasis and its absence causes pneumonia. J Clin Invest 101:982–992

Lemke, G, and Lu, Q (2003) Macrophage regulation by Tyro 3 family receptors. Curr Opin Immunol 15:31–36

Leonardi-Essmann F, Emig M, Kitamura Y et al (2005) Fractalkine-upregulated milk-fat globule EGF factor-8 protein in cultured rat microglia. J Neuroimmunol 160:92–101

Li, MO Wan YY, Sanjabi S et al (2006) Transforming growth factor-beta regulation of immune responses. Annu Rev Immunol 24:99–146

Lichanska AM, Browne CM, Henkel GW et al (1999) Differentiation of the mononuclear phagocyte system during mouse embryogenesis: the role of transcription factor PU.1. Blood 94:127–138

Lorimore SA, Coates PJ, Wright EG (2003) Radiation-induced genomic instability and bystander effects: inter-related nontargeted effects of exposure to ionizing radiation. Oncogene 22:7058–7069

Lucas M, Stuart LM, Savill J et al (2003) Apoptotic cells and innate immune stimuli combine to regulate macrophage cytokine secretion. J Immunol 171:2610–2615

Lucas M, Stuart LM, Zhang A et al (2006) Requirements for apoptotic cell contact in regulation of macrophage responses. J Immunol 177:4047–4054

Martin SJ, Reutelingsperger CPM, McGahon AJ et al (1995) Early redistribution of plasma membrane phosphatidylserine is a general feature of apoptosis regardless of the initiating stimulus: inhibition by overexpression of Bcl-2 and Abl. J Exp Med 182:1545–1556

McDonald PP, Fadok VA, Bratton D et al (1999) Transcriptional and translational regulation of inflammatory mediator production by endogenous TGF-beta in macrophages that have ingested apoptotic cells. J Immunol 163:6164–6172

McPhillips K, Janssen WJ, Ghosh M et al (2007) TNF-alpha inhibits macrophage clearance of apoptotic cells via cytosolic phospholipase A2 and oxidant-dependent mechanisms. J Immunol 178:8117–8126

Meagher LC, Savill JS, Baker A et al (1992) Phagocytosis of apoptotic neutrophils does not induce macrophage release of thromboxane-b2. J Leukocyte Biol 52:269–273

Medzhitov R, Janeway CA (1997) Innate immunity: The virtues of a nonclonal system of recognition. Cell 91:295–298

Mevorach D, Mascarenhas JO, Gershov D et al (1998) Complement-dependent clearance of apoptotic cells by human macrophages. J Exp Med 188:2313–2320

Miksa M, Amin D, Wu R et al (2007) Fractalkine-induced MFG-E8 leads to enhanced apoptotic cell clearance by macrophages. Mol Med 13:553–560

Mitchell DA, Pickering MC, Warren J et al (2002) C1q deficiency and autoimmunity: The effects of genetic background on disease expression. J Immunol 168:2538–2543

Miyanishi M, Tada K, Koike M et al (2007) Identification of Tim4 as a phosphatidylserine receptor. Nature 450:435–439

Miyasaka K, Hanayama R, Tanaka M et al (2004) Expression of milk fat globule epidermal growth factor 8 in immature dendritic cells for engulfment of apoptotic cells. Eur J Immunol 34:1414–1422

Morimoto K, Amano H, Sonoda F et al (2001) Alveolar macrophages that phagocytose apoptotic neutrophils produce hepatocyte growth factor during bacterial pneumonia in mice. Am J Respir Cell Mol Biol 24:608–615

Muhl H, Nold M, Chang JH et al (1999) Expression and release of chemokines associated with apoptotic cell death in human promonocytic U937 cells and peripheral blood mononuclear cells. Eur J Immunol 29:3225–3235

Nakai Y, Shiratsuchi A, Manaka J et al (2005) Externalization and recognition by macrophages of large subunit of eukaryotic translation initiation factor 3 in apoptotic cells. Exp Cell Res 309:137–148

Nauta AJ, Trouw LA, Daha MR et al (2002) Direct binding of C1q to apoptotic cells and cell blebs induces complement activation. Eur J Immunol 32:1726–1736

Nauta AJ, Raaschou-Jensen N, Roos A et al (2003) Mannose-binding lectin engagement with late apoptotic and necrotic cells. Eur J Immunol 33:2853–2863

Navratil JS, Watkins SC, Wisnieski JJ et al (2001) The globular heads of C1q specifically recognize surface blebs of apoptotic vascular endothelial cells. J Immunol 166:3231–3239

Norsworthy PJ, Fossati-Jimack L, Cortes-Hernandez J et al (2004) Murine CD93 (C1qRp) contributes to the removal of apoptotic cells in vivo but is not required for C1q-mediated enhancement of phagocytosis. J Immunol 172:3406–3414

Ogden CA, de Cathelineau A, Hoffmann PR et al (2001) C1q and mannose binding lectin engagement of cell surface calreticulin and CD91 initiates macropinocytosis and uptake of apoptotic cells. J Exp Med 194:781–795

Ogden CA, Pound JD, Batth BK et al (2005) Enhanced Apoptotic Cell Clearance Capacity and B Cell Survival Factor Production by IL-10-Activated Macrophages: Implications for Burkitt's Lymphoma. J Immunol 174:3015–3023

Palaniyar N, Nadesalingam J, Clark H et al (2004) Nucleic acid is a novel ligand for innate, immune pattern recognition collectins surfactant proteins A and D and mannose-binding lectin. J Biol Chem 279:32728–32736

Parente L, Solito E (2004) Annexin 1: more than an anti-phospholipase protein. Inflamm Res 53:125–132

Park D, Tosello-Trampont AC, Elliott MR et al (2007) BAI1 is an engulfment receptor for apoptotic cells upstream of the ELMO/Dock180/Rac module. Nature 450:430–434

Park SY, Jung MY, Kim HJ et al (2008) Rapid cell corpse clearance by stabilin-2, a membrane phosphatidylserine receptor. Cell Death Differ 15:192–201

Pickering MC, Fischer S, Lewis MR et al (2001) Ultraviolet-radiation-induced keratinocyte apoptosis in C1q-deficient mice. J Invest Dermatol 117:52–58

Platt N, Suzuki H, Kurihara Y et al (1996) Role for the class A macrophage scavenger receptor in the phagocytosis of apoptotic thymocytes in vitro. Proc Natl Acad Sci U S A 93:12456–12460

Platt N, Suzuki H, Kodama T et al (2000) Apoptotic thymocyte clearance in scavenger receptor class A-deficient mice is apparently normal. J Immunol 164:4861–4867

Quartier P, Potter PK, Ehrenstein MR et al (2005) Predominant role of IgM-dependent activation of the classical pathway in the clearance of dying cells by murine bone marrow-derived macrophages in vitro. Eur J Immunol 35:252–260

Sano H, Hsu DK, Apgar JR et al (2003) Critical role of galectin-3 in phagocytosis by macrophages. J Clin Invest 112:389–397

Santiago C, Ballesteros A, Martinez-Munoz L et al (2007) Structures of T cellimmunoglobulin mucin protein 4 show a metal-Ion-dependent ligand binding site where phosphatidylserine binds. Immunity 27:941–951

Savill J, Dransfield I, Gregory C et al (2002) A blast from the past: clearance of apoptotic cells regulates immune responses. Nat Rev Immunol 2:965–975

Schiemann B, Gommerman JL, Vora K et al (2001) An essential role for BAFF in the normal development of B cells through a BCMA-independent pathway. Science 293:2111–2114

Scott RS, McMahon EJ, Pop SM et al (2001) Phagocytosis and clearance of apoptotic cells is mediated by MER. Nature 411:207–211

Shiratsuchi A, Watanabe I, Takeuchi O et al (2004) Inhibitory effect of toll-like receptor 4 on fusion between phagosomes and endosomes/lysosomes in macrophages. J Immunol 172:2039–2047

Stern M, Savill J, Haslett C (1996) Human monocyte-derived macrophage phagocytosis of senescent eosinophils undergoing apoptosis -mediation by alpha(v)beta(3)/CD36/thrombospondin recognition mechanism and lack of phlogistic response. Am J Pathol 149:911–921

Stuart LM, Takahashi K, Shi L et al (2005) Mannose-binding lectin-deficient mice display defective apoptotic cell clearance but no autoimmune phenotype. J Immunol 174:3220–3226

Su HP, Nakada-Tsukui K, Tosello-Trampont AC et al (2002) Interaction of CED-6/GULP, an adapter protein involved in engulfment of apoptotic cells with CED-1 and CD91/low density lipoprotein receptor-related protein (LRP). J Biol Chem 277:11772–11779

Surh CD, and Sprent J (1994) T-cell apoptosis detected in-situ during positive and negative selection in the thymus. Nature 372:100–103

Svensson PA, Johnson MSC, Ling C et al (1999) Scavenger receptor class B type Iin the rat ovary: Possible role in high density lipoprotein cholesterol uptake and in the recognition of apoptotic granulosa cells. Endocrinology 140:2494–2500

Szondy Z, Sarang Z, Molnar P et al (2003) Transglutaminase 2$^{-/-}$mice reveal a phagocytosis-associated crosstalk between macrophages and apoptotic cells. Proc Natl Acad Sci U S A 100:7812–7817

Taylor PR, Carugati A, Fadok VA et al (2000) A hierarchical role for classical pathway complement proteins in the clearance of apoptotic cells in vivo. J Exp Med 192:359–366

Tennant I, Pound J, Bournazou I et al (2008) Interaction of apoptotic cells with the innate immune system: exposure of apoptotic cell-associated molecular patterns (ACAMPs) that permit binding of anti-LPS antibodies and lactoferrin. Submitted

Tosello-Trampont AC, Brugnera E, Ravichandran KS (2001) Evidence for a conserved role for CrkII and Rac in engulfment of apoptotic cells. J Biol Chem 276:13797–13802

Travis MA, Reizis B, Melton AC et al (2007) Loss of integrin alpha(v)beta8 on dendritic cells causes autoimmunity and colitis in mice. Nature 449:361–365

Truman L, Ford C, Pazikowska M et al (2008) CX3CL1 (fractalkine) is released from apoptotic lymphocytes to stimulate macrophage chemotaxis. Blood. In press.

Uchimura E, Watanabe N, Niwa O et al (2000) Transient infiltration of neutrophils into the thymus in association with apoptosis induced by whole-body X-irradiation. J Leukocyte Biol 67:780–784

Van Den Eijnde SM, Boshart L, Reutelingsperger CPM et al (1997) Phosphatidylserine plasma membrane asymmetry in vivo: A pancellular phenomenon which alters during apoptosis. Cell Death Differ 4:311–316

Voll RE, Herrmann M, Roth EA et al (1997) Immunosuppressive effects of apoptotic cells. Nature 390:350–351

Weihua Z, Tsan R, Schroit AJ et al (2005) Apoptotic cells initiate endothelial cells prouting via electrostatic signalling. Cancer Res 65:11529–11535

Wood W, Turmaine M, Weber R et al (2000) Mesenchymal cells engulf and clear apoptotic footplate cells in macrophageless PU.1 null mouse embryos. Development 127:5245–5252

Wu YC, Horvitz HR (1998) *C. elegans* phagocytosis and cell-migration protein CED-5 is similar to human DOCK180. Nature 392:501–504

Wyllie AH (1992) Apoptosis and the regulation of cell numbers in normal and neoplastic tissues -an overview. Cancer Metast Rev 11:95–103298

Chapter 10
Clearance of Dying Cells and Systemic Lupus Erythematosus

Dror Mevorach

Abstract: Death-Associated Molecular Patterns (DAMPs) maintain peripheral tolerance and immune suppression following binding and phagocytosis of apoptotic cells. In systemic lupus erythematosus (SLE), a multisystem autoimmune disease of unknown etiology, alteration in cell death patterns, apoptotic cell recognition and DAMP signalling generate the characteristic pathogenic autoantibodies to a diverse group of autoantigens.

The normal innate immune response to cell death and the abnormalities identified in SLE are presented, along with possible relations to mechanisms of autoantibody generation in SLE, the phenomenon of drug-induced lupus, and the paradoxical role of complement in the clearance of dying cells and in disease progression.

Keywords: Apoptosis • Necrosis • Clearance • Autoimmunity • Dendritic cells • SLE

10.1 Introduction

Systemic lupus erythematosus (SLE) is a multisystem autoimmune disease of unknown etiology characterized by the presence of high titers of autoantibodies to a diverse group of autoantigens. Autoantibodies are present in 88% of patients an average of 3.3 years before diagnosis (Arbuckle et al. 2003). Of the characteristic panel of autoantibodies, antinuclear, anti-Ro, anti-La, and antiphospholipid antibodies appear first, followed by anti-dsDNA, anti-Sm and anti-RNP. These autoantibodies have features of an antigen-driven, T cell-dependent immune response (Diamond et al. 1992; Radic and Weigert 1994). Once these antibodies are present,

D. Mevorach (✉)
The Laboratory for Cellular and Molecular Immunology
Department of Medicine, Hadassah and the Hebrew University
POB 12000, Kiryat Hadassah
Jerusalem 91200, Israel
Tel.: 972-2-677-7317
Fax: 972-2-646-8801
E-mail: mevorachd@hadassah.org.il

D.V. Krysko, P. Vandenabeele (eds.), *Phagocytosis of Dying Cells,*
DOI: 10.1007/978-1-4020-9292-3_10, © Springer Science+Business Media B.V. 2009

the course of SLE is characterized by disease flares and autoimmune dysregulation. Treatment is based on immunosuppressive drugs, such as corticosteroids, azathioprine, mycophenolate mofetil, and cyclophosphamide. Prognosis depends on the organs involved, the outcome of treatment, and the extent of adverse effects inflicted by the immunosuppressive drugs.

Programmed cell death (PCD), an essential developmental and homeostatic mechanism (Krysko et al., Chap. 1; Diez-Fraile et al., Chap. 2, this Vol.), is the preferred physiological death process for cells and also an important regulator of the immune response. Appropriate clearance of apoptotic material completes the PCD process, and is essential for regulating inflammation and maintaining self-tolerance (Ucker, Chap. 6; Lacy-Hulbert, Chap. 7; Gregory and Pound, Chap. 9; Divito and Morelli, Chap. 11, this Vol.).

In this chapter we will discuss altered mechanisms for clearance of dying material, which represent a central pathogenic process in the development and acceleration of SLE.

10.2 Cell Death in the Human Body

It estimated that every day more then one billion cells/kg of the human body undergo apoptosis (Fig. 10.1). Granulocytes, with a half lifetime of 6–7 hours, provide about 50 billion apoptotic cells each day in an average adult (Dancey et al. 1976). Apoptotic cells, which carry cellular components that could be antigens, present 'find me' and 'eat-me' signals and are engulfed by phagocytes such as macrophages and immature dendritic cells (iDCs; Peter et al., Chap. 3; Napirei and Mannherz, Chap. 4; Lacy-Hulbert, Chap. 7; Divito and Morelli, Chap. 11, this Vol.). Endosomes containing the engulfed dead cells fuse with lysosomes, where the dead cell components are degraded into amino acids and nucleotides. The degradation of chromosomal DNA occurs in two steps (Napirei and Mannherz, Chap. 4, this Vol.). First, the chromosomal DNA of apoptotic cells is cell-autonomously cleaved into nucleosomal units (Wyllie 1980) by caspase-activated DNase (CAD; Enari et al. 1998; Sakahira et al. 1998). Then, following engulfment by macrophages, the DNA of the dead cells is further degraded into nucleotides by DNase II, an acid DNase located in the lysosomes of macrophages (Kawane et al. 2003).

The terms apoptosis and PCD were once used interchangeably. However, PCD today can be further subdivided into at least three death pathways: apoptosis, necrosis (Krysko et al., Chap. 1; Diez-Fraile et al., Chap. 2, this Vol.; Chiarugi 2005; Kroemer and Martin 2005; Zong and Thompson 2006), and autophagy (Levine and Deretic 2007; Maiuri et al. 2007). Cell necrosis is morphologically different from apoptosis and is characterized by cellular swelling, chromatin digestion, organelle vacuolation, and plasma membrane disruption (Darzynkiewicz et al. 1997). Primary necrosis was for a long time considered as an 'accidental cell death,' in contrast to 'programmed cell death'. However, there is now extensive evidence that necrosis can be due to either accidental or programmed cell death (Chiarugi 2005; Kroemer

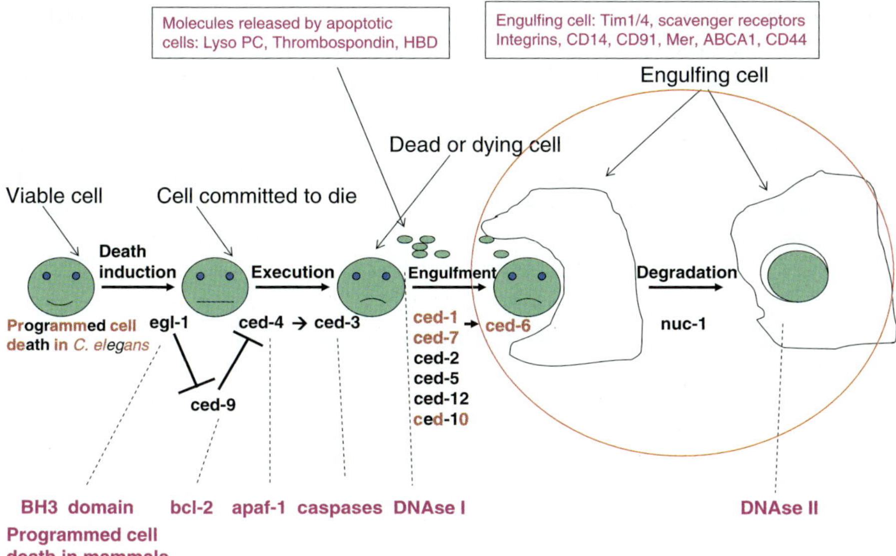

Fig. 10.1 Programmed cell death in *C. elegans* and mammals We can identify four different steps of programmed cell death in *C. elegans* and mammals: death induction, execution, engulfment, and degradation. Engulfment can be further subdivided into adhesion and engulfment. The main genes involved in *C. elegans* and mammals (pink) are presented. HBD, heparin binding domain; Lyso PC, lysophosphatidylcholine.

and Martin 2005; Zong and Thompson 2006). In general, necrosis is believed to occur when the cell cannot maintain membrane integrity. In the case of primary necrosis, this is due to abrupt and rapid cellular insults that overwhelm the cell's processes (Zong and Thompson 2006). Thus, necrosis proceeds down energetic and thermodynamic gradients, usually without needing active generation of energy generation for completion (Zong and Thompson 2006). All cells undergoing necrosis will, by definition, eventually end in lysis and spillage of the cellular contents, and in that regard primary necrosis and even secondary necrosis can be sometimes pro-inflammatory.

10.3 The Normal Innate Immune Response to Cell Death

10.3.1 Receptors

Different receptors, including integrins, scavenger receptors, phosphatidylserine (PS) receptor (PsR), CD14, ABC1 cassette transporter, CD44, C1q/CD91, stabilin-2, TIM 1 and 4, have been shown to have specific roles in tethering and/or uptake of apoptotic cells by macrophages and immature dendritic cells (iDCs) in mammals (Napirei and Mannherz, Chap. 3; Gregory and Pound, Chap. 9, this Vol.;

Miyanishi et al. 2007; Nagata 2007, Park et al. 2008; Savill and Fadok 2000). As indicated by the many receptors involved (Savill and Fadok 2000; Serhan and Savill 2005), the mechanisms underlying the recognition, engulfment, and phagocytosis of apoptotic cells by macrophages are complex. In addition, serum factors have been shown to increase the uptake of apoptotic cells by mammalian macrophages and iDCs by three to tenfold (Mevorach et al. 1998a; Verbovetski et al. 2002). The roles of these receptors in different patterns of PCD and in necrotic cell clearance are less understood.

10.3.2 Signalling Cascade Following Interaction with Apoptotic Cells

Apoptotic cells have been shown to signal their neighbours in a variety of ways. 'Eat me' signals, which can appear on the membranes of apoptotic cells, serve as markers for phagocytes to recognize specific cells and subsequently ingest them (Napirei and Mannherz, Chap. 4; Gregory and Pound, Chap. 9, this Vol.). Direct signals include alteration in cell surface phospholipid composition (Fadok et al. 1992), and changes in cell surface glycoproteins or in surface charge (Henson et al. 2001). In an alternative and complementary manner, certain serum proteins can opsonize an apoptotic cell surface and signal to phagocytes to engulf the opsonized apoptotic cells (Mevorach et al. 1998a; Mevorach 2000). Similarly, viable cells express 'do not eat me' signals by restricting phosphatidylserine to the inner leaflet of the membrane (Verhoven et al. 1995) or by CD31 expression (Brown et al. 2002). Apoptotic cells can also secrete molecules that are important for recruitment of phagocytic cells, phagocytosis, and modulation of the immune response. TGF-β (Chen et al. 2001), phosphoisocholine (Lauber et al. 2003), annexin 1 (Scannell et al. 2007), thrombospondin 1, and heparin binding domain (Krispin et al. 2006) are examples of immune-suppressive and phagocyte-recruiting molecules secreted by cells undergoing apoptosis (Peter et al., Chap. 3, this Vol.). Most of these mechanisms enable accurate identification of a cell undergoing apoptosis and efficient clearance, as well as appropriate non-inflammatory and non-autoimmune consequences (Ucker, Chap. 6; Lacy-Hulbert, Chap. 7; Divito and Morelli, Chap. 11, this Vol.; Fadok et al. 1998; Huynh et al. 2002; Verbovetski et al. 2002; Tassiulas et al. 2007). Phagocytosis of apoptotic cells is considered non-inflammatory, and has been shown to suppress the release of GM-CSF, IL-1β, IL-8, TNF-α and thromboxane B2, but not TGF-β and PGE2 (Fadok et al. 1998). IL-10 has anti-inflammatory properties manifested by suppression of the release of pro-inflammatory cytokines such as IL-1, IL-6, IL-8, and TNF-α The involvement of NF-κB as an inhibitory mechanism has been suggested (Tassiulas et al. 2007; Amarilyo et al., unpublished data). Contradictory results were obtained in two studies that evaluated the production of IL-10 following phagocytosis of apoptotic cells. One study reported increased production of IL-10 (Voll et al. 1997) and the second its suppressed release (Fadok et al. 1998) following ingestion of apoptotic cells by human macrophages. These studies were performed in the absence of serum; therefore, the effect of complement and other opsonins on IL-10 release could not have been evaluated. We found that IL-10 is secreted in the presence of serum and complement opsonization (Amarilyo et al., unpublished data).

The general immune suppression or immune non-activation consequences that are induced following the binding of apoptotic cells by both specific receptors indicates there are at least some partly shared downstream pathways. We propose to call these common pathways by the general name Death-Associated Molecular Patterns (DAMPs), because they bear some resemblance to Pathogen-Associated Molecular Patterns (PAMPs, see below). We prefer the term Death-Associated Molecular Patterns to Apoptosis-Associated Molecular Patterns because, contrary to 'accepted' assumptions, some forms of necrotic cell death result in cell clearance processes that resemble those for apoptotic cells (Chung et al. 2000), whereas some apoptotic patterns induce inflammation (see below). DAMPs (not to be mixed with Danger-Associated Molecular Patterns) and PAMPs are associated with different receptors, and lead in general to opposing immune consequences.

Apoptotic cells generally induce immune suppression, but exceptions exist. In some conditions, cells undergoing apoptosis produce pro-inflammatory cytokines. Earlier studies led to the hypothesis that Fas ligand expression enables cells to counterattack the immune system, and that transplant rejection, for example, could be prevented by expressing FasL on transplanted organs. More recent studies have indicated that the notion of Fas ligand as a mediator of immune privilege should be reconsidered, and that Fas ligation may even be pro-inflammatory (Restifo 2000). In fact, Fas was proposed to mediate the processing and secretion of pro-inflammatory cytokines such as IL-1β (Miwa et al. 1998). Park et al. (2003) suggested that following anti-Fas (CH11)-induced apoptosis, Fas-dependent secretion of IL-8 and TNF-α was triggered in human monocytes. This was shown to occur in macrophages even in the absence of apoptosis and was associated with NF-κB activation.

10.4   Immune Response to Cell Death in SLE

10.4.1   *Apoptotic and Necrotic Cells Are the Sources of Antigens That Drive an Autoimmune Response in SLE*

SLE is characterized by the production of a variety of heterogeneous autoantibodies, such as against nucleosomes, U1-RNP, Sm, SSA, SSB, PARP, NuMA, cardiolipin. These autoantibodies are targeted against a variety of structures, such as DNA, RNA, and acidic phospholipids, which are found in diverse locations within the cell, including the nucleus, cytoplasm, and membrane. It has long been appreciated that DNA and histones are major autoantigens in SLE. However, recently it has become clear that DNA-histone complexes, i.e. nucleosomes, are preferred targets for autoantibodies in SLE (Mohan et al. 1993).

How do nucleosomes and several other intracellular antigens become immunogenic in SLE patients? Casciola-Rosen et al. (1994) showed in their pioneering work that exposure of keratinocytes to UV-B-mediated apoptosis induces cell surface expression of Ro, La, nucleosomes, and ribosomes, possibly due to translocation of certain intracellular particles to the apoptotic surface blebs (Casciola-Rosen et al. 1994). Subsequent work showed that phosphatidylserine, which is restricted to the inner membrane leaflet in viable cells, translocates to the cell surface during apoptosis (Verhoven et al. 1995). and that β2GPI, a major autoantigen for anti-phospholipid antibodies in SLE, binds to phosphatidylserine on apoptotic cells (Price et al. 1996). In vivo experiments in mice showed transient elevations of anti-phospholipid, ANA, and anti-ssDN autoantibodies following high-dose injections of apoptotic cells (Mevorach et al. 1998b). It has therefore been hypothesized that SLE is triggered by immune responses to exposed intracellular macromolecules that are translocated to cell surfaces during apoptosis (Casciola-Rosen et al. 1994). We argue that necrotic cell death is also a source for autoantigens because during necrosis DNAse1 enters the cell and cleaves its chromatin to nucleosomes that are the antigens against which anti-DNA is generated (Napirei and Mannherz, Chap. 4, this Vol.; Napirei et al. 2000).

10.4.2 Abnormal Clearance of Apoptotic or Necrotic Material in SLE

Why do SLE patients mount an immune response to the apoptotic material? One possible explanation is their inability to clear apoptotic cells in the normal manner. Apoptosis usually occurs while apoptotic cell membranes are intact so that nucleosomes can not leak out. In SLE patients, however, we find high levels of nucleosomes circulating in peripheral blood (Rumore and Steinman 1990), indicating accelerated apoptosis, inefficient clearance (Herrmann et al. 1998; Shoshan et al. 2001) or both. The nucleosomes are cleaved and derived from apoptotic cells. Necrotic cell death is another source of nucleosomes, and accelerated primary necrosis may occur in SLE. Indeed, reduction of DNAse1 and increase in blood nucleosome levels are suggested to play a critical role in the initiation of SLE (Napirei et al. 2000).

Increased spontaneous apoptosis of lymphocytes (Emlen et al. 1994; Perniok et al. 1998) and higher rates of spontaneous monocyte apoptosis (Shoshan et al. 2001) have been documented in patients with SLE. These observations point to accelerated mononuclear apoptosis rate in SLE as a possible source of increased nucleosome levels. However, these observations are ex vivo, and in vivo these cells may be live cells undergoing accelerated apoptosis due to withdrawal of cytokines or growth factors. Nevertheless, leukopenia, typical in SLE patients, may result from this accelerated apoptosis, as it does in AIDS patients (Jaworowski and Crowe 1999). In addition, SLE sera may induce leukocyte apoptosis in vitro.

Is there convincing evidence for impaired apoptotic or necrotic cell clearance in SLE? In vitro studies have indicated that apoptotic cell clearance by monocyte-derived macrophages is impaired in SLE (Herrmann et al. 1998; Shoshan et al. 2001; Ren et al. 2003), and that this impairment is at least partially related to accelerated in vitro monocyte apoptosis (Shoshan et al. 2001). This accelerated apoptosis causes an increase in the amount of apoptotic material and an impairment of the ability of the remaining diluted macrophages to phagocytose apoptotic cells due to loss of 'a community effect' (Shoshan et al. 2001). Dendritic cells have a similar, albeit perhaps, smaller decrease in their ability to phagocytose apoptotic cells in these patients (Berkun et al. 2008).

It is difficult to conclude solely on these in vitro studies that clearance of apoptotic cells in SLE patients is altered. However, Baumann et al. (2002) have shown that in one subgroup of SLE patients, apoptotic cells are not properly cleared by tingible body macrophages of the germinal centers. So we may conclude that impaired clearance of dying cells probably occurs in SLE, and that this may have been documented as early as 1948. In the LE cell, nuclei are opsonized by antichromatin/anti-histones antibodies (Hargraves 1949; Haserick and Bortz 1949), then phagocytosed and digested by polymorphonuclear leukocytes (Hargraves et al. 1948). Description of the LE cell represents the first documentation of abnormal chromatin clearance, opsonization by an autoantibody, and clearance by a polymorphonuclear cell.

10.4.3 Apoptotic Cell Opsonins in SLE

SLE patients have reduced levels not only of complement, but also of other important apoptotic cell opsonins. Complement maybe reduced due to a genetic defect in some patients, or to complement consumption in most of them. C-reactive protein (CRP), was recently shown to be genetically controlled (Russell et al. 2004), is another example. SLE patients also have increased levels of potentially dangerous opsonin-like autoantibodies (Rovere et al. 1999) that can inhibit normal clearance of apoptotic cells (Reefman et al. 2007). Abnormal monocytes (Blanco et al. 2001; Verbovetski et al. 2002; Verbovetski et al., unpublished data) and iDCs (Blanco et al. 2001) as well as tolerogenic DC generation following apoptotic cell-DC interaction, may represent additional mechanisms (Berkun et al. 2008).

10.4.4 The Complement Story

How is complement involved in general in clearance of apoptotic cells and in SLE particularly? The complement system is activated by human and murine apoptotic material, as reviewed by Fishelson et al. (2001). We first showed that activation of both the classical and alternative complement pathways by apoptotic cells leads to their opsonization by iC3b (Mevorach et al. 1998a). Taylor et al. (2000) confirmed our observation and suggested an hierarchical role for C1q in the ex vivo clearance of apoptotic thymocytes by thioglycollate-derived peritoneal murine macrophages.

Ogden et al. (2001) showed a role for CD91 and calreticulin in the uptake of apoptotic cells opsonised by mannose-binding lectin. C1q and complement factors were suggested to mediate late apoptotic or necrotic cell clearance (Gaipl et al. 2001). Clearance by iDCs was also shown to be facilitated by opsonization with iC3b (Verbovetski et al. 2002) In addition, apoptotic cells rapidly lose the key complement regulator, CD46, which enables complement activation (Elward et al. 2005). Thus, complement opsonins and degradation products such as C1q, mannose binding lectin, and iC3b, appear to have a role in the uptake of apoptotic cells and cell debris, possibly by interacting with C1qR/CD91, CR3 (CD11b/CD18), and/or CR4 (CD11c/CD18) on phagocytes (Mevorach et al. 1998a; Mevorach et al. 1998b; Godson et al. 2000; Mevorach 2000; Ogden et al. 2001; Verbovetski et al. 2002). Other opsonins may also play a role in apoptotic cell clearance, whether or not they are involved in complement activation (Peter et al., Chap. 3, this Vol.). These include IgM, IgG, serum amyloid protein (SAP), C-reactive protein (CRP), protein S, thrombospondin, and other serum proteins (Mevorach et al. 1998b; Bickerstaff et al. 1999; Gershov et al. 2000; Kim et al. 2002; Anderson et al. 2003; Krispin et al. 2006).

Tagging the surfaces of apoptotic cells with C3b and iC3b may not only promote efficient clearance of these cells, it may also induce anti-inflammatory responses. Binding and phagocytosis via macrophage CD11b/CD18 does not trigger release of leukotriene (Aderem et al. 1985) or a respiratory burst (Wright and Silverstein 1983; Yamamoto and Johnston 1984). Furthermore, ligation of CD11b/CD18 and other complement receptors may actually be immunosuppressive due to the resultant down-regulation of IL-12 and IFN-γ production by human monocytes (Marth et al. 1997; Sutterwala et al. 1997).

It seems likely, therefore, that the pro- and anti-inflammatory consequences of complement activation depend on the specific ligands involved and the co-receptors engaged. Macrophages ingesting iC3b-opsonized apoptotic cells, after being triggered by zymosan (which stimulates Toll-like receptor4), down-regulated their secretion of IL-1β and IL-6 (Amarilyo et al., unpublished data).

Susceptibility to SLE is heritable, and studies in humans and mice indicate that multiple genetic loci are involved (Kono and Theofilopoulos 2006). A possible exception to polygenic predisposition to this disease seems to be single-gene deficiency in either C1 (C1q, C1r or C1s) or C4, but it is possible that confounding genes are hidden in these individuals. Those with C3 deficiency are less susceptible to SLE compared with individuals, who are deficient in C1 or C4, which are considered the major susceptibility factors. Genetic deficiencies in individual loci are rare, but acquired deficiencies are possible. For example, they may result from drugs that inactivate C4 (Speirs et al. 1989), or from development of antibodies against C1q. Additional mechanisms can also lead to SLE. Partial deficiency in C4– of which there are two human isotypes, C4A and C4B is relatively common and may be an important factor in SLE susceptibility among Scandinavians (Steinsson et al. 1998; Kristjansdottir et al. 2000). Mice deficient in C1q or C4 are also predisposed to an SLE-like condition (Botto et al. 1998; Chen et al. 2000; Paul et al. 2002), but the frequency and severity of the disease is strain dependent. Mice deficient in C1q or

C4 on a mixed background of 129 and C57BL/6 develop SLE autoantibodies and evidence of renal disease, whereas the same deficiency on a C57BL/6 background has little effect.

Taken together, these observations indicate that early complement proteins are important in protecting humans against the development of SLE. These observations raise a paradox, because the complement system is also an important effector of innate immunity and of the inflammation that mediates the pathogenesis of SLE. Decreased serum C3 levels and kidney deposition of C3, C1q and other complement components are common in patients with necrotizing lesions in class III and IV lupus nephritis. Analysis of immune complexes from humans with lupus nephritis shows that complement activation via the classical or alternative pathway occurs in proliferative lupus nephritis and thus complement was considered for years as an important factor in the effector mechanism of lupus nephritis. Several studies in animals demonstrate that kidney inflammation and glomerular necrosis is prevented when C3 or C1q is knocked out, which indicates the possibility of using complement inhibition in the treatment of lupus glomerulonephritis (Manzi et al. 1996; Niculescu et al. 1997; Makrides 1998).

Thus, we are faced with a paradox: complement may play a protective role, a destructive role, or even both in SLE. It has been suggested that complement plays an intrinsic protective role in the rapid clearance of the IgG-containing immune complexes that are characteristic of SLE (Takahashi et al. 1977; Paul et al. 1988). Rapid clearance of these complexes limits their pathogenicity (Lachmann and Walport 1987).

We have suggested that during apoptotic cell clearance, complement is needed for adhesion and phagocytosis as well as for the anti-inflammatory effect (Mevorach et al. 1998a; Verbovetski et al. 2002; Amarilyo et al., unpublished data). Thus, complement supports a tolerogenic pathway in interactions between dying cells and macrophages, dendritic cells, and even B cells (Carroll 2004). Taken together, complement may play a protective role at the initial formation of autoantibodies and in patients in remission, by supporting a tolerogenic pathway, but may participate in the destruction and inflammation characteristic of a lupus flare.

10.4.5 Dendritic Cells in SLE

Dendritic cells (DCs) are professional antigen-presenting cells of the immune system. They are continuously produced from hematopoietic stem cells in the bone-marrow, and they are widely distributed in both lymphoid and non-lymphoid tissues (Banchereau and Steinman 1998; Reis e Sousa 2001). DCs are initially immature (iDC), but when pathogens trigger pattern-recognition receptors and other receptors on their surface, they phagocytose the pathogen and transform into mature DCs (mDCs). Transformation into mDCs is accompanied by loss of phagocytic capacity and acquisition of antigen presentation capability. These mDCs present peptides derived from the pathogen they had ingested on their surface, together with large quantities of cell surface major histocompatibility complexes (MHC) and co-stimulatory molecules. In this way, they can initiate primary T cell-mediated immune responses. It has become clear that DCs are not only excellent immune response stimulators, but are also potent immune inhibitors. Immature dendritic cells, including epidermal Langerhans cells, splenic marginal zone DCs, and interstitial DCs within non-lymphoid tissues, continuously sample self-antigen to maintain T cell self-tolerance. For a review, see Steinman et al. (2003).

Do iDCs use specific receptors for engulfing apoptotic cells? iDCs engulf apoptotic cells and are able to acquire antigens present in the dying cells (Lacy-Hulbert, Chap. 7; Divito and Morelli, Chap. 11, this Vol.; Albert et al. 1998a; Albert et al. 1998b; Berard et al. 2000; Nouri-Shirazi et al. 2000; Yrlid and Wick 2000). Rubartelli et al. (1997) suggested that iDCs may use an integrin, the vitronectin receptor $\alpha v\beta 3$, for apoptotic cell uptake. Albert et al. (1998b) showed that another integrin, $\alpha v\beta 5$, and the scavenger receptor CD36 both mediate apoptotic cell uptake by iDCs. Later, Nouri-Shirazi et al. (2000) suggested that different subpopulations of iDCs may use different integrins. We showed that iDCs increase apoptotic cell uptake in the presence of complement degradation products (Verbovetski et al. 2002), and that interaction of DCs with iC3b-opsonised apoptotic cells leads to generation of tolerizing DCs (Verbovetski et al. 2002). Tolerizing DCs were able to migrate to lymph nodes, because they express CCR7, they were resistant to toll-like and CD40 stimulation, and they inhibited NF-κB activation (Amarilyo et al., unpublished data). Thus clearance of apoptotic cells was shown to down-regulate the DC immune response following PAMP triggering.

A mechanism that may contribute to the generation of autoantibodies may be related to the failure to generate tolerogenic dendritic cells and/or B cells due to the fact that their interaction with apoptotic or necrotic cells occurs in the presence of autoantibodies that allow toll-like receptor triggering and B cell differentiation (for review see Mevorach 2004; Marshak-Rothstein 2006). Complement factors were shown to mediate efficient clearance of apoptotic cells by macrophages (Botto et al. 1998; Mevorach et al. 1998a), and also the generation of tolerogenic dendritic cells (Verbovetski et al. 2002). Complement deficiencies trigger the development of SLE in both in mice and humans.

In humans, two major DCs types were identified: myeloid DCs, including Langerhans cells and interstitial DCs, and plasmacytoid DCs (pDCs; Banchereau et al. 2000). It was suggested that patients with SLE have increased differentiation of precursor cells into monocyte-derived DCs due to the effect of IFN-γ produced by pDCs (Blanco et al. 2001). Myeloid DCs were found to be either spontaneously hypoactivated (Scheinecker et al. 2001; Koller et al. 2004) or hyperactivated (Decker et al. 2006; Ding, et al. 2006), with a decreased (Scheinecker et al. 2001; Koller et al. 2004) or increased (Ding et al. 2006) mixed lymphocyte reaction (MLR). However, their capacity for presentation and their function following interaction with apoptotic cells were not examined. We showed the generation of monocyte-derived DCs, their capacity to phagocytose iC3b-opsonised apoptotic cells, and their ability to function as tolerogenic DCs in SLE patients (Verbovetski et al. 2002; Berkun et al. 2008).

High quantities of MHC and costimulatory molecules trigger T cell proliferation. In contrast, when Death-Associated Molecular Patterns are triggered (Mevorach et al. 1998a), tolerogenic DCs are generated (Verbovetski et al. 2002), and interaction with autoreactive T cells will trigger T cell apoptosis or anergy.

The clearance of dying cells in SLE patients is probably altered (Herrmann et al. 1998; Shoshan et al. 2001; Baumann et al. 2002). For a review see Mevorach (2004). Accelerated leukocyte apoptosis, genetic or functional deficiencies of natural opsonins such as complement and CRP, the presence of abnormal opsonin-like T cell-derived autoantibodies, as well as phagocyte, monocyte, and macrophage dysfunction, may lead to accumulation of apoptotic debris and immune complexes of chromatin and other components of dying cells. In healthy individuals, apoptotic cells are tolerogenic and not immunogenic under most conditions (Fadok et al. 1998; Verbovetski et al. 2002; Mevorach 2004). However, when normal clearance is altered, e.g. in patients with SLE, B cells are dysfunctional and autoantibodies are present. Clearance of apoptotic cells may trigger auto-antibody production via co-stimulation of toll-like receptors 9 and 7, and possibly other pro-inflammatory receptors (Marshak-Rothstein 2006). In the current study, we demonstrate alterations in both iDC phenotype and tolerizing DC generation following apoptotic cell clearance.

Abnormally strong expression of MHC class II and co-stimulatory molecule were not consistently found in previous studies on SLE patients. Blood-derived myeloid DCs were reported to have low expression (Scheinecker et al. 2001). Koller et al. (2004) reported comparable expression of CD86 and even lower expression of MHC class II.

However, recent data obtained by Ding et al. (2006) and Decker et al. (2006) were more comparable to our findings with respect to MHC class II and co-stimulatory molecule expression. The results of the more recent studies by Ding et al. (2006) and Berkun et al. (2008) may differ from those of the Koller study due tothe presence of younger DC populations (day 5–7 vs. day 8 in the Koller study), or to maturation effects resulting from the use of non-autologous serum (FCS) to generate cells in the Koller study. In addition, as shown by Berkun et al. (2008) a significant portion of patients may have CD86 and MHC class II levels that are lower or comparable to those in healthy individuals. We believe that derivation of DCs in autologous human plasma, which was used only in the Berkun et al. (2008) study, is a more accurate method for generating DCs, because it resembles in vivo physiological processes.

On the basis of these and the previous studies, we conclude that abnormal CD86 and MHC class II expression occur often in SLE patients. In most patients, CD86 and MHC class II are up-regulated, but in some patients it is down-regulated. Up-regulation may imply accelerated spontaneous iDC maturation, analogous to maturation rates observed when monocyte-derived DCs are left in media for an extra day or two, and as reported in experiments with monocytes from SLE patients (Blanco et al. 2001).

This 'mature' phenotype may be partially responsible for the significantly decreased apoptotic cell clearance in some SLE patients, and specifically in patients with active SLE, as judged by high SLEDAI-2K (Systemic Lupus Erythematosus

Disease Activity Index 2000) score. Thus, accelerated acquisition of a mature phenotype, in which cells become antigen presenters rather than remaining phagocytic, may disrupt the natural generation of tolerogenic DCs. This hypothesis is consistent with the accelerated autologous mixed lymphocyte reaction (AMLR) and heterologous mixed lymphocytic reaction (HMLR) response seen in most SLE patients. Again, Koller et al. (2004) saw an AMLR decrease in lupus patients. However, they used relatively 'old' iDCs that may have exhausted their extra maturation capacity. Alternatively, some patients may have a lower AMLR response. In our experiments, and in studies performed by Ding et al. (2006), the majority of patients had accelerated AMLR.

Mammalian chromatin is considered a very weak adjuvant (Pisetsky 1992). Li et al. (2003) have shown that developing B cells expressing phosphatidylserine-reactive 3H9 are regulated by receptor editing in a manner similar to the anti-DNA-reactive 3H9. The authors proposed that the apoptotic cell surface is an active B cell tolerogen in vivo. However, triggering autoimmune BCR with a chromatin and anti-chromatin complex leads to a marked autoimmune response that involves BCR as well as TLR9 (Leadbetter et al. 2002). Vigilanti et al. (2003) showed that the apoptotic mammalian chromatin immune complex is an effective ligand in the context of BCR-mediated delivery to TLR9, and can stimulate B cell proliferation. Thus, whereas apoptotic cells and chromatin are tolerogenic for APCs such as B cells in healthy individuals, they may become immunogenic in the context of the immune complex of chromatin-antichromatin by triggering BCR and TLR9 (Viglianti et al. 2003). TLR9 is absent in monocyte-derived DCs, and immunogenicity via chromatin-antichromatin was shown only for plasmacytoid DCs (Means et al. 2005) that express TLR9. Thus, the trigger for immunogenicity of myeloid DCs in patients with SLE may be something other than B cells or plasmacytoid dendritic cells that express TLR9, and may be related to TLR7 or TLR8. Barrat et al. (2005) showed that mammalian DNA and RNA in the form of immune complexes are potent self antigens for TLR9 and TLR7, respectively, and induce IFN-α production by pDCs. Thus, impaired clearance of apoptotic material in SLE patients may play critical role in promotion of the disease through the induction of IFN-α by pDCs.

Vollmer et al. (2005) showed that specific, highly conserved RNA sequences within small nuclear ribonuclearprotein particles (snRNPs) can stimulate TLR7 and TLR8, as well as activate pDCs to secrete large amounts of IFN-α. SLE patient sera contain autoantibodies to snRNPs from immune complexes that are taken up through Fc RII, and that efficiently stimulate pDCs to secrete IFN-α. They demonstrated that prototype autoantigen, the snRNP, can directly stimulate innate immunity, and suggested that autoantibodies against snRNP may initiate SLE by stimulating TLR7/8.

Interestingly, clearance of apoptotic cells correlated with SLEDAI-2K score. The finding that ingestion of latex beads was not as disturbed as phagocytosis of apoptotic cells suggests a rather specific defect in apoptotic cell phagocytosis, rather then a general defect in phagocytosis, as we had suggested for macrophages (Shoshan et al. 2001). A cautionary note must be raised, as apoptotic cell clearance studies are conducted in vitro. The situation in vivo may be worse, considering that apoptotic cells may be opsonised by autoantibodies, that IFN-α is present, and that there is a decrease in both complement and CRP, both of which are important for appropriate apoptotic cell clearance (Mevorach et al. 1998a; Gershov et al. 2000).

10.5 Drug Induced Lupus (DIL)

Despite a considerable body of research, the mechanism by which multiple chemically diverse medications cause DIL remains unknown. It was once assumed that medications may complex with cell membranes, inducing autoantibody production, in a mechanism analogous with penicillin-induced hemolytic anemia (Alarcon-Segovia 1976). Though appealing in its simplicity, later research has not substantiated this hypothesis.

Considerable attention was later directed toward drug-induced DNA hypomethylation, which leads to T cell activation, as a possible mechanism of DIL. This effect has been described for various drugs, including procainamide and hydralazine (Cornacchia et al. 1988). It was demonstrated in a murine model that treating CD4 T cells with DNA methylation inhibitors, such as procainamide, is sufficient to induce a lupus-like syndrome (Quddus et al. 1993). Various medications may inhibit DNA methylation by different mechanisms, and this inhibition could serve as a final common pathway leading to autoimmune activation culminating in clinical DIL.

As shown in this review, recent research has implicated impaired handling and removal of apoptotic cells in the pathogenesis of SLE. The assumption is that appropriate mononuclear cells must by phagocytosis efficiently clear cells undergoing apoptosis, a non-inflammatory process that depends on the presence of complement components. Disruption of this process may lead to accumulation of apoptotic cell remnants at various stages, thus instigating an inflammatory, autoimmune reaction.

Following treatment with an established drug that induces DIL, such as infliximab (D'Auria et al. 2004), there is accumulation of apoptotic material, such as nucleosomes, which indicates that DIL may be linked to impaired clearance of such material secondary to anti-TNFα treatment.

On the basis of this knowledge, we have raised the possibility that DIL could be the result of drugs impairing the uptake of apoptotic cells. In a murine model we have shown that both procainamide and quinidine inhibit uptake of syngeneic apoptotic thymocytes by peritoneal macrophages (Ablin et al. 2005). Similar results were achieved when macrophages were presented with necrotic thymocytes, while uptake of innate latex beads was not decreased, thus implying a relatively specific effect of these drugs on the uptake of dying cells.

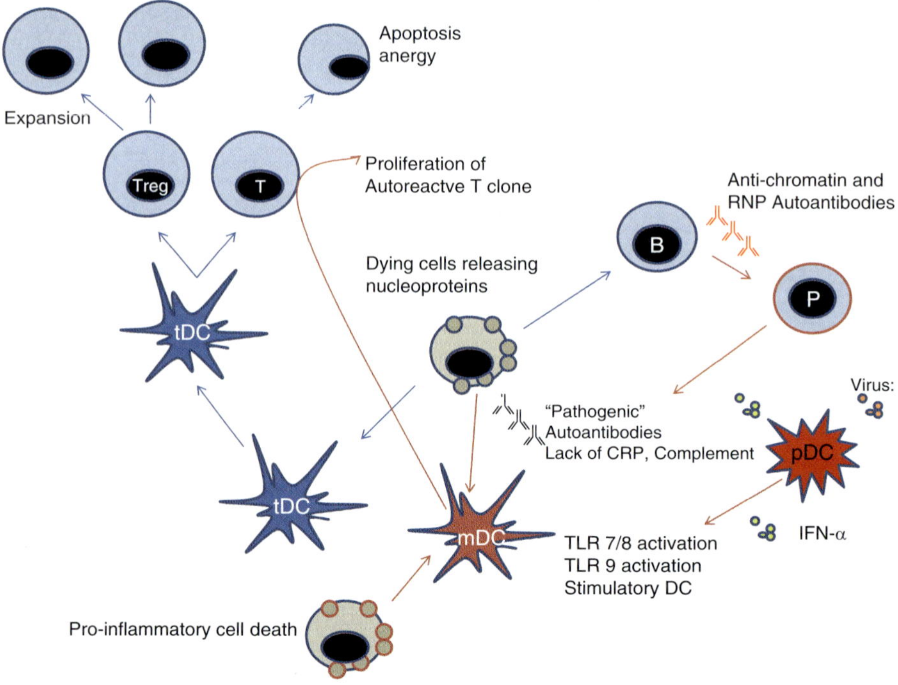

Fig. 10.2 Abnormal clearance of dying cells in SLE Altered clearance of apoptotic cells or proinflammtory death in the context of antichromatin, RNP, and other nucleoproteins and autoantibodies activates immature dendritic cells (TRL3 and TRL7) and/or B cells (TLR9), resulting in T cell proliferation and autoantibody production. Viral infection and other environmental factors can serve as additional triggers. The normal tolerogenic dendritic cell phenotype (tDC) is avoided, and peripheral tolerance leading to autoreactive T cells and Tregs expansion is disturbed. B, B cell; mDC, mature dendritic cells; P, plasma cell; pDCs, plasmacytoid dendritic cells; mDC, mature dendritic cells; T, T cell; tDC, tolerogenic dendritic cells; Treg, T regulatory cells.

10.6 Conclusions

Clearance of dying cells is most likely altered in SLE patients, and it might have a role in drug-induced lupus as well. Many different factors may lead to increased levels of apoptotic debris and immune complexes of chromatin and other components of dying cells. These potential factors include accelerated leukocyte apoptosis, genetic or functional deficiencies of natural opsonins such as complement and CRP, the presence of abnormal opsonin-like T cell-derived autoantibodies, as well as phagocyte, monocyte, macrophage, and dendritic cell dysfunction. These events may result in an autoimmune response and persistence of inflammation (summarized in Fig. 10.2). Environmental factors, such as viral infections, as well as intrinsic factors may play a triggering role in the exacerbation of SLE.

Acknowledgments The author wishes to thank Mrs. Shifra Fraifeld for her assistance in the preparation.

References

Ablin J, Verbovetski I, Trahtemberg U et al (2005) Quinidine and procainamide inhibit murine macrophage uptake of apoptotic and necrotic cells: A novel contributing mechanism of drug-induced-lupus. Apoptosis 10(5):1009–1018

Aderem AA, Wright SD, Silverstein SC et al (1985) Ligated complement receptors do not activate the arachidonic acid cascade in resident peritoneal macrophages. J Exp Med 161(3):617–622

Alarcon-Segovia D (1976) Drug-induced antinuclear antibodies and lupus syndromes. Drugs 12(1):69–77

Albert ML, Darnell JC, Bender A et al (1998a) Tumour-specific killer cells in paraneoplastic cerebellar degeneration. Nat Med 4(11):1321–1324

Albert ML, Pearce SF, Francisco LM et al (1998b) Immature dendritic cells phagocytose apoptotic cells via alphavbeta5 and cd36, and cross-present antigens to cytotoxic t lymphocytes. J Exp Med 188(7):1359–1368

Amarilyo G, Verbovetski I, Atallah M et al Cd11b/cd18 mediates a distinct anti-inflammatory response and transcriptional nf-kappab-dependent blockade following interaction, but not engulfment, with ic3b-opsonized apoptotic cells. Submitted 2008

Anderson HA, Maylock CA, Williams JA et al (2003) Serum-derived protein s binds to phosphatidylserine and stimulates the phagocytosis of apoptotic cells. Nat Immunol 4(1):87–91

Arbuckle MR, McClain MT, Rubertone MV et al (2003) Development of autoantibodies before the clinical onset of systemic lupus erythematosus. N Engl J Med 349(16):1526–1533

Banchereau J, Pulendran B, Steinman R et al (2000) Will the making of plasmacytoid dendritic cells in vitro help unravel their mysteries? J Exp Med 192(12):F39–F44

Banchereau J, Steinman RM (1998) Dendritic cells and the control of immunity. Nature 392(6673):245–252

Barrat FJ, Meeker T, Gregorio J et al (2005) Nucleic acids of mammalian origin can act as endogenous ligands for toll-like receptors and may promote systemic lupus erythematosus. J Exp Med 202(8):1131–1139

Baumann I, Kolowos W, Voll RE et al (2002) Impaired uptake of apoptotic cells into tingible body macrophages in germinal centers of patients with systemic lupus erythematosus. Arthritis Rheum 46(1):191–201

Berard F, Blanco P, Davoust J et al (2000) Cross-priming of naive cd8 t cells against melanoma antigens using dendritic cells loaded with killed allogeneic melanoma cells. J Exp Med 192(11):1535–1544

Berkun Y, Verbovetsky I, Ben-Ami A et al (2008) Altered tolerizing dendritic cells in patients with systemic lupus erythematosus. Eur J Immunol. In print.

Bickerstaff MC, Botto M, Hutchinson WL et al (1999) Serum amyloid p component controls chromatin degradation and prevents antinuclear autoimmunity. Nat Med 5(6):694–697

Blanco P, Palucka AK, Gill M et al (2001) Induction of dendritic cell differentiation by ifn-alpha in systemic lupus erythematosus. Science 294(5546):1540–1543

Botto M, Dell Agnola C, Bygrave AE et al (1998) Homozygous c1q deficiency causes glomerulonephritis associated with multiple apoptotic bodies. Nat Genet 19(1):56–59

Brown S, Heinisch I, Ross E et al (2002) Apoptosis disables cd31-mediated cell detachment from phagocytes promoting binding and engulfment. Nature 418(6894):200–203

Carroll MC (2004) A protective role for innate immunity in systemic lupus erythematosus. Nat Rev Immunol 4(10):825–831

Casciola-Rosen LA, Anhalt G, Rosen A (1994) Autoantigens targeted in systemic lupus erythematosus are clustered in two populations of surface structures on apoptotic keratinocytes. J Exp Med 179(4):1317–1330

Chen W, Frank ME, Jin W et al (2001) Tgf-beta released by apoptotic t cells contributes to an immunosuppressive milieu. Immunity 14(6):715–725

Chen Z, Koralov SB, Kelsoe G (2000) Complement c4 inhibits systemic autoimmunity through a mechanism independent of complement receptors cr1 and cr2. J Exp Med 192(9):1339–1352

Chiarugi A (2005) Simple but not simpler Toward a unified picture of energy requirements in cell death. FASEB J 19(13):1783–1788

Chung S, Gumienny TL, Hengartner MO et al (2000) A common set of engulfment genes mediates removal of both apoptotic and necrotic cell corpses in c. Elegans. Nat Cell Biol 2(12):931–937

Cornacchia E, Golbus J, Maybaum J et al (1988) Hydralazine and procainamide inhibit t cell DNA methylation and induce autoreactivity. J Immunol 140(7):2197–2200

D Auria F, Rovere-Querini P, Giazzon M et al (2004) Accumulation of plasma nucleosomes upon treatment with anti-tumour necrosis factor-alpha antibodies. J Intern Med 255(3):409–418

Dancey JT, Deubelbeiss KA, Harker LA et al (1976) Neutrophil kinetics in man. J Clin Invest 58(3):705–715

Darzynkiewicz Z, Juan G, Li X et al (1997) Cytometry in cell necrobiology: Analysis of apoptosis and accidental cell death (necrosis). Cytometry 27(1):1–20

Decker P, Kotter I, Klein R et al (2006) Monocyte-derived dendritic cells over-express cd86 in patients with systemic lupus erythematosus. Rheumatology (Oxford) 45(9):1087–1095

Diamond B, Katz JB, Paul E et al (1992) The role of somatic mutation in the pathogenic anti-DNA response. Annu Rev Immunol 10731–10757

Ding D, Mehta H, McCune WJ et al (2006) Aberrant phenotype and function of myeloid dendritic cells in systemic lupus erythematosus. J Immunol 177(9):5878–5889

Duperrier K, Eljaafari A, Dezutter-Dambuyant C et al (2000) Distinct subsets of dendritic cells resembling dermal dcs can be generated in vitro from monocytes, in the presence of different serum supplements. J Immunol Methods 238(1–2):119–131

Elward K, Griffiths M, Mizuno M et al (2005) Cd46 plays a key role in tailoring innate immune recognition of apoptotic and necrotic cells. J Biol Chem 280(43):36342–36354

Emlen W, Niebur J, Kadera R (1994) Accelerated in vitro apoptosis of lymphocytes from patients with systemic lupus erythematosus. J Immunol 152(7):3685–3692

Enari M, Sakahira H, Yokoyama H et al (1998) A caspase-activated dnase that degrades DNA during apoptosis, and its inhibitor icad. Nature 391(6662):43–50

Fadok VA, Bratton DL, Konowal A et al (1998) Macrophages that have ingested apoptotic cells in vitro inhibit pro-inflammatory cytokine production through autocrine/paracrine mechanisms involving tgf-beta, pge2, and paf. J Clin Invest 101(4):890–898

Fadok VA, Voelker DR, Campbell PA et al (1992) Exposure of phosphatidylserine on the surface of apoptotic lymphocytes triggers specific recognition and removal by macrophages. J Immunol 148(7):2207–2216

Fishelson Z, Attali G, Mevorach D (2001) Complement and apoptosis. Mol Immunol 38(2–3):207–219

Gaipl US, Kuenkele S, Voll RE et al (2001) Complement binding is an early feature of necrotic and a rather late event during apoptotic cell death. Cell Death Differ 8(4):327–334

Gershov D, Kim S, Brot N et al (2000) C-reactive protein binds to apoptotic cells, protects the cells from assembly of the terminal complement components, and sustains an anti-inflammatory innate immune response: Implications for systemic autoimmunity. J Exp Med 192(9):1353–1364

Godson C, Mitchell S, Harvey K et al (2000) Cutting edge: Lipoxins rapidly stimulate non-phlogistic phagocytosis of apoptotic neutrophils by monocyte-derived macrophages. J Immunol 164(4):1663–1667

Hargraves MM (1949) Production in vitro of the le cell phenomenon: Use of normal bone marrow elements and blood plasma from patients with acute disseminated lupus erythematosus. Proc Staff Meet Mayo Clin 24234–24237

Hargraves MM, Richmond H, Morton R (1948) Presentation of two bone marrow elements; the tart cell and le cell. Proc Staff Meet Mayo Clin 2325–2328

Haserick JR, Bortz DW (1949) Normal bone marrow inclusion phenomena induced by lupus erythematosus plasma. J Invest Dermatol 1347–1349

Henson PM, Bratton DL, Fadok VA (2001) Apoptotic cell removal. Curr Biol 11(19):R795–R805

Herrmann M, Voll RE, Zoller OM et al (1998) Impaired phagocytosis of apoptotic cell material by monocyte-derived macrophages from patients with systemic lupus erythematosus. Arthritis Rheum 41(7):1241–1250

Huynh ML, Fadok VA, Henson PM (2002) Phosphatidylserine-dependent ingestion of apoptotic cells promotes tgf-beta1 secretion and the resolution of inflammation. J Clin Invest 109(1):41–50

Jaworowski A, Crowe SM (1999) Does hiv cause depletion of cd4+ t cells in vivo by the induction of apoptosis? Immunol Cell Biol 77(1):90–98

Kawane K, Fukuyama H, Yoshida H et al (2003) Impaired thymic development in mouse embryos deficient in apoptotic DNA degradation. Nat Immunol 4(2):138–144

Kim SJ, Gershov D, Ma X et al (2002) I-pla(2) activation during apoptosis promotes the exposure of membrane lysophosphatidylcholine leading to binding by natural immunoglobulin m antibodies and complement activation. J Exp Med 196(5):655–665

Koller M, Zwolfer B, Steiner G et al (2004) Phenotypic and functional deficiencies of monocyte-derived dendritic cells in systemic lupus erythematosus (sle) patients. Int Immunol 16(11):1595–1604

Kono DH, Theofilopoulos AN (2006) Genetics of sle in mice. Springer Semin Immunopathol 28(2):83–96

Krispin A, Bledi Y, Atallah M et al (2006) Apoptotic cell thrombospondin-1 and heparin-binding domain lead to dendritic-cell phagocytic and tolerizing states. Blood 108(10):3580–3589

Kristjansdottir H, Bjarnadottir K, Hjalmarsdottir IB et al (2000) A study of c4aq0 and mhc haplotypes in icelandic multicase families with systemic lupus erythematosus. J Rheumatol 27(11):2590–2596

Kroemer G, Martin SJ (2005) Caspase-independent cell death. Nat Med 11(7):725–730

Lachmann PJ, Walport MJ (1987) Deficiency of the effector mechanisms of the immune response and autoimmunity. Ciba Found Symp 129149–129171

Lauber K, Bohn E, Krober SM et al (2003) Apoptotic cells induce migration of phagocytes via caspase-3-mediated release of a lipid attraction signal. Cell 113(6):717–730

Leadbetter EA, Rifkin IR, Hohlbaum AM et al (2002) Chromatin-igg complexes activate b cells by dual engagement of igm and toll-like receptors. Nature 416(6881):603–607

Levine B, Deretic V (2007) Unveiling the roles of autophagy in innate and adaptive immunity. Nat Rev Immunol 7(10):767–777

Li H, Jiang Y, Cao H et al (2003) Regulation of anti-phosphatidylserine antibodies. Immunity 18(2):185–192

Maiuri MC, Zalckvar E, Kimchi A et al (2007) Self-eating and self-killing: Crosstalk between autophagy and apoptosis. Nat Rev Mol Cell Biol 8(9):741–752

Makrides SC (1998) Therapeutic inhibition of the complement system. Pharmacol Rev 50(1):59–87

Manzi S, Rairie JE, Carpenter AB et al (1996) Sensitivity and specificity of plasma and urine complement split products as indicators of lupus disease activity. Arthritis Rheum 39(7):1178–1188

Marshak-Rothstein A (2006) Toll-like receptors in systemic autoimmune disease. Nat Rev Immunol 6(11):823–835

Marth T, Strober W, Seder RA et al (1997) Regulation of transforming growth factor-beta production by interleukin-12. Eur J Immunol 27(5):1213–1220

Means TK, Latz E, Hayashi F et al (2005) Human lupus autoantibody-DNA complexes activate dcs through cooperation of cd32 and tlr9. J Clin Invest 115(2):407–417

Mevorach D (2000) Opsonization of apoptotic cells. Implications for uptake and autoimmunity. Ann N Y Acad Sci 926226–926235

Mevorach D (2004) The role of death-associated molecular patterns in the pathogenesis of systemic lupus erythematosus. Rheum Dis Clin North Am 30(3):487–504, viii

Yamamoto K, Johnston RB Jr. (1984) Dissociation of phagocytosis from stimulation of the oxidative metabolic burst in macrophages. J Exp Med 159(2):405–416

Yrlid U, Wick MJ (2000) Salmonella-induced apoptosis of infected macrophages results in presentation of a bacteria-encoded antigen after uptake by bystander dendritic cells. J Exp Med 191(4):613–624

Zong WX, Thompson CB (2006) Necrotic death as a cell fate. Genes Dev 20(1):1–15

Chapter 11
Apoptotic Cells for Therapy of Transplant Rejection

Sherrie J. Divito and Adrian E. Morelli

Abstract: The ability of apoptotic cells to serve as a source of alloantigen and to exert immune-regulatory effects on the innate and adaptive immune responses make apoptotic cell-based therapies a promising tool to down-regulate the anti-donor response that causes rejection of transplanted organs. Advances in understanding the mechanisms of interaction between apoptotic cells and antigen-presenting cells, including dendritic cells and macrophages, have provided the basis for the therapeutic harnessing of cells in the early stages of apoptosis to restrain the immune response or promote immunological tolerance. In this chapter, we compiled the available information on the mechanism(s) by which apoptotic cells interact with cells of the immune system and down-regulate the adaptive immune response, the effect of apoptotic cells on organ allograft survival in experimental models, and the potential application of apoptotic cell-based therapies for prevention or treatment of graft rejection and autoimmune disorders.

Keywords: Apoptotic cells • Transplantation • Graft rejection • Dendritic cells

11.1 Introduction

Organ transplantation is becoming an increasingly important and common surgical procedure with nearly 30,000 performed each year in the U.S. (based on OPTN data as of Jan. 4, 2008). In most cases, transplantation surgery serves as the only life-saving treatment available for end-stage organ disease. Although the constant develop-

---

A. E. Morelli (✉) · S. J. Divito
Thomas E. Starzl Transplantation Institute and
Departments of Surgery and Immunology
University of Pittsburgh Medical Center
W1556. Biomedical Science Tower
Pittsburgh, PA 15261, USA
Tel.: 412-624-2193
Fax: 412-624-1172
E-mail: morelliae@upmc.edu

D.V. Krysko, P. Vandenabeele (eds.), *Phagocytosis of Dying Cells*,
DOI: 10.1007/978-1-4020-9292-3_11, © Springer Science+Business Media B.V. 2009

ment of new and more effective immunosuppressive drugs and the better knowledge of their therapeutic application have revolutionized the prevention and treatment of acute graft rejection, these drugs have significant toxicity and greatly increase patient susceptibility to malignant neoplasias and infections. Further, the implementation of immunosuppressive agents has exerted little impact on the incidence of chronic graft rejection, and therefore overall long-term graft survival has only improved modestly. Thus, it is necessary to develop novel cell- or vesicle-based therapies able to down-regulate, in a donor-specific manner, the immune response that causes transplant rejection, without inducing generalized immune-suppression and its harmful side-effects. The discovery that apoptotic cells exert potent anti-inflammatory and immune-regulatory effects under certain conditions has paved the way for a new field of research specialized in the impact of apoptotic cells on the immune system in physiologic and pathologic conditions and the potential therapeutic use of apoptotic cells in autoimmune disorders and transplant rejection.

The concept of using apoptotic cells therapeutically to prevent, delay or ameliorate the anti-donor immune response that causes transplant rejection has received great attention over the past decade. This idea is based on the facts that (i) apoptotic cells are a source of self- or foreign-antigen (Ag) for Ag-presenting cells (APC), particularly for the quiescent (semi-mature) dendritic cells (DC) of secondary lymphoid organs; and (ii) cells in early stages of apoptosis deliver an inhibitory signal that interferes with maturation/activation of APC and up-regulation of the co-stimulatory signals required for full activation of T cells. Significant research efforts have partially illuminated how APC reprocess apoptotic cell fragments into peptides for presentation to T cells, the receptor-ligand interactions between APC and apoptotic cells, the resulting down-stream signalling events that follow and the mechanisms by which apoptotic cells down-regulate the immune response. Importantly, a few studies in murine models have demonstrated the efficacy of apoptotic cell-based therapies to prolong allograft survival and have partially unveiled the mechanisms by which apoptotic cells down-regulate the anti-donor T cell response. Further, there is indirect evidence from clinical studies that suggests that administration of cells in early stages of apoptosis may delay rejection of solid organ allografts in humans. Although more investigation is needed to elucidate the mechanism(s) of action of apoptotic cell-based therapies and to compare their efficacy with other forms of cell-based and pharmacological treatments, the use of apoptotic cells appears promising as a therapeutic against transplant rejection.

11.2 Apoptotic Cells and the Immune System

As a consequence of the normal cell turnover that occurs daily in our body, billions of dead cells are generated without causing inflammation. Conditions that induce apoptosis such as moderate ultraviolet-B light (UV-B) irradiation, certain viral infections and malignant tumours are also accompanied by lack of inflammation and adaptive immunity. Originally, this was attributed to the rapid clearance of apoptotic cells that would prevent the release of toxic cellular components into the micro-environment. However, Voll and colleagues (1997) demonstrated that internalization of apoptotic cells suppresses "actively" the inflammatory response by delivering inhibitory signals to the phagocytes (Ucker, Chap. 6; Lacy-Hulbert, Chap. 7; Gregory and Pound, Chap. 9, this Vol.). It was further shown that endocytosis of early apoptotic cells by immature DC prevents DC activation in both humans and rodents, as the APC fail to up-regulate expression of MHC class-II and co-stimulatory molecules CD80, CD86, CD40 and (in humans) CD83, despite subsequent stimulation with DC-activating factors such as lipopolysaccharride (LPS), CD40-ligation, TNF-α and monocyte-conditioned medium (Gallucci et al. 1999; Urban et al. 2001; Stuart et al. 2002; Verbovestski et al. 2002; Takahashi and Kobayashi 2003). The inhibitory effect of apoptotic cells was not simply a result of the process of phagocytosis, as activation of immature DC was not impaired after ingestion of control latex beads that were similar in size to apoptotic cell fragments (Stuart et al. 2002).

11.2.1 *Apoptotic Cells and Peripheral T Cell Tolerance*

Although most thymocytes that recognize self-peptides with high affinity are eliminated centrally in the thymus through negative selection, a percentage of self-reactive T cells escape thymic deletion and access the periphery. An efficient mechanism of peripheral T cell tolerance is therefore necessary to prevent activation of self-reactive T cells and avoid autoimmunity. In the periphery, the "two-signal" hypothesis states that APC presenting MHC molecules loaded with antigenic peptides (signal 1) together with low levels (or absence) of co-stimulation (signal 2) triggers defective activation of Ag-specific T cells (reviewed in Steinman and Nussenzweig 2002; Steinman et al. 2003). Incomplete activation of T cells results in poor cellular proliferation followed by deletion, anergy, and likely differentiation/expansion of regulatory T cells (Treg), all mechanisms that lead to T cell hypo-response or tolerance.

Self-Ag derived from apoptotic cells (resulting from the cell turnover that takes place in the steady-state) is constantly sampled by quiescent DC that migrate constitutively from peripheral tissues to lymph nodes (LN) and spleen. Quiescent DC are defined as semi-mature APC with intermediate levels of co-stimulatory molecule expression (CD86^{int}) and therefore are unable to properly activate T cells. In other words, quiescent DC present in a sub-optimal manner apoptotic cell-derived (self-) peptides to auto-reactive T cells (Chernysheva et al. 2002; reviewed in Steinman and Nussenzweig 2002; Steinman et al. 2003). In support of this idea, it was reported that ingestion of apoptotic cells causes DC to decrease expression of the chemokine receptor CCR5 and increase levels of CCR7, which indicates that the DC remain semi-mature but acquire the ability to home to draining secondary lymphoid organs in response to the chemokine MIP-3β (CCL19; Verbovetski et al. 2002; Ip and Lau 2004). This idea has been confirmed in vivo by Huang et al. (2000) who demonstrated that intestinal DC that have internalized apoptotic cell fragments (derived from intestinal epithelial cells) migrate to mesenteric LN. Using gnotobiotic rats under germ-free conditions, they further showed that intestinal DC migration occurred independently of the presence of DC-maturation stimuli derived from the intestinal bacterial flora.

In summary, since quiescent DC that migrate from peripheral tissues express low levels of co-stimulatory signals, it is believed that auto-reactive T cells in secondary lymphoid organs receive sub-threshold stimulation resulting in defective activation that leads to T cell tolerance (reviewed in Steinman and Nussenzweig 2002; Steinman et al. 2003).

11.2.2 Apoptotic Cells Are Phagocytosed by Different Subsets of Antigen-Presenting Cells

In normal conditions, apoptotic cells are removed rapidly by phagocytes that include different subpopulations of APC (Ucker, Chap. 6; Lacy-Hulbert, Chap. 7; Gregory and Pound, Chap. 9, this Vol.). DC are unique "professional" APC able to prime naïve T cells. By contrast, "amateur or non-professional" APC like macrophages, B cells and endothelial cells can only activate effector and memory T lymphocytes, which have fewer requirements for re-stimulation than naïve cells.

Most of the research on apoptotic cell-based therapies in transplantation has been conducted using young inbred mice maintained in clean or nearly pathogen-free conditions, which therefore may contain significantly lower numbers of memory T cells compared to outbred animals. However, transplant rejection in humans is mediated by both naïve and memory T cells and as such, the ability of apoptotic cells to tolerize not only DC but also other non-professional APC capable of activating anti-donor memory T cells may be critical for successful therapy.

A further complication to investigating the impact of apoptotic cell-derived peptide presentation by DC and the resulting outcome on the T cell response is the existence of different DC subpopulations, whose roles in induction/maintenance of T cell tolerance are still unclear. In mice, three subpopulations of DC (all CD11c[hi]) have been described in the spleen and LN: CD8α^+ DC, CD4$^+$CD8$^-$ DC and CD4$^-$ CD8$^-$ DC. In the steady-state, the CD8α^+ DC (known as lymphoid-related DC) represent 20–30% of the DC population and are quiescent/semi-mature APC (CD86[int]) located in the T cell-dependent areas (Vremec et al. 2000). The CD8$^-$ DC (known as myeloid-DC) consist of immature DC located in the marginal zone and throughout the red pulp of the spleen. LN are also populated by two other subsets of migrating DC derived from epidermal Langerhans cells from the skin and interstitial DC from peripheral tissues, which traffic from periphery to LN through lymphatic circulation (Henri et al. 2001).

More than two decades ago, Fossum and Rolstad (1986) demonstrated that DC located in T cell areas of LN and spleen were capable of internalizing fragments derived from allogeneic lymphocytes killed by host natural killer (NK) cells. Further work showed that DC located in peripheral tissues were also capable of internalizing apoptotic cells, such is the case of Langerhans cells of the vaginal epithelium that internalize apoptotic epithelial cells during late metestrus and early diestrus in mice (Parr et al. 1991). These peripheral tissue-resident DC can easily migrate and transport the internalized apoptotic cell fragments and their antigenic load to secondary lymphoid organs for presentation to T cells.

Spleen-resident DC have received the most attention for their ability to capture circulating apoptotic cells and down-regulate the T cell response against apoptotic cell-derived peptides. However, the individual contribution of each subpopulation of splenic DC to the capture, processing, and presentation of blood-borne apoptotic cells is not completely understood. Iyoda and colleagues (2002) have shown that apoptotic leukocytes generated by osmotic shock and administered intravenously are internalized exclusively by $CD8\alpha^+$ DC of the T cell areas of the spleen. Conversely, our work has demonstrated that UV-B-induced apoptotic leukocytes are captured mainly by splenic $CD8^-$ DC of the marginal zone and to a lesser extent by $CD8\alpha^+$ DC of the T cell areas (Morelli et al. 2003). Intriguingly, we have shown that intravenous (i.v.) treatment of naïve recipients with splenic DC isolated from mice pre-treated (i.v., 24-26 hours before) with donor apoptotic cells, prolongs heart allograft survival in those treated mice, but that depletion of $CD8\alpha^+$ DC from the transferred inoculum abrogates the effect (Wang et al. 2006). This finding suggests that presentation of donor allogeneic Ag (alloAg) by splenic $CD8\alpha^+$ DC is necessary for the beneficial effect of donor apoptotic cells in allograft survival. It is likely that, following i.v. administration, apoptotic cells are trapped by $CD8^-$ DC and other phagocytes of the splenic marginal zone rather than actively migrating to the T cell areas for internalization by $CD8\alpha^+$ DC. Alternatively, marginal zone $CD8^-$ DC with internalized apoptotic cells could migrate to the T cell zone and transfer the apoptotic cell fragments to $CD8\alpha^+$ DC, or (although less likely) $CD8^-$ DC with the apoptotic cell fragments could acquire $CD8\alpha$ expression during their mobilization to the T cell areas. Another possibility is that the stage of apoptosis in the injected cells dictates what subset of DC internalize the apoptotic cell in vivo. In peripheral blood circulation, dying cells or living cells that have been primed to die may be capable of actively migrating to the T cell areas for internalization by $CD8\alpha^+$ DC, whereas blood-borne apoptotic cells unable to actively traffic are captured by marginal zone $CD8^-$ DC.

Plasmacytoid DC (pDC), a more recently described subpopulation of APC, could potentially interact with apoptotic cells to induce T cell tolerance. In mice, pDC or their precursors facilitate bone marrow engraftment (Fugier-Vivier et al. 2005) and together with CD154 monoclonal antibody (Ab) therapy prolong significantly solid organ allograft survival (Abe et al. 2005; Bjorck et al. 2005). In autoimmunity, administration of pDC expressing transgenic proinsulin prevented onset of type 1 diabetes in NOD mice (Steptoe et al. 2005) and pDC promoted generation of Treg that prevented autoimmunity in lupus-prone mice (Kang et al. 2007). However, it is unclear whether pDC interact with apoptotic cells to down-regulate the T cell response. In this regard, Dalgaard and colleagues (2005) have shown that, unlike the classic myeloid CD11c^{hi} DC, pDC (CD11c^{int}) do not phagocytose apoptotic or necrotic cells in vitro.

(Kobayashi et al. 2007). This study also demonstrated that human and mouse DC express TIM-4, although whether TIM-4 expression contributes to apoptotic cell phagocytosis by DC remains unexplored. Other receptors such as CD14 and soluble opsonins also bind externalized PS and other ligands expressed on apoptotic cells (Gregory 2000), and such binding triggers internalization of apoptotic cells by phagocytes.

Although most work studying PS receptors has been performed in macrophages, mouse splenic DC have been shown to transcribe mRNA for SR-AII, SR-BI and CD14 (Iyoda et al. 2002) although it is unknown whether these transcripts are all translocated into functional proteins. However, human monocyte-derived DC do alter their phenotype and function following incubation with liposomes containing PS, suggesting that DC must express receptors for PS (Chen et al. 2004). Likewise, opsonization of apoptotic cells with β2-glycoprotein I (β2-GPI), an aminophospholipid ligand that recognizes externalized PS, facilitates apoptotic cell phagocytosis by DC (Balasubramanian and Schroit 1998) and enhances presentation of MHC class-II-restricted apoptotic cell-derived peptides by DC to CD4 T cells (Rovere et al. 1999).

Additional ligands are expressed by or attached to the surface of apoptotic cells to facilitate their phagocytosis. Apoptotic cells expose anionic sites on their surface that allows for deposition of thrombospondin-1 (TSP-1), a platelet-derived multifunctional trimeric glycoprotein. TSP-1 serves as a bridge linking the apoptotic cell to integrins $\alpha_v\beta_3$ and $\alpha_v\beta_5$, and to CD36 expressed by the phagocyte. CD36 actually amplifies $\alpha_v\beta_3$ - TSP-1 recognition and thus increases phagocytosis of apoptotic cells (Ren et al. 1995; Savill et al. 1992). Interestingly, human monocyte-derived DC decrease surface expression of $\alpha_v\beta_5$ and CD36 following maturation in vitro, co-relating with the reduced capacity of mature DC to phagocytose apoptotic cells (Albert et al. 1998a).

Complement factors bound to the surface of apoptotic cells increase apoptotic cell phagocytosis (Napirei and Mannherz, Chap. 4; Mevorach, Chap. 10, this Vol.;Takizawa et al. 1996; Mevorach et al. 1998). IgM Ab bind lysophosphatidylcholine on the surface of apoptotic cells (Kim et al. 2002, Zwart et al. 2004), and C1q, the first component of the classical complement pathway then binds IgM. The classical pathway generates C3-derived opsonins that are recognized by phagocytes through complement receptors (CR). In vitro, human monocyte-derived DC and mouse splenic DC use the iC3b receptors CR3 and CR4 to take up iC3b-opsonized apoptotic cells (Verbovetski et al. 2002; Morelli et al. 2003). Alternatively, the globular head of C1q may directly bind to ligands clustered on the apoptotic cell surface (Korb and Ahearn 1997; Navratil et al. 2001), which could trigger complement activation (Nauta et al. 2002) or the collagen-like tail of C1q could bind calreticulin that is attached to the endocytic receptor CD91 on the surface of the phagocyte (Ogden et al. 2001). Similarly, mannose-binding lectin (MBL) and lung surfactant proteins A and D are collectins structurally related to C1q. One study showed that MBL enhances uptake of apoptotic cells by immature DC (Nauta et al. 2003a; Nauta et al. 2004). This is likely due to the ability of MBL and surfactant proteins to attach their globular domains directly to the apoptotic cell surface and their collagen-like tails to calreticulin/CD91 on the phagocyte surface (Ogden et al. 2001; Schagat et al. 2001; Vandivier et al. 2002a). Mice rendered hypocomplementeric following injection of cobra venom factor had a significantly impaired ability of splenic DC to phagocytose circulating apoptotic cells (Morelli et al. 2003). Cobra venom factor results in severe inactivation of C3, and thus affects all downstream factors in the complement cascade, so complement factors deriving from C3 or later in the cascade are central to apoptotic cell phagocytosis by DC in vivo (Morelli et al. 2003).

Pentraxins including C reactive protein (CRP), serum amyloid P (SAP) and pentraxin 3 (PTX3) are acute-phase proteins secreted in response to inflammation that bind the surface of apoptotic cells (Gershov et al. 2000; Rovere et al. 2000; Familian et al. 2001; reviewed in Nauta et al. 2003b). In addition to binding anionic phospholipids, CRP binds small nuclear ribonucleoproteins and SAP binds chromatin/nucleolar components that mobilize to the cell surface during apoptosis (Pepys and Butler 1987; Butler et al. 1990; Hicks et al. 1992). Both SAP and CRP increase apoptotic cell phagocytosis by macrophages in the absence of complement, supporting the existence of specific receptors for pentraxins, and evidence also suggests that pentraxins may bind Fcγ receptors (Bharadwaj et al. 2001; Mold et al. 2002).

Milk-fat globule protein epidermal growth factor 8 (MFG-E8), also known as lactadherin, participates in the apoptosis that takes place during mammary gland involution. MFG-E8 binds to anionic phospholipids and serves as a bridge between externalized PS and the $\alpha_v\beta_3$ integrin expressed by phagocytes (reviewed in Wu et al. 2006). It has been shown that mouse immature DC and Langerhans cells synthesize MFG-E8 and that mice lacking MFG-E8 have a markedly reduced ability of DC to phagocytose apoptotic cells (Miyasaka et al. 2004). Further, MFG-E8 expression is reduced in LPS-matured DC (Miyasaka et al. 2004).

The surface expression of apoptotic cell receptors by phagocytes does offer insight into the types of DC responsible for apoptotic cell phagocytosis. Only splenic CD8α$^+$ DC express high levels of CD36 on their surface (Belz et al. 2002), and in humans, cross-presentation of apoptotic cell-derived Ag in vitro depends on phagocytosis of apoptotic cell fragments by DC via CD36 and $\alpha_v\beta_3$ /β_5 integrins (Rubartelli et al. 1997; Albert et al. 1998a), thus explaining the ability of CD8α$^+$ DC to cross-present apoptotic cell-derived Ag. However, splenic CD8α$^+$ DC from mice lacking CD36 or $\alpha_v\beta_3$ /β_5 integrins are capable of cross-presenting extracellular self- or foreign-Ag (associated with either apoptotic or viable cells) (Belz et al. 2002; Schulz et al. 2002), suggesting that receptors necessary for apoptotic cell clearance exist at a high level of redundancy.

Importantly, phagocytes discriminate between viable and apoptotic cells not only by the presence of 'eat me' signals but also by the expression of detachment signals such as the CD31 molecule (Brown et al. 2002). While both viable and apoptotic cells express CD31, apoptotic cells alter the function of CD31 from repulsive to adhesive to promote the binding of dying cells (Brown et al. 2002), although the mechanism by which this occurs remains unknown.

11.2.4   Presentation of Apoptotic Cell-Derived Antigens

There is evidence that apoptotic cells dock on the surface of DC and macrophages through binding to the $\alpha_v\beta_5$ integrin, that recruits the CrkII-Dock180 molecular complex and in turn triggers Rac1 activation and phagosome formation (Gronski and Ravichandran, Chap. 7, this Vol.; Albert et al. 2000; Akakura et al. 2004). Once internalized, the apoptotic cells are processed within MHC class-II rich compartments (MHC class-II$^+$ LAMP$^+$ H-2M$^+$ cytoplasmic vesicles) for presentation as peptides loaded in MHC class-II molecules (Inaba et al. 1998) to CD4 T cells. Alternatively, apoptotic cell-derived Ag can be routed out of the endosomal compartment into the lumen of the endoplasmic reticulum for loading into MHC class-I for cross-presentation to CD8 T lymphocytes (Albert et al. 1998b; Rodriguez et al. 1999). The ultimate mechanism by which apoptotic cell-derived Ag are shuttled into the endoplasmic reticulum is unknown, however the ability of lactacystin, a 26S proteasome inhibitor to partially block cross-presentation of apoptotic cell-derived Ag in DC suggests that both classical and non-classical MHC class-I pathways participate in this process (Albert et al. 1998b).

Significant work has established in vitro and in vivo the ability of DC to cross-present, via MHC class-I molecules, peptides derived from internalized apoptotic cells to CD8 T cells. Although macrophages phagocytose apoptotic cells more efficiently than DC, only the latter APC cross-present efficiently the apoptotic cell-derived peptides to CD8 T cells (Albert et al. 1998b). In this regard, it has been shown that DC, unlike macrophages, cross-present Ag derived from salmonella-infected apoptotic cells to CD8 T cells (Yrlid and Wick 2000). In vivo, i.v. injected apoptotic leukocytes pre-loaded with the model Ag ovalbumin (OVA) were internalized by splenic DC for presentation to OVA-specific TCR transgenic CD8 T cells (Liu et al. 2002). Ag derived from apoptotic cells can also be presented through MHC class-II molecules to CD4 T cells. In fact, murine bone-marrow-derived DC present Ag from internalized apoptotic cells to CD4 T cells 1-10,000 times better than pre-processed peptide (Inaba et al. 1998).

11.2.5   Phagocytosis of Early Apoptotic Cells and Tolerogenic Antigen-Presenting Cells

There is accumulated evidence that DC that have phagocytosed early apoptotic cells exhibit a decreased capacity to stimulate Ag-specific TCR transgenic T cells, allogeneic T cells (Lacy-Hulbert, Chap. 7, this Vol.; Stuart et al. 2002; Takahashi and Kobayashi 2003) and even T cell clones (which do not require co-stimulation) (Urban et al. 2001). This is likely attributable to reduced expression of surface MHC-peptide complexes and co-stimulatory molecules by those DC that have interacted with apoptotic cells, rather than to a deficiency in the Ag-processing capacity of DC (Urban et al. 2001). As stated above, if phagocytosis of early apoptotic cells prevents activation of immature/semi-mature DC, this should result in presentation of apoptotic cell-derived peptides in the context of insufficient co-stimulation, which according to the two-signal hypothesis of T cell activation, leads to T cell hypo-response or tolerance. Besides the decreased expression of co-stimulatory molecules, other immune-suppressive changes in the DC induced by the apoptotic cells could also contribute to down-regulation of the T cell response. Phagocytosis of apoptotic cells decreases in DC (and macrophages) activation of NF-κB, a transcription factor required for DC maturation/activation and synthesis of several pro-inflammatory cytokines. This may explain why DC that have interacted with early apoptotic cells secrete significantly lower levels of IL-1α, IL-1β, IL-6, IL-12p70, IL-23 and TNF-α (Urban et al. 2001; Savill et al. 2002; Morelli et al. 2003; Stark et al. 2005), a phenomenon that is maintained even in the presence of LPS (Morelli et al. 2003). Interestingly, phagocytosis of apoptotic cells does not interfere with secretion of TGF-β1 by mouse DC (Morelli et al. 2003), and even increases release of IL-10 by human DC (Urban et al. 2001), both immuno-suppressive cytokines. The effect of apoptotic cells on cytokine production by DC is at least partly due to altered cytokine mRNA transcription or stabilization (Morelli et al. 2003).

Following internalization of cells in early apoptosis, macrophages increase secretion of additional anti-inflammatory mediators including prostaglandin E_2 (PGE_2), platelet-activating factor (PAF), IL-1α receptor-antagonist and hepatocyte growth factor (Fadok et al. 1998b; Morimoto et al. 2001; Craciun et al. 2005). The release of TGF-β1, PGE_2 and PAF can actually inhibit bystander macrophages that have not phagocytosed apoptotic cells by means of a paracrine mechanism (Fadok et al. 1998b). Unlike macrophages, down-modulation of DC function by apoptotic cells is not paracrine and requires physical contact of the DC with the dead cell (Stuart et al. 2002). More recently, other inhibitory mediators have been discovered to have pervasive anti-inflammatory effects on the immune responses, including the tryptophan catabolizing enzyme indoleamine 2,3-dioxygenase (IDO) and the cell surface receptor programmed cell death (PD)-1 and its ligands PD-L1 and PD-L2 (Munn et al. 1998; Nishimura et al. 1999; Mellor et al. 2002; Ozkaynak et al. 2002; Ansari et al. 2003; Gao et al. 2003; Mellor et al. 2003; Salama et al. 2003; Bauer et al. 2005). The contribution of these mediators to tolerance induction following phagocytosis of apoptotic cells remains to be explored.

Significant evidence supports that phagocytosis of early apoptotic cells by APC down-regulates the T cell response in vivo through T cell deletion, anergy and/or de novo differentiation/expansion of Treg in secondary lymphoid organs. By using mice expressing model Ag controlled by tissue-specific promoters, it has been shown that in the steady state, constitutively migrating DC transport and process tissue-specific Ag from periphery to LN and spleen. These migrating immature/semi-mature DC silence rather than stimulate self-reactive T cells (Kurts et al. 1997; Adler et al. 1998). A similar mechanism seems to operate when foreign-Ag are delivered directly to DC in secondary lymphoid organs by i.v. administration of early apoptotic cells. After i.v. injection of OVA-loaded dying cells in mice, DC cross-presented OVA-peptides and induced abortive proliferation and deletion of OVA-specific CD8 T cells and subsequent T cell tolerance to OVA-challenge (Liu et al. 2002). In the transplantation setting, human monocyte-derived DC loaded with allogeneic apoptotic cells down-regulated specifically the anti-donor T cell response in vitro (Nouri-Shirazi and Guinet 2002). More importantly in vivo, i.v. administration of donor splenocytes in early stages of apoptosis induced by UV-B irradiation led to defective activation of donor-reactive TCR transgenic CD4 T cells and their peripheral deletion (Wang et al. 2006). When combined with blockade of the CD40-CD154 pathway, systemic administration of donor apoptotic splenocytes promoted expansion of donor-specific CD4 Treg in mice (Wang et al. 2006). These findings have important implications for transplantation since, although opinions vary, induction of donor-specific tolerance will ultimately depend on both deletion of allo-reactive T cells and generation/expansion of Treg.

11.3 Apoptotic Cell-Based Therapies Against Transplant Rejection

11.3.1 Mechanisms of Allo-recognition and Transplant Rejection

To understand why apoptotic cell-based therapies prolong transplant survival, it is essential to have a basic knowledge of the mechanisms of allo-recognition and rejection of allografts, the latter defined as organs/tissues/cells transplanted between MHC-mismatched individuals of the same species. Based on the tempo of onset, there are three types of allograft rejection: hyperacute, acute and chronic. Hyperacute rejection occurs within minutes or hours after transplantation surgery and is mediated by deposition of pre-formed circulating Ab on the graft and the consequent activation of the complement and coagulation cascades. Hyperacute rejection of allografts is currently a preventable phenomenon that very rarely occurs. Acute rejection takes place within weeks or months after transplantation, is mediated by the innate and adaptive immune responses and constitutes the main immediate threat for allograft survival. However, with the advent of immuno-suppressive drugs, acute rejection is largely preventable. Chronic rejection develops in months or in general years after transplantation and results from immune and non-immune mechanisms. Unfortunately, the immuno-suppressive drugs currently used are not effective to prevent or treat chronic rejection.

In order to reject an allograft, recipient's T cells must recognize the donor alloAg (mostly derived from mismatched-MHC molecules) through the direct and indirect pathways of allo-recognition (reviewed in Game and Lechler 2002). By the direct pathway, recipient's T cells recognize donor "intact" MHC molecules expressed by the donor APC transplanted along with the graft (i.e. donor DC, macrophages, endothelial cells). Among these donor APC, only DC are capable of migrating as "passenger leukocytes" to secondary lymphoid organs of the recipient, where they activate donor-reactive naïve T cells. Unlike naïve T cells, memory T cells that cross-react with donor alloAg can recirculate through the graft and be re-activated in the periphery by professional or non-professional APC. The precursor frequency of direct pathway donor-reactive T cells is extremely high, roughly 1–10%.

By the indirect pathway, recipient's T cells recognize donor-derived allopeptides presented by recipient's APC in self-MHC molecules. The precursor frequency of indirect pathway T cells is extremely low (1:100,000-200,000), the same as that for any other nominal Ag. Due to the high precursor frequency of direct pathway T cells, it is assumed that the direct pathway is the more significant contributor of acute rejection (Talmage et al. 1976; Lechler and Batchelor 1982; Larsen et al. 1990). However, as the supply of donor APC within the graft decreases over time, the contribution of the indirect pathway becomes the main mechanism of allo-recognition during chronic rejection (Valujskikh et al. 1998).

11.3.2 *Apoptotic Leukocytes Promote Engraftment of Allogeneic Bone-Marrow*

Bittencourt and colleagues (2001) have demonstrated that i.v. infusion of apoptotic leukocytes enhances allogeneic bone-marrow engraftment following a non-myelo-ablative conditioning regimen in mice. The beneficial effect of apoptotic cells was independent of the stimulus used to trigger cell death. Interestingly, donor, recipient and even xenogeneic apoptotic leukocytes were capable of promoting allogeneic bone-marrow engraftment, suggesting that the therapeutic effect of apoptotic cells was not donor-specific. However, since the apoptotic cells were administered simultaneously with the bone-marrow graft, it is possible that the apoptotic cells had exerted a bystander pro-tolerogenic effect on the recipient's DC when they were presenting the donor alloAg from the graft. Further, co-administration of apoptotic leukocytes with allogeneic hematopoietic stem cells has been proven to expand Treg able to delay onset of graft vs. host disease (Kleinclauss et al. 2006), and also to prevent the anti-donor humoral response (Perruche et al. 2004).

Apoptotic cell therapy has been shown to improve bone marrow engraftment by inducing mixed chimerism, defined as the presence of donor-derived cells (normally of hematopoietic origin) in the tissues of allograft recipients (reviewed in Sykes and Sachs 2001). Micro- and macro-chimerism refers to the detection of <1.0 and >1.0%, respectively, of donor cells in the recipient. The reason(s) that mixed chimerism correlates with a state of anti-donor hypo-response/tolerance and the consequent prolongation of graft survival has not been entirely elucidated.

11.3.3 *Donor Apoptotic Leukocytes Prolong Solid Organ Allograft Survival*

A limited number of studies in murine models have proven the efficacy of apoptotic cell-based therapies for prolongation of solid organ allograft survival. Our group has demonstrated in mice that i.v. administration of donor-derived UV-B-irradiated apoptotic splenocytes seven days prior to transplantation significantly prolongs survival of heart allografts in the absence of immunosuppression (Wang et al. 2006). Moreover, combination of donor apoptotic splenocytes with suboptimal blockade of the CD40-CD154 pathway with a single dose of the anti-CD154 monoclonal Ab MR1 resulted in long-term survival of cardiac transplants for more than 100 days (which is considered indefinite graft survival in mice; Wang et al. 2006). The therapeutic effect of donor apoptotic cells was donor-specific and required interaction of the apoptotic cells with recipient's DC in secondary lymphoid organs. It also depended on the physical properties of the apoptotic leukocytes, since administration of donor necrotic cells did not affect graft survival (Wang et al. 2006). Similarly, infusion of donor apoptotic leukocytes prolonged significantly cardiac allograft survival in a rat model (Sun et al. 2004). By using a more indirect approach, Xu and colleagues (2004) showed that intra-portal administration of recipient-derived immature DC pre-loaded ex vivo with donor apoptotic splenocytes prolongs survival of cardiac allografts in otherwise untreated mice. In humans, there is indirect evidence suggesting that apoptotic leukocytes may have a beneficial effect in graft recipients treated with extracorporeal photopheresis (ECP), a technique by which blood leukocytes are UV-B-irradiated ex vivo and then re-infused systemically (Barr et al. 1998). The ultimate mechanism of action of ECP is still unknown. However, it is believed that the UV-B-irradiation primes the leukocytes to become apoptotic when re-infused into the bloodstream with consequent anti-inflammatory and down-regulatory effects on the immune system.

11.3.4 Donor Apoptotic Leukocytes Down-regulate the Anti-donor T Cell Response

Some mechanisms by which donor apoptotic leukocytes prolong allograft survival are becoming elucidated. We have demonstrated that i.v. administered donor-apoptotic cells are phagocytosed rapidly by recipient's splenic DC, which present apoptotic cell-derived allopeptides in recipient's MHC molecules to indirect pathway T cells (Wang et al. 2006). Importantly, internalization of apoptotic cells did not induce maturation/activation of recipient's DC in vivo, as reflected in similar expression of MHC molecules, CD40, CD80 and CD86 compared to splenic DC from untreated mice.

Using a model of C57BL/10 (B10) mice reconstituted with 1H3.1 TCR transgenic CD4 T cells [specific for the BALB/c allopeptide $IE\alpha_{52-68}$ loaded in IA^b molecules (MHC class-II of B10)], we characterized the in vivo effect of apoptotic cell-derived allopeptide presentation on indirect pathway T cells (Wang et al. 2006). Interestingly, splenic 1H3.1 CD4 T cells proliferated in response to injection of BALB/c apoptotic splenocytes, but did not up-regulate expression of the T cell activation markers CD25, CD44, CD69 and CD152 and secreted lower amounts of IL-2 and IFN-γ upon ex vivo re-stimulation with $IE\alpha_{52-68}$, when compared to controls. Importantly, the defective activation of anti-donor 1H3.1 CD4 T cells resulted in their peripheral deletion, as their numbers decreased significantly in spleen, LN, blood and peripheral tissues, 14 days after administration of apoptotic cells. Deletion of donor-reactive CD4 T cells was likely due to deficient presentation of apoptotic cell-derived allopeptides by immature/semi-mature DC in the spleen, leading to generation of anti-donor T cell clones unfit for survival. In fact, after administration of BALB/c apoptotic splenocytes, proliferating 1H3.1 CD4 T cells failed to increase levels of the anti-apoptotic protein $Bcl-X_L$ and of receptors for IL-7 and IL-15, which are cytokines required for homeostatic survival/proliferation of T cells.

Besides inducing peripheral deletion of donor-reactive T cells, administration of donor apoptotic splenocytes in combination with suboptimal CD40-CD154 blockade promoted differentiation/expansion of donor-specific CD4 Treg (Wang et al. 2006). Indeed, adoptive transfer of CD4 T cells from B10 mice with long-term surviving BALB/c cardiac grafts (>100 days) following treatment with BALB/c apoptotic splenocytes plus CD154 monoclonal Ab, into naïve B10 recipients prolonged survival of BALB/c hearts, but not third-party control grafts. Accordingly, long-term heart allografts of mice treated with donor apoptotic splenocytes plus MR1 Ab were infiltrated with a high number of CD4 T cells expressing the Treg marker Foxp3 and containing intracellular IL-10 and TGF-β.

In theory, processing of donor apoptotic cells by recipient's APC should down-modulate exclusively indirect pathway T cells, since there is no evidence that dead cells interact directly with T cells. However, more recent evidence indicates that indirect pathway Treg can suppress direct pathway T cells (through a mechanism known as "bystander suppression"; Spadafora-Ferreira et al. 2007). This phenomenon could explain why administration of donor apoptotic leukocytes down-regulates both direct and indirect pathway T cells and delays significantly acute rejection of hearts in mice, a phenomenon elicited mainly by direct pathway T cells.

11.3.5 Donor Apoptotic Leukocytes Decrease the Anti-donor B Cell Response

Following transplantation, the recipient develops anti-donor plasma cells able to synthesize Ab against donor MHC molecules. These alloAb play an important role in graft rejection. In order to produce alloAb, recipient's B cells must specifically recognize, internalize and process donor intact MHC molecules and then present donor MHC-derived peptides loaded in recipient's MHC molecules to already activated CD4 T cells (helper cells) recognizing the same allopeptides through the indirect pathway. As a result of this T cell-B cell interaction, B cells differentiate into plasma cells and secrete alloAb. Simply put, activation of indirect pathway CD4 T cells helps anti-donor B cells to develop alloAb. Therefore, the deletion of indirect pathway CD4 T cells that follows administration of donor apoptotic leukocytes should indirectly reduce generation of alloAb. Indeed, we have shown in mice that therapy with donor apoptotic splenocytes reduced significantly the level of circulating alloAb in recipients of cardiac allografts (Wang et al. 2006).

11.3.6   Effect of Donor Apoptotic Leukocytes on Chronic Rejection

Although administration of donor apoptotic cells has been proven to down-regulate the anti-donor response that causes acute rejection, to our knowledge the impact of apoptotic cell-based therapies on chronic rejection has not been explored. This is of particular importance, since presentation of apoptotic cell-derived alloAg by recipient's APC should down-regulate mainly the indirect pathway, which, as stated above, is the principal mechanism of allo-recognition in chronic rejection. Typical features of chronic rejection include steady decline of organ function, interstitial fibrosis, chronic inflammatory infiltrate (i.e. lymphocytes, plasma cells), atrophy and gradual loss of parenchymal cells and chronic vascular arteriopathy (CVA), the latter a condition manifested by endothelitis, intimal proliferation, elastic fiber disruption, fibrosis and leukocyte infiltration of medium- and small-size arteries of the graft. An experimental approach to investigate the onset of CVA in mice is the model of aortic (abdominal) allografts, where the histological features of CVA develop in the transplanted aorta 30–60 days after surgery. By using this model, we found that administration of BALB/c apoptotic splenocytes to recipient B6 mice seven days before transplant of BALB/c aortic grafts, results in significant inhibition of the indirect pathway T cell response and the histopathological features of CVA (manuscript in preparation).

Therapies with early apoptotic cells may be beneficial not only for their ability to down-regulate adaptive immunity, but also for their anti-inflammatory effects. There is evidence that stem cell transplantation after myocardial infarction restores cardiac function. Interestingly, the effect has been attributed to the ability of apoptotic stem cells to resolve the local inflammation, and by doing so, reducing scar formation and further damage of myocardial fibers (Thum et al. 2005).

11.3.7   Apoptotic Cells Down-regulate Contact Hypersensitivity and Autoimmunity

In addition to transplantation, apoptotic cells have been used in mouse models to regulate the immune response in contact hypersensitivity and autoimmunity. Ferguson et al. (2002) demonstrated in mice that following i.v. administration, trinitrophenyl (TNP)-coupled splenocytes become apoptotic (via a Fas-Fas ligand dependent pathway) and down-regulate the contact hypersensitivity response in a hapten-specific manner.

In autoimmunity, induction of apoptosis of pancreatic β-cells in vivo decreased the incidence of type 1 diabetes in NOD mice (Hugues et al. 2002). In agreement with this observation, Xia and colleagues (2007) demonstrated that administration of UV-B-irradiated apoptotic β-cells (NIT-1 cell line) delayed significantly onset of type 1 diabetes in NOD mice. The beneficial effect of the apoptotic β-cells was accompanied with decreased T cell proliferation, Th2 bias, expansion of IL-10-producing Treg, and reduced levels of Ab against β-cells.

11.4 Clinical Utility of Apoptotic Cells

Although apoptotic cells have been proven to prolong allograft survival in experimental models, there are some characteristics that make apoptotic cells a promising tool for their potential application for therapy of transplant rejection in humans. Early apoptotic cells: i) are a rich source of alloAg; ii) are easy and fast to prepare in the laboratory compared to soluble MHC molecules, allopeptides, chimeric MHC molecules, alloAg tagged to engineered Ab, and in vitro-generated tolerogenic DC; iii) are relatively safe when administered i.v. and iv) can be prepared from living or deceased organ donors.

11.4.1 Apoptotic Cells Compared to Alternative Cell-based Therapies for Transplantation Tolerance

Interestingly, there is growing evidence that alternative cellular therapies to control the anti-donor response such as administration of donor living cells like donor specific transfusion (DST) of splenocytes or blood mononuclear cells, and infusion of donor tolerogenic DC, may prolong allograft survival simply by generating donor apoptotic cells. Based on the "missing self" hypothesis, recipient's NK cells would recognize as non-self and eliminate the i.v. administered donor cells. Even though i.v. administered recipient-derived tolerogenic DC loaded ex vivo with alloAg would be spared from NK cell recognition, they would have a limited lifespan and become apoptotic in the recipient (Kamath et al. 2000). Supporting this concept, i.v. administration of donor apoptotic leukocytes, DST and tolerogenic DC have resulted in a relatively similar prolongation of cardiac allograft survival in murine models (O'Connell et al. 2002; Xu et al. 2004; Peche et al. 2005; Taner et al. 2005; Emmer et al. 2006; Lan et al. 2006; Wang et al. 2006; Turnquist et al. 2007). Further, Quezada et al. (2003) have shown that, similar to i.v. infused donor apoptotic splenocytes, DST induces peripheral deletion of indirect pathway CD4 T cells and increases the percentage of Treg. Importantly in this model, there was no direct pathway allo-recognition of the injected cells by CD4 T cells, indicating the absence of direct contact between the DST cells and the recipient's T cells. By contrast, this finding suggests that the cells of the DST become apoptotic in vivo and were processed by recipient's APC for indirect allo-recognition. Our unpublished work demonstrates that the same is true following i.v administration of donor-derived DC rendered tolerogenic with vitamin D_3 treatment (manuscript in preparation).

Although there are multiple similarities between the mechanisms of action of donor apoptotic and living leukocytes injected i.v., dead or dying cells behave in some aspects differently than living cells. Whereas donor apoptotic cells injected i.v. induced activation of donor-reactive T cells only in the spleen, DST is followed by T cell stimulation in spleen and LN (Wang et al. 2006). This difference is likely attributable to the ability of living splenocytes to traffic actively to both spleen and LN, as opposed to the passive transport of apoptotic cells by the bloodstream into the spleen, where they are trapped mainly by marginal zone phagocytes. Further, whereas donor apoptotic cells induce defective activation of donor-reactive T cells, DST is followed by up-regulation of the activation marker CD69 and CD44 by T cells (Wang et al. 2006).

Previous findings in the field of hematopoietic stem cell transplantation may also provide some insight into the mechanisms by which apoptotic cells down-regulate the anti-donor response. Multiple studies have shown that allogeneic bone-marrow transplantation induces tolerance to solid organ transplants from the same donor (Sayegh et al. 1991; Jacobsen et al. 1994; Helg et al. 1994). Further, apoptotic cell infusion not only facilitates donor bone marrow engraftment, but also promotes tolerance to donor, but not to third-party, skin grafts in a murine model (Kleinclauss et al. 2001). Therefore it is likely that, once infused, the donor hematopoietic stem cells become apoptotic and decrease the anti-donor response to prolong allograft survival.

As all cell-based therapies to promote transplantation tolerance in the clinic are inherently complex, potentially risky, expensive and difficult to implement, it will be essential to compare side by side therapies based on apoptotic vs. living cells to identify the approach with the greatest clinical utility.

11.4.2　Considerations for Apoptotic Cell-based Therapies

Besides their anti-inflammatory and immune-regulatory properties, under specific circumstances apoptotic cells can promote immunity (Bartunkova and Spisek, Chap. 12, this Vol.). It has been shown that DC loaded with apoptotic cells can efficiently activate CD4 T cells (Rovere et al. 1998; Yrlid and Wick 2000) and cross-prime CD8 T cells (Albert et al. 1998b; Rovere et al. 1998; Yrlid and Wick 2000). In fact, apoptotic cells can trigger immune responses in numerous infectious and autoimmune disorders (Mevorach, Chap. 10, this Vol.). In this regard, DC have been shown to prime Ag-specific T cells following internalization of apoptotic monocytes infected by influenza A virus (Albert et al. 1998b) or apoptotic macrophages infected by *Salmonella typhimurium* (Yrlid and Wick 2000). This is likely attributable to the ability of certain pathogens to trigger host cell apoptosis and release of pro-inflammatory mediators. Several virus-infected cells release the DC-activating cytokine IFN-α and then become apoptotic, inducing activation of those DC that internalize the apoptotic cells (Gallucci et al. 1999). Likewise, other cytokines including IL-1β and TNF-α or Toll-like receptor ligands (i.e. double stranded RNA, CpG DNA motifs, LPS) are released in response to pathogens and contribute to activation of APC during phagocytosis of apoptotic infected cells.

Apoptotic cells are also implicated in the onset of autoimmunity. Accumulation of cells in late apoptosis due to deficient apoptotic cell clearance plays a key role in the pathogenesis of systemic lupus erythematosus (SLE; Gregory and Pound, Chap. 9; Mevorach, Chap. 10, this Vol.; Bijl et al. 2006), rheumatoid arthritis (reviewed in Peng 2006) and type 1 diabetes (O'Brien et al. 2002) and in exacerbation of cystic fibrosis (Vandivier et al. 2002b) and chronic obstructive pulmonary disease (Hodge et al. 2003). An increase in the ratio between apoptotic cells and phagocytes can promote inflammation and immunity. DC incubated with an excess of apoptotic cells resulted in DC maturation and efficient Ag presentation to both MHC class-I and class-II-restricted T cells (Rovere et al. 1998), an effect that was apoptotic cell dose-dependent and accompanied by secretion of IL-1β and TNF-α (Rovere et al. 1998). Further, immunization with DC exposed to high numbers of apoptotic cells resulted in priming of tumour-specific cytotoxic T cells and protection against tumour challenge in mice (Ronchetti et al. 1999). More recently, Maree and colleagues (2007) provided insight into the long-held hypothesis that type 1 diabetes results from deficient apoptotic cell clearance. They showed that macrophages from NOD mice clear and digest apoptotic cells slower than control BALB/c macrophages. DeFranco et al. (2001) found that Fas-resistant individuals have an earlier age of onset of type 1 diabetes than Fas-sensitive patients, supporting the notion that variations in the ability to induce and clear apoptotic cells is a factor in pathogenesis of autoimmune diabetes.

Deficits in apoptotic cell clearance can also be caused by deficiencies in the receptors or ligands involved in apoptotic cell recognition/internalization by phagocytes (Gregory and Pound, Chap. 9, this Vol.). Mice deficient in the complement factors C1q and C4, IgM, SAP, and the receptor Mer develop systemic autoimmunity (Botto et al. 1998; Bickerstaff et al. 1999; Boes et al. 2000; Chen et al. 2000; Ehrenstein et al. 2000; Cohen et al. 2002). In humans, SLE is associated with genetic deficiencies in the complement factors C1q, C1r, C1s, C4 and C2 (Mevorach, Chap. 10, this Vol.; reviewed in Pickering and Walport 2000) and phagocytes from SLE patients exhibit impaired apoptotic cell uptake. Interestingly, there is a hierarchical association between SLE susceptibility and disease severity based on the position of the deficient complement factor in the activation pathway, as the highest disease incidence occurs in patients with C1q deficiency (reviewed in Pickering and Walport 2000).

Molecules bound to the apoptotic cell surface can stimulate the immune response. For example, IgG opsonized to the apoptotic cell surface exert pro-inflammatory effects on phagocytes. Patients with SLE exhibit in serum anti-phospholipid autoantibodies that recognize externalized phospholipids on the apoptotic cell surface (Manfredi et al. 1998). Internalization of IgG-opsonized apoptotic cells stimulate secretion of TNF-α by macrophages (Manfredi et al. 1998). Endocytosis of IgG-opsonized apoptotic cells by macrophages does not prevent LPS-induced secretion of pro-inflammatory cytokines (Fadok et al. 1998b).

Apoptotic cells may, by themselves, affect the function of APC. Various pro- and anti-inflammatory mediators can be released by dying cells (Peter et al., Chap. 3, this vol.). Secondary necrotic (late apoptotic) cells are known to passively release pro-inflammatory signals that stimulate macrophage and DC maturation/activation (Sauter et al. 2000; Chernysheva et al. 2002; Ip and Lau 2004), including uric acid, heat shock proteins (HSP) and the high mobility group box 1 (HMGB1) molecule (Basu et al. 2000; Scaffidi et al. 2002; Shi et al. 2003). Interestingly, apoptotic cells do not normally release HSP, but when stressed, up-regulate its expression and promote APC activation. For example, vaccination with DC pulsed with heat-stressed apoptotic tumour cells elicits tumour-specific immunity in mice (Bartunkova and Spisek, Chap. 12, this Vol.; Feng et al. 2001). Additionally, a report by Lauber and colleagues (2003) showed that apoptotic cells release lysophosphatidylcholine, a lipid chemotactic for monocytes and macrophages, that activates pro-inflammatory transcription factors like NF-κB and signalling through mitogen-activated protein kinase (MAKs) pathways (Peter et al., Chap. 3, this Vol.). Alternatively, in some cases apoptotic cells release the immunosuppressive cytokines IL-10 and bio-active TGF-β1 (Gao et al. 1998; Chen et al. 2001).

In summary, multiple factors determine the ability of apoptotic cells to promote tolerance or trigger immunity. With so many variables in play and the devastating consequences at stake, the conditions to use apoptotic cells for therapeutic purposes must be carefully optimized.

11.5 Conclusions

From the initial findings that APC internalize, process and present apoptotic cell-derived Ag to T cells with regulatory effects on the immune response, apoptotic cell-based therapies have been a promising candidate for treatment of transplant rejection and autoimmune disorders. Increasing evidence in murine models supports the efficacy of apoptotic cell-based therapies against allograft rejection and provides significant insight into the mechanisms by which prolongation of graft survival is achieved. Increasing evidence suggest that therapies with living leukocytes for prolongation of allograft survival may function through generation of apoptotic cells. Importantly, apoptotic cell therapy is not without risk, and clearly therapeutic regimens need to be optimized for any realistic hope that therapies based on administration of exogenous apoptotic cells achieve clinical use. Based on the tremendous need for donor-specific immuno-suppression/tolerance in transplantation, further investigation in this fascinating area of research is warranted.

Acknowledgments The work of the authors is supported by grants from the National Institutes of Health: T32 AI074490 (S.J.D.) and by R01 HL075512 and HL077545 (A.E.M.).

References

Abe M, Wang Z, de Creus A et al (2005) Plasmacytoid dendritic cell precursors induce allogeneic T cell hyporesponsiveness and prolong heart graft survival. Am J Transplant 5:1808–1819

Adler AJ, Marsh DW, Yochum GS et al (1998) CD4+ T cell tolerance to parenchymal self-antigens requires presentation by bone marrow-derived antigen-presenting cells. J Exp Med 187:1555–1564

Akakura S, Singh S, Spataro M et al (2004) The opsonin MFG-E8 is a ligand for the alphavbeta5 integrin and triggers DOCK180-dependent Rac1 activation for the phagocytosis of apoptotic cells. Exp Cell Res 292:403–416

Albert ML, Pearce SF, Francisco LM et al (1998a) Immature dendritic cells phagocytose apoptotic cells via alphavbeta5 and CD36, and cross-present antigens to cytotoxic T lymphocytes. J Exp Med 188:1359–1368

Albert ML, Sauter B, Bhardwaj N (1998b) Dendritic cells acquire antigen from apoptotic cells and induce class I-restricted CTLs. Nature 392:86–89

Albert ML, Kim JI, Birge RB (2000) alphavbeta5 integrin recruits the CrkII-Dock180-rac1 complex for phagocytosis of apoptotic cells. Nat Cell Biol 2:899–905

Ansari MJ, Salama AD, Chitnis T et al (2003) The programmed death-1 (PD-1) pathway regulates autoimmune diabetes in non-obese diabetic (NOD) mice. J Exp Med 198:63–69

Arur S, Uche UE, Rezaul K et al (2003) Annexin I is an endogenous ligand that mediates apoptotic cell engulfment. Dev Cell 4:587–598

Balasubramanian K, Schroit AJ (1998) Characterization of phosphatidylserine-dependent beta2-glycoprotein I macrophage interactions. Implications for apoptotic cell clearance by phagocytes. J Biol Chem 273:29272–29277

Barr ML, Meiser BM, Eisen HJ et al (1998) Photopheresis for the prevention of rejection in cardiac transplantation. Photopheresis Transplantation Study Group. N Engl J Med 339:1744–1751

Basu S, Binder RJ, Suto R et al (2000) Necrotic but not apoptotic cell death releases heat shock proteins, which deliver a partial maturation signal to dendritic cells and activate the NF-kappa B pathway. Int Immunol 12:1539–1546

Bauer TM, Jiga LP, Chuang JJ et al (2005) Studying the immunosuppressive role of indoleamine 2,3-dioxygenase: tryptophan metabolites suppress rat allogeneic T cell responses in vitro and in vivo. Transpl Int 18:95–100

Belz GT, Vremec D, Febbraio M et al (2002) CD36 is differentially expressed by CD8+ splenic dendritic cells but is not required for cross-presentation in vivo. J Immunol 168:6066–6070

Bevers EM, Comfurius P, Dekkers DW et al (1999) Lipid translocation across the plasma membrane of mammalian cells. Biochim Biophys Acta 1439:317–330

Bharadwaj D, Mold C, Markham E et al (2001) Serum amyloid P component binds to Fc gamma receptors and opsonizes particles for phagocytosis. J Immunol 166:6735–6741

Bickerstaff MC, Botto M, Hutchinson WL et al (1999) Serum amyloid P component controls chromatin degradation and prevents antinuclear autoimmunity. Nat Med 5:694–697

Bijl M, Reefman E, Horst G et al (2006) Reduced uptake of apoptotic cells by macrophages in systemic lupus erythematosus: correlates with decreased serum levels of complement. Ann Rheum Dis 65:57–63

Bittencourt MC, Perruche S, Contassot E et al (2001) Intravenous injection of apoptotic leukocytes enhances bone marrow engraftment across major histocompatibility barriers. Blood 98:224–230

Bjorck P, Coates PT, Wang Z et al (2005) Promotion of long-term heart allograft survival by combination of mobilized donor plasmacytoid dendritic cells and anti-CD154 monoclonal antibody. J Heart Lung Transplant 24:1118–1120

Boes M, Schmidt T, Linkemann K et al (2000) Accelerated development of IgG autoantibodies and autoimmune disease in the absence of secreted IgM. Proc Natl Acad Sci U S A 97:1184–1189

Botto M, Dell'Agnola C, Bygrave AE et al (1998) Homozygous C1q deficiency causes glomerulonephritis associated with multiple apoptotic bodies. Nat Genet 19:56–59

Brown S, Heinisch I, Ross E et al (2002) Apoptosis disables CD31-mediated cell detachment from phagocytes promoting binding and engulfment. Nature 418:200–203

Butler PJ, Tennent GA, Pepys MB (1990) Pentraxin-chromatin interactions: serum amyloid P component specifically displaces H1-type histones and solubilizes native long chromatin. J Exp Med 172:13–18

Carrasco YR, Batista FD (2007) B cells acquire particulate antigen in a macrophage-rich area at the boundary between the follicle and the subcapsular sinus of the lymph node. Immunity 27:160–171

Chang MK, Bergmark C, Laurila A et al (1999) Monoclonal antibodies against oxidized low-density lipoprotein bind to apoptotic cells and inhibit their phagocytosis by elicited macrophages: evidence that oxidation-specific epitopes mediate macrophage recognition. Proc Natl Acad Sci U S A 96:6353–6358

Chen Z, Koralov SB, Kelsoe G (2000) Complement C4 inhibits systemic autoimmunity through a mechanism independent of complement receptors CR1 and CR2. J Exp Med 192:1339–1352

Chen W, Frank ME, Jin W et al (2001) TGF-beta released by apoptotic T cells contributes to an immunosuppressive milieu. Immunity 14:715–725

Chen X, Doffek K, Sugg SL et al (2004) Phosphatidylserine regulates the maturation of human dendritic cells. J Immunol 173:2985–2994

Chen Q, Stone PR, McCowan LM et al (2006) Phagocytosis of necrotic but not apoptotic trophoblasts induces endothelial cell activation. Hypertension 47:116–121

Chernysheva AD, Kirou KA, Crow MK (2002) T cell proliferation induced by autologous non-T cells is a response to apoptotic cells processed by dendritic cells. J Immunol 169:1241–1250

Cohen PL, Caricchio R, Abraham V et al (2002) Delayed apoptotic cell clearance and lupus-like autoimmunity in mice lacking the c-mer membrane tyrosine kinase. J Exp Med 196:135–140

Craciun LI, DiGiambattista M, Schandene L et al (2005) Anti-inflammatory effects of UV-irradiated lymphocytes: induction of IL-1Ra upon phagocytosis by monocyte/macrophages. Clin Immunol 114:320–326

Daleke DL, Lyles JV (2000) Identification and purification of aminophospholipid flippases. Biochim Biophys Acta 1486:108–127

Dalgaard J, Beckstrom KJ, Jahnsen FL et al (2005) Differential capability for phagocytosis of apoptotic and necrotic leukemia cells by human peripheral blood dendritic cell subsets. J Leukoc Biol 77:689–698

DeFranco S, Bonissoni S, Cerutti F et al (2001) Defective function of Fas in patients with type 1 diabetes associated with other autoimmune diseases. Diabetes 50:483–488

Ehrenstein MR, Cook HT, Neuberger MS (2000) Deficiency in serum immunoglobulin (Ig)M predisposes to development of IgG autoantibodies. J Exp Med 191:1253–1258

Emmer PM, Van Der Vlag J, Adema GJ et al (2006) Dendritic cells activated by lipopolysaccharide after dexamethasone treatment induce donor-specific allograft hyporesponsiveness. Transplantation 81:1451–1459

Fadok VA, Bratton DL, Frasch SC (1998a) The role of phosphatidylserine in recognition of apoptotic cells by phagocytes. Cell Death Differ 5:551–562

Fadok VA, Bratton DL, Konowal A et al (1998b) Macrophages that have ingested apoptotic cells in vitro inhibit pro-inflammatory cytokine production through autocrine/paracrine mechanisms involving TGF-beta, PGE2, and PAF. J Clin Invest 101:890–898

Fadok VA, Bratton DL, Henson PM (2001) Phagocyte receptors for apoptotic cells: recognition, uptake, and consequences. J Clin Invest 108:957–962

Familian A, Zwart B, Huisman HG et al (2001) Chromatin-independent binding of serum amyloid P component to apoptotic cells. J Immunol 167:647–654

Feng H, Zeng Y, Whitesell L et al (2001) Stressed apoptotic tumour cells express heat shock proteins and elicit tumour-specific immunity. Blood 97:3505–3512

Ferguson TA, Herndon J, Elzey B et al (2002) Uptake of apoptotic antigen-coupled cells by lymphoid dendritic cells and cross-priming of CD8(+) T cells produce active immune unresponsiveness. J Immunol 168:5589–5595

Finnemann SC, Bonilha VL, Marmorstein AD et al (1997) Phagocytosis of rod outer segments by retinal pigment epithelial cells requires alpha(v)beta5 integrin for binding but not for internalization. Proc Natl Acad Sci U S A 94:12932–12937

Fossum S, Rolstad B (1986) The roles of interdigitating cells and natural killer cells in the rapid rejection of allogeneic lymphocytes. Eur J Immunol 16:440–450

Fugier-Vivier IJ, Rezzoug F, Huang Y et al (2005) Plasmacytoid precursor dendritic cells facilitate allogeneic hematopoietic stem cell engraftment. J Exp Med 201:373–383

Gallucci S, Lolkema M, Matzinger P (1999) Natural adjuvants: endogenous activators of dendritic cells. Nat Med 5:1249–1255

Game DS, Lechler RI (2002) Pathways of allorecognition: implications for transplantation tolerance. Transpl Immunol 10:101–108

Gao W, Demirci G, Strom TB et al (2003) Stimulating PD-1-negative signals concurrent with blocking CD154 co-stimulation induces long-term islet allograft survival. Transplantation 76:994–999

Gao Y, Herndon JM, Zhang H (1998) Anti-inflammatory effects of CD95 ligand (FasL)-induced apoptosis. J Exp Med 188:887–896

Gershov D, Kim S, Brot N et al (2000) C-Reactive protein binds to apoptotic cells, protects the cells from assembly of the terminal complement components, and sustains an anti-inflammatory innate immune response: implications for systemic autoimmunity. J Exp Med 192:1353–1364

Gregory CD (2000) CD14-dependent clearance of apoptotic cells: relevance to the immune system. Curr Opin Immunol 12:27–34

Hamon Y, Broccardo C, Chambenoit O et al (2000) ABC1 promotes engulfment of apoptotic cells and transbilayer redistribution of phosphatidylserine. Nat Cell Biol 2:399–406

Helg C, Chapuis B, Bolle JF et al (1994) Renal transplantation without immunosuppression in a host with tolerance induced by allogeneic bone marrow transplantation. Transplantation 58:1420–1422

Henri S, Vremec D, Kamath A et al (2001) The dendritic cell populations of mouse lymph nodes. J Immunol 167:741–748

Hicks PS, Saunero-Nava L, Du Clos TW et al (1992) Serum amyloid P component binds to histones and activates the classical complement pathway. J Immunol 149:3689–3694

Hodge S, Hodge G, Scicchitano R et al (2003) Alveolar macrophages from subjects with chronic obstructive pulmonary disease are deficient in their ability to phagocytose apoptotic airway epithelial cells. Immunol Cell Biol 81:289–296

Huang FP, Platt N, Wykes M et al (2000) A discrete subpopulation of dendritic cells transports apoptotic intestinal epithelial cells to T cell areas of mesenteric lymph nodes. J Exp Med 191:435–444

Hugues S, Mougneau E, Ferlin W et al (2002) Tolerance to islet antigens and prevention from diabetes induced by limited apoptosis of pancreatic beta cells. Immunity 16:169–181

Inaba K, Turley S, Yamaide F et al (1998) Efficient presentation of phagocytosed cellular fragments on the major histocompatibility complex class II products of dendritic cells. J Exp Med 188:2163–2173

Ip WK, Lau YL (2004) Distinct maturation of, but not migration between, human monocyte-derived dendritic cells upon ingestion of apoptotic cells of early or late phases. J Immunol 173:189–196

Iyoda T, Shimoyama S, Liu K et al (2002) The CD8 +dendritic cell subset selectively endocytoses dying cells in culture and in vivo. J Exp Med 195:1289–1302

Jacobsen N, Taaning E, Ladefoged J et al (1994) Tolerance to an HLA-B,DR disparate kidney allograft after bone-marrow transplantation from same donor. Lancet 343:800

Kamath AT, Pooley J, O'Keeffe MA (2000) The development, maturation, and turnover rate of mouse spleen dendritic cell populations. J Immunol 165:6762–6770

Kang HK, Liu M, Datta SK (2007) Low-dose peptide tolerance therapy of lupus generates plasmacytoid dendritic cells that cause expansion of auto-antigen-specific regulatory T cells and contraction of inflammatory Th17 cells. J Immunol 178:7849–7858

Khera TK, Martin J, Riley SG et al (2007) Glucose modulates handling of apoptotic cells by mesangial cells: involvement of TGF-beta1. Lab Invest 87:690–701

Kim SJ, Gershov D, Ma X et al (2002) I-PLA(2) activation during apoptosis promotes the exposure of membrane lysophosphatidylcholine leading to binding by natural immunoglobulin M antibodies and complement activation. J Exp Med 196:655–665

Kleinclauss F, Perruch S, Angonin R et al (2001) Intravenous infusion of apoptotic leukocytes facilitates allogeneic hematopoietic cell engraftment after a non-myeloablative conditioning regimen and induces tolerance to donor antigens (abstract). Blood 98:735a

Kleinclauss F, Perruche S, Masson E et al (2006) Intravenous apoptotic spleen cell infusion induces a TGF-beta-dependent regulatory T cell expansion. Cell Death Differ 13:41–52

Kobayashi N, Karisola P, Pena-Cruz V et al (2007) TIM-1 and TIM-4 glycoproteins bind phosphatidylserine and mediate uptake of apoptotic cells. Immunity 27:927–940

Korb LC, Ahearn JM (1997) C1q binds directly and specifically to surface blebs of apoptotic human keratinocytes: complement deficiency and systemic lupus erythematosus revisited. J Immunol 158:4525–4528

Kurts C, Kosaka H, Carbone FR et al (1997) Class I-restricted cross-presentation of exogenous self-antigens leads to deletion of autoreactive CD8(+) T cells. J Exp Med 186:239–245

Lan YY, Wang Z, Raimondi G et al (2006) "Alternatively activated" dendritic cells preferentially secrete IL-10, expand Foxp3+CD4+ T cells, and induce long-term organ allograft survival in combination with CTLA4-Ig. J Immunol 177:5868–5877

Larsen CP, Steinman RM, Witmer-Pack M et al (1990) Migration and maturation of Langerhans cells in skin transplants and explants. J Exp Med 172:1483–1493

Lauber K, Bohn E, Krober SM et al (2003) Apoptotic cells induce migration of phagocytes via caspase-3-mediated release of a lipid attraction signal. Cell 113:717–730

Lechler RI, Batchelor JR (1982) Restoration of immunogenicity to passenger cell-depleted kidney allografts by the addition of donor strain dendritic cells. J Exp Med 155:31–41

Lechler RI, Garden OA, Turka LA (2003) The complementary roles of deletion and regulation in transplantation tolerance. Nat Rev Immunol 3:147–158

Liu K, Iyoda T, Saternus M et al (2002) Immune tolerance after delivery of dying cells to dendritic cells in situ. J Exp Med 196:1091–1097

Manfredi AA, Rovere P, Galati G et al (1998) Apoptotic cell clearance in systemic lupus erythematosus. I. Opsonization by antiphospholipid antibodies. Arthritis Rheum 41:205–214

Maree AF, Komba M, Finegood DT et al (2008) A quantitative comparison of rates of phagocytosis and digestion of apoptotic cells by macrophages from normal (BALB/c) and diabetes-prone (NOD) mice. J Appl Physiol 104:157–169

Marguet D, Luciani MF, Moynault A et al (1999) Engulfment of apoptotic cells involves the redistribution of membrane phosphatidylserine on phagocyte and prey. Nat Cell Biol 1:454–456

Marino E, Batten M, Groom J et al (2008) Marginal zone B cells of Non-obese diabetic mice expand with diabetes onset, invade the pancreatic lymph nodes and present auto-antigen to diabetogenic T cells. Diabetes 57:395–404

Mellor AL, Keskin DB, Johnson T et al (2002) Cells expressing indoleamine 2,3-dioxygenase inhibit T cell responses. J Immunol 168:3771–3776

Mellor AL, Baban B, Chandler P et al (2003) Cutting edge: induced indoleamine 2,3 dioxygenase expression in dendritic cell subsets suppresses T cell clonal expansion. J Immunol 171:1652–1655

Menges M, Rossner S, Voigtlander C et al (2002) Repetitive injections of dendritic cells matured with tumour necrosis factor alpha induce antigen-specific protection of mice from autoimmunity. J Exp Med 195:15–21

Mevorach D, Mascarenhas JO, Gershov D et al (1998) Complement-dependent clearance of apoptotic cells by human macrophages. J Exp Med 188:2313–2320

Miyasaka K, Hanayama R, Tanaka M et al (2004). Expression of milk fat globule epidermal growth factor 8 in immature dendritic cells for engulfment of apoptotic cells. Eur J Immunol 34:1414–1422

Mold C, Baca R, Du Clos TW (2002) Serum amyloid P component and C-reactive protein opsonize apoptotic cells for phagocytosis through Fcgamma receptors. J Autoimmun 19:147–154

Morelli AE, Larregina AT, Shufesky WJ et al (2003) Internalization of circulating apoptotic cells by splenic marginal zone dendritic cells: dependence on complement receptors and effect on cytokine production. Blood 101:611–620

Morimoto K, Amano H, Sonoda F et al (2001) Alveolar macrophages that phagocytose apoptotic neutrophils produce hepatocyte growth factor during bacterial pneumonia in mice. Am J Respir Cell Mol Biol 24:608–615

Munn DH, Zhou M, Attwood JT et al (1998) Prevention of allogeneic fetal rejection by tryptophan catabolism. Science 281:1191–1193

Nauta AJ, Trouw LA, Daha MR et al (2002) Direct binding of C1q to apoptotic cells and cell blebs induces complement activation. Eur J Immunol 32:1726–1736

Nauta AJ, Raaschou-Jensen N, Roos A et al (2003a) Mannose-binding lectin engagement with late apoptotic and necrotic cells. Eur J Immunol 33:2853–2863

Nauta AJ, Daha MR, van Kooten C et al (2003b) Recognition and clearance of apoptotic cells: a role for complement and pentraxins. Trends Immunol 24:148–154

Nauta AJ, Castellano G, Xu W et al (2004) Opsonization with C1q and mannose-binding lectin targets apoptotic cells to dendritic cells. J Immunol 173:3044–3050

Navratil JS, Watkins SC, Wisnieski JJ et al (2001) The globular heads of C1q specifically recognize surface blebs of apoptotic vascular endothelial cells. J Immunol 166:3231–3239

Nishimura H, Nose M, Hiai H et al (1999) Development of lupus-like autoimmune diseases by disruption of the PD-1 gene encoding an ITIM motif-carrying immunoreceptor. Immunity 11:141–151

Nouri-Shirazi M, Guinet E (2002) Direct and indirect cross-tolerance of alloreactive T cells by dendritic cells retained in the immature stage. Transplantation 74:1035–1044

O'Brien BA, Huang Y, Geng X et al (2002) Phagocytosis of apoptotic cells by macrophages from NOD mice is reduced. Diabetes 51:2481–2488

O'Connell PJ, Li W, Wang Z et al (2002) Immature and mature CD8alpha+ dendritic cells prolong the survival of vascularized heart allografts. J Immunol 168:143–154

Ogden CA, de Cathelineau A, Hoffmann PR et al (2001) C1q and mannose binding lectin engagement of cell surface calreticulin and CD91 initiates macropinocytosis and uptake of apoptotic cells. J Exp Med 194:781–795

Ozkaynak E, Wang L, Goodearl A et al (2002) Programmed death-1 targeting can promote allograft survival. J Immunol 169:6546–6553

Parr MB, Kepple L, Parr EL (1991) Langerhans cells phagocytose vaginal epithelial cells undergoing apoptosis during the murine estrous cycle. Biol Reprod 45:252–260

Pearson AM (1996) Scavenger receptors in innate immunity. Curr Opin Immunol 8:20–28

Peche H, Trinite B, Martinet B et al (2005) Prolongation of heart allograft survival by immature dendritic cells generated from recipient type bone marrow progenitors. Am J Transplant 5:255–267

Peng SL (2006) Fas (CD95)-related apoptosis and rheumatoid arthritis. Rheumatology (Oxford) 45:26–30

Pepys MB, Butler PJ (1987) Serum amyloid P component is the major calcium-dependent specific DNA binding protein of the serum. Biochem Biophys Res Commun 148:308–313

Perruche S, Kleinclauss F, Bittencourt M de C et al (2004) Intravenous infusion of apoptotic cells simultaneously with allogeneic hematopoietic grafts alters anti-donor humoral immune responses. Am J Transplant 4:1361–1365

Pickering MC, Walport MJ (2000) Links between complement abnormalities and systemic lupus erythematosus. Rheumatology (Oxford) 39:133–141

Quezada SA, Fuller B, Jarvinen LZ et al (2003) Mechanisms of donor-specific transfusion tolerance: preemptive induction of clonal T cell exhaustion via indirect presentation. Blood 102:1920–1926

Ren Y, Silverstein RL, Allen J et al (1995) CD36 gene transfer confers capacity for phagocytosis of cells undergoing apoptosis. J Exp Med 181:1857–1862

Rodriguez A, Regnault A, Kleijmeer M et al (1999) Selective transport of internalized antigens to the cytosol for MHC class I presentation in dendritic cells. Nat Cell Biol 1:362–368

Ronchetti A, Rovere P, Iezzi G et al (1999) Immunogenicity of apoptotic cells in vivo: role of antigen load, antigen-presenting cells, and cytokines. J Immunol 163:130–136

Rovere P, Vallinoto C, Bondanza A et al (1998) Bystander apoptosis triggers dendritic cell maturation and antigen-presenting function. J Immunol 161:4467–4471

Rovere P, Sabbadini MG, Vallinoto C et al (1999) Dendritic cell presentation of antigens from apoptotic cells in a pro-inflammatory context: role of opsonizing anti-beta2-glycoprotein I antibodies. Arthritis Rheum 42:1412–1420

Rovere P, Peri G, Fazzini F et al (2000) The long pentraxin PTX3 binds to apoptotic cells and regulates their clearance by antigen-presenting dendritic cells. Blood 96:4300–4306

Rubartelli A, Poggi A, Zocchi MR (1997) The selective engulfment of apoptotic bodies by dendritic cells is mediated by the alpha(v)beta3 integrin and requires intracellular and extracellular calcium. Eur J Immunol 27:1893–1900

Salama AD, Chitnis T, Imitola J et al (2003) Critical role of the programmed death-1 (PD-1) pathway in regulation of experimental autoimmune encephalomyelitis. J Exp Med 198:71–78

Sauter B, Albert ML, Francisco L et al (2000) Consequences of cell death: exposure to necrotic tumour cells, but not primary tissue cells or apoptotic cells, induces the maturation of immunostimulatory dendritic cells. J Exp Med 191:423–434

Savill J, Hogg N, Ren Y et al (1992) Thrombospondin cooperates with CD36 and the vitronectin receptor in macrophage recognition of neutrophils undergoing apoptosis. J Clin Invest 90:1513–1522

Savill J, Dransfield I, Gregory C et al (2002) A blast from the past: clearance of apoptotic cells regulates immune responses. Nat Rev Immunol 2:965–975

Sayegh MH, Fine NA, Smith JL et al (1991) Immunologic tolerance to renal allografts after bone marrow transplants from the same donors. Ann Intern Med 114:954–955

Scaffidi P, Misteli T, Bianchi ME (2002) Release of chromatin protein HMGB1 by necrotic cells triggers inflammation. Nature 418:191–195

Schagat TL, Wofford JA, Wright JR (2001) Surfactant protein A enhances alveolar macrophage phagocytosis of apoptotic neutrophils. J Immunol 166:2727–2733

Schulz O, Pennington DJ, Hodivala-Dilke K et al (2002) CD36 or alphavbeta3 and alphavbeta5 integrins are not essential for MHC class I cross-presentation of cell-associated antigen by CD8 alpha+ murine dendritic cells. J Immunol 168:6057–6065

Shi Y, Evans JE, Rock KL (2003) Molecular identification of a danger signal that alerts the immune system to dying cells. Nature 425:516–521

Sims PJ, Wiedmer T (2001) Unraveling the mysteries of phospholipid scrambling. Thromb Haemost 86:266–275

Spadafora-Ferreira M, Caldas C, Fae KC et al (2007) CD4+CD25+Foxp3+ indirect alloreactive T cells from renal transplant patients suppress both the direct and indirect pathways of allorecognition. Scand J Immunol 66:352–361

Stark MA, Huo Y, Burcin TL et al (2005) Phagocytosis of apoptotic neutrophils regulates granulopoiesis via IL-23 and IL-17. Immunity 22:285–294

Steinman RM, Nussenzweig MC (2002) Avoiding horror autotoxicus: the importance of dendritic cells in peripheral T cell tolerance. Proc Natl Acad Sci U S A 99:351–358

Steinman RM, Hawiger D, Nussenzweig MC (2003) Tolerogenic dendritic cells. Annu Rev Immunol 21:685–711

Steptoe RJ, Ritchie JM, Jones LK et al (2005) Autoimmune diabetes is suppressed by transfer of proinsulin-encoding Gr-1 +myeloid progenitor cells that differentiate in vivo into resting dendritic cells. Diabetes 54:434–442

Stuart LM, Lucas M, Simpson C et al (2002) Inhibitory effects of apoptotic cell ingestion upon endotoxin-driven myeloid dendritic cell maturation. J Immunol 168:1627–1635

Sun E, Gao Y, Chen J et al (2004) Allograft tolerance induced by donor apoptotic lymphocytes requires phagocytosis in the recipient. Cell Death Differ 11:1258–1264

Sykes M, Sachs DH (2001) Mixed chimerism. Philos Trans R Soc Lond B Biol Sci 356:707–726

Takahashi M, Kobayashi Y (2003) Cytokine production in association with phagocytosis of apoptotic cells by immature dendritic cells. Cell Immunol 226:105–115

Takizawa F, Tsuji S, Nagasawa S (1996) Enhancement of macrophage phagocytosis upon iC3b deposition on apoptotic cells. FEBS Lett 397:269–272

Talmage DW, Dart G, Radovich J et al (1976) Activation of transplant immunity: effect of donor leukocytes on thyroid allograft rejection. Science 191:385–388

Taner T, Hackstein H, Wang Z et al (2005) Rapamycin-treated, alloantigen-pulsed host dendritic cells induce ag-specific T cell regulation and prolong graft survival. Am J Transplant 5:228–236

Thum T, Bauersachs J, Poole-Wilson PA et al (2005) The dying stem cell hypothesis: immune modulation as a novel mechanism for progenitor cell therapy in cardiac muscle. J Am Coll Cardiol 46:1799–1802

Turnquist HR, Raimondi G, Zahorchak AF et al (2007) Rapamycin-conditioned dendritic cells are poor stimulators of allogeneic CD4+ T cells, but enrich for antigen-specific Foxp3+ T regulatory cells and promote organ transplant tolerance. J Immunol 178:7018–7031

Urban BC, Willcox N, Roberts DJ (2001) A role for CD36 in the regulation of dendritic cell function. Proc Natl Acad Sci U S A 98:8750–8755

U.S. Organ Procurement and Transplantation Network, Health Resources and Services Administration, and Healthcare Systems Bureau, Rockville, MD; United Network for Organ Sharing, Richmond, VA. Transplants by Donor Type; U.S. Transplants Performed: Jan. 1, 1988—Oct. 31, 2007. Retrieved Jan. 10, 2008 from http://www.optn.org/latestData/rptData.asp

Valujskikh A, Matesic D, Gilliam A et al (1998) T cells reactive to a single immunodominant self-restricted allopeptide induce skin graft rejection in mice. J Clin Invest 101:1398–1407

Valujskikh A, Lantz O, Celli S et al (2002) Cross-primed CD8(+) T cells mediate graft rejection via a distinct effector pathway. Nat Immunol 3:844–851

Vandivier RW, Ogden CA, Fadok VA et al (2002a) Role of surfactant proteins A, D, and C1q in the clearance of apoptotic cells in vivo and in vitro: calreticulin and CD91 as a common collectin receptor complex. J Immunol 169:3978–3986

Vandivier RW, Fadok VA, Ogden CA et al (2002b) Impaired clearance of apoptotic cells from cystic fibrosis airways. Chest 121:89S

Verbovetski I, Bychkov H, Trahtemberg U et al (2002) Opsonization of apoptotic cells by autologous iC3b facilitates clearance by immature dendritic cells, down-regulates DR and CD86, and up-regulates CC chemokine receptor 7. J Exp Med 196:1553–1561

Verhasselt V, Vosters O, Beuneu C et al (2004) Induction of FOXP3-expressing regulatory CD4pos T cells by human mature autologous dendritic cells. Eur J Immunol 34:762–772

Voll RE, Herrmann M, Roth EA et al (1997) Immunosuppressive effects of apoptotic cells. Nature 390:350–351

Vremec D, Pooley J, Hochrein H et al (2000) CD4 and CD8 expression by dendritic cell subtypes in mouse thymus and spleen. J Immunol 164:2978–2986

Wang Z, Larregina AT, Shufesky WJ et al (2006) Use of the inhibitory effect of apoptotic cells on dendritic cells for graft survival via T cell deletion and regulatory T cells. Am J Transplant 6:1297–1311

Wood W, Turmaine M, Weber R et al (2000) Mesenchymal cells engulf and clear apoptotic footplate cells in macrophageless PU.1 null mouse embryos. Development 127:5245–5252

Wu Y, Tibrewal N, Birge RB (2006) Phosphatidylserine recognition by phagocytes: a view to a kill. Trends Cell Biol 16:189–197

Xia CQ, Peng R, Qiu Y et al (2007) Transfusion of apoptotic beta-cells induces immune tolerance to beta-cell antigens and prevents type 1 diabetes in NOD mice. Diabetes 56:2116–2123

Xu DL, Liu Y, Tan JM et al (2004) Marked prolongation of murine cardiac allograft survival using recipient immature dendritic cells loaded with donor-derived apoptotic cells. Scand J Immunol 59:536–544

Yrlid U, Wick MJ (2000) Salmonella-induced apoptosis of infected macrophages results in presentation of a bacteria-encoded antigen after uptake by bystander dendritic cells. J Exp Med 191:613–624

Zwart B, Ciurana C, Rensink I et al (2004) Complement activation by apoptotic cells occurs predominantly via IgM and is limited to late apoptotic (secondary necrotic) cells. Autoimmunity 37:95–102

Chapter 12
Impact of Tumour Cell Death on the Activation of Anti-tumour Immune Response

Jiřina Bartůňková and Radek Špíšek

Abstract: The primary function of the immune system is to protect the host against pathogens. Success of vaccination against infectious diseases has led to efforts to develop strategies to activate a specific immune response against cancer. However, this task has proved to be very challenging. In the last few years there has been rapid progress in the identification of the molecular mechanisms by which the immune system detects pathogens and initiates immune responses. By contrast, recognition of neoplastic cells lags behind, and induction of specific anti-tumour immunity is even less understood. In this chapter we discuss evidence for the role of the immune system in the control of tumour growth. We also review recent advances in the understanding of mechanisms that can alert the immune system to the presence of neoplastic lesions, and we discuss recent data on the immunogenicity of tumour cells and their interaction with antigen presenting cells.

Keywords: Cancer immune-editing • Dendritic cells • Heat shock proteins • Immunogenicity of tumour cells • Immunotherapy • Tumour immunology

12.1 Introduction

A combination of surgery, chemotherapy and radiotherapy greatly improves prognosis of cancer patients. Though this approach results in significant reduction of tumour mass, a small population of precursor tumour cells or cancer stem cells

J. Bartůňková (✉) · R. Špíšek
Department of Immunology
Charles University, Second Medical School,
University Hospital Motol, V Úvalu 84,
Prague 5, 150 06, Czech Republic
Tel.: +420 224 435 961
Fax: +420 224 435 962
E-mail: jirina.bartunkova@lf2.cuni.cz; radek.spisek@lf2.cuni.cz

D.V. Krysko, P. Vandenabeele (eds.), *Phagocytosis of Dying Cells,*
DOI: 10.1007/978-1-4020-9292-3_3, © Springer Science+Business Media B.V. 2009

often survives and subsequently gives rise to a new population of tumour cells and consequent relapse. Clearly, other treatment modalities should be sought to further improve clinical outcome (Spisek 2006; Spisek and Dhodapkar 2006).

The remarkable success of vaccination programs against infectious diseases is one of the most impressive triumphs of modern medicine (Ada 1990, 2001). Though the aim of vaccination is protection against infectious diseases, its success prompted efforts to develop strategies to activate a specific immune response against cancer (Spisek and Dhodapkar 2006). However, this task has proved to be very challenging. It soon became clear that vaccination strategies used against infectious diseases can not be applied directly to cancer because of several fundamental differences between the two situations (Rosenberg et al. 2004; Lollini et al. 2005; Lollini et al. 2006). During evolution the immune system has evolved powerful mechanisms to recognize pathogens and to elicit specific immune responses against them. First during the last few years there has been a rapid progress in the identification of the molecular mechanisms deployed by the immune system to detect pathogens and initiate pathogen-specific immune responses. By contrast, the recognition of neoplastic cells is not as well understood, and knowledge of the induction of specific anti-tumour immunity lags even further behind. Second, the protective efficacy vaccines against pathogens has been evaluated and proven mostly in healthy individuals (Lollini and Forni 2002; Lollini and Forni 2003). By contrast, vaccination against cancer has been tested almost exclusively as a therapy in patients who already had advanced tumours (Gilboa 2004).

In this chapter we will discuss evidence for the role of the immune system in the control of tumour growth. We will also review recent advances in the understanding of mechanisms that can alert the immune system to the presence of neoplastic lesions, and we will discuss recent data on the immunogenicity of tumour cells and their interaction with antigen presenting cells.

12.2 Role of the Immune System in Protection Against Cancer

A pre-requisite for the development of successful protocols for cancer immunotherapy is the notion that the immune system can recognize transformed cells and destroy them before the onset of clinically apparent disease. This idea was first suggested in the late 19th century by William B. Coley, who noted that rare events of spontaneous tumour regression were sometimes preceded by severe infections. Thomas and Burnet demonstrated in the 1950s that inbred mice can be immunized against carcinogen-induced transplantable tumours and that the tumour rejection antigens are tumour-specific. These findings led them to formulate the cancer immunosurveillance hypothesis. They postulated that "sentinel thymus dependent cells of the body constantly surveyed host tissues for nascent transformed cells"

(Burnet 1970). It took over 30 years to convincingly validate the prediction made in the cancer immunosurveillance hypothesis that immunodeficient individuals would exhibit an increased incidence of tumours (Stutman 1975). Definitive proof came later from studies performed in genetically modified immune-deficient mice.

12.2.1 Effector Mechanisms in Tumour Protection

One of the most definitive studies employed gene-targeted mice lacking the recombinase activating genes RAG-1 or RAG-2. These mice lack T and B lymphocytes and natural killer T (NKT) cells (Shinkai et al. 1992). RAG$^{-/-}$ mice not only developed chemically induced tumours more rapidly and with greater frequency than immunocompetent controls, they also formed many more spontaneous tumours (Shankaran et al. 2001).

Studies in other mice models identified the role of various components of the immune system in cancer prevention. Selective depletion of NK cells by anti-asialo-GM1 rendered mice three times more susceptible to chemically induced tumours (Smyth et al. 2001). Similarly, Jα281$^{-/-}$ mice lacking NKT cells had a higher incidence of chemically induced tumours (Smyth et al. 2000; Smyth et al. 2001; Street et al. 2001). Administration of an NKT cell-activating ligand (α-galactosylceramide) in models of chemically induced tumourigenesis decreased the incidence of tumours and prolonged the latency period to tumour formation compared to control mice. These studies identified both innate and adaptive immunity as important effectors in the control of tumour formation.

Pioneering studies in the laboratory of Dr. Robert Schreiber at Washington University demonstrated that IFNγ plays a critical role in immune-surveillance (Kaplan et al. 1998). Abrogation of IFNγ function by neutralization with monoclonal antibodies, or by genetic manipulation of its expression or its signalling pathways enhanced the susceptibility to both chemically induced and spontaneous tumours (Dighe et al. 1994). It is likely that IFNγ acts both on the host and on the tumour. It promotes the generation of antigen-specific CD4 Th1 T cells and cytotoxic T cells (CTLs), it enhances activation of dendritic cells, NK cells, and macrophages, and it inhibits angiogenesis (Bach et al. 1997; Hayakawa et al. 2002). At the level of tumour cells IFNγ induces increased expression of major histocompatibility complex (MHC) class I components and enhances antigen processing and presentation (Wong et al. 1997). Perforin was also shown to be important for anti-tumour immune responses because its deficiency makes mice significantly more susceptible to chemically induced tumours (van den Broek et al. 1996). Untreated perforin$^{-/-}$ mice also showed an increased incidence of spontaneous disseminated lymphomas.

monly breast or ovarian adenocarcinomas or small-cell lung cancer. These tumours usually show very limited spread, and often only microscopic foci of tumour can be found, often with inflammatory infiltration (Darnell and Posner 2003b; Albert and Darnell 2004). Spontaneous tumour regression has also been described. The identities of some of the neuron-specific antigens were revealed by screening of expression libraries with patient sera and subsequent cloning. This showed that PND antigens are ectopically expressed in tumours of PND patients. Analysis of the immune response against PND antigens revealed the presence of antibodies and PND antigens reactive T cells (Albert et al. 1998). Although PND can not be considered as a pre-neoplasia, these studies offer evidence of naturally occurring tumour immunity in humans.

12.2.5 Mechanisms of Immune Escape

Despite immunosurveillance, tumour cells have evolved several strategies to evade the immune system (Zitvogel et al. 2006). Tumours often induce the production of factors that may suppress or attenuate the anti-tumour immune response, such as transforming growth factor-beta (TGF-β) and interleukin-10 (IL-10; Gilboa 2004). Tumours may also escape the immune system due to mutations in the antigen processing pathway, such as those in β2-microglobulin, TAP, or proteasome components (Marincola et al. 2000). These mutations confer resistance to CD8 T lymphocytes that is very difficult to overcome although down-regulation of MHC class I expression can make tumour cells more susceptible to NK cell-mediated death. Hopefully, as shown in murine studies, the induction of a broad immune response, including the activation of CD4 T cells and antibody production by B cells, could compensate for this defect (Hung et al. 1998; Mumberg et al. 1999).

Tumour cells can also express B7-H1, B7-H4 or indolamine 2,3 oxygenase that can induce anergy of antigen-specific T cells (Curiel et al. 2003; Uyttenhove et al. 2003). Recent studies also identified frequent activation of STAT3 in both tumour cells and dendritic cells; this leads to immune suppression by inhibiting production of pro-inflammatory cytokines and costimulatory molecules (Burdelya et al. 2005; Cheng et al. 2003; Wang et al. 2004b).

12.2.6 Antigen Presenting Cells in Tumour Immunity vs. Tolerance

Selective pressure exerted on tumour cells by specific anti-tumour immune responses, as proposed by the cancer immunoediting hypothesis, may not be the only mechanism accounting for the escape of a tumour from the control of the immune system. Induction of tolerance to tumour cells, possibly by their recognition in the absence of appropriate stimulatory signals, has also been shown to result in the progressive growth of previously immunogenic tumours (Willimsky and Blankenstein 2005).

Tumour cells themselves are poor antigen presenting cells. To induce an immune response, tumour cells need to be captured, processed and their antigens presented by professional antigen presenting cells, especially dendritic cells (Pardoll 1998). Mice depleted of dendritic cells failed to generate protective anti-tumour immunity (Zaft et al. 2005; Frenzel et al. 2006; Winau et al. 2006). Dendritic cells are professional antigen presenting cells capable of activating naive T cells and initiating primary immune responses (Banchereau and Steinman 1998; Steinman et al. 2005). Different subsets of dendritic cells exist in mice and in humans. In humans, two major subsets of dendritic cells, plasmacytoid and myeloid dendritic cells, can be identified by the surface markers they express and by their functional characteristics (Shortman and Liu 2002). Myeloid dendritic cells in the skin can be further subtyped into Langerhans cells in the epidermis and interstitial dendritic cells in dermis. Large numbers of monocyte-derived dendritic cells can also be generated in vitro from monocytes in the presence of IL-4 and GM-CSF (Bender et al. 1996).

Precursors of dendritic cells enter tissues as immature dendritic cells where they continuously sample the environment. Upon encountering a pathogen or other danger signals, they undergo a process of activation commonly termed maturation and leading to the acquisition of the ability to activate specific immune response (Divito and Morelli, Chap. 11, this Vol.; Banchereau and Steinman 1998; Steinman and Banchereau 2007). Dendritic cell maturation is a complex process characterized by increased antigen processing, translocation of MHC:peptide complexes to the cell surface, increased expression of co-stimulatory molecules, cytokines production, and abrogation of phagocytic activity. Due to a change in their expression of chemokine receptors, activated dendritic cell migrate via afferent lymphatics to the secondary lymphoid organs, where they present antigens to antigen-specific T cells and initiate immune responses. Moreover, dendritic cells can present captured exogenous antigens not only on MHC class II but also on MHC class I molecules. This process is called cross-presentation and allows dendritic cells to activate efficient CTL responses against virally infected cells and tumour cells.

Importantly, immature or incompletely activated dendritic cells efficiently generate IL-10 producing regulatory T cells (Tregs) and FoxP3$^+$ CD4$^+$ CD25$^+$ Tregs (Dhodapkar et al. 2001; Dhodapkar and Steinman 2002; Zou 2006). Expansion of Tregs is significantly enhanced in the presence of TGFβ and IL-10, which are often produced by tumour cells or tumour stroma (Yamazaki et al. 2007). Presentation of tumour cells by dendritic cells affected by immunosuppressive cytokine milieu in the tumour tissue may thus promote generation of Tregs (Wang et al. 2004a). A number of studies correlated the frequency of Tregs with an adverse prognosis for cancer patients (Woo et al. 2001; Liyanage et al. 2002; Curiel et al. 2004).

Dendritic cell activation during infection has been extensively studied. These studies led to the identification of various families of pattern recognition receptors (e.g. Toll-like receptors) that recognize evolutionarily conserved pathogen-associated molecular patterns (PAMPs; Medzhitov 2001; Iwasaki and Medzhitov 2004; Pasare and Medzhitov 2004a; Pasare and Medzhitov 2004b). In his self-non-self theory, Charles Janeway predicted the existence of such patterns (Janeway et al. 1996; Medzhitov and Janeway 1997a; Medzhitov and Janeway 1997b; Medzhitov and Janeway 2000a; Medzhitov and Janeway 2000b). Dendritic cell activation can be further modified by other danger signals, such as pro-inflammatory cytokines, innate lymphocytes, and CD40L on activated T cells (Edwards et al. 2002; Reis e Sousa et al. 2003; Sporri and Reis e Sousa 2005). Identification of potent activatory signals expressed by tumour cells would significantly contribute to elucidating the interactions between tumour cells and the immune system and would facilitate the design of more effective immunotherapeutic strategies.

12.3 Cancer Immunotherapy Using Dendritic Cell-based Tumour Vaccines

Discovery of the key role of dendritic cells in initiating anti-tumour responses initiated intensive efforts to design immunotherapeutic trials in which activated mature dendritic cells presenting tumour antigens would induce specific anti-tumour responses (Figdor et al. 2004). Use of whole killed tumour cells as a source of tumour antigens is one of the most often tested strategies for several reasons (Divito and Morelli, Chap. 11, this Vol.):

- this approach does not require prior knowledge of potent tumour rejection antigens;
- presentation of peptides derived from tumour cells on both MHC class I and class II leads to the induction of broad immune response;
- use of autologous tumour cells provides a full spectrum of tumour-specific antigens, including patient-specific antigens resulting from random mutations.

The first dendritic cell-based approach consisted of the administration of in vitro generated dendritic cells loaded with tumour antigens (Figs. 12.1 and 12.2). Dendritic cells can be activated in vitro by exogenous stimuli, such as TLR ligands (e.g. Poly I:C or clinically used bacterial lysates; Spisek et al. 2004). A number of studies convincingly showed that mature dendritic cells pulsed with killed tumour cells can efficiently process and cross-present tumour antigens and expand tumour-specific CTLs as well as CD4 Th cells (Dhodapkar et al. 2002; Labarriere et al. 2002; Schnurr et al. 2002; Spisek et al. 2002; Tobiasova et al. 2007). Migration of dendritic cells from the tissues to the lymph nodes is tightly controlled and involves many different mediators and their receptors. Inefficient migration of dendritic cells generated in vitro can be partially enhanced by pre-conditioning of the vaccine injection site with inflammatory cytokines.

An alternative approach that is currently being explored is the targeting tumour antigens directly to dendritic cells in situ, thereby exploiting the intricate migratory ability of dendritic cells in vivo. This can be done either by injecting dead tumour cells or by killing tumour cells in situ by chemotherapy. A pre-requisite for this approach is the induction of a type of cell death that provides activating signals to dendritic cells. In other words, there is an urgent need to identify signals and drugs that induce an immunogenic type cell death (Adema et al. 2005; De Vries et al. 2007; Tacken et al. 2007; Verdijk et al. 2007).

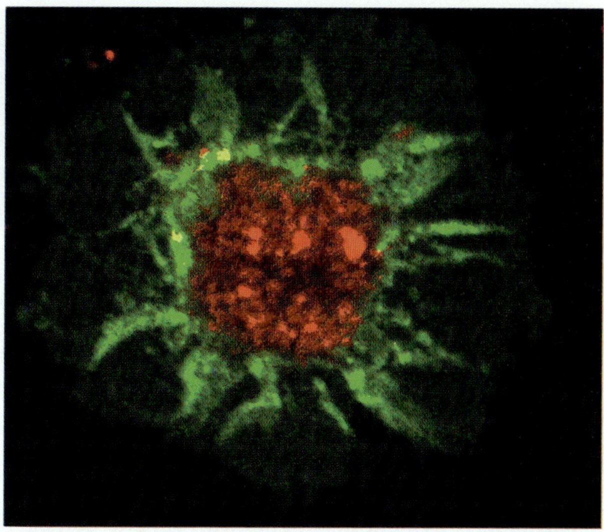

Fig. 12.1 Confocal microscopy (Olympus, Czech Republic) of a dendritic cell (green) phagocytosing apoptotic human ovarian cancer cells isolated from ascites (red).

12.4 Interactions Between Dead Tumour Cells and Dendritic Cells

12.4.1 Various Forms of Cell Death

Tumour cells dying either naturally or as a result of anti-tumour chemotherapy are rapidly cleared by phagocytes, including dendritic cells. How exogenous PAMPs initiate dendritic cell activation when these cells encounter pathogens is understood, but activation of dendritic cells after encounter with dead tumour cells remains elusive. The danger signal hypothesis presumes the existence of specific molecules (damage-associated molecular patterns, DAMPs) that alert the immune system to the presence of damaged or injured tissue even in the absence of infection. Extracellular matrix components, such as hyaluronan and heparan sulfate, have been identified as markers of tissue injury (Scheibner et al. 2006). The existence of tumour-specific DAMPs has been deduced from the results confirming the existence of cancer immunosurveillance. The first series of studies aimed to link the outcome of the interaction between tumour cell and dendritic cell to the type of cell death. Although these initial studies often yielded contradictory results, it was considered logical that apoptotic cells would not induce dendritic cell maturation (Basu et al. 2000; Sauter et al. 2000; Somersan et al. 2001). During homeostatic turnover, billions of cells undergo apoptosis each day and are replaced by newly differentiated progeny. Dying cells are readily removed by phagocytes without inducing autoimmunity (Ucker, Chap. 6; Lacy-Hulbert, Chap. 7, this Vol.; Savill et al. 1993; Savill and Fadok 2000; Savill et al. 2002). However, an aberrant immune response against an excess of apoptotic cells may be harmful, as in

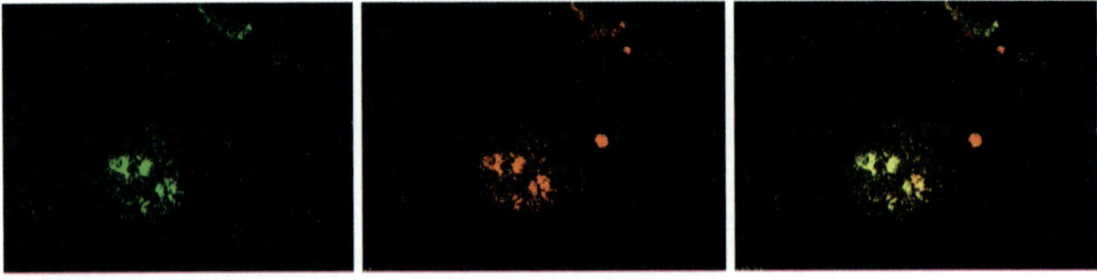

Fig. 12.2 Phagocytosis of apoptotic ovarian cancer cells by immature dendritic cells. Confocal microscopy of PKH-26-labelled (red) irradiated tumour cells engulfed by HLA-DR (FITC) labelled (green) DC after 4 h of co-cultivation. The co-localization of FITC anti-LAMP1/2 antibody with red-labeled engulfed tumour cells shows that at this time the apoptotic cells reach the late endosomal and lysosomal compartments. Pictures reproduced from Tobiasová et al. 2007, with the permission of publisher.

systemic lupus erythematosus (Mevorach, Chap. 10, this Vol.; Walport 2000). This is particularly evident in patients with genetic defects in the complement pathway, in whom defective and delayed clearance of apoptotic cells leads to the development of autoantibodies against intracellular antigens displayed on the surface of apoptotic cells.

Apoptosis as a "physiological type of cell death" was considered non-immunogenic or immunologically silent. By contrast, necrosis often leads to the disruption of the cell membrane and release of the cellular contents, including inflammatory mediators such as IL-8 and TNF (Fadok et al. 2001). However, subsequent studies challenged this over-simplified and misleading concept, as neither killed tumour cells with apoptotic morphology nor necrotic cells induced protective immunity against subsequent challenge with live tumour cells. Recent studies in murine and human models have brought some novel insights into the mechanisms of cell death that activate specific immunity.

Confusion resulting from earlier studies can be attributed mostly to incorrect terminology. A shared terminology is a pre-requisite for a debate about the relationship between the tumour cell death type and the activation of the immune system. Current nomenclature is based on the morphological appearance of a dying cell, and it distinguishes four types of cell death (Krysko et al., Chap. 1; Diez-Fraile et al., Chap. 2, this Vol.; Kroemer et al. 2005; Galluzzi et al. 2007):

- **Apoptosis:** type I cell death is characterized by chromatin condensation, nuclear fragmentation and formation of apoptotic bodies;
- **Autophagy:** type II cell death characterized by sequestration of cellular organelles in double-membraned vacuoles and degradation by lysosomal hydrolases (Maiuri et al. 2007);
- **Necrosis:** type III cell death usually characterized by cytoplasmic swelling followed by loss of membrane integrity and subsequent release of cellular contents;

- **Mitotic catastrophe** is the last distinct type of cell death occurring after failed mitosis. Hallmarks of mitotic catastrophe are micronucleation and multinucleation (Castedo et al. 2004a; Castedo et al. 2004b; Castedo et al. 2004c).

Thus description of a dead cell as apoptotic is only a description of its morphology; it does not reflect its immunogenicity or for the outcome of the immune response. Morphological criteria are probably not relevant for the debate about the consequences of cell death for the activation of the immune system. The immunogenicity of tumour cells rather than their morphology should be taken into account.

12.4.2 Recognition of Tumour Cells by the Immune System— Molecular Markers of Immunogenicity

In this section, we discuss novel findings supporting the concept that chemotherapy can, under certain circumstances, contribute to the induction of measurable anti-tumour immune responses and that a combination of chemotherapy and immuno-therapy could represent another step towards improving anti-cancer therapy. We will discuss recently identified molecular mechanisms underlying immunogenic cell death.

It is important to emphasize that the interaction between dying tumour cells and dendritic cells is mediated by multiple receptors that often engage opposing activatory and inhibitory signalling pathways (Peter et al., Chap. 3; Napirei and Mannherz, Chap. 4; Gronski and Ravichandran, Chap. 5; Gregory and Pound, Chap. 9, this Vol.). The ultimate outcome of cellular interaction thus depends on the downstream integration of positive and negative signals from several pathways and receptors, including uptake and pattern recognition receptors, and signals from inflammatory/anti-inflammatory cytokines (Ucker, Chap. 6; Lacy-Hulbert, Chap. 7, this Vol.; Dhodapkar et al. 2008). Despite this complexity, recent studies identified several molecules and mechanisms that alert the immune system of the presence of neoplastic cells. These molecules can serve as markers of immunogenic tumour cell death.

12.4.2.1 DNA Damage Response to Pre-neoplastic Cells

Tumourigenesis is an evolutionary process that selects for genetic and epigenetic changes and thereby enables evasion of anti-proliferative and cell-death inducing mechanisms that normally limit clonal expansion of somatic cells. During oncogenesis most tumours acquire genetic instability, but how early this occurs and whether it drives tumour development is unknown. A recent report has shown that in clinical specimens from different stages of human tumours, including urinary bladder, breast, lung and colon cancer, the early pre-neoplastic lesions commonly express markers of an activated DNA damage response. DNA damage response is also activated by DNA damaging agents, such as ionizing irradiation or topoisomerase inhibitors. DNA damage response results in phosporylation of kinases ATM and Chk2 and tumour suppressor p53 (Bartkova et al. 2005; Gorgoulis et al. 2005). It was also induced in cultured cells upon over-expression of several protooncogenes known to de-regulate DNA replication, such as Cyclin E, Cdc25A or S phase-promoting transcription factor E2F1, all of which are commonly over-expressed in many carcinomas. It was further shown that activation of the DNA damage response cascade in human cells occurs very early in tumourigenesis, even before genomic instability and malignant conversion. It thus appears that tumourigenic events early in the progression of major human cancers activate an inducible barrier against tumour progression and genetic instability caused by de-regulated DNA replication or DNA damage. Some of the candidate inducers of this response include de-regulation of growth factors pathways, mutations in DNA repair genes such as BRCA2, and over-expression of regulatory proteins that advance the cell cycle, such as cyclin E and Cdc25A. Activation of DNA damage response pathway leads to p53-dependent cell-cycle blockade, senescence and apoptosis, and is thus likely to contribute to long latency periods or failure of early pre-neoplastic lesions to ever become malignant. Tumour progression in the presence of this and possibly other mechanisms relies on selection of cells defective in DNA damage components (such as ATM and p53) with compromised cell cycle arrest and apoptosis. Interestingly, the DNA damage response not only activates cell cycle arrest but also provides a link to the activation of the immune system by inducing the expression of NKG2D ligands on tumour cells (Gasser et al. 2005). Induction of NKG2D ligand expression on cells in the process of transformation is dependent on ATM and Chk1. NKG2D is an activating receptor expressed on NK cells, NKT cells and activated CD8 cells. Activation of cytotoxic effector mechanisms by NKG2D ligands on tumour cells further potentiates defence against cells in the early stages of malignant transformation and limits the formation of tumour escape variants.

12.4.2.2   Cell Membrane-Bound Markers: HSP70, HSP90 and Calreticulin

Heat Shock Proteins

While the DNA damage response can ensure from both intrinsic and extrinsic signals, a second group of markers of immunogenic cell death was identified on tumour cells killed by certain chemotherapeutics. Changes in the spectrum of proteins expressed on the plasma membrane are typical hallmarks of cell death (Napirei and Mannherz, Chap. 4, this Vol.; Gardai et al. 2006; Krysko et al. 2006). Therefore, chemotherapy-induced cell death can be accompanied by a different spectrum of proteins on the plasma membrane, and signals delivered by dying tumour cells can influence the functional characteristics of dendritic cells, especially antigen uptake, processing and presentation, and dendritic cell maturation status. Induction of heat shock proteins (HSPs) is a typical response to cellular stress. Inducible HSPs are chaperone proteins that protect cells against death by refolding damaged, misfolded proteins and by guiding them to proteasomes for degradation. They also inhibit apoptosis by interfering with various components of the apoptotic signalling pathway, such as Apaf-1 (Beere et al. 2000). HSPs are typically located in the endoplasmic reticulum, but at least two members of the HSP family, HSP70 and HSP90, are expressed on the cell surface, although their translocation mechanism is unclear (Udono and Srivastava 1993). It is important to note that over-expression of HSPs has been detected in many tumours (Jaattela 1995; Garrido et al. 1998). However, it seems that cytoplasmic over-expression of HSPs has no effect on tumour cell immunogenicity or on their cytoprotective and anti-apoptotic function as opposed to cell surface expression. HSP70 expression was detected in PC12 tumour cells and its ligation by phosphatidylserine accelerated tumour cell apoptosis (Arispe et al. 2004).

Recent studies reported induction of immunogenic cell death by bortezomib (Velcade), a specific inhibitor of the 26S proteasome subunit. It induced immunogenic cell death in primary myeloma tumour cells by delivering activation signals to dendritic cells. The immunogenicity of the tumour cells correlated with the expression of HSP90 on the surface of myeloma cells killed by bortezomib (Spisek et al. 2007). Cells treated with dexamethasone or gamma irradiation failed to up-regulate HSP90. Cell surface expression of HSP90 was critical for immunogenicity of killed tumour cells, as activation of dendritic cells was dependent on cell contact, and specific blockade of HSP90 abolished immunogenicity of myeloma cells. Interestingly, uptake of bortezomib-killed tumour cells by dendritic cells was sufficient to induce myeloma-specific T cells in culture without the need for an exogenous activation signal. This finding is in accordance with murine studies showing that cell surface HSPs represent a potent immunogenic signal (Dai et al. 2003). Moreover, up-regulation of maturation-associated markers was reported on murine dendritic cells that phagocytosed bortezomib-killed 67NR colon carcinoma cells (Demaria et al. 2005). This report also showed increased the immunogenicity of bortezomib-killed tumour cells in tumour protection experiments. Even clinical observation of better clinical responses in patients who developed vasculitis rush after bortezomib administration as well as the observations that the clinical responses may be delayed by several months in follicular lymphoma, support the possibility that clinical effect of bortezomib may be partially mediated by induction of cell mediated immunity (Gerecitano et al. 2006). Immunostimulatory activity and increased immunogenicity of tumour cells enriched in HSPs after induction of heat shock was also reported in various animal models of tumour vaccination as well as in a human in vitro model (Masse et al. 2004; Shi et al. 2006).

Calreticulin

The group of L. Zitvogel and G. Kroemer screened an array of chemotherapeutic drugs for their ability to induce immunogenic cell death in a mouse model of colon carcinoma (CT26 cell line). Using tumour cells killed by tested drugs for their capacity to serve as a protective vaccine in tumour protection experiments, they identified anthracyclins as compounds that induced immunogenic cell death even in the absence of external activation signals (Zitvogel et al. 2004; Casares et al. 2005). Rapid translocation of the chaperone protein calreticulin, which resides in the endoplasmic reticulum, to the cell surface of dying tumour cells was identified as a mechanism underlying the increased immunogenicity of tumour cells (Obeid et al. 2007a; Obeid et al. 2007b). Calreticulin translocation enhanced phagocytosis of tumour cells by dendritic cells, and blockade of calreticulin abolished anthracyclin-induced immunogenicity of killed tumour cells in mice. However, as in the case of bortezomib-induced HSP90 expression, the mechanisms accounting for calreticulin translocation are not yet known. Upon exposure to anthracyclins or ionizing irradiation, calreticulin translocates very quickly to the outer leaflet of the cell membrane, whereas the other tested chemotherapeutics (mitomycin C and etoposide) failed to induce calreticulin translocation and thus do not induce immunogenic cell death.

Questions remain concerning the mechanisms that cell membrane-bound chaperones stimulate the immune system. There are three possible scenarios based on published studies.

1. Beside repairing or transporting misfolded proteins, chaperones bound to the cell surface may transfer tumour antigens to MHC class I and class II processing pathways. Indeed, HSP-peptide complexes were reported as particularly efficient in providing antigens for cross-presentation (Binder et al. 2007; Binder and Srivastava 2004; Binder and Srivastava 2005).
2. Chaperones could also provide an activating stimulus to dendritic cells, as seen for HSPs (Binder et al. 2000a; Singh-Jasuja et al. 2000).
3. Besides providing an activating signal, chaperones also have an important role as "eat me" signals and increase the rate of tumour cell uptake by dendritic cells.

Receptors on dendritic cells proposed to specifically mediate interaction with chaperones include CD91 and scavenger receptors LOX-1 and SR-A, although this is still a matter of considerable debate (Binder et al. 2000b; Binder et al. 2000c; Basu et al. 2001; Becker et al. 2002).

12.4.2.3 Soluble Endogenous Danger Signals: Uric Acid and HMGB1

Upon disintegration of the plasma membrane, dying tumour cells (necrotic or late apoptotic) can release factors that act as endogenous DAMPs and alert the immune system to danger (Peter et al., Chap. 3, this Vol.).

Uric acid

In an elegant study, Shi et al. identified uric acid released from injured cells as an endogenous danger signal that activates dendritic cells (Shi and Rock 2002; Shi et al. 2003; Rock et al. 2005). The concentration of uric acid increases in stressed

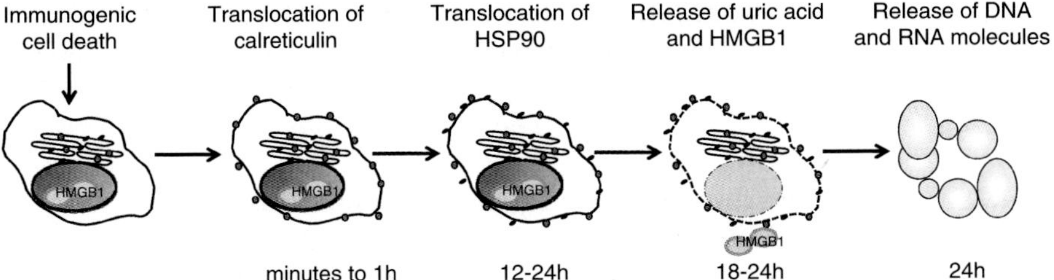

Fig. 12.3 Schematic representation of immunogenic cell death markers after induction of cell death.

cells due to rapid degradation of RNA and DNA molecules and conversion of the generated purines to uric acid. An increase in uric acid concentration was reported after heat shock treatment, cycloheximide treatment of EL4 cells, and UV irradiation of fibroblasts. Inhibition of uric acid production by allopurinol was shown to inhibit induction of antigen-specific CTL responses. Uric acid is specifically recognized in antigen presenting cells by inflammasome cryopyrin/NALP3, a large multiprotein complex that also recruits inflammatory caspases and triggers their activation (Martinon et al. 2006).

High Mobility Group Box 1 Protein (HMGB1)

Recently, Apetoh et al. performed elegant studies that led to the discovery of another soluble endogenous danger signal (Apetoh et al. 2007a; Apetoh et al. 2007b; Tesniere et al. 2008). They found that TLR4 deficiency compromised immunogenicity of tumour cells and identified HMGB1 as a specific TLR4 ligand released from dying tumour cells late in apoptosis. HMGB1 is a non-histone chromatin binding protein that influences transcription and other cell functions. HMGB1 is actively secreted from inflammatory cells and released from necrotic cells (Wang et al. 1999; Scaffidi et al. 2002). Identity of its receptor is still controversial but it signals through TLR2 and TLR4 and through receptor for advanced glycosylation products (RAGE; Park et al. 2004; Rovere-Querini et al. 2004; Park et al. 2006). Depletion of HMGB1 from tumour cells abolished TLR4-dependent dendritic cell-mediated presentation of tumour antigens. Relevance of this study is further illustrated by the finding that breast cancer patients with a TLR4 allele variant that reduces the affinity of TLR4 for HMGB1 had increased incidence of metastases after conventional treatment than patients with wild type allele (Apetoh et al. 2007b).

12.4.2.4 End Stage Degradation Products

Disintegration of the cell plasma membrane in the final stages of apoptosis is characterized by the release of RNA and DNA molecules that can activate the immune system through TLR3 and TLR4, respectively (Ishii et al. 2001; Krysko et al. 2005). The kinetics of the expression of immunogenic cell death markers is summarized in Fig. 12.3.

12.4.2.5 Cellular Markers of Immunogenicity: Future Directions

The newly found markers of immunogenicity reviewed in the previous paragraphs illustrate the substantial improvement in our understanding of the events involved in the interaction of tumour cells with the immune system. It is very likely that other markers of immunogenic cell death will be identified. However, the signals that have been identified have to be elucidated further. For example, most calreticulin experiments were performed in murine models using colon cancer cell line, and it remains be seen whether similar results will be observed in humans and for other types of cancer. Surface expression of HSP90 on myeloma cells was only detected after bortezomib treatment, but not after treatment with dexamethasone or gamma irradiation. It is necessary to test a wider spectrum of chemotherapeutic agents. Although it has been shown that bortezomib induces surface expression of HSP90 in other tumour cells lines, immunogenicity of these cells, i.e. their capacity to stimulate tumour specific CTLs, has not been tested. However, it is remarkable that two independent studies using different drugs and experimental approaches identified very similar mechanisms of tumour immunogenicity. Both calreticulin and HSPs are chaperone proteins that participate in the folding of newly synthesized proteins and in the elimination of unfolded and misfolded proteins. In both cases translocation of proteins to the cell surface is required for the delivery of activation signal. However, killing of myeloma cells by bortezomib does not significantly alter the kinetics of their phagocytosis by dendritic cells, as opposed to enhanced phago-cytosis by expression of calreticulin. It is evident that mechanisms of interaction between tumour cells and the immune system are very complex, and many studies are needed to clarify them in detail.

12.5 New Concepts in the Management of Cancer Patients

We reviewed experimental findings that convincingly document the existence of cancer immunosurveillance that can be exploited for the development of immu-notherapeutic strategies. Cells of the immune system constantly survey the body for the presence of cells in early stages of malignant transformation. Several mechanisms that alert dendritic cells to the presence of neoplastic cells have been identified. Importantly, even fundamental cellular defense responses to DNA dam-age are linked to the activation of immune response. These findings justify clinical studies based on the activation of specific immune responses against tumour cells. However, breakthrough studies that identified markers of immunogenic tumour cell death after chemotherapy treatment challenge the long-held view of chemotherapy and immunotherapy as conflicting and incompatible treatment modalities. Introduction of chemotherapy regimens and their diligent testing in well-designed clinical trials represents one of the greatest triumphs of modern medicine. For example, in childhood acute lymphoblastic leukemias were invariably fatal in 1960s, but introduction and subsequent improvement of chemotherapy protocols has led to a cure rate of almost 90% in the last decade. Despite the continuous introduction of new drugs and the improvement of chemotherapy protocols, it is likely that the efficacy of chemotherapy will eventually plateau. Moreover, despite undeniable success in treating some malignancies, chemotherapy rarely cures some tumours, particularly solid tumours. Combined treatment modalities have been a standard strategy for cancer treatment, as exemplified by surgery combined with chemo- or radiotherapy. We believe that our rapidly increasing knowledge of the biology of the immune response and the importance of an anti-tumour immune response for long-term survival of cancer patients has led to the emergence of immunotherapy as another treatment modality. It would be unwise to ignore new findings about suppressor mechanisms used by tumour cells (suppressive cytokines, Tregs) and about the immunogenic modification of tumour cells (HPSs, calreticulin, post-apoptotic HMGB1 production). New immunotherapeutic strategies are needed. Furthermore, chemotherapy and immunotherapy are not antagonistic forms of therapy, and their rational combination (Table 12.1) could substantially improve prognosis of cancer patients. This can be done at several levels.

As discussed above, clinically used drugs should be tested for their ability to induce immunogenic cell death in specific cancers. The degree of immunosupression induced by chemotherapy depends on the nature of the drug. Thorough studies are needed to evaluate the kinetics of immune reconstitution, not only in terms of the absolute cell numbers but also with respect to the breadth of the T cell repertoire and the proportion of T cell subsets known to affect antitumour immune responses, such as regulatory T cells. Monitoring of immune reconstitution after chemotherapy cycles would allow precise timing of active immunotherapy that should be administered to patients with low tumour burden. Experimental data indicate that induction of cancer-specific CTLs after active immunotherapy sensitizes tumour cells to subsequent chemotherapeutic treatment, although the underlying mechanism remains unknown.

Cancer patients should also be tested for polymorphisms of genes that are important for the clinical outcome (Zitvogel et al. 2008). Preliminary data are available on some genes, such as those coding for TLR4, IL-10 and IL-18 and on their role in therapeutic responses in breast cancer, lymphoma and ovarian cancer, respectively (Bushley et al. 2004; Lech-Maranda et al. 2004; Apetoh et al. 2007a). This field of immunoepidemiology is expanding and it may play an increasingly important role in the management of cancer patients.

Table 12.1 Rationale design of combined chemo-immunotherapy therapeutic protocols.

- Reduction of tumor mass by surgery or conventional chemotherapy
- Monitoring of immune reconstitution
- Chemotherapy regimens targeting suppresor populations, such as Tregs
- Immunotherapy at the stage of minimal residual disease

Patients should also be monitored for the presence of tumour-induced immunosuppressive mechanisms (Tregs, myeloid suppressor cells, IL-10, TGFβ) because an increase in these cells and cytokines correlates with adverse prognosis. Thus, specific therapies aimed at eliminating suppressor cells should also be investigated. Immunotherapy will certainly be optimized. For instance, identification of important rejection antigens in patients with pre-neoplastic diseases or patients who experience significant clinical improvement after immunotherapy could yield better target antigens than studies in patients with advanced disease. The long-term goal of a successful combination of chemotherapy and immunotherapy should not only be the complete eradication of tumour cell population, but also the establishment of a long-lived tumour-cell-specific immunological memory. Such complex interdisciplinary management of tumour diseases would likely represent a major improvement in long-term clinical prognosis of cancer patients.

Acknowledgments   Research on tumour cells at the Department of Immunology, Second Medical School, Charles University is supported by MSM 0021620812 grant from The Czech Ministry of Education and by grant GACR 310/08/0838.

References

Ada G (2001) Vaccines and vaccination. N Engl J Med 345:1042–1053

Ada GL (1990) The immunological principles of vaccination. Lancet 335:523–526

Adema GJ, de Vries IJ, Punt CJ et al (2005) Migration of dendritic cell based cancer vaccines: in vivo veritas? Curr Opin Immunol 17:170–174

Albert ML, Darnell JC, Bender A et al (1998) Tumour-specific killer cells in paraneoplastic cerebellar degeneration. Nat Med 4:1321–1324Albert ML, Darnell RB (2004) Paraneoplastic neurological degenerations: keys to tumour immunity. Nat Rev Cancer 4:36–44

Apetoh L, Ghiringhelli F, Tesniere A et al (2007a) Toll-like receptor 4-dependent contribution of the immune system to anti-cancer chemotherapy and radiotherapy. Nat Med 13:1050–1059

Apetoh L, Obeid M, Tesniere A et al (2007b) Immunogenic chemotherapy: discovery of a critical protein through proteomic analyses of tumour cells. Cancer Genomics Proteomics 4:65–70

Arispe N, Doh M, Simakova O et al (2004) Hsc70 and Hsp70 interact with phosphatidylserine on the surface of PC12 cells resulting in a decrease of viability. FASEB J 18:1636–1645

Bach EA, Aguet M, Schreiber RD (1997) The IFN gamma receptor: a paradigm for cytokine receptor signaling. Annu Rev Immunol 15:563–591

Banchereau J, Steinman RM (1998) Dendritic cells and the control of immunity. Nature 392:245–252

Bartkova J, Horejsi Z, Koed K et al (2005) DNA damage response as a candidate anti-cancer barrier in early human tumourigenesis. Nature 434:864–870

Basu S, Binder RJ, Ramalingam T et al (2001) CD91 is a common receptor for heat shock proteins gp96, hsp90, hsp70, and calreticulin. Immunity 14:303–313

Basu S, Binder RJ, Suto R et al (2000) Necrotic but not apoptotic cell death releases heat shock proteins, which deliver a partial maturation signal to dendritic cells and activate the NF-kappa B pathway. Int Immunol 12:1539–1546

Becker T, Hartl FU, Wieland F (2002) CD40, an extracellular receptor for binding and uptake of Hsp70-peptide complexes. J Cell Biol 158:1277–1285

Beere HM, Wolf BB, Cain K et al (2000) Heat-shock protein 70 inhibits apoptosis by preventing recruitment of procaspase-9 to the Apaf-1 apoptosome. Nat Cell Biol 2:469–475

Bender A, Sapp M, Schuler G et al (1996) Improved methods for the generation of dendritic cells from non-proliferating progenitors in human blood. J Immunol Methods 196:121–135

Binder RJ, Anderson KM, Basu S et al (2000a) Cutting edge: heat shock protein gp96 induces maturation and migration of CD11c+ cells in vivo. J Immunol 165:6029–6035

Binder RJ, Han DK, Srivastava PK (2000b) CD91: a receptor for heat shock protein gp96. Nat Immunol 1:151–155

Binder RJ, Harris ML, Menoret A et al (2000c) Saturation, competition, and specificity in interaction of heat shock proteins (hsp) gp96, hsp90, and hsp70 with CD11b+ cells. J Immunol 165:2582–2587

Binder RJ, Kelly JB, 3rd, Vatner RE et al (2007) Specific immunogenicity of heat shock protein gp96 derives from chaperoned antigenic peptides and not from contaminating proteins. J Immunol 179:7254–7261

Binder RJ, Srivastava PK (2004) Essential role of CD91 in re-presentation of gp96-chaperoned peptides. Proc Natl Acad Sci U S A 101:6128–6133

Binder RJ, Srivastava PK (2005) Peptides chaperoned by heat-shock proteins are a necessary and sufficient source of antigen in the cross-priming of CD8+ T cells. Nat Immunol 6:593–599

Burdelya L, Kujawski M, Niu G et al (2005) Stat3 activity in melanoma cells affects migration of immune effector cells and nitric oxide-mediated anti-tumour effects. J Immunol 174:3925–3931

Burnet FM (1970) The concept of immunological surveillance. Prog Exp Tumour Res 13:1–27

Bushley AW, Ferrell R, McDuffie K et al (2004) Polymorphisms of interleukin (IL)-1alpha, IL-1beta, IL-6, IL-10, and IL-18 and the risk of ovarian cancer. Gynecol Oncol 95:672–679

Casares N, Pequignot MO, Tesniere A et al (2005) Caspase-dependent immunogenicity of doxorubicin-induced tumour cell death. J Exp Med 202:1691–1701

Castedo M, Perfettini JL, Roumier T et al (2004a) Cell death by mitotic catastrophe: a molecular definition. Oncogene 23:2825–2837

Castedo M, Perfettini JL, Roumier T et al (2004b) Mitotic catastrophe constitutes a special case of apoptosis whose suppression entails aneuploidy. Oncogene 23:4362–4370

Castedo M, Perfettini JL, Roumier T et al (2004c) The cell cycle checkpoint kinase Chk2 is a negative regulator of mitotic catastrophe. Oncogene 23:4353–4361

Curiel TJ, Coukos G, Zou L et al (2004) Specific recruitment of regulatory T cells in ovarian carcinoma fosters immune privilege and predicts reduced survival. Nat Med 10:942–949

Curiel TJ, Wei S, Dong H et al (2003) Blockade of B7-H1 improves myeloid dendritic cell-mediated anti-tumour immunity. Nat Med 9:562–567

Dai J, Liu B, Caudill MM et al (2003) Cell surface expression of heat shock protein gp96 enhances cross-presentation of cellular antigens and the generation of tumour-specific T cell memory. Cancer Immun 3:1

Darnell RB, Posner JB (2003a) Observing the invisible: successful tumour immunity in humans. Nat Immunol 4:201

Darnell RB, Posner JB (2003b) Paraneoplastic syndromes involving the nervous system. N Engl J Med 349:1543–1554

De Vries IJ, Bernsen MR, van Geloof WL et al (2007) In situ detection of antigen-specific T cells in cryo-sections using MHC class I tetramers after dendritic cell vaccination of melanoma patients. Cancer Immunol Immunother 56:1667–1676

Demaria S, Santori FR, Ng B et al (2005) Select forms of tumour cell apoptosis induce dendritic cell maturation. J Leukoc Biol 77:361–368

Dhodapkar MV (2005) Harnessing host immune responses to preneoplasia: promise and challenges. Cancer Immunol Immunother 54:409–413

Dhodapkar MV, Bhardwaj N (2000) Active immunization of humans with dendritic cells. J Clin Immunol 20:167–174

Dhodapkar MV, Dhodapkar KM, Palucka AK (2008) Interactions of tumour cells with dendritic cells: balancing immunity and tolerance. Cell Death Differ 15:39–50

Dhodapkar MV, Krasovsky J, Olson K (2002) T cells from the tumour microenvironment of patients with progressive myeloma can generate strong, tumour-specific cytolytic responses to autologous, tumour-loaded dendritic cells. Proc Natl Acad Sci U S A 99:13009–13013

Dhodapkar MV, Krasovsky J, Osman K et al (2003) Vigorous premalignancy-specific effector T cell response in the bone marrow of patients with monoclonal gammopathy. J Exp Med 198:1753–1757

Dhodapkar MV, Steinman RM (2002) Antigen-bearing immature dendritic cells induce peptide-specific CD8(+) regulatory T cells in vivo in humans. Blood 100:174–177

Dhodapkar MV, Steinman RM, Krasovsky J et al (2001) Antigen-specific inhibition of effector T cell function in humans after injection of immature dendritic cells. J Exp Med 193:233–238

Dighe AS, Richards E, Old LJ et al (1994) Enhanced in vivo growth and resistance to rejection of tumour cells expressing dominant negative IFN gamma receptors. Immunity 1:447–456

Dunn GP, Bruce AT, Ikeda H et al (2002) Cancer immunoediting: from immunosurveillance to tumour escape. Nat Immunol 3:991–998

Dunn GP, Old LJ, Schreiber RD (2004a) The immunobiology of cancer immunosurveillance and immunoediting. Immunity 21:137–148

Dunn GP, Old LJ, Schreiber RD (2004b) The three Es of cancer immunoediting. Annu Rev Immunol 22:329–360

Edwards AD, Manickasingham SP, Sporri R et al (2002) Microbial recognition via Toll-like receptor-dependent and -independent pathways determines the cytokine response of murine dendritic cell subsets to CD40 triggering. J Immunol 169:3652–3660

Fadok VA, Bratton DL, Guthrie L et al (2001) Differential effects of apoptotic vs. lysed cells on macrophage production of cytokines: role of proteases. J Immunol 166:6847–6854

Figdor CG, de Vries IJ, Lesterhuis WJ et al (2004) Dendritic cell immunotherapy: mapping the ay. Nat Med 10:475–480

Frenzel H, Hoffmann B, Brocks C et al (2006) Toll-like receptor interference in myeloid dendritic cells through head and neck cancer. Anticancer Res 26:4409–4413

Galluzzi L, Maiuri MC, Vitale I et al (2007) Cell death modalities: classification and pathophysiological implications. Cell Death Differ 14:1237–1243

Gardai SJ, Bratton DL, Ogden CA et al (2006) Recognition ligands on apoptotic cells: a perspective. J Leukocyte Biol 79:896–903

Garrido C, Fromentin A, Bonnotte B et al (1998) Heat shock protein 27 enhances the tumourigenicity of immunogenic rat colon carcinoma cell clones. Cancer Res 58:5495–5499

Gasser S, Orsulic S, Brown EJ et al (2005) The DNA damage pathway regulates innate immune system ligands of the NKG2D receptor. Nature 436:1186–1190

Gerecitano J, Goy A, Wright J et al (2006) Drug-induced cutaneous vasculitis in patients with non-Hodgkin lymphoma treated with the novel proteasome inhibitor bortezomib: a possible surrogate marker of response? Br J Haematol 134:391–398

Gilboa E (2004) The promise of cancer vaccines. Nat Rev Cancer 4:401–411

Gorgoulis VG, Vassiliou LV, Karakaidos P et al (2005) Activation of the DNA damage checkpoint and genomic instability in human precancerous lesions. Nature 434:907–913

Hayakawa Y, Takeda K, Yagita H et al (2002) IFN-gamma-mediated inhibition of tumour angiogenesis by natural killer T-cell ligand, alpha-galactosylceramide. Blood 100:1728–1733

Hung K, Hayashi R, Lafond-Walker A et al (1998) The central role of CD4(+) T cells in the antitumour immune response. J Exp Med 188:2357–2368

Cheng F, Wang HW, Cuenca A et al (2003) A critical role for Stat3 signaling in immune tolerance. Immunity 19:425-436

Ishii KJ, Suzuki K, Coban C et al (2001) Genomic DNA released by dying cells induces the maturation of APCs. J Immunol 167:2602–2607

Iwasaki A, Medzhitov R (2004) Toll-like receptor control of the adaptive immune responses. Nat Immunol 5:987–995

Jaattela M (1995) Over-expression of hsp70 confers tumourigenicity to mouse fibrosarcoma cells. Int J Cancer 60:689–693

Janeway CA, Jr., Goodnow CC, Medzhitov R (1996) Danger - pathogen on the premises! Immunological tolerance. Curr Biol 6:519–522

Kaplan DH, Shankaran V, Dighe AS et al (1998) Demonstration of an interferon gamma-dependent tumour surveillance system in immunocompetent mice. Proc Natl Acad Sci U S A 95:7556–7561

Kroemer G, El-Deiry WS, Golstein P et al (2005) Classification of cell death: recommendations of the Nomenclature Committee on Cell Death. Cell Death Differ 12 Suppl 2:1463–1467

Krysko DV, D'Herde K, Vandenabeele P (2006) Clearance of apoptotic and necrotic cells and its immunological consequences. Apoptosis 11:1709–1726

Krysko DV, Leybaert L, Vandenabeele P et al (2005) Gap junctions and the propagation of cell survival and cell death signals. Apoptosis 10:459–469

Kyle RA, Rajkumar SV (1999) Monoclonal gammopathies of undetermined significance. Hematol Oncol Clin North Am 13:1181–1202

Labarriere N, Bretaudeau L, Gervois N et al (2002) Apoptotic body-loaded dendritic cells efficiently cross-prime cytotoxic T lymphocytes specific for NA17-A antigen but not for Melan-A/MART-1 antigen. Int J Cancer 101:280–286

Lech-Maranda E, Baseggio L, Bienvenu J et al (2004) Interleukin-10 gene promoter polymorphisms influence the clinical outcome of diffuse large B-cell lymphoma. Blood 103:3529–3534

Liyanage UK, Moore TT, Joo HG et al (2002) Prevalence of regulatory T cells is increased in peripheral blood and tumour microenvironment of patients with pancreas or breast adenocarcinoma. J Immunol 169:2756–2761

Lollini PL, Cavallo F, Nanni P et al (2006) Vaccines for tumour prevention. Nat Rev Cancer 6:204–216

Lollini PL, De Giovanni C, Pannellini T et al (2005) Cancer immunoprevention. Future Oncol 1:57–66

Lollini PL, Forni G (2002) Anti-tumour vaccines: is it possible to prevent a tumour? Cancer Immunol Immunother 51:409–416

Lollini PL, Forni G (2003) Cancer immunoprevention: tracking down persistent tumour antigens. Trends Immunol 24:62–66

Maiuri MC, Zalckvar E, Kimchi A et al (2007) Self-eating and self-killing: crosstalk between autophagy and apoptosis. Nat Rev Mol Cell Biol 8:741–752

Marincola FM, Jaffee EM, Hicklin DJ et al (2000) Escape of human solid tumours from T-cell recognition: molecular mechanisms and functional significance. Adv Immunol 74:181–273

Martinon F, Petrilli V, Mayor A et al (2006) Gout-associated uric acid crystals activate the NALP3 inflammasome. Nature 440:237–241

Masse D, Ebstein F, Bougras G et al (2004) Increased expression of inducible HSP70 in apoptotic cells is correlated with their efficacy for anti-tumour vaccine therapy. Int J Cancer 111:575–583

Medzhitov R (2001) Toll-like receptors and innate immunity. Nat Rev Immunol 1:135–145

Medzhitov R, Janeway C, Jr. (2000a) Innate immune recognition: mechanisms and pathways. Immunol Rev 173:89–97

Medzhitov R, Janeway C, Jr. (2000b) Innate immunity. N Engl J Med 343:338–344

Medzhitov R, Janeway CA, Jr. (1997a) Innate immunity: impact on the adaptive immune response. Curr Opin Immunol 9:4–9

Medzhitov R, Janeway CA, Jr. (1997b) Innate immunity: the virtues of a non-clonal system of recognition. Cell 91:295–298

Mumberg D, Monach PA, Wanderling S et al (1999) CD4(+) T cells eliminate MHC class II-negative cancer cells in vivo by indirect effects of IFN-gamma. Proc Natl Acad Sci U S A 96:8633–8638

Obeid M, Tesniere A, Ghiringhelli F et al (2007a) Calreticulin exposure dictates the immunogenicity of cancer cell death. Nat Med 13:54–61

Obeid M, Tesniere A, Panaretakis T et al (2007b) Ecto-calreticulin in immunogenic chemotherapy. Immunol Rev 220:22–34

Pardoll DM (1998) Cancer vaccines. Nat Med 4:525–531

Park JS, Gamboni-Robertson F, He Q et al (2006) High mobility group box 1 protein interacts with multiple Toll-like receptors. Am J Physiol Cell Physiol 290:C917–C924

Park JS, Svetkauskaite D, He Q et al (2004) Involvement of toll-like receptors 2 and 4 in cellular activation by high mobility group box 1 protein. J Biol Chem 279:7370–7377

Pasare C, Medzhitov R (2004a) Toll-like receptors and acquired immunity. Semin Immunol 16:23–26

Pasare C, Medzhitov R (2004b) Toll-like receptors: linking innate and adaptive immunity. Microbes Infect 6:1382–1387

Reis e Sousa C, Diebold SD, Edwards AD et al (2003) Regulation of dendritic cell function by microbial stimuli. Pathol Biol (Paris) 51:67–68

Rock KL, Hearn A, Chen CJ et al (2005) Natural endogenous adjuvants. Springer Semin Immunopathol 26:231–246

Rosenberg SA, Yang JC, Restifo NP (2004) Cancer immunotherapy: moving beyond current vaccines. Nat Med 10:909–915

Rovere-Querini P, Capobianco A, Scaffidi P et al (2004) HMGB1 is an endogenous immune adjuvant released by necrotic cells. EMBO Rep 5:825–830

Sauter B, Albert ML, Francisco L et al (2000) Consequences of cell death: exposure to necrotic tumour cells, but not primary tissue cells or apoptotic cells, induces the maturation of immunostimulatory dendritic cells. J Exp Med 191:423–434

Savill J, Dransfield I, Gregory C et al (2002) A blast from the past: clearance of apoptotic cells regulates immune responses. Nat Rev Immunol 2:965–975

Savill J, Fadok V (2000) Corpse clearance defines the meaning of cell death. Nature 407:784–788

Savill J, Fadok V, Henson P et al (1993) Phagocyte recognition of cells undergoing apoptosis. Immunol Today 14:131–136

Scaffidi P, Misteli T, Bianchi ME (2002) Release of chromatin protein HMGB1 by necrotic cells triggers inflammation. Nature 418:191–195

Shankaran V, Ikeda H, Bruce AT et al (2001) IFNgamma and lymphocytes prevent primary tumour development and shape tumour immunogenicity. Nature 410:1107–1111

Shi H, Cao T, Connolly JE et al (2006) Hyperthermia enhances CTL cross-priming. J Immunol 176:2134–2141

Shi Y, Evans JE, Rock KL (2003) Molecular identification of a danger signal that alerts the immune system to dying cells. Nature 425:516–521

Shi Y, Rock KL (2002) Cell death releases endogenous adjuvants that selectively enhance immune surveillance of particulate antigens. Eur J Immunol 32:155–162

Shinkai Y, Rathbun G, Lam KP et al (1992) RAG-2-deficient mice lack mature lymphocytes owing to inability to initiate V(D)J rearrangement. Cell 68:855–867

Shortman K, Liu YJ (2002) Mouse and human dendritic cell subtypes. Nat Rev Immunol 2:151–161

Scheibner KA, Lutz MA, Boodoo S et al (2006) Hyaluronan fragments act as an endogenous danger signal by engaging TLR2. J Immunol 177:1272–1281

Schnurr M, Scholz C, Rothenfusser S et al (2002) Apoptotic pancreatic tumour cells are superior to cell lysates in promoting cross-priming of cytotoxic T cells and activate NK and gammadelta T cells. Cancer Res 62:2347–2352

Singh-Jasuja H, Toes RE, Spee P et al (2000) Cross-presentation of glycoprotein 96-associated antigens on major histocompatibility complex class I molecules requires receptor-mediated endocytosis. J Exp Med 191:1965–1974

Smyth MJ, Crowe NY, Godfrey DI (2001) NK cells and NKT cells collaborate in host protection from methylcholanthrene-induced fibrosarcoma. Int Immunol 13:459–463

Smyth MJ, Thia KY, Street SE et al (2000) Differential tumour surveillance by natural killer (NK) and NKT cells. J Exp Med 191:661–668

Somersan S, Larsson M, Fonteneau JF et al (2001) Primary tumour tissue lysates are enriched in heat shock proteins and induce the maturation of human dendritic cells. J Immunol 167:4844–4852

Spisek R (2006) Immunoprevention of cancer: time to reconsider timing of vaccination against cancer. Expert Rev Anticancer Ther 6:1689–1691

Spisek R, Brazova J, Rozkova D et al (2004) Maturation of dendritic cells by bacterial immunomodulators. Vaccine 22:2761–2768

Spisek R, Dhodapkar MV (2006) Immunoprevention of cancer. Hematol Oncol Clin North Am 20:735–750

Spisek R, Charalambous A, Mazumder A et al (2007) Bortezomib enhances dendritic cell (DC) mediated induction of immunity to human myeloma via exposure of cell surface heat shock protein 90 on dying tumour cells: therapeutic implications. Blood

Spisek R, Chevallier P, Morineau N et al (2002) Induction of leukemia-specific cytotoxic response by cross-presentation of late-apoptotic leukemic blasts by autologous dendritic cells of non-leukemic origin. Cancer Res 62:2861–2868

Sporri R, Reis e Sousa C (2005) Inflammatory mediators are insufficient for full dendritic cell activation and promote expansion of CD4+ T cell populations lacking helper function. Nat Immunol 6:163–170

Steinman RM, Banchereau J (2007) Taking dendritic cells into medicine. Nature 449:419–426

Steinman RM, Bonifaz L, Fujii S et al (2005) The innate functions of dendritic cells in peripheral lymphoid tissues. Adv Exp Med Biol 560:83–97

Street SE, Cretney E, Smyth MJ (2001) Perforin and interferon-gamma activities independently control tumour initiation, growth, and metastasis. Blood 97:192–197

Stutman O (1975) Immunodepression and malignancy. Adv Cancer Res 22:261–422

Tacken PJ, de Vries IJ, Torensma R et al (2007) Dendritic-cell immunotherapy: from ex vivo loading to in vivo targeting. Nat Rev Immunol 7:790–802

Tesniere A, Panaretakis T, Kepp O et al (2008) Molecular characteristics of immunogenic cancer cell death. Cell Death Differ 15:3–12

Tobiasova Z, Pospisilova D, Miller AM et al (2007) In vitro assessment of dendritic cells pulsed with apoptotic tumour cells as a vaccine for ovarian cancer patients. Clin Immunol 122:18–27

Udono H, Srivastava PK (1993) Heat shock protein 70-associated peptides elicit specific cancer immunity. J Exp Med 178:1391–1396

Uyttenhove C, Pilotte L, Theate I et al (2003) Evidence for a tumoural immune resistance mechanism based on tryptophan degradation by indoleamine 2,3-dioxygenase. Nat Med 9:1269–1274

Van Den Broek ME, Kagi D, Ossendorp F et al (1996) Decreased tumour surveillance in perforin-deficient mice. J Exp Med 184:1781–1790

Verdijk P, Scheenen TW, Lesterhuis WJ et al (2007) Sensitivity of magnetic resonance imaging of dendritic cells for in vivo tracking of cellular cancer vaccines. Int J Cancer 120:978–984

Walport MJ (2000) Lupus, DNase and defective disposal of cellular debris. Nat Genet 25:135–136

Wang H, Bloom O, Zhang M et al (1999) HMG-1 as a late mediator of endotoxin lethality in mice. Science 285:248–251

Wang HY, Lee DA, Peng G et al (2004a) Tumour-specific human CD4+ regulatory T cells and their ligands: implications for immunotherapy. Immunity 20:107–118

Wang T, Niu G, Kortylewski M et al (2004b) Regulation of the innate and adaptive immune responses by Stat-3 signaling in tumour cells. Nat Med 10:48–54

Willimsky G, Blankenstein T (2005) Sporadic immunogenic tumours avoid destruction by inducing T-cell tolerance. Nature 437:141–146

Winau F, Weber S, Sad S et al (2006) Apoptotic vesicles crossprime CD8 T cells and protect against tuberculosis. Immunity 24:105–117

Wong LH, Krauer KG, Hatzinisiriou I et al (1997) Interferon-resistant human melanoma cells are deficient in ISGF3 components, STAT1, STAT2, and p48-ISGF3gamma. J Biol Chem 272:28779–28785

Woo EY, Chu CS, Goletz TJ et al (2001) Regulatory CD4(+)CD25(+) T cells in tumours from patients with early-stage non-small cell lung cancer and late-stage ovarian cancer. Cancer Res 61:4766–4772

Yamazaki S, Bonito AJ, Spisek R et al (2007) Dendritic cells are specialized accessory cells along with TGF-{beta} for the differentiation of Foxp3+ CD4+ regulatory T cells from peripheral Foxp3- precursors. Blood

Zaft T, Sapoznikov A, Krauthgamer R et al (2005) CD11chigh dendritic cell ablation impairs lymphopenia-driven proliferation of naive and memory CD8+ T cells. J Immunol 175:6428–6435

Zitvogel L, Apetoh L, Ghiringhelli F et al (2008) Immunological aspects of cancer chemotherapy. Nat Rev Immunol 8:59–73

Zitvogel L, Casares N, Pequignot MO et al (2004) Immune response against dying tumour cells. Adv Immunol 84:131–179

Zitvogel L, Tesniere A, Kroemer G (2006) Cancer despite immunosurveillance: immunoselection and immunosubversion. Nat Rev Immunol 6:715–727

Zou W (2006) Regulatory T cells, tumour immunity and immunotherapy. Nat Rev Immunol 6:295–307

Chapter 13
Phagocytosis of Dying Cells in the Pathogenesis of Atherosclerosis

Wim Martinet, Dorien M. Schrijvers and Guido R.Y. De Meyer

Abstract: Atherosclerosis is a progressive inflammatory disease characterized by the formation of atheromatous plaques in the intima of medium- and large-sized arteries. A large body of evidence indicates that apoptosis is a prominent feature of advanced plaques. The phagocytic clearance of apoptotic cells by macrophages in these lesions is severely impaired, which is at least partly attributed to oxidative stress and cytoplasmic saturation with indigestible material. The consequences of this defect are likely to be substantial because it promotes three main processes—inflammation, necrosis and thrombosis—that are thought to play an important role in plaque disruption and its acute clinical consequences such as myocardial infarction and stroke. Moreover, the lack of lesional phagocytes to safely clear apoptotic cells undermines the relevance of recent therapeutic approaches to stabilize vulnerable plaques via selective induction of macrophage death. To limit the detrimental effects of defective phagocytic clearance of apoptotic cells, suppression of apoptosis in atherosclerotic plaques is not recommended because reduced levels of macrophage apoptosis promotes plaque development. A more promising approach would be to therapeutically enhance selective phagocytosis of apoptotic cells, on condition that this intervention does not alter the uptake of other plaque components such as lipoproteins, erythrocytes or platelets, an event that is considered to be pro-atherogenic.

Keywords: Apoptosis • Atherosclerosis • Macrophages • Phagocytosis • Plaque destabilization

W. Martinet (✉) · D. M. Schrijvers · G. R. Y. De Meyer
Division of Pharmacology
University of Antwerp
Universiteitsplein 1
B-2610 Antwerp (Wilrijk), Belgium
Tel.: +32 3 820 26 79
Fax.: +32 3 820 25 67
E-mail: wim.martinet@ua.ac.be

D. V. Krysko, P. Vandenabeele (eds.), *Phagocytosis of Dying Cells*,
DOI: 10.1007/978-1-4020-9292-3_13, © Springer Science+Business Media B.V. 2009

13.1   Atherosclerosis and Plaque Destabilization

Atherosclerosis is a long-term inflammatory disease of large- and medium-sized arteries and the leading cause of death and morbidity among adults in developed countries (Ross 1999; Lusis 2000; Hansson 2005). It slowly progresses over a period of decades before clinical symptoms become manifest. Progression of the disease is characterized by the formation of atherosclerotic plaques (Fig. 13.1) and depends on genetic make-up, gender and certain well-recognized risk factors such as hypercholesterolemia, diabetes mellitus, hypertension, obesity and smoking (Linton and Fazio 2003; Kannel et al. 2004). Because most of these risk factors

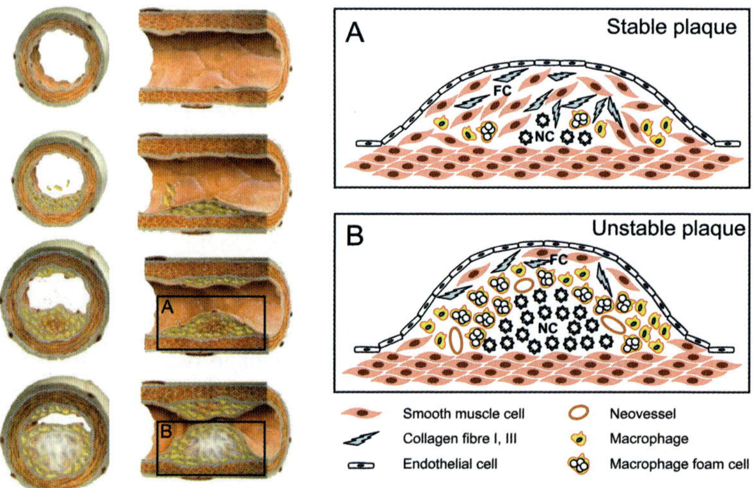

Fig. 13.1   Schematic representation of atherosclerosis and plaque development. Atherosclerosis is a chronic inflammatory disease of the arterial wall of large and medium-sized arteries, and is characterized by the formation of an atherosclerotic plaque. Until recently, it was thought that the major pathophysiological mechanism by which atherosclerosis contributes to various coronary syndromes was a slow progressive luminal obstruction by the atherosclerotic plaque that eventually resulted in decreased coronary blood flow reserve and subsequent myocardial ischemia. Recent evidence, however, indicates that also the composition and stability of the plaque are important aspects that determine acute clinical events. During plaque development, monocytes are recruited from the circulation in the sub-endothelial space, where they differentiate into macrophages. These recruited macrophages endocytose modified forms of LDL via scavenger receptors to form foam cells. The death of foam cells by necrosis leads to the formation of a necrotic core (NC), rich in cholesterol and cell debris. However, before lipid-laden macrophages undergo cell death, many pro-inflammatory cytokines are released which in turn promote progression of atherosclerosis. Moreover, activated macrophages in the atherosclerotic plaque release matrix degrading enzymes. This leads to collagen breakdown and thinning of the fibrous cap (FC). Therefore, lesions with a thick fibrous cap and many smooth muscle cells are stable (box A) and do not pose a clinical threat. Lesions characterized by a thin fibrous cap, a large necrotic core, significant neovascularization, and the presence of numerous inflammatory cells, in particular macrophages and macrophage-derived foam cells, are considered unstable and risk-bearing (box B).

are inherent to modern life, atherosclerosis transformed from a rare and sporadic disease in the beginning of the 20th century to a disease of pandemic dimensions by the mid-century. Initially, plaque formation was viewed as an inevitable degenerative process. Nowadays, our knowledge about plaque development emerged to a much better defined scenario of molecular and cellular events so that we can begin to approach atherogenesis as a modifiable rather than an inevitable process. Indeed, changes in diet and exercise as well as the development of certain drugs that may improve plaque stability (e.g. lipid-lowering agents) have made significant inroads in preventing acute atherothrombotic events (Franklin and Kahn 1996; Getz and Reardon 2006; Grines 2006).

The first signs of atherosclerosis occur after birth at specific orifices and branching points in the vascular tree, in the form of physiological adaptations to changes in flow and wall tension (Stary 1987). Although not yet considered as atherosclerosis per se, these adaptive *intimal thickenings*, consisting of smooth muscle cells (SMCs) and connective tissue, often characterize the predilection sites for further plaque development. The non-random distribution of early lesions suggests that hemodynamic forces might be acting as local 'biomechanical risk factors' (VanderLaan et al. 2004). Sites of disturbed flow (i.e. arterial bifurcations and curvatures) typically are 'atherosclerosis-prone regions', whereas atherosclerosis-resistant regions' (i.e. the straight tubular parts of the arteries) are associated with uniform laminar flow, which exhibits atheroprotective mechanisms (Cunningham and Gotlieb 2005).

Excess amounts of low-density lipoproteins (LDL) infiltrate atherosclerosis-prone regions of the artery and are retained in the intima (Skalen et al. 2002). Modification of LDL, through oxidation or enzymatic attack, leads to inflammatory lipids, which can activate endothelial cells (Chisolm and Steinberg 2000). Furthermore, disturbed oscillary blood flow alters the expression of endothelial adhesion molecules, resulting in increased adhesion of monocytes to the endothelium (Chappell et al. 1998; Hsiai et al. 2003). Once trapped in the arterial wall, monocytes undergo differentiation. Some monocytes differentiate in dendritic cells, which represent the most potent antigen-presenting cells. The majority of the monocytes convert, however, into macrophages, which accumulate lipids in their cytoplasm and become foam cells (Bobryshev 2006). Clusters of these lipid-laden macrophages are macroscopically visible as 'fatty streaks'. Macrophage-derived foam cells and other inflammatory cells that invaded the vessel wall such as T lymphocytes and mast cells produce cytokines and growth factors (Hansson 2005), and together with the continuous presence of oxidative stress (Chisolm and Steinberg 2000), this environment causes vascular SMCs to migrate, proliferate and synthesize extracellular matrix components on the luminal side of the vessel wall, forming the fibrous cap of the atherosclerotic lesion (Dzau et al. 2002).

In advanced stages of atherosclerosis, plaques that have formed can partially or totally occlude the blood vessel (Fig. 13.1). Arterial narrowings (or stenoses) are at first opposed by a process called positive remodeling, in which lumen size is maintained by outward growth of the vessel (Pasterkamp and Smits 2002). Moreover, stenoses by themselves seldom cause acute unstable angina or myocardial infarction. Sizeable atheroma with a thick fibrous cap may remain silent for decades or produce symptoms such as angina pectoris precipitated by increased oxygen demand during exercise. However, seemingly without warning, such lesions may cause dreaded acute manifestations of atherosclerosis (Libby 2000). Indeed, when atheroma accumulate more lipids, also more inflammatory mediators (e.g. interferon-γ) and proteolytic enzymes (e.g. matrix metalloproteinases [MMPs]) are produced, that contribute to thinning of the established fibrous cap, by digesting its components (Lee and Libby 1997). Therefore, not only the degree of a stenosis, but also the plaque composition seems to be clinically important. A stable (macrophage-poor) plaque has a small necrotic core separated from the lumen by a thick fibrous cap containing SMCs and collagen fibers (Fig. 13.1, box A). A plaque with an unstable phenotype, on the other hand, is characterized by a relatively large lipid core, a high macrophage density in the shoulder region and a thin fibrous cap (Fig. 13.1, box B). When the fibrous cap looses its strength, the plaque can disrupt, which often results in the formation of a mural thrombus. Non-occlusive thrombi are subsequently incorporated into the plaque and organized by invading macrophages and SMCs, thereby further compromising the lumen (Libby et al. 2002). This sequence of fissure, thrombus formation, organization and incorporation into the plaque may occur repeatedly (Kolodgie et al. 2004). Occlusive thrombi cause acute manifestations of atherosclerosis such as myocardial infarction and stroke (Klein 2005). In man, bleeding from small blood vessels inside the plaque (hemorrhages) is another important mechanism leading to plaque rupture and/or accelerated plaque growth (Kockx et al. 2003; Levy and Moreno 2006).

13.2   Cell Death in Atherosclerosis

Numerous studies have identified apoptosis as a prominent feature of advanced, unstable lesions (Kockx and Herman 2000; Mallat and Tedgui 2000; Martinet and Kockx 2001; Fig. 13.2). However, the distribution of apoptosis within advanced plaques as determined by in situ end-labelling techniques (TUNEL and ISNT) is heterogeneous, being more frequent (1–2%) in regions with a high density of inflammatory cells (Kockx and Herman 2000). Indeed, monocytes and T lymphocytes release cytokines with pro-apoptotic potential including IFN-γ, TNF-α and IL-2. TNF-α stimulates the migration and replication of human SMCs in vitro, but when given together with IFN-γ and IL-1β, TNF-α may be involved in the local induction of apoptosis (Geng et al. 1996). Mast cells are also present in the inflammatory infiltrate and induce SMCs apoptosis by releasing chymase in the extracellular matrix (Leskinen et al. 2001). Furthermore, inducible nitric oxide synthase (iNOS) is over-expressed in macrophage-derived foam cells so that high amounts of NO are produced that are toxic for SMCs (Cromheeke et al. 1999). NO may contribute to oxidative stress and tissue damage through formation of peroxynitrite. Once formed, peroxynitrite can potentially oxidize and/or damage a variety of biomolecules, including polyunsaturated fatty acids, sulfhydryl groups, DNA and RNA (Martinet et al. 2002a, 2004a; Kamat, 2006). Oxidative DNA damage is

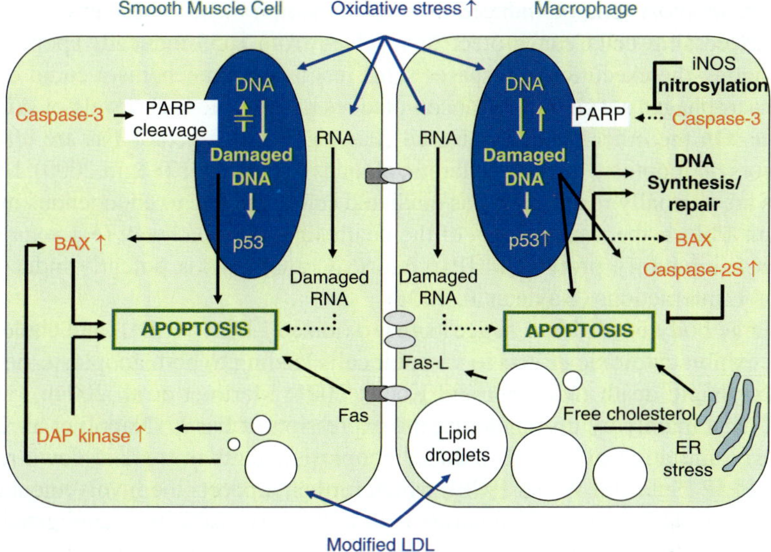

Fig. 13.2 Overview of the cell death mechanisms in atherosclerosis. High levels of reactive oxygen species (ROS) are produced in advanced atherosclerotic plaques, especially in and around macrophages. Increased levels of ROS may lead to oxidative damage of DNA and RNA. DNA damage is followed by increased DNA repair activity (up-regulation of PARP and p53 which is most prominent in macrophages) so that initial damage is efficiently repaired and, more importantly, remains sublethal. However, some cells may reach a point of no return, after which the DNA repair systems can no longer cope with the extensive damage and apoptotic cell death becomes imminent. This may occur through up-regulation of the pro-apoptotic factor Bax and activation of the caspase cascade. Overwhelming oxidative damage of RNA may also affect cell viability, although thorough factual evidence is lacking. In contrast to SMCs, plaque macrophages show high levels of DNA synthesis/repair and low levels of Bax. Since plaque macrophages express inducible nitric oxide synthase (iNOS), it is plausible to assume that they inactivate their caspases by nitrosylation. This inactivation mechanism suggests that macrophages are capable of repairing their DNA continuously (no caspase-mediated cleavage of PARP and other repair enzymes). Moreover, macrophages express elevated levels of the anti-apoptotic isoform of caspase-2 (caspase-2S) after genomic DNA damage. These data could explain why macrophages do not undergo apoptosis, despite the high levels of ROS that they produce. Apart from oxidative stress, uptake of modified LDL and foam cell formation are important additional features of advanced atherosclerotic plaques. Macrophages internalize large amounts of modified LDL. A portion of free cholesterol, derived from internalized LDL, stimulates ER-stress-induced apoptosis. Unlike macrophages, SMCs over-express the pro-apoptotic proteins DAP kinase and Bax after uptake of aggregated LDL, which make them more susceptible to cell death. Finally, it is worth mentioning that Fas/Fas-L interactions are a major cause of SMC death in atherosclerotic plaques.

associated with the up-regulation of DNA repair enzymes such as PARP-1, Ref-1 and DNA-PK as well as p53 and caspase-2S, the anti-apoptotic isoform of caspase-2 (Martinet et al. 2002a, 2003). These proteins are over-expressed predominantly in plaque macrophages, suggesting that only this type of cells and not SMCs are capable of repairing their DNA adequately. In case of severe oxidative DNA damage, however, apoptosis can be induced through activation of p53 (Mercer et al. 2007). iNOS expressing cells can protect themselves from (p53-induced) apoptosis by nitrosylating the executioner caspases. This inactivation mechanism could explain why macrophages survive in advanced plaques despite the high levels of NO they produce. On the other hand, Fas-ligand (Fas-L) and its receptor Fas are effective mediators of apoptosis in atherosclerotic plaques (Mallat and Tedgui 2000). Macrophages are normally resistant to Fas mediated cell death due to endogenous mechanisms that block the transmission of the death signalling pathway (e.g. expression of FLICE-inhibitory protein [FLIP]), but SMC apoptosis is potently induced via Fas/Fas-L interactions (Boyle et al. 2001).

A large body of evidence suggests that oxidized LDL (oxLDL) and cholesterol oxides exhibit cytotoxic effects to vascular cells leading to both apoptosis, necrosis and autophagic death (Martinet and Kockx 2001; Martinet et al. 2004b). Mildly oxLDL acts mainly by up-regulating the expression of Fas-L (Napoli et al. 2000). Moreover, initiation of oxLDL-induced apoptosis results in marked down-regulation of FLIP (Sata and Walsh 1998), which further supports the involvement of the Fas pathway. This is in sharp contrast with both native LDL and aggregated LDL (agLDL) which are known to be non-apoptotic (Martinet and Kockx 2001). It should be noted, however, that the uptake of agLDL by SMCs is associated with enhanced expression of Bax and death-associated protein kinase (DAPK) without affecting the viability of the cells (Martinet et al. 2002b). This is consistent with immunohistochemical findings showing that most of the SMCs present in early atherosclerotic lesions (fatty streaks) express high amounts of Bax as compared with normal SMCs without undergoing the execution phase of apoptosis (Kockx et al. 1998). We therefore believe that SMCs over-expressing Bax and/or DAPK are sensitized to die in a more advanced stage of atherosclerosis by additional pro-apoptotic stimuli.

All cell types present in atherosclerotic plaques, including endothelial cells, SMCs, lymphocytes and macrophages, are known to undergo apoptosis with varying outcome (Kockx and Herman 2000; Mallat and Tedgui 2000; Martinet and Kockx, 2001). However, in advanced lesions, up to 50% of the apoptotic cells are macrophages (Lutgens et al. 1999). A likely cause of macrophage death is the progressive accumulation of free cholesterol in the endoplasmic reticulum (ER) membrane (Tabas 2004), which is normally cholesterol poor and highly fluid. Probably by altering the function of integral ER membrane proteins, this event induces the ER stress signal transduction pathway, known as the unfolded protein response (UPR). A branch of the UPR eventually leads to an apoptotic response in macrophages, which is mediated by both Fas and mitochondrial pathways (Yao and Tabas 2000, 2001).

Despite all the efforts in determining the potential mechanisms and inducers of apoptosis as described above, the significance of apoptotic cell death in atherosclerosis remains unclear. SMCs apoptosis can be harmful because it may weaken the fibrous cap by decreasing the synthesis of interstitial collagen fibers (Clarke et al. 2006). In contrast, apoptosis of macrophages and T lymphocytes can be beneficial as removal of inflammatory cells from the plaque could attenuate the inflammatory response and decrease the synthesis of matrix degrading proteases. However, the presence of macrophages may not be entirely detrimental for the structure of the plaque. Macrophages are also professional phagocytes and thus designed to remove pathogens, small particles as well as dead or damaged cells from the atherosclerotic tissue via phagocytosis (Schrijvers et al. 2007). Although the concept of this process is known for decades and the mechanisms underlying phagocytosis are well defined, we are only beginning to acknowledge the importance of phagocytic clearance of apoptotic cells in atherosclerosis as well as its potential consequences.

13.3 Phagocytosis of Apoptotic Cells by Macrophages in Atherosclerosis

13.3.1 General Principles

When cells undergo apoptosis, they are rapidly recognized and phagocytized by professional phagocytes, such as macrophages and dendritic cells (Ucker, Chap. 6; Lacy-Hulbert, Chap. 7, this Vol.). This event is essential for normal development and prevention of inflammation and disease as phagocytosis of apoptotic cells inhibits the production of pro-inflammatory cytokines by macrophages such as IL-1β, IL-8 and TNF-α through a suppressive mechanism involving the autocrine/paracrine secretion of transforming growth factor-β (TGF-β; Fadok et al. 1998; Erwig and Henson 2007). TGF-β secretion in turn inhibits recruitment of circulating monocytes (Grainger et al. 2004). In addition, phagocytosis of apoptotic cells strongly inhibits expression of the pro-inflammatory IL-12 family of cytokines (IL-12, IL-23, IL-27) whereas the production of the anti-inflammatory cytokine IL-10 is stimulated (Chung et al. 2006). In this way, tissues are protected from harmful exposure to inflammatory responses as well as from the immunogenic content of the dying cells after undergoing post-apoptotic necrosis.

Apoptotic cells express many cell surface changes, which may serve as "eat me" signals for recognition by macrophages (Napirei and Mannherz, Chap. 4; Gregory and Pound, Chap. 9, this Vol.; Ravichandran 2003). A large number of surface molecules, ligands and receptors have also been identified on the macrophage (Aderem and Underhill 1999; Maderna and Godson 2003). The best characterized recognition mechanism is the exposure of phosphatidylserine (PS) on dying cells (Krahling et al. 1999; Callahan et al. 2000), directly followed by binding of PS to the PS receptor on the macrophage membrane (Fadok et al. 2000; Li et al. 2003; Park et al. 2007a, 2008). Interestingly, a number of PS-binding proteins exist that can act as a bridge between apoptotic cells and macrophages. Among these bridge molecules, it is worth mentioning (i) plasma-protein β_2-glycoprotein 1 (β_2-GPI), (ii) growth arrest-specific gene 6 (Gas6) protein that binds to the Mer kinase and (iii) milk-fat globule epidermal growth factor 8 (MFG-E8, also known as lactadherin) that connects vitronectin receptor integrin ($\alpha_v\beta_3$) to PS. This multiplicity of ways in which phagocytes recognize and engulf apoptotic cells suggests that a hierarchy of engulfment mechanisms and back-up systems may exist. Indeed, upon blocking PS with a specific antibody or annexin V, phagocytosis of apoptotic cells decreases but this inhibition is never complete, suggesting cooperation between different recognition mechanisms (Trahtemberg and Mevorach, Chap. 8, this Vol.; Pradhan et al. 1997). It should be noted that exposure of PS has been documented on many different cell types undergoing apoptosis. Some membrane changes, however, have only been described in certain cell types (e.g. expression of altered carbohydrates on apoptotic thymocytes and hepatocytes, ICAM-3 expression on apoptotic B-cells and loss of CD16 on apoptotic neutrophils).

13.3.2 Phagocytosis of Apoptotic Cells in Early vs. Advanced Plaques

It has been proposed that phagocytic clearance of apoptotic cells in early lesions is very efficient and physiologically beneficial (Tabas 2005). Consistent with this theory, cell culture studies have shown that cholesterylester-loaded macrophages, the most prominent type of phagocytes in early lesions, can effectively recognize and ingest apoptotic macrophages (Tabas 2005). However, cell death in these early lesions is almost undetectable (<0.1% TUNEL positive cells), so that it may seem obvious that few apoptotic cells are rapidly and safely cleared by neighbouring macrophages. In contrast with early lesions, phagocytic clearance of apoptotic cells in advanced plaques is far from efficient (Schrijvers et al. 2005). The ratio of free vs. phagocytosed apoptotic cells in advanced human atherosclerotic plaques is substantially higher as compared to the germinal centers of tonsils, in which a high turnover of B lymphocytes and consequently a high apoptosis rate occurs (Fig. 13.3A–G). Moreover, macrophages in tonsils can engulf several apoptotic bodies at the same time to ensure effective and swift clearance, whereas the majority of phagocytic macrophages in atherosclerotic plaques engulf one single apoptotic cell (Fig. 13.3H). Importantly, apoptotic bodies from different cell types may expose different ligands to act as "eat-me" signals, so that we cannot rule out the possibility that recognition of apoptotic lymphocytes (the predominant cell type undergoing apoptosis in tonsil) may occur differently as compared with recognition of apoptotic bodies (derived from macrophages and SMCs) in human plaques (Fadok 1999). However, in both tissues apoptotic cells are found in the vicinity of macrophages, which rules out a possible geographic problem. Moreover, atherosclerotic plaques are highly inflamed tissue and contain large areas of macrophage infiltration, whereas in tonsils usually no signs of inflammation are detected. Therefore, it is not the quantity of the phagocytes that is important for phagocytosis, but their functionality. Examination of plaques from cholesterol-fed rabbits confirm the findings obtained in human plaques, but also show that impaired phagocytosis of apoptotic cells remains constant during plaque progression (Schrijvers et al. 2005).

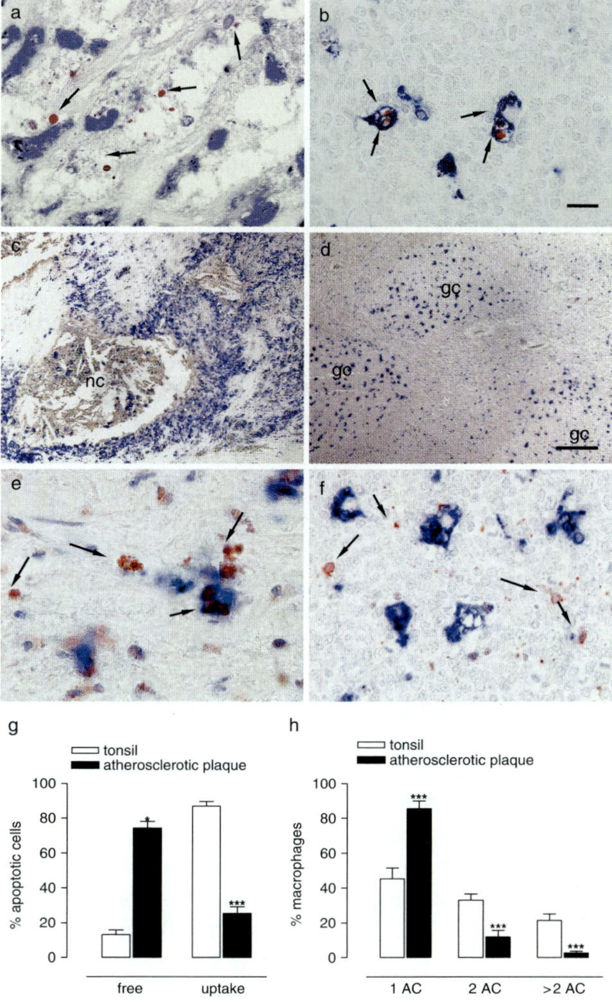

Fig. 13.3 Phagocytosis efficiency of apoptotic cells (AC) in human tonsils and advanced human atherosclerotic plaques (A, B). Immunohistochemical detection of CD68 (macrophages, blue) combined with TUNEL (AC, red) in atherosclerotic plaques (A) and tonsils (B). Arrows in panel B show sequestration of AC in macrophages. Despite presence of many CD68 positive macrophages in panel A, most apoptotic cells in the plaque remain unremoved (arrows). (C,D) Immunohistochemical staining for CD68 (macrophages, blue) in human plaques (C) and tonsils (D). nc = necrotic core, gc = germinal center. (E, F) Immunohistochemical detection of cleaved caspase-3 (AC, red) and CD68 (macrophages, blue) in atherosclerotic plaques (E) and tonsils (F). Arrows show cleaved caspase-3-positive apoptotic cells, either free or sequestrated in macrophages. (G) Quantification of free and phagocytized apoptotic cells in tonsils (open bars) and atherosclerotic plaques (solid bars) after immunostaining of both tissues for CD68 combined with TUNEL. (H) Quantification of macrophages engulfing one or more apoptotic cells in tonsils compared to atherosclerotic plaques. *p<0.05; ***p<0.001 vs. tonsil (Mann-Whitney test). Bar = 10 μm (A, B, E, F) or 200 μm (C, D). Reproduced with permission from Schrijvers et al., Arterioscler Thromb Vasc Biol. 2005; 25: page 1258 of 1256-1261.

13.3.3 Possible Mechanisms of Defective Phagocytic Clearance of Apoptotic Cells in Advanced Atherosclerotic Plaques

One of the first discoveries in the quest for possible mechanisms that contribute to defective clearance of apoptotic cells in atherosclerosis was made by Steinberg and co-workers, who showed that oxLDL, but not acetylated LDL, competes with apoptotic and oxidatively damaged cells for macrophage binding (Sambrano and Steinberg 1995; Ramprasad et al. 1995). Apoptotic cells, oxLDL as well as oxidized red blood cells share oxidatively modified moieties on their surfaces that serve as ligands for macrophage recognition, thereby profoundly inhibiting the phagocytosis of apoptotic cells (Fig. 13.4, mechanism 1). Moreover, oxLDL is immunogenic and anti-oxLDL autoantibodies are commonly found in atherosclerotic lesions of both animals and patients (Palinski et al. 1995; Shaw et al. 2001). These autoantibodies also specifically bind to the surface of apoptotic cells, but not to normal cells, thereby inhibiting their phagocytosis by macrophages (Chang et al. 1999; Fig. 13.4, mechanism 2). Another mechanism related to uptake of oxLDL is based on cell culture data showing that minimally oxidized LDL induces actin polymerization and spreading of macrophages, which results in inhibition of phagocytosis of apoptotic cells but enhancement of oxLDL uptake (Miller et al. 2003). Furthermore, secretion of MFG-E8 (lactadherin), a macrophage-derived molecule that facilitates phagocytosis of apoptotic cells, is downregulated upon free cholesterol-loading in macrophages (Su et al. 2005; Fig. 13.4, mechanism 3).

Electron microscopy images from macrophage-derived foam cells in atherosclerotic plaques show that these cells are often crammed with large lipid droplets. It is therefore tempting to speculate that these macrophages are no longer able to engulf apoptotic cells ("full is full," Fig. 13.4, mechanism 4). Uptake of apoptotic cells in macrophages treated in vitro with an excess of platelets or aggregated LDL is not inhibited, most likely because foam cells in culture rarely reach the same degree of foam cell formation as seen in human plaques, but phagocytosis of beads results in a nearly complete inhibition of phagocytosis of apoptotic cells (Schrijvers et al. 2005). These findings confirm a study by Moller et al. (2002) showing that uptake of indigestible, rigid particles by macrophages induces cellular 'stiffening' so that changes in cellular shape, which are needed to form pseudopodia for phagocytosis, are inhibited.

In addition to foam cell formation, increased oxidative stress is a hallmark of advanced atherosclerotic plaques. Although several reactive oxygen species (ROS) can be found in human plaques, peroxynitrite (ONOO$^-$) plays a central role in the pathophysiology of atherosclerosis as it induces protein nitrosylation and oxidative DNA damage (Martinet et al. 2002a; Kamat 2006). The peroxynitrite donor SIN1A concentration-dependently decreases phagocytosis of apoptotic cells in vitro (Schrijvers et al. 2005), suggesting that PS or other factors present on macrophages or dying cells required for phagocytosis may be sensitive to oxidative treatment (Fig. 13.4, mechanism 5).

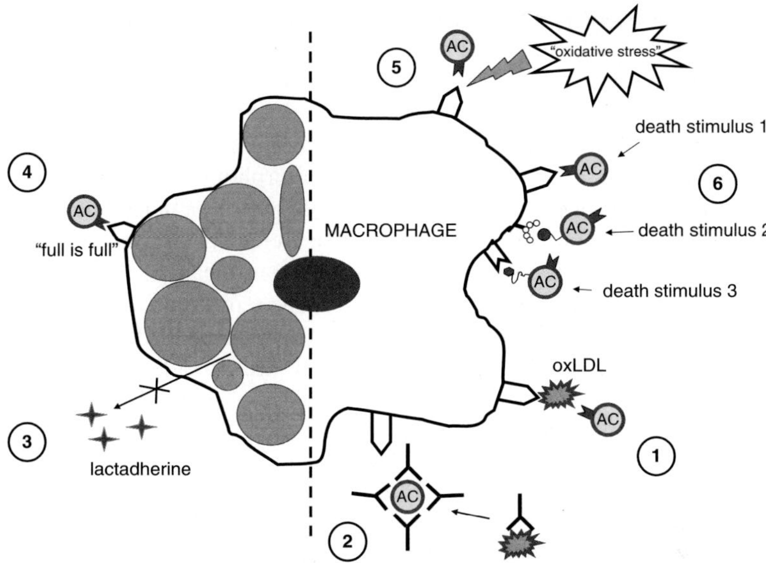

Fig. 13.4 Schematic overview of the different mechanisms that are responsible for diminished phagocytosis of apoptotic cells (AC) by macrophages in advanced atherosclerotic plaques (1) oxLDL competes with AC for the same epitopes on the surface of macrophages, thereby profoundly inhibiting AC phagocytosis. (2) Monoclonal autoantibodies to various epitopes of oxLDL bind specifically to the surface of AC and not to normal cells. Once bound, these antibodies profoundly inhibit uptake of AC by macrophages. (3) Free cholesterol loading in macrophages downregulates the secretion of lactadherin, a macrophage-derived molecule that facilitates phagocytosis of AC. (4) Accumulation of indigestible material (e.g. lipid droplets in foam cells) causes macrophage 'stiffening', thereby decreasing macrophage phagocytosis capacity. (5) Oxidative stress exerted by peroxynitrite (and probably also by other reactive oxygen species) inhibits phagocytosis of AC by macrophages. (6) The cell death stimulus that results in the formation of AC determines the extent to which dying cells are recognized and removed by phagocytes.

Another factor that may contribute to diminished phagocytosis of apoptotic cells is macrophage heterogeneity within atherosclerotic plaques. Some cell surface markers are expressed differently on macrophages close to the endothelial layer vs. those in the deeper areas of the plaque, due to variable stages of monocyte-to-macrophage differentiation, different degrees of lipid loading, and/or effects of focal extracellular molecules (van der Wal et al. 1992; Wintergerst et al. 1998). Given that some of the differentially expressed cell surface markers can be involved in the uptake of apoptotic cells (Wintergerst et al. 1998), and that phagocytosis efficiency changes with the stage of differentiation (Neumann and Sorg 1980), the relative proportion of phagocytic-defective subpopulations of macrophages in lesions may affect the clearance of apoptotic cells and the sites within plaques where lesional necrosis occurs.

Finally, it is worth mentioning that success or failure of adequate phagocytosis of apoptotic cells is determined not only by phagocyte availability, but also by the nature of the cell death trigger (Wiegand et al. 2001). Indeed, cells undergoing apoptosis are not equally marked for safe clearance, which implies that clearance of apoptotic bodies depends on the cell death initiator that is used (Fig. 13.4, mechanism 6). Future studies may reveal new mechanisms for defective apoptotic clearance in advanced atherosclerosis based on other diseases with defective phagocytosis. For example, impairment of serum complement activation and reduced expression of CD44 on apoptotic polymorphonuclear neutrophils seem to render reduced phagocytosis of apoptotic polymorphonuclear neutrophils in systemic lupus erythematosis (Ren et al. 2003). Defective clearance of apoptotic cells in cystic fibrosis is caused by elastase-mediated cleavage of the phosphatidylserine receptor and contributes to ongoing airway inflammation (Vandivier et al. 2002). In this context, it is interesting to note that vascular SMCs, in particular those isolated from abdominal aortic aneurysms, are capable of secreting large amounts of elastase, and therefore may contribute to impaired phagocytosis of apoptotic cells by macrophages (Cohen et al. 1992).

13.3.4 *Potential Consequences of Impaired Apoptotic Cell Clearance for Atherogenesis*

The accumulation of uncleared apoptotic cells has a number of consequences that promote plaque progression (Fig. 13.5). Firstly, impaired phagocytosis of apoptotic cells results in enhanced secretion of pro-inflammatory cytokines including TNF-α and IL-6 by the phagocyte, and a reduced release of anti-inflammatory proteins such as TGF-β and IL-10 (Fadok et al. 1998; Khan et al. 2003; Erwig and Henson 2007). De-regulated expression of these cytokines may result in inflammatory autoimmune responses, as seen in systemic lupus erythematosus and rheumatoid arthritis, two non-atherosclerotic inflammatory disorders with a similar defect in phagocytosis of dying cells by macrophages. Secondly, PS often becomes oxidized during apoptosis (Kadl et al. 2004). These oxidized phospholipids are known to induce secretion of pro-inflammatory proteins such as monocyte chemotactic protein-1 (MCP-1) and IL-8, thereby contributing to a persistent state of chronic inflammation. Thirdly, tissue factor (TF) expression co-localizes with apoptotic cells, especially around the lipid core of human atherosclerotic plaques, suggesting that uncleared apoptotic cells are an important source of this molecule (Tedgui and Mallat 2001). TF is a key element in the initiation of the coagulation cascade and mediates thrombus formation after rupture of an unstable plaque, when free apoptotic cells are exposed to the blood stream. It binds coagulation factor VII and its activated form factor VIIa. This complex proteolytically activates factors IX and X, which in turn leads to thrombin generation. TF activity is highly dependent on the presence of PS because PS increases the catalytic activity of the TF/factor VIIa complex (Tedgui and Mallat 2001). Fourthly, macrophages that have ingested cholesterol-loaded apoptotic cells show a remarkable set of survival responses, some of which are not present during the loading of macrophages with lipoprotein-derived cholesterol (Cui et al. 2007). These responses include cholesterol esterification and massive cholesterol efflux

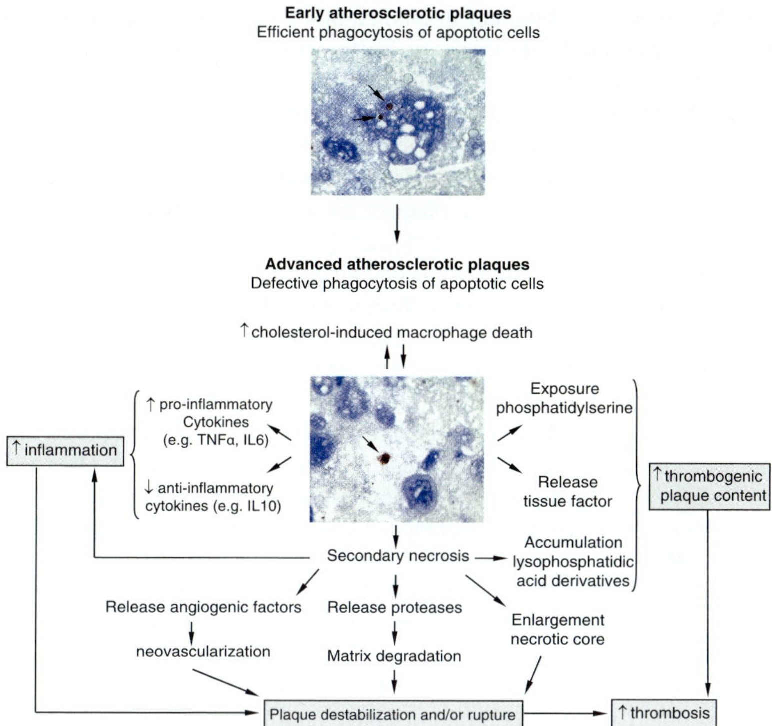

Fig. 13.5 Potential consequences of defective phagocytosis of apoptotic bodies in advanced atherosclerotic plaques. Early lesions reveal efficient phagocytosis of apoptotic cells (TUNEL, arrows) by macrophages (CD68, blue), which prevents the release of toxic intracellular contents and the formation of an inflammatory infiltrate. Apoptotic bodies in advanced plaques, however, are not efficiently removed by phagocytosis and may accumulate in the plaque. Lack of scavenging of apoptotic bodies in advanced plaques underlies plaque progression as well as plaque destabilization and/or rupture through different, well-defined mechanisms. Moreover, free apoptotic cells could lead to an increased level of thrombogenicity and a higher risk for plaque complications. Finally, impaired phagocytosis of apoptotic cells promotes cholesterol-induced macrophage death, and therefore amplifies the accumulation of dead cells as well as its consequences.

as well as the triggering of cell survival signal transduction pathways involving PI3-kinase/Akt and NF-κB. Consequently, impaired phagocytosis of apoptotic cells would render the phagocyte more susceptible to cholesterol-induced death. Finally, expansion of the necrotic core is stimulated due to secondary necrosis of free apoptotic cells. Necrotic core formation is associated with the accumulation of lysophosphatidic acid derivatives, which enhance platelet aggregation (Siess and Tigyi 2004) and may prevent the emigration of macrophages from atherosclerotic lesions (Llodra et al. 2004). On the other hand, the interaction of macrophages with necrotic cells or compounds released from necrotic cells often results in an additional inflammatory response or stimulate secretion of matrix-degrading enzymes in the plaque (Galis et al. 1994; Scaffidi et al. 2002). Necrotic macrophages also passively release matrix degrading proteases as well as the potent angiogenic factor thymidine phosphorylase, which is abundant in the necrotic core of advanced atherosclerotic lesions (Boyle et al. 2000). In this way, post-apoptotic necrosis may contribute to plaque instability through enlargement of the necrotic core, matrix breakdown and stimulation of inflammation and neovascularization (Fig. 13.5).

The exact number of unphagocytosed apoptotic cells required for plaque progression is presently unknown. It is therefore unclear whether free apoptotic cells in advanced human atherosclerotic plaques are frequent enough to induce significant inflammation or other plaque destabilizing effects. Nonetheless, the impact of impaired phagocytosis of apoptotic cells on atherogenesis was demonstrated recently in different mouse models. Aortic valve lesions in LDL receptor (LDLR)$^{-/-}$ recipients of transglutaminase 2 (TG2)$^{-/-}$ bone-marrow, for example, are larger than in recipients of TG2$^{+/+}$ bone-marrow (Boisvert et al. 2006). TG2 deficiency fundamentally impairs the capacity of macrophages to ingest apoptotic cells (Szondy et al. 2003), which supports the general hypothesis that accumulation of undigested apoptotic cells may function as a pro-atherogenic factor that stimulates expansion of atherosclerotic plaques. Besides TG2, apolipoprotein E (apoE), a structural component of all lipoproteins other than LDL, modulates clearance of apoptotic cells both in vitro and in vivo, resulting in a systemic pro-inflammatory state in apoE$^{-/-}$ mice, independent of its role in lipoprotein metabolism (Grainger et al. 2004). This finding is highly relevant in terms of atherogenesis because apoE$^{-/-}$ mice, unlike wild type strains, spontaneously develop atherosclerotic plaques (Zhang et al. 1992). In this light, they are widely used as a model for experimental atherosclerosis. Gld.apoE$^{-/-}$ mice, lacking the genes for both Fas-ligand and apolipoprotein E, have high levels of free apoptotic cells in tissues and in the circulation which was due, at least in part, to an impaired ability to scavenge apoptotic debris (Aprahamian et al. 2004). These mice display enhanced atherosclerosis as compared to apoE$^{-/-}$ mice. More recently, Ait-Oufella et al. (2007) demonstrated that lack of lactadherin in bone-marrow-derived cells enhances the accumulation of apoptotic cell corpses in plaques from LDL receptor-deficient mice and alters the protective immune response (a reduction in IL-10 in the spleen, but an increase in IFN-γ in both the spleen and atherosclerotic plaques). These effects in turn lead to the formation of a large acellular core and a marked acceleration of atherogenesis. All together, the present results suggest that defective phagocytosis of apoptotic cells in atherosclerotic lesions may represent a major risk in patients with cardiovascular disease and a novel target for disease modulation.

13.3.5 Strategies to Limit the Detrimental Effects of Impaired Phagocytosis of Apoptotic Cells in Atherosclerosis

A potential strategy targeted to the problem of free apoptotic cells and its detrimental consequences would be to inhibit apoptosis in advanced atherosclerotic plaques. Liu et al. (2005) demonstrated that reconstitution of lethally irradiated LDLR$^{-/-}$ mice with Bax$^{-/-}$ bone-marrow reduces the levels of macrophage apoptosis but promotes the development of atherosclerosis, indicating that macrophage apoptosis provides a critical self-defence mechanism in suppressing atherogenesis. Moreover, using APOE*3-Leiden mice as atherosclerotic background, van Vlijmen et al. (2001) showed that macrophage p53 deficiency leads to a decrease in apoptosis, but a 2.3-fold increase in mean total lesion area and significantly more necrosis in the plaque, which is consistent with the hypothesis that macrophage apoptosis is important in suppressing the progression of atherosclerosis. A decreased level of macrophage apoptosis might also cause expansion of the atherosclerotic plaque via increased growth factor production and stimulation of SMC growth. Suppression of apoptosis is therefore not a valuable approach for the prevention and treatment of atherosclerosis.

An alternative strategy targeted to the problem of defective phagocytosis of apoptotic cells in advanced plaques would be to therapeutically enhance the phagocytosis capacity of macrophages in these lesions. However, it is important to note that macrophage phagocytosis in atherogenesis might act as a double edged sword (Schrijvers et al. 2007). Macrophages are necessary for removal of apoptotic cells from plaques, but exert strong pro-atherogenic properties upon phagocytosis of lipoproteins, erythrocytes and platelets, which are abundantly present in advanced lesions. Given this complex role of macrophage phagocytosis in atherosclerosis, it is unclear whether an increase in the rate of phagocytosis has a beneficial or detrimental effect on plaque progression and whether the ultimate effect of phagocytosis depends on the type and/or complexity of the lesion. Direct in vivo evidence from studies using phagocytosis inhibitors such as cytochalasin B or latrunculin A is lacking. Therefore, we can only draw preliminary conclusions based on indirect observations. Early lesions do not contain large amounts of infiltrated platelets or erythrocytes and reveal undetectable levels of apoptosis (Kockx and Herman 2000). Therefore, phagocytosis in early lesions is mainly focused on lipoproteins and might be pro-atherogenic. In advanced lesions, where different particles or cellular debris can be phagocytosed including platelets, erythrocytes, apoptotic cells and lipoproteins, only phagocytosis of apoptotic cells seems beneficial for plaque stability (Schrijvers et al. 2007). Because phagocytosis of apoptotic cells by macrophages in advanced plaques is significantly impaired, the benefit resulting from enhanced uptake of apoptotic cells might not overrule the adverse effects associated with phagocytosis of lipoproteins, platelets or erythrocytes, especially in complicated lesions with significant neovascularization or in ruptured plaques. As a consequence, both in early and advanced lesions, enhancement of phagocytosis may stimulate rather than inhibit plaque formation. Thus, non-specific enhancement of phagocytosis in atherosclerosis seems not desirable. Nonetheless, apoptotic clearance is a beneficial process and selectively enhancing uptake of apoptotic cells might help to prevent the progression of advanced atherosclerotic lesions, on condition that the pro-clearance intervention would not alter the fundamental property of phagocytes to selectively recognize and ingest only apoptotic cells. Glucocorticoids are known to stimulate phagocytic clearance of apoptotic cells, but the serious adverse effects of glucocorticoid therapy render this approach impractical (Maderna and Godson 2003). Thiazolidinediones, which affect cells through activation of the nuclear receptor peroxisome proliferator-activated receptor (PPAR)-γ and through other 'off-target' mechanisms, selectively improve phagocytosis of apoptotic cells in vitro, but at the same time enhance macrophage apoptosis by a number of stimuli, including those thought to be important in advanced atherosclerosis (Thorp et al. 2007). Consequently, the net in vivo effect of these drugs is increased plaque necrosis and subsequent plaque destabilization. Because oxLDL or oxLDL antibodies interfere with phagocytosis of apoptotic cells, it is tempting to speculate that antioxidants can improve clearance of dying cells. However, clinical trials in humans with antioxidants showed only limited success in preventing coronary artery disease, although it is possible that more encouraging results will be obtained in the future through the use of drugs that are targeted to specific oxidation reactions in atherosclerosis (Tardif 2005; Cherubini et al. 2005). It should also be noted that certain antioxidants may not promote but inhibit recognition of apoptotic cells by phagocytes by inhibiting oxidation of externalized PS (Tyurina et al. 2004). Examples of other drugs that might promote phagocytic clearance of apoptotic cells are the cholesterol-lowering agent lovastatin (Morimoto et al. 2006), the macrolide antibiotic azithromycin (Reynolds and Hodge, Chap. 14, this Vol.; Hodge et al. 2006) and members of the lipoxin family (Godson et al. 2000). Future studies are needed to determine whether these drugs may provide the basis for a novel therapeutic strategy to prevent the progression of advanced atherosclerotic plaques via modulation of phagocytosis. Indeed, it has been proposed that abundant phagocytosis of apoptotic cells might be associated with the production of reactive oxygen species and tissue injury which has prompted the search for attenuation mechanisms of phagocytosis (de Almeida and Linden 2005).

13.3.6 *Implications for Plaque-stabilizing Interventions*

Because macrophages play a central role in atherosclerotic plaque destabilization, selective induction of macrophage death now gains increasing attention in cardiovascular medicine to stabilize "vulnerable", rupture-prone lesions (Martinet et al. 2007). Several successful strategies have recently been reported to induce macrophage cell death in atherosclerotic plaques without affecting SMC viability. For example, local administration of the mammalian target of rapamycin (mTOR) inhibitor everolimus or the protein synthesis inhibitor cycloheximide selectively depletes macrophages in plaques from cholesterol-fed rabbits via autophagic or apoptotic cell death, respectively, without significant adverse effects (Verheye et al. 2007; Croons et al. 2007). It remains, however, unclear what type of death is preferred to deplete macrophages in atherosclerotic plaques. Given that the most abundant phagocytes in advanced lesions are macrophages themselves, stimulation of macrophage apoptosis would deplete the pool of phagocytes in these lesions and, at the same time, enhance the number of free apoptotic corpses that amplify plaque progression, as explained above. Stoneman et al. (2007) recently proposed a novel transgenic mouse model to study more in depth the role of monocyte/macrophage apoptosis in atherogenesis and established plaques. This mouse strain expresses the human diphtheria toxin receptor (hDTR) from a monocyte/macrophage-specific CD11b promoter sequence so that conditional induction of monocyte/macrophage apoptosis can be achieved by intraperitoneal injection of diphtheria toxin (DT). Using this model, Stoneman et al. (2007) showed that DT-induced macrophage death does not induce inflammation or markers of plaque rupture in established plaques, which justifies therapeutic means to selectively remove plaque macrophages via induction of apoptosis. It should be noted, however, that DT was administered systemically, not locally, and thus resulted in a 50% reduction of circulating monocytes. This effect is obviously not favorable in clinical applications and makes the approach not representative for local therapy. Possibly, selective induction of macrophage autophagy may offer better perspectives. During autophagy, cells literally digest themselves to death. As a consequence, the cytoplasmic content progressively decreases so that activation of inflammatory responses, the release of matrix degrading proteases and the deposition of necrotic debris after post-autophagic necrosis is minimal.

13.4 Conclusions

Unstable rupture-prone plaques are characterized by the accumulation of macrophages and many free, non-phagocytosed apoptotic cells, suggesting inefficient removal of the latter. Several factors present in unstable plaques, such as indigestible material in the macrophage cytoplasm, oxidative stress, and the presence of oxLDL or oxidized erythrocytes may contribute to the defective phagocytic clearance of apoptotic cells. Lack of lesional macrophages to safely clear apoptotic cells may contribute to the formation of the necrotic core through accumulation of necrotic debris, which in concert with pro-atherogenic effects of residual surviving macrophages, promotes further inflammation, plaque instability and thrombosis. Although these detrimental effects fit with current in vitro data as well as with several mouse models for impaired phagocytosis of apoptotic cells, it is important to emphasize that more evidence is needed. For example, it remains unclear whether the number of free apoptotic cells in advanced human plaques is sufficiently high to evoke the proposed plaque-destabilizing events. On the other hand, a thorough understanding of the mechanisms and consequences of apoptosis and phagocytosis of apoptotic cells is required in order to translate these concepts into novel therapeutic strategies supporting plaque stability.

Acknowledgments Research was supported by the Fund for Scientific Research (FWO)-Flanders (Belgium) (project N G.0308.04 and G.0113.06), the University of Antwerp (NOI-BOF) and the Bekales Foundation. Wim Martinet and Dorien M. Schrijvers are postdoctoral fellows of the FWO-Flanders.

References

Aderem A, Underhill DM (1999) Mechanisms of phagocytosis in macrophages. Annu Rev Immunol 17:593–623

Ait-Oufella H, Kinugawa K, Zoll J et al (2007) Lactadherin deficiency leads to apoptotic cell accumulation and accelerated atherosclerosis in mice. Circulation 115:2168–2177

Aprahamian T, Rifkin I, Bonegio R et al (2004) Impaired clearance of apoptotic cells promotes synergy between atherogenesis and autoimmune disease. J Exp Med 199:1121–1131

Bobryshev YV (2006) Monocyte recruitment and foam cell formation in atherosclerosis. Micron 37:208–222

Boisvert WA, Rose DM, Boullier A et al (2006) Leukocyte transglutaminase 2 expression limits atherosclerotic lesion size. Arterioscler Thromb Vasc Biol 26:563–569

Boyle JJ, Bowyer DE, Weissberg PL et al (2001) Human blood-derived macrophages induce apoptosis in human plaque-derived vascular smooth muscle cells by Fas-ligand/Fas interactions. Arterioscler Thromb Vasc Biol 21:1402–1407

Boyle JJ, Wilson B, Bicknell R et al (2000) Expression of angiogenic factor thymidine phosphorylase and angiogenesis in human atherosclerosis. J Pathol 192:234–242

Callahan MK, Williamson P, Schlegel RA (2000) Surface expression of phosphatidylserine on macrophages is required for phagocytosis of apoptotic thymocytes. Cell Death Differ 7:645–653

Chang MK, Bergmark C, Laurila A et al (1999) Monoclonal antibodies against oxidized low-density lipoprotein bind to apoptotic cells and inhibit their phagocytosis by elicited macrophages: evidence that oxidation-specific epitopes mediate macrophage recognition. Proc Natl Acad Sci U S A 96:6353–6358

Chappell DC, Varner SE, Nerem RM et al (1998) Oscillatory shear stress stimulates adhesion molecule expression in cultured human endothelium. Circ Res 82:532–539

Cherubini A, Vigna GB, Zuliani G et al (2005) Role of antioxidants in atherosclerosis: epidemiological and clinical update. Curr Pharm Des 11:2017–2032

Chisolm GM, Steinberg D (2000) The oxidative modification hypothesis of atherogenesis: an overview. Free Radic Biol Med 28:1815–1826

Chung EY, Kim SJ, Ma XJ (2006) Regulation of cytokine production during phagocytosis of apoptotic cells. Cell Res 16:154–161

Clarke MC, Figg N, Maguire JJ et al (2006) Apoptosis of vascular smooth muscle cells induces features of plaque vulnerability in atherosclerosis. Nat Med 12:1075–1080

Cohen JR, Sarfati I, Danna D et al (1992) Smooth muscle cell elastase, atherosclerosis, and abdominal aortic aneurysms. Ann Surg 216:327–330

Cromheeke KM, Kockx MM, De Meyer GRY et al (1999) Inducible nitric oxide synthase colocalizes with signs of lipid oxidation/peroxidation in human atherosclerotic plaques. Cardiovasc Res 43:744–754

Croons V, Martinet W, Herman AG et al (2007) Selective clearance of macrophages in atherosclerotic plaques by the protein synthesis inhibitor cycloheximide. J Pharmacol Exp Ther 320:986–993

Cui D, Thorp E, Li Y et al (2007) Pivotal Advance: Macrophages become resistant to cholesterol-induced death after phagocytosis of apoptotic cells. J Leukocyte Biol 82:1040–1050

Cunningham KS, Gotlieb AI (2005) The role of shear stress in the pathogenesis of atherosclerosis. Lab Invest 85:9–23

de Almeida CJ, Linden R (2005) Phagocytosis of apoptotic cells: a matter of balance. Cell Mol Life Sci 62:1532–1546

Dzau VJ, Braun-Dullaeus RC, Sedding DG (2002) Vascular proliferation and atherosclerosis: new perspectives and therapeutic strategies. Nat Med 8:1249–1256

Erwig LP, Henson PM (2007) Immunological consequences of apoptotic cell phagocytosis. Am J Pathol 171:2–8

Fadok VA (1999) Clearance: the last and often forgotten stage of apoptosis. J Mammary Gland Biol Neoplasia 4:203–211

Fadok VA, Bratton DL, Konowal A et al (1998) Macrophages that have ingested apoptotic cells in vitro inhibit pro-inflammatory cytokine production through autocrine/paracrine mechanisms involving TGF-beta, PGE2, and PAF. J Clin Invest 101:890–898

Fadok VA, Bratton DL, Rose DM et al (2000) A receptor for phosphatidylserine-specific clearance of apoptotic cells. Nature 405:85–90

Franklin BA, Kahn JK (1996) Delayed progression or regression of coronary atherosclerosis with intensive risk factor modification. Effects of diet, drugs, and exercise. Sports Med 22:306–320

Galis ZS, Sukhova GK, Lark MW et al (1994) Increased expression of matrix metalloproteinases and matrix degrading activity in vulnerable regions of human atherosclerotic plaques. J Clin Invest 94:2493–2503

Geng YJ, Wu Q, Muszynski M et al (1996) Apoptosis of vascular smooth muscle cells induced by in vitro stimulation with interferon-gamma, tumour necrosis factor-alpha, and interleukin-1 beta. Arterioscler Thromb Vasc Biol 16:19–27

Getz GS, Reardon CA (2006) Diet and murine atherosclerosis. Arterioscler Thromb Vasc Biol 26:242–249

Godson C, Mitchell S, Harvey K et al (2000) Cutting edge: lipoxins rapidly stimulate non-phlogistic phagocytosis of apoptotic neutrophils by monocyte-derived macrophages. J Immunol 164:1663–1667

Grainger DJ, Reckless J, McKilligin E (2004) Apolipoprotein E modulates clearance of apoptotic bodies in vitro and in vivo, resulting in a systemic pro-inflammatory state in apolipoprotein E-deficient mice. J Immunol 173:6366–6375

Grines CL (2006) The role of statins in reversing atherosclerosis: what the latest regression studies show. J Interv Cardiol 19:3–9

Hansson GK (2005) Inflammation, atherosclerosis, and coronary artery disease. N Engl J Med 352:1685–1695

Hodge S, Hodge G, Brozyna S et al (2006) Azithromycin increases phagocytosis of apoptotic bronchial epithelial cells by alveolar macrophages. Eur Respir J 28:486–495

Hsiai TK, Cho SK, Wong PK et al (2003) Monocyte recruitment to endothelial cells in response to oscillatory shear stress. FASEB J 17:1648–1657Kadl A, Bochkov VN, Huber J et al (2004) Apoptotic cells as sources for biologically active oxidized phospholipids. Antioxid Redox Signal 6:311–320

Kamat JP (2006) Peroxynitrite: a potent oxidizing and nitrating agent. Indian J Exp Biol 44:436–447

Kannel WB, D'Agostino RB, Sullivan L et al (2004) Concept and usefulness of cardiovascular risk profiles. Am Heart J 148:16–26

Khan M, Pelengaris S, Cooper M et al (2003) Oxidised lipoproteins may promote inflammation through the selective delay of engulfment but not binding of apoptotic cells by macrophages. Atherosclerosis 171:21–29

Klein LW (2005) Clinical implications and mechanisms of plaque rupture in the acute coronary syndromes. Am Heart Hosp J 3:249–255

Kockx MM, Cromheeke KM, Knaapen MW et al (2003) Phagocytosis and macrophage activation associated with hemorrhagic microvessels in human atherosclerosis. Arterioscler Thromb Vasc Biol 23:440–446

Kockx MM, De Meyer GRY, Muhring J et al (1998) Apoptosis and related proteins in different stages of human atherosclerotic plaques. Circulation 97:2307–2315

Kockx MM, Herman AG (2000) Apoptosis in atherosclerosis: beneficial or detrimental? Cardiovasc Res 45:736-746

Kolodgie FD, Virmani R, Burke AP et al (2004) Pathologic assessment of the vulnerable human coronary plaque. Heart 90:1385–1391

Krahling S, Callahan MK, Williamson P et al (1999) Exposure of phosphatidylserine is a general feature in the phagocytosis of apoptotic lymphocytes by macrophages. Cell Death Differ 6:183–189

Lee RT, Libby P (1997) The unstable atheroma. Arterioscler Thromb Vasc Biol 17:1859–1867

Leskinen M, Wang Y, Leszczynski D et al (2001) Mast cell chymase induces apoptosis of vascular smooth muscle cells. Arterioscler Thromb Vasc Biol 21:516–522

Levy AP, Moreno PR (2006) Intraplaque hemorrhage. Curr Mol Med 6:479–488

Li MO, Sarkisian MR, Mehal WZ et al (2003) Phosphatidylserine receptor is required for clearance of apoptotic cells. Science 302:1560–1563

Libby P (2000) Changing concepts of atherogenesis. J Intern Med 247:349–358

Libby P, Ridker PM, Maseri A (2002) Inflammation and atherosclerosis. Circulation 105:1135-1143

Linton MF, Fazio S (2003) A practical approach to risk assessment to prevent coronary artery disease and its complications. Am J Cardiol 92:19i–26i

Liu J, Thewke DP, Su YR et al (2005) Reduced macrophage apoptosis is associated with accelerated atherosclerosis in low-density lipoprotein receptor-null mice. Arterioscler Thromb Vasc Biol 25:174–179

Llodra J, Angeli V, Liu J et al (2004) Emigration of monocyte-derived cells from atherosclerotic lesions characterizes regressive, but not progressive, plaques. Proc Natl Acad Sci U S A 101:11779–11784

Lusis AJ (2000) Atherosclerosis. Nature 407:233–241

Lutgens E, de Muinck ED, Kitslaar PJ et al (1999) Biphasic pattern of cell turnover characterizes the progression from fatty streaks to ruptured human atherosclerotic plaques. Cardiovasc Res 41:473–479

Maderna P, Godson C (2003) Phagocytosis of apoptotic cells and the resolution of inflammation. Biochim Biophys Acta 1639:141–151

Mallat Z, Tedgui A (2000) Apoptosis in the vasculature: mechanisms and functional importance. Br J Pharmacol 130:947–962

Martinet W, De Meyer GRY, Herman AG et al (2004a) Reactive oxygen species induce RNA damage in human atherosclerosis. Eur J Clin Invest 34:323–327

Martinet W, De Bie M, Schrijvers DM et al (2004b) 7-ketocholesterol induces protein ubiquitination, myelin figure formation, and light chain 3 processing in vascular smooth muscle cells. Arterioscler Thromb Vasc Biol 24:2296–2301

Martinet W, Knaapen MW, De Meyer GRY et al (2002a) Elevated levels of oxidative DNA damage and DNA repair enzymes in human atherosclerotic plaques. Circulation 106:927–932

Martinet W, Knaapen MW, De Meyer GRY et al (2003) Over-expression of the anti-apoptotic caspase-2 short isoform in macrophage-derived foam cells of human atherosclerotic plaques. Am J Pathol 162:731–736

Martinet W, Kockx MM (2001) Apoptosis in atherosclerosis: focus on oxidized lipids and inflammation. Curr Opin Lipidol 12:535–541

Martinet W, Schrijvers DM, De Meyer GRY et al (2002b) Gene expression profiling of apoptosis-related genes in human atherosclerosis: up-regulation of death-associated protein kinase. Arterioscler Thromb Vasc Biol 22:2023–2029

Martinet W, Verheye S, De Meyer GRY (2007) Selective depletion of macrophages in atherosclerotic plaques via macrophage-specific initiation of cell death. Trends Cardiovasc Med 17:69–75

Mercer J, Mahmoudi M, Bennett M (2007) DNA damage, p53, apoptosis and vascular disease. Mutat Res 621:75–86

Miller YI, Viriyakosol S, Binder CJ et al (2003) Minimally modified LDL binds to CD14, induces macrophage spreading via TLR4/MD-2, and inhibits phagocytosis of apoptotic cells. J Biol Chem 278:1561–1568

Moller W, Hofer T, Ziesenis A et al (2002) Ultrafine particles cause cytoskeletal dysfunctions in macrophages. Toxicol Appl Pharmacol 182:197–207

Morimoto K, Janssen WJ, Fessler MB et al (2006) Lovastatin enhances clearance of apoptotic cells (efferocytosis) with implications for chronic obstructive pulmonary disease. J Immunol 176:7657–7665

Napoli C, Quehenberger O, De Nigris F et al (2000) Mildly oxidized low density lipoprotein activates multiple apoptotic signalling pathways in human coronary cells. FASEB J 14:1996–2007

Neumann C, Sorg C (1980) Sequential expression of functions during macrophage differentiation in murine bone marrow liquid cultures. Eur J Immunol 10:834–840

Palinski W, Tangirala RK, Miller E et al (1995) Increased auto-antibody titers against epitopes of oxidized LDL in LDL receptor-deficient mice with increased atherosclerosis. Arterioscler Thromb Vasc Biol 15:1569–1576

Park D, Tosello-Trampont AC, Elliott MR et al (2007a) BAI1 is an engulfment receptor for apoptotic cells upstream of the ELMO/Dock180/Rac module. Nature 450:430–434

Park SY, Jung MY, Kim HJ et al (2008) Rapid cell corpse clearance by stabilin-2, a membrane phosphatidylserine receptor. Cell Death Differ 15:192–201

Pasterkamp G, Smits PC (2002) Imaging of atherosclerosis. Remodelling of coronary arteries. J Cardiovasc Risk 9:229–235

Pradhan D, Krahling S, Williamson P, Schlegel RA (1997) Multiple systems for recognition of apoptotic lymphocytes by macrophages. Mol Biol Cell 8:767–778

Ramprasad MP, Fischer W, Witztum JL et al (1995) The 94- to 97-kDa mouse macrophage membrane protein that recognizes oxidized low density lipoprotein and phosphatidylserine-rich liposomes is identical to macrosialin, the mouse homologue of human CD68. Proc Natl Acad Sci U S A 92:9580–9584

Ravichandran KS (2003) "Recruitment signals" from apoptotic cells: invitation to a quiet meal. Cell 113:817–820

Ren Y, Tang J, Mok MY et al (2003) Increased apoptotic neutrophils and macrophages and impaired macrophage phagocytic clearance of apoptotic neutrophils in systemic lupus erythematosus. Arthritis Rheum 48:2888–2897

Ross R (1999) Atherosclerosis--an inflammatory disease. N Engl J Med 340:115–126

Sambrano GR, Steinberg D (1995) Recognition of oxidatively damaged and apoptotic cells by an oxidized low density lipoprotein receptor on mouse peritoneal macrophages: role of membrane phosphatidylserine. Proc Natl Acad Sci U S A 92:1396–1400

Sata M, Walsh K (1998) Endothelial cell apoptosis induced by oxidized LDL is associated with the down-regulation of the cellular caspase inhibitor FLIP. J Biol Chem 273:33103–33106

Scaffidi P, Misteli T, Bianchi ME (2002) Release of chromatin protein HMGB1 by necrotic cells triggers inflammation. Nature 418:191–195

Schrijvers DM, De Meyer GRY, Herman AG et al (2007) Phagocytosis in atherosclerosis: Molecular mechanisms and implications for plaque progression and stability. Cardiovasc Res 73:470–480

Schrijvers DM, De Meyer GRY, Kockx MM et al (2005) Phagocytosis of apoptotic cells by macrophages is impaired in atherosclerosis. Arterioscler Thromb Vasc Biol 25:1256–1261

Shaw PX, Horkko S, Tsimikas S et al (2001) Human-derived anti-oxidized LDL auto-antibody blocks uptake of oxidized LDL by macrophages and localizes to atherosclerotic lesions in vivo. Arterioscler Thromb Vasc Biol 21:1333–1339

Siess W, Tigyi G (2004) Thrombogenic and atherogenic activities of lysophosphatidic acid. J Cell Biochem 92:1086–1094

Skalen K, Gustafsson M, Rydberg EK et al (2002) Subendothelial retention of atherogenic lipoproteins in early atherosclerosis. Nature 417:750–754

Stary HC (1987) Macrophages, macrophage foam cells, and eccentric intimal thickening in the coronary arteries of young children. Atherosclerosis 64:91–108

Stoneman V, Braganza D, Figg N et al (2007) Monocyte/macrophage suppression in CD11b diphtheria toxin receptor transgenic mice differentially affects atherogenesis and established plaques. Circ Res 100:884–893

Su YR, Dove DE, Major AS et al (2005) Reduced ABCA1-mediated cholesterol efflux and accelerated atherosclerosis in apolipoprotein E-deficient mice lacking macrophage-derived ACAT1. Circulation 111:2373–2381

Szondy Z, Sarang Z, Molnar P et al (2003) Transglutaminase 2-/- mice reveal a phagocytosis-associated crosstalk between macrophages and apoptotic cells. Proc Natl Acad Sci U S A 100:7812–7817

Tabas I (2004) Apoptosis and plaque destabilization in atherosclerosis: the role of macrophage apoptosis induced by cholesterol. Cell Death Differ 11 Suppl 1:S12–S16

Tabas I (2005) Consequences and therapeutic implications of macrophage apoptosis in atherosclerosis: the importance of lesion stage and phagocytic efficiency. Arterioscler Thromb Vasc Biol 25:2255–2264

Tardif JC (2005) Antioxidants and atherosclerosis: emerging drug therapies. Curr Atheroscler Rep 7:71–77

Tedgui A, Mallat Z (2001) Apoptosis as a determinant of atherothrombosis. Thromb Haemost 86:420–426

Thorp E, Kuriakose G, Shah YM et al (2007) Pioglitazone increases macrophage apoptosis and plaque necrosis in advanced atherosclerotic lesions of non-diabetic low-density lipoprotein receptor-null mice. Circulation 116:2182–2190

Tyurina YY, Serinkan FB, Tyurin VA et al (2004) Lipid antioxidant, etoposide, inhibits phosphatidylserine externalization and macrophage clearance of apoptotic cells by preventing phosphatidylserine oxidation. J Biol Chem 279:6056–6064

Van Der Wal AC, Das PK, Tigges AJ et al (1992) Macrophage differentiation in atherosclerosis. An in situ immunohistochemical analysis in humans. Am J Pathol 141:161–168

van Vlijmen BJ, Gerritsen G, Franken AL et al (2001) Macrophage p53 deficiency leads to enhanced atherosclerosis in APOE*3-Leiden transgenic mice. Circ Res 88:780–786

VanderLaan PA, Reardon CA, Getz GS (2004) Site specificity of atherosclerosis: site-selective responses to atherosclerotic modulators. Arterioscler Thromb Vasc Biol 24:12–22

Vandivier RW, Fadok VA, Hoffmann PR et al (2002) Elastase-mediated phosphatidylserine receptor cleavage impairs apoptotic cell clearance in cystic fibrosis and bronchiectasis. J Clin Invest 109:661–670

Verheye S, Martinet W, Kockx MM et al (2007) Selective clearance of macrophages in atherosclerotic plaques by autophagy. J Am Coll Cardiol 49:706–715

Wiegand UK, Corbach S, Prescott AR et al (2001) The trigger to cell death determines the efficiency with which dying cells are cleared by neighbours. Cell Death Differ 8:734–746

Wintergerst ES, Jelk J, Asmis R (1998) Differential expression of CD14, CD36 and the LDL receptor on human monocyte-derived macrophages. A novel cell culture system to study macrophage differentiation and heterogeneity. Histochem Cell Biol 110:231–241

Yao PM, Tabas I (2000) Free cholesterol loading of macrophages induces apoptosis involving the fas pathway. J Biol Chem 275:23807–23813

Yao PM, Tabas I (2001) Free cholesterol loading of macrophages is associated with widespread mitochondrial dysfunction and activation of the mitochondrial apoptosis pathwayzz. J Biol Chem 276:42468–42476

Zhang SH, Reddick RL, Piedrahita JA et al (1992) Spontaneous hypercholesterolemia and arterial lesions in mice lacking apolipoprotein E. Science 258:468–471392

Chapter 14
The Impact of Defective Clearance of Apoptotic Cells in the Pathogenesis of Chronic Lung Diseases: Chronic Obstructive Pulmonary Disease, Asthma and Cystic Fibrosis

Paul N. Reynolds and Sandra J. Hodge

Abstract: Apoptosis is an important process in the regulation of cell turnover and inflammatory responses in the lungs. Efficient clearance of apoptotic cells is a critical component of the process to avoid accumulation of apoptotic material and the subsequent development of secondary necrosis and incitement or perpetuation of inflammation. Airway and alveolar macrophages have the major role in ensuring effective clearance, but other cells including airway epithelial cells also have a role. Several recent studies have identified defects in the phagocytic capacity of macrophages obtained from subjects with chronic pulmonary diseases including chronic obstructive pulmonary disease, asthma and cystic fibrosis. There is evidence that these defects relate to levels of expression of various cell surface receptors including the mannose receptor and components of the collectin system. The defects in macrophage function are a potential target for new therapeutic interventions which may complement existing therapies. Approaches currently being pursued include the use of macrolide antibiotics, statins, exogenous surfactant proteins and mannose binding lectin. Pre-clinical studies and early phase human studies show some promise but further work is needed before these strategies emerge as established therapies.

Keywords: Airways • Collectins • Macrolides • Smoking • Statins

P. N. Reynolds (✉) · S. J. Hodge
Department of Thoracic Medicine
Royal Adelaide Hospital, Hanson Institute and University of Adelaide
Chest Clinic, 275 North Tce
South Australia 5000, Australia
Tel.: +61 8 82225376
Fax: +61 8 82225957
E-mail: paul.reynolds@adelaide.edu.au

D. V. Krysko, P. Vandenabeele (eds.), *Phagocytosis of Dying Cells,*
DOI: 10.1007/978-1-4020-9292-3_14, © Springer Science+Business Media B.V. 2009

14.1   Introduction

The regulation of apoptosis and the effective clearance of apoptotic material has emerged as an important area of research in the pathogenesis of pulmonary disease (Vandivier et al. 2006). Our group has predominantly focussed on the role of these processes in the pathogenesis of chronic obstructive pulmonary disease (COPD). COPD is projected to become the third leading cause of death by the year 2020 and the costs of the disease are enormous both in economic terms and in the impact on quality of life. From a clinical standpoint, interest in the management of this disease has increased significantly in recent years. There is now good evidence that quality of life and survival for COPD sufferers can be improved with modern medical regimens based on the use of long-acting bronchodilators and inhaled steroids. That having been said, the efficacy of treatments in absolute terms is modest and there is an accepted need for improvements to therapy. Smoking cessation is of course critical, but patients with established disease will continue to have some degree of symptoms even after smoking cessation.

14.2   Increased Apoptosis in COPD

As outlined in other chapters, apoptosis and subsequent phagocytosis of the apoptotic cells provides a mechanism for dealing with cell turnover and death in a manner that avoids inflammation. Perturbation of this mechanism may occur either via an excess of apoptosis per se, a defect in clearance mechanisms or a combination of these factors. Several studies have identified evidence of increased apoptosis in COPD (or more specifically emphysema) through the analysis of resected lung specimens (Dini and Vergallo, Chap. 15, this Vol.; Kasahara et al. 2001; Henson and Tuder 2008). Histological analysis using standard techniques such as TUNEL assay have identified increased rates of both endothelial and epithelial apoptosis in this condition. Further evidence is derived from animal studies where the induction of alveolar endothelial apoptosis via inhibition of VEGF was shown to induce apoptosis with subsequent emphysema development (Kasahara et al. 2000) and VEGF levels are reduced in clinical samples of emphysematous lung (Kasahara et al. 2001). We initially hypothesized that excess apoptosis may be present in the airways of COPD patients and if so, that this might contribute to the ongoing inflammation in the airways via secondary necrosis of uncleared material.

To enable assessment of subjects with COPD and for comparison to healthy subjects, samples for analysis were obtained from COPD patients and normal volunteers by bronchoscopy and brush biopsy of airway epithelium. The brushing technique yields a population of over 90% epithelial cells, and effectively samples a wider area than is possible with forceps biopsy. This approach may improve the robustness of results obtained as previous histological studies had found that apoptotic cells were detected at low frequency in sections. A number of techniques were used to analyse apoptosis in these cells including techniques reliant on membrane changes (7AAD and Annexin V staining) and those reflecting intracellular changes (single stranded DNA and caspase activation; Trahtemberg and Mevorach, Chap. 8, this Vol.; Hodge et al. 2005). These studies identified an increase in apoptosis of airway epithelial cells of 87% compared to normal controls (Fig. 14.1). Particularly notable was the finding that the increased apoptosis was also apparent in subjects with COPD that had ceased smoking for over 12 months. This finding indicates the failure of the airway epithelium to return to normal after years of noxious insult, a finding consistent with others who have noted ongoing inflammation in the airways of COPD subjects after smoking cessation (Gamble et al. 2007). Recent theories have also proposed that uncleared apoptotic material might promote the generation of autoantibodies giving rise to an "auto-immune" mechanism to help explain the perpetuation of disease after smoking cessation (Curtis et al. 2007).

In studies using human subjects it can be difficult to confirm cause-effect relationships. In an effort to provide some evidence for an association between excess apoptosis and secondary necrosis we measured lactate dehydrogenase activity in the bronchoalveolar lavage samples obtained from the subjects. An increase was seen in the COPD subjects which is at least consistent with the hypothesis. These findings suggest that tackling the presence of excess apoptotic material in the airways might be a therapeutic option for future COPD treatment. This could be achieved either by reducing the rate of apoptosis, or by evaluating and addressing the clearance mechanisms.

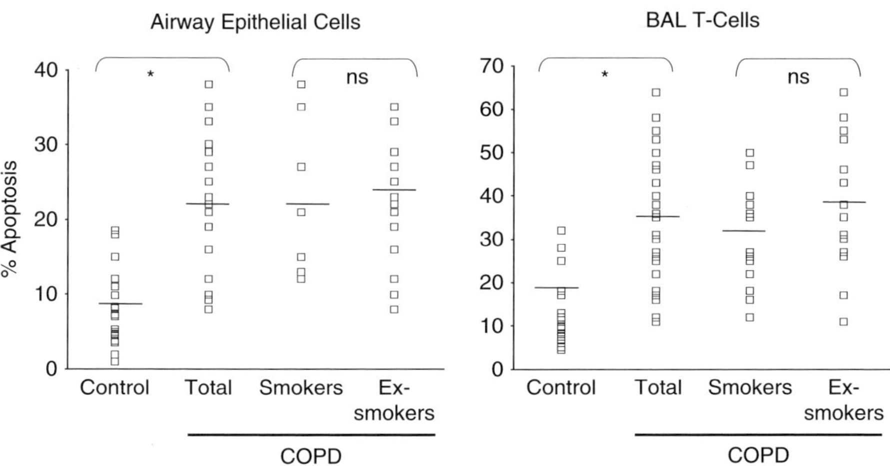

Fig. 14.1 Apoptosis of brushing derived AEC and BAL-derived T cells measured by 7AAD staining and flow cytometry, in controls, COPD subjects (total group) and smokers and ex-smokers with COPD. *Significant difference between COPD and control groups, p≤0.05; ns no significant difference between smokers and ex-smokers with COPD. Adapted from Hodge et al. (2006) Eur Respir J 28:486–495.

14.3 Macrophage Dysfunction in COPD

Airway and alveolar macrophages are key cells involved in the clearance of apoptotic material, as well as a host of other functions including the phagocytosis of invading pathogens. To ascertain whether AM dysfunction might be contributing to the accumulation of apoptotic cells in COPD, a novel flow cytometry assay was developed (Hodge et al. 2003; Hodge et al. 2004). AM cells were obtained from COPD and healthy subjects by bronchoalveolar lavage (BAL), then placed in short term culture. Airway epithelial cells (from an established immortalized cell line) were rendered apoptotic in a controlled manner by exposure to UV irradiation. These apoptotic cells were labelled with a fluorescent dye, then incubated with the AMs. The cells were then analysed by flow cytometry. By using parameters to determine cell size and co-staining with the epithelial cell dye as well as a fluorescing AM marker (anti-CD-33) the percentage of AMs that had ingested epithelial cells could be determined. Using this approach, a significant defect in phagocytic capacity was seen in cells obtained from COPD subjects compared to non-COPD controls. In subsequent studies, further analysis has shown that even smokers without clinically evident COPD have a defect in phagocytic capacity compared to never smokers (Hodge et al. 2007). The degree of abnormality in the smokers was similar to that seen in ex-smokers with COPD whereas continuing smokers with COPD were more affected, even though the level of physiological impairment as determined by spirometry was similar to the ex-smoker COPD group (Fig. 14.2). These findings are consistent with there being a multifactorial basis to the problem—some of which appears to be related to smoking per se, but some effect is seen in established COPD even without ongoing smoking. Others have also described a phagocytosis defect in subjects with COPD but in their

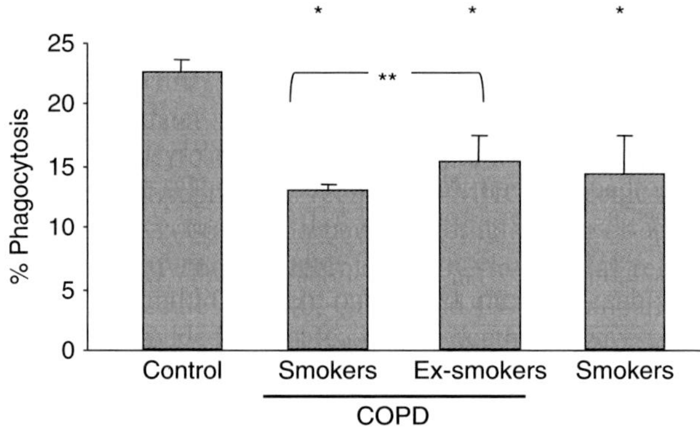

Fig. 14.2 Phagocytosis of apoptotic bronchial epithelial cells by AM from never smoker controls (n = 12), smokers with COPD (n = 12), ex-smokers with COPD (n = 12) and smokers without COPD (n = 10). Results are expressed as the percentage of AM ingesting apoptotic cells (mean ± SEM). *significantly reduced AM phagocytic ability compared to never smoker controls. **Significant increase in phagocytic ability of AM from ex-smoker COPD subjects compared to current smoker COPD subjects ($p < 0.05$ respectively). Adapted from Hodge et al. (2007) Am J Respir Cell Mol Biol 37:748–755.

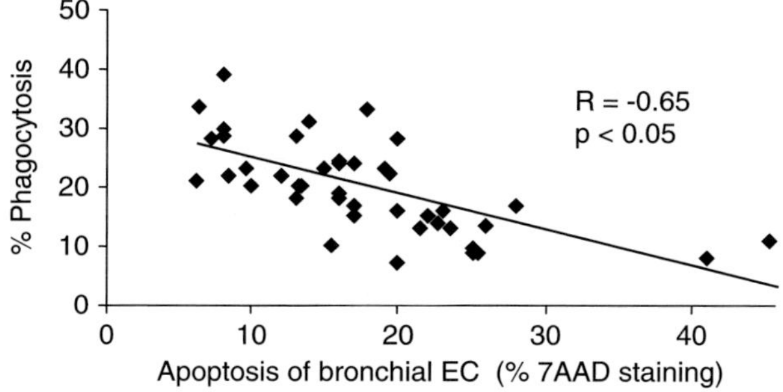

Fig. 14.3 Significant correlation between apoptosis of bronchial epithelial cells and AM phagcoytosis of apoptotic epithelial cells (%). Bronchial epithelial cells and BAL-derived AM were obtained from 40 volunteers (20 never-smoker controls and 20 subjects with COPD of varying severity) at fibreoptic bronchoscopy. (Spearman's rank correlation, p<0.05). Adapted from Hodge et al. (2007) Am J Respir Cell Mol Biol 37:748–755.

study the abnormality appeared to relate to the presence of certain subtypes of lung cancer (Pouniotis et al. 2006). These latter findings imply an effect of the cancer per se might be a further contributing factor. We have analyzed cells from COPD and smoking cohorts in which cancer has been excluded (by CXR and bronchoscopy) thereby confirming a cancer-independent effect. In the human in vivo setting the link between defective macrophage function and accumulation of apoptotic material is difficult to prove directly. However, across a number of subjects we noted a significant correlation between phagocytosis and airway epithelial apoptosis, which is at least supportive of such a link (Hodge et al. 2007; Fig. 14.3)

14.4 Mechanisms Underlying AM Dysfunction in COPD

A large array of surface receptors and secreted factors play a role in AM phagocytic function. Identification of the key defects in COPD could help to more precisely target therapies. The precise molecular basis of the defective phagocytic ability of AM in COPD is currently unknown. The interaction between the AM and apoptotic cells is mediated by a variety of macrophage membrane-associated proteins and intracellular signalling pathways. Several of these recognition molecules on AM include thrombospondin receptor [CD36], integrin $\alpha V\beta 3$ [CD51/CD61] (Savill et al. 1992) phosphatidylserine receptor [PSR] (Fadok et al. 1992) class A scavenger receptors (SR-As) and PECAM [CD31] (Brown et al. 2002) and the lung collectin associated receptor, LDL receptor related protein [CD91] (Gardai et al. 2003). We have begun to characterize the level of expression of these molecules in our subject cohorts. Strong associations between current smoking (both in "healthy" smokers and those with COPD) and reduced expression of CD91, CD44, and CD31 have been seen. However, the ex-smoker COPD cohort had similar levels of expression to the controls indicating that smoking per se rather than COPD was the key factor involved. Further support for these findings was derived from in vitro studies in which AMs were exposed in culture to cigarette smoke. Significant falls in CD91, CD44, and CD31 were again noted (Hodge et al. 2007).

Monocytes from the peripheral blood undergo maturation into AM upon migration into lung tissue at sites of inflammation. The maturation process is associated with an increase in phagocytic capacity, thus an excess of newly arrived relatively immature cells could result in a net reduction in AM phagocytic capacity. To address this question the level of expression of a maturation marker (transferring receptor, CD71) was assessed on AMs. Significant reductions were seen in smokers both with and without COPD, but not in COPD alone. Conversely, the monocyte marker CD14 was increased in the smokers. Further, local AM proliferation was assessed using staining for Ki-67, with increased levels again seen in the smokers (Hodge et al. 2007). Taken together, these results indicate that smoking is associated with an increase in relatively immature AMs, which may reflect increased influx of blood-derived monocytes, or in situ proliferation. These studies of surface makers and maturation do not however explain the phagocytic defect in ex-smoker COPD subjects.

Further studies have been conducted to identify markers that may more closely associate with COPD rather than smoking alone. These studies have included investigations relating to the collagenous lectin family of molecules (collectins) (Stuart et al. 2006). Collectins include mannose binding lectin (MBL), surfactant proteins A (SP-A) and D (SP-D). These molecules have long been known to have critical roles in host defence, specifically in regard to the phagocytosis of invading pathogens. More recently, their role in phagocytosis of apoptotic cells has also been recognized. Links between this group of molecules and the pathogenesis of COPD are suggested by the associations between MBL polymorphisms and exacerbation rates (Yang et al. 2003) and by the development of emphysema in SP-D deficient mice (Vandivier et al. 2002b; Clark et al. 2003). However, the majority of attention has been directed towards the role of infecting pathogens on disease progression rather than any attention to clearance of apoptotic material.

To begin to address the role of collectins in COPD, we assessed AM expression of mannose receptor as well as BAL levels of MBL and SP-D. Importantly, we found reduced levels of SP-D, MBL and AM expression of MR in COPD subjects who were ex-smokers as well as current smokes compared to non-smoking controls (Hodge et al. 2008; Fig. 14.4). There was a trend toward lower MBL levels in the COPD smokers than the ex-smokers but this did not reach statistical significance. Thus, unlike the surface makers and maturation markers discussed above, the collectin abnormalities are seen in COPD in the absence of ongoing smoking. This has important implications for using this system as a potential therapeutic target in the broad range of COPD patients.

To further explore the link between collectins and AM function, in vitro analyses have been performed. AM macrophages obtained by BAL were incubated with a mannose receptor blocking antibody, then phagocytic capacity was assessed for apoptoctic airway epithelial cells as discussed above. These studies showed a greater than 50% reduction in phagocytosis was induced by MR blocking. Conversely, when cells were incubated with recombinant MBL, a substantial increase of over 100% in phagocytosis could be achieved (Hodge et al. 2008).

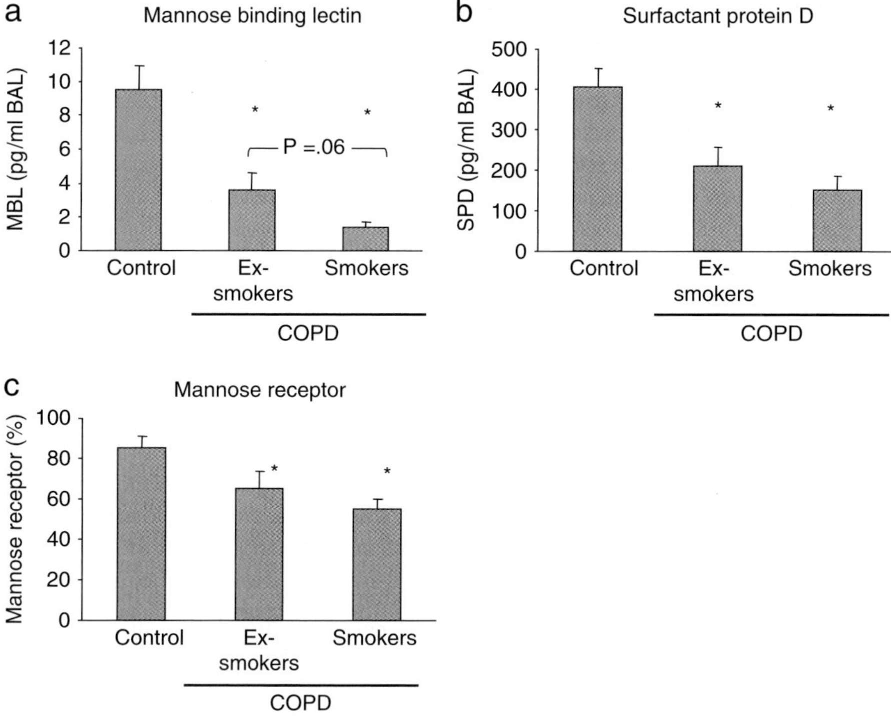

Fig. 14.4 Expression of collectin–associated mediators in control subjects and current- and ex-smoker COPD patients a). Mannose binding lectin (MBL) in BAL. b) Surfactant protein - D (SP-D) in BAL. c) Percentage of AM expressing mannose receptor (MR). Data shows mean ± SEM. MBL and SP-D were investigated by ELISA in samples of BAL from 10 in each group of current- and ex-smoker COPD subjects and never-smoker controls. MR was measured by flow cytometry in AM collected by BAL from the 10 never-smoker controls and 10 COPD subjects (6 current- and 4 ex-smoker). Note * significantly (p<0.05) lower expression of MBL, SP-D and MR in both COPD groups compared to controls. Adapted from Hodge et al. (2008) Am J Respir Crit Care Med 178(2):139–148.

14.5 Macrophage Phenotype in COPD

Investigation and understanding of phenotypic differences in macrophages has lead to the concept of M1 and M2—type cells (Mantovani et al. 2002). The former are activated by the classical pathway and principally have a role as effectors against pathogens, with relatively higher inflammatory cytokine production and phagocytic capacity. Exposure to cigarette smoke in a mouse model resulted in a shift away from the M1 phenotype in as much as cytokine production (TNFα, IL6 and RANTES) in response to viral and bacterial mimics (pI:C and LPS respectively) was reduced (Gaschler et al. 2008). Conceivably this could increase susceptibility to infection. In COPD we have noted an increase in inflammatory cytokines but a reduction in

CD86 (M1 marker; unpublished observations) and also a reduction in MR (M2 marker). Thus the phenotypic shifts in COPD vs. smoking warrant further clarification, but early results suggest a mixed phenotype.

14.6 Therapeutic Implications

14.6.1 Collectins

The use of exogenous MBL as a potential therapy has already been explored in animal models of infection, and benefits have been confirmed (Jensenius et al. 2003; Kaur et al. 2007). The potential links between collectins and COPD raises the possibility that this mechanism could be a therapeutic target here also. Further work is needed in this area, including assessment of SP-A and SP-D. It should be noted that until now the surfactant therapies that have been trialled clinically in diseases such as ARDS have not contained SP-A and SP-D, thus the generally disappointing results with commercially available surfactant therapies to date do not preclude further consideration of using surfactant proteins as therapies. It has been shown that a recombinant fragment of human SP-D could ameliorate the inflammatory pulmonary phenotype seen in SP-D deficient mice, and recently also the structural emphysematous changes seen in these animals (Clark et al. 2003; Knudsen et al. 2007).

14.6.2 Statins

Statins are now known to have a range of anti-inflammatory properties in addition to their lipid lowering roles. With respect to phagocytosis of apoptotic cells, the effect of statins on AM function are mediated at least in part by their inhibitory effects on RhoA (itself an inhibitor of phagocytosis of apoptotic cells) (Morimoto et al. 2006b). Morimoto et al conducted studies using lovostatin on human monocyte derived macrophages (HMDMs) and AMs from COPD subjects, using apoptotic human neutophils and Jurkat T cells as targets (Morimoto et al. 2006a). They found that Lovostatin significantly increased the phagocytosis of apoptotic neutrophils. This effect was not associated with any detectable changes in surface receptors ($\alpha_v\beta_3$, β_3, CD91, calreticulin, CD36, CD44 or CD14). No change in staining with mAb 217 was detected suggesting the PS pathway may not be involved. No effect was seen on apoptotic cell binding either. The effect was dependant lovostatin's ability to block HMG-CoA reductase. Using specific inhibitors of gernylgeranyltransferase I (GGTI) and farnesyltransferase (FTI), involvement of GGTI was confirmed and FTI strongly implicated. The inhibitory effect on prenylation appeared to be greater for RhoA (a negative regulator of phagocytosis of apoptotic cells) than Rac-1 (a positive regulator of apoptosis), based on greater observed reduction on membrane localization of RhoA, which thus is consistent with the enhancement in phagocytosis observed. The effects of lovostatin were further assessed using a mouse model, where enhanced phagocytosis of labelled apoptotic cells instilled into the trachea was seen in those animals given the drug. Whether the effects of statins on phagocytosis will translate to human clinical benefit is as yet uncertain because the doses used to achieve thein vitro effects are somewhat higher than those used clinically.

14.6.3 Macrolides

There has been a great deal of interest in the use of macrolide antibiotics (e.g. erythromycin, clarithromycin, azithromycin) as therapeutic agents beyond their role as anti-microbials. This interest stems from their now widely recognized anti-inflammatory effects and more recently their effect on phagocytic capacity. These drugs are now first line therapy for diffuse panbronchiolitis and have had a major impact in that disease. Small trials have also shown some benefits in cystic fibrosis and in bronchiolitis obliterans syndrome in lung transplant recipients, although larger controlled studies are needed. Macrolides have been shown in vitro to improve AM phagocytosis of apoptotic neutrophils. In a study by Yamaryo et al. (2003), AMs obtained from healthy volunteers by BAL demonstrated enhanced phagocytic capacity for apoptotic neutrophils in response to 14-member macrolides clarithromycin and erythromycin as well as the 15-member macrolides azithromycin, but not 16-member macrolides. The phosphatidylserine pathway was involved in this response.

We assessed the effects of a macrolide on the phagocytic capacity of AMs, particularly in regard to the phagocytosis of apoptotic airway epithelial cells. In vitro studies confirmed that low concentrations of azithromycin (levels below that recommended for antimicrobial effects) significantly improved AM phagocytic capacity (Hodge et al. 2006; Fig. 14.5). We also noted that AM expression of pro-inflammatory cytokines was reduced in the presence of azithromycin, although interestingly the concentrations required to achieve this were higher than those needed for the phagocytosis effect.

In vitro analyses were done to try to identify the mechanism by which azithromycin was exerting its beneficial effects. Analysis of AM expression of several recognition molecules including CD31, CD36, CD91, CD44 $\alpha_v\beta_3$ integrin as well as expression of GMCSF revealed no significant changes in response to incubation with azithromycin. However, the effect was partially blocked by the inhibition of phosphatidylserine (PS) pathway with PS-containing liposomes, indicating involvement of that pathway (Hodge et al. 2006). Further studies then showed that incubation of AMs with azithromycin lead to a dose-dependent increase in expression of the mannose receptor, suggesting this pathway is also involved in the beneficial effects seen with macrolides (Hodge et al. 2008). Murphy et al. (2008) have more extensively characterized the phenotypic changes induced by azithromycin using a mouse macrophage cell line J774. They found, as did we that azithromycin increased

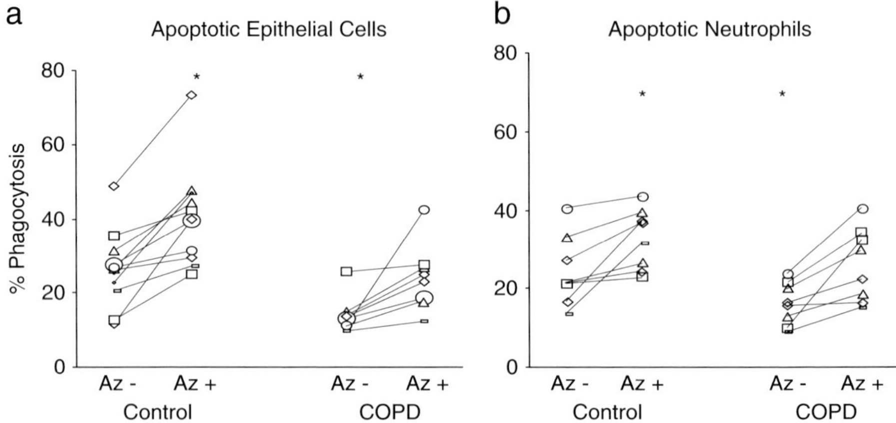

Fig. 14.5 Effect of azithromycin (500 ng/mL) on the phagocytosis of apoptotic (a) bronchial epithelial cells, and (b) neutrophils by alveolar macrophages (AMs)AMs from normal volunteers or subjects with chronic obstructive pulmonary disease (COPD) were incubated with (Az+) or without (Az-) 500 ng/mL azithromycin for 24 h prior to phagocytosis assay. Each value represents the mean of triplicate determinations. *: $p<0.05$ vs. control with no azithromycin added. Adapted from Hodge et al. (2006) Eur Respir J 28: 486–495.

MR expression and reduced inflammatory cytokines. They also found increased production of the "anti-inflammatory" cytokine IL-10 and arginase and reduced iNOS. These changes are all consistent with a shift toward M2 phenotype.

14.6.4   *In Vivo Human Studies*

A number of reports have now been published concerning the potential role for statins in COPD. There is some evidence from a population-based study of improved survival, but further studies to account for the potential confounding factors of cardio-vascular disease are needed (Ishida et al. 2007). As yet there is no controlled study of statins in this disease, but some evidence that statin use is associated with a slower decline in lung function has been found (Keddissi et al. 2007). No study has specifically addressed the question of statin effects on AM function in COPD patients in vivo.

Our in vitro analyses suggested that macrolides may have benefits in subjects with COPD, at least in part via an effect on phagocytosis. As noted previously, a number of studies in other disease settings have provided evidence of benefits for macrolides, which have generally been attributed to the reduction in expression of inflammatory mediators. This mechanism could be of benefit in COPD, and formed the rationale for a controlled clinical trial which showed a reduced rate of exacerbations with 1 year of macrolide therapy (reported in abstract form Seemungal et al. 2007). Our particular interest, however, was to determine whether azithromycin would have an impact on phagocytic function.

To address this question, a small clinical study was conducted involving nine stable COPD subjects. Bronchoscopy was performed at baseline primarily to quantify AM phagocytic capacity, then subjects were treated with low dose azithromycin (250mg daily for 5 days, then 250mg twice weekly for 12 weeks). The subjects then underwent a further bronchoscopy for follow-up assessment of AM function. A number of additional measurements of inflammatory mediators were also made, using BAL fluid and cells, breath condensate and peripheral blood.

The major outcome of the study was that azithromycin treatment led to a significant increase in AM phagocytic function (Fig. 14.6; Hodge et al. 2008). There was a concurrent fall in the number of apoptotic airway epithelial cells detected by 7AAD analysis of brushing samples. No changes were detected in CD91, CD31, CD36 or CD44. However, there was a significant increase in AM expression of the mannose receptor. No changes were seen in BAL concentrations of MBL or SP-D. Some systemic effects were noted with a fall in peripheral blood total white cell count and C–reactive protein levels. The study was not powered to detect changes in physiological parameters (spirometry) or other clinical outcomes such as quality of life.

The principal finding of the study is that the in vitro effects of azithromycin on AM function can be translated to the in vivo situation. To our knowledge this is the first study to show that macrophage function can indeed be pharmacologically modified in vivo in COPD. The study provides further biological rationale for developing macrolides as therapies in this disease. Of course the phagocytic improvements might not only have a beneficial impact in the context of clearance of apoptotic cells, but may improve host defence against infection. Such an effect could plausibly be considered to lead to a reduction in exacerbations of the disease, which has indeed been found in the recent large clinical study using erythromycin referred to above (Seemungal et al. 2007). In that study however, the relatively high

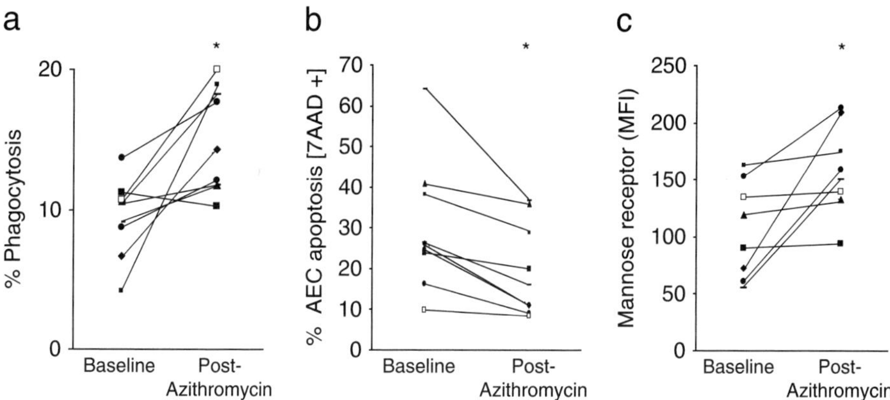

Fig. 14.6 Effect of 12 weeks azithromycin treatment on subjects with COPD a) *3. Phagocytosis of apoptotic airway epithelial cells by AMs. b) *3. Airway epithelial apoptosis determined by analysis of airway brushing samples. c) *3. Mean fluorescence intensity (MFI) of mannose receptor expression on BAL derived macrophages. *Significant change in median value post-treatment (p < 0.05). Adapted from Hodge et al. (2008) Am J Respir Crit Care Med 178(2)139–148.

dose of 500mg bd of erythromycin was used, which may also raise the problem of the development of resistant strains of bacteria.

14.7 Asthma

Less work has been done to examine the role of AM function in asthma than in COPD. In a study by Huynh et al, phagocytic function of AMs was compared between normal volunteers, mild to moderate asthmatics, and severe, oral steroid dependant asthmatics (Huynh et al. 2005). They initially examined AMs obtained by BAL for the presence of phagocytic bodies (PBs), finding a reduced number of such bodies in the severe asthmatics but not in the mild-moderate group compared to normals. The number of PBs significantly increased in the severe group following a burst of high-dose oral steroid. A series of studies on ex vivo cells was then undertaken which confirmed that the AMs from severe asthmatics had reduced ability to phagocytose apoptotic Jurkat cells. An interesting further finding was that the cells from the severe group were resistant to the phagocytosis—stimulating effect of LPS, but were responsive to dexamethasone. The cells from the mild to moderate asthmatics responded in a similar fashion to the normals. It was noted in the normals that phagocytosis was associated with release of the anti-inflammatory and anti-fibrotic mediators PGE2 and 15-HETE. The relative lack of these mediators in the severe group could thus be contributing to the inflammation and airway remodelling seen in the severe disease.

A recent study has shown that in persistent neutrophilic asthma, clarithromycin 500mg twice daily for 8 weeks reduced sputum IL8 levels, neutophil accumulation and activation, but AM phagocytic capacity was not specifically assessed (Simpson et al. 2008).

Airway and alveolar macrophages are not the only pulmonary cells involved in the clearance of apoptotic bodies. Several studies have shown that airway epithelial cells (ECs) have the capacity to phagocytose apoptotic cells, and in the context of asthma, phagocytosis of eosinophils may be important in the resolution of inflammation. Using cell culture assays Sexton et al evaluated the phagocytic capacity of both large and small airway ECs for apoptotic eosinophils, finding that both cells types had this capacity and that the phagocytosis was increased by stimulation with IL-1 (Sexton et al. 2004). Interestingly, these same cells were not capable of ingesting apoptotic neutrophils. This deficiency may be due to a lack of mannose receptor expression on the ECs. In direct comparison with monocyte-derived macrophages the ECs had about one third of the macrophage phagocytic capability for eosinophils. Unlike the epithelial cells, the macrophages showed good capability for ingesting apoptotic neutrophils, although their avarice for eosinophils was greater. In asthma, the airway inflammation frequently leads to EC cell loss, which could thus impair clearance of apoptotic eosinophils leading to secondary necrosis and perpetuation of inflammation. As yet there are no reports of EC phagocytic capacity in vivo in asthma and this phenomenon has not been studied to any extent in COPD.

14.8 Cystic Fibrosis

CF is known to be associated with an increased accumulation of apoptotic material, particularly neutrophils, in the airways. AMs from cystic fibrosis patients appear to be defective in engulfing bacteria, and also have a defect in phagosome acidification which leads to defective killing of internalized pathogens (Di et al. 2006). Sputum from CF patients has been shown to inhibit phagocytosis of apoptotic cells by HMDMs and AMs. This effect appears to be due to the high levels of neutophil elastase in the sputum, which causes cleavage of the PS receptor (Vandivier et al. 2002a). This study evaluated subjects with CF, non-CF bronchiectasis and subjects with chronic bronchitis (defined as current or ex-smokers with sputum production for 3 months of 2 consecutive years). Phagocytosis was lowest in the CF group, with an intermediate level in the non-CF bronchiectasis patients although this was not significant different to those with chronic bronchitis. No normal comparator group was used.

The airways of CF patients are typically chronically infected with Pseudomonas Aeruginosa, which has been shown to induce neutrophil apoptosis. Further work has now also shown that a bacterial product, pyocyanin, can inhibit the phagocytosis of apoptotic cells by HMDMs in in vitro assays (Bianchi et al. 2008). The effect was associated with the generation of reactive oxygen species and an increase in RHO GTPase signalling. Phagocytosis of latex beads was not affected and the effects were not due to a reduction in HMDM viability. The effects were confirmed in vivo in a mouse model, where animals infected with a pyocyanin producing strain of pseudomonas were seen to have a reduced proportion of AMs containing apoptotic neutrophils, despite an excess of apoptotic cells per se.

With regard to therapeutic interventions, there is now a significant emerging body of evidence that macrolides are of benefit in CF. Generally however this has been attributed to their anti-inflammatory properties. An improvement in phagocytic function is plausible although this has not been directly assessed.

14.9 Conclusion

The role of defective phagocytosis of apoptotic cells is clearly an evolving area in lung disease. The basic biology of the process and studies to date in a limited range of pulmonary conditions indicate that this process is likely to be of pathological relevance and ultimately clinical importance. At this stage however, the precise link between defective apoptotic cell clearance and lung disease requires a lot more work to establish its real potential as a target pathway for therapy. Nevertheless, pre-clinical and early clinical studies in the area are already underway and providing support for continued development of this approach.

References

Bianchi SM, Prince LR, McPhillips K et al (2008) Impairment of apoptotic cell engulfment by pyocyanin, a toxic metabolite of Pseudomonas aeruginosa. Am J Respir Crit Care Med 177:35–43

Brown S, Heinisch I, Ross E et al (2002) Apoptosis disables CD31-mediated cell detachment from phagocytes promoting binding and engulfment. Nature 418:200–203

Clark H, Palaniyar N, Hawgood S et al (2003) A recombinant fragment of human surfactant protein D reduces alveolar macrophage apoptosis and pro-inflammatory cytokines in mice developing pulmonary emphysema. Ann N Y Acad Sci 1010:113–116

Curtis JL, Freeman CM, Hogg JC (2007) The immunopathogenesis of chronic obstructive pulmonary disease: insights from recent research. Proc Am Thorac Soc 4:512–521

Di A, Brown ME, Deriy LV et al (2006) CFTR regulates phagosome acidification in macrophages and alters bactericidal activity. Nat Cell Biol 8:933–944

Fadok VA, Voelker DR, Campbell PA et al (1992) Exposure of phosphatidylserine on the surface of apoptotic lymphocytes triggers specific recognition and removal by macrophages. J Immunol 148:2207–2216

Gamble E, Grootendorst DC, Hattotuwa K et al (2007) Airway mucosal inflammation in COPD is similar in smokers and ex-smokers: a pooled analysis. Eur Respir J 30:467–471

Gardai SJ, Xiao YQ, Dickinson M et al (2003) By binding SIRPalpha or calreticulin/CD91, lung collectins act as dual function surveillance molecules to suppress or enhance inflammation. Cell 115:13–23

Gaschler GJ, Zavitz CC, Bauer CM et al (2008) Cigarette smoke exposure attenuates cytokine production by mouse alveolar macrophages. Am J Respir Cell Mol Biol 38:218–226

Henson PM and Tuder RM (2008) Apoptosis in the Lung: Induction, Clearance and Detection. Am J Physiol Lung Cell Mol Physiol EPub

Hodge S, Hodge G, Scicchitano R et al (2003) Alveolar macrophages from subjects with chronic obstructive pulmonary disease are deficient in their ability to phagocytose apoptotic airway epithelial cells. Immunol Cell Biol 81:289–296

Hodge S, Hodge G, Holmes M et al (2005) Increased airway epithelial and T-cell apoptosis in COPD remains despite smoking cessation. Eur Respir J 25:447–454

Hodge S, Hodge G, Brozyna S et al (2006) Azithromycin increases phagocytosis of apoptotic bronchial epithelial cells by alveolar macrophages. Eur Respir J 28:486–495

Hodge S, Hodge G, Ahern J et al (2007) Smoking alters alveolar macrophage recognition and phagocytic ability: implications in chronic obstructive pulmonary disease. Am J Respir Cell Mol Biol 37:748–755

Hodge S, Hodge G, Jersmann H et al (2008) Azithromycin improves macrophage phagocytic function and expression of mannose receptor on COPD. Am J Respir Crit Care Med 178(2):139–148

Hodge SJ, Hodge GL, Holmes M et al (2004) Flow cytometric characterization of cell populations in bronchoalveolar lavage and bronchial brushings from patients with chronic obstructive pulmonary disease. Cytometry 61B:27–34

Huynh ML, Malcolm KC, Kotaru C et al (2005) Defective apoptotic cell phagocytosis attenuates prostaglandin E2 and 15-hydroxyeicosatetraenoic acid in severe asthma alveolar macrophages. Am J Respir Crit Care Med 172:972–979

Ishida W, Kajiwara T, Ishii M et al (2007) Decrease in mortality rate of chronic obstructive pulmonary disease (COPD) with statin use: a population-based analysis in Japan. Tohoku J Exp Med 212:265–273

Jensenius JC, Jensen PH, McGuire K et al (2003) Recombinant mannan-binding lectin (MBL) for therapy. Biochem Soc Trans 31:763–767

Kasahara Y, Tuder RM, Taraseviciene-Stewart L et al (2000) Inhibition of VEGF receptors causes lung cell apoptosis and emphysema. J Clin Invest 106:1311–1319

Kasahara Y, Tuder RM, Cool CD et al (2001) Endothelial cell death and decreased expression of vascular endothelial growth factor and vascular endothelial growth factor receptor 2 in emphysema. Am J Respir Crit Care Med 163:737–744.

Kaur S, Gupta VK, Thiel S et al (2007) Protective role of mannan-binding lectin in a murine model of invasive pulmonary aspergillosis. Clin Exp Immunol 148:382–389

Keddissi JI, Younis WG, Chbeir EA et al (2007) The use of statins and lung function in current and former smokers. Chest 132:1764–1771

Knudsen L, Ochs M, Mackay R et al (2007) Truncated recombinant human SP-D attenuates emphysema and type II cell changes in SP-D deficient mice. Respir Res 8:70

Mantovani A, Sozzani S, Locati M et al (2002) Macrophage polarization: tumour-associated macrophages as a paradigm for polarized M2 mononuclear phagocytes. Trends Immunol 23:549–555

Morimoto K, Janssen WJ, Fessler MB et al (2006a) Lovastatin enhances clearance of apoptotic cells (efferocytosis) with implications for chronic obstructive pulmonary disease. J Immunol 176:7657–7665

Morimoto K, Janssen WJ, Fessler MB et al (2006b) Statins enhance clearance of apoptotic cells through modulation of Rho-GTPases. Proc Am Thorac Soc 3:516–517

Murphy BS, Sundareshan V, Cory TJ et al (2008) Azithromycin alters macrophage phenotype. J Antimicrob Chemother 61: 554–560

Pouniotis DS, Plebanski M, Apostolopoulos V et al (2006) Alveolar macrophage function is altered in patients with lung cancer. Clin Exp Immunol 143:363–372

Savill J, Hogg N, Ren Y et al (1992) Thrombospondin cooperates with CD36 and the vitronectin receptor in macrophage recognition of neutrophils undergoing apoptosis. J Clin Invest 90:1513–1522

Seemungal TAR, Wilkinson TMA, Perera W et al (2007) Long term macrolide therapy decreases exacerbations in COPD. Am J Respir Crit Care Med 175:A764

Sexton DW, Al-Rabia M, Blaylock MG et al (2004) Phagocytosis of apoptotic eosinophils but not neutrophils by bronchial epithelial cells. Clin Exp Allergy 34:1514–1524

Simpson JL, Powell H, Boyle MJ et al (2008) Clarithromycin targets neutrophilic airway inflammation in refractory asthma. Am J Respir Crit Care Med 177:148–155

Stuart LM, Henson PM, Vandivier RW (2006) Collectins: opsonins for apoptotic cells and regulators of inflammation. Curr Dir Autoimmun 9:143–161

Vandivier RW, Fadok VA, Hoffmann PR et al (2002a) Elastase-mediated phosphatidylserine receptor cleavage impairs apoptotic cell clearance in cystic fibrosis and bronchiectasis. J Clin Invest 109:661–670.

Vandivier RW, Ogden CA, Fadok VA et al (2002b) Role of surfactant proteins A, D, and C1q in the clearance of apoptotic cells in vivo and in vitro: calreticulin and CD91 as a common collectin receptor complex. J Immunol 169:3978–3986

Vandivier RW, Henson PM, Douglas IS (2006) Burying the dead: the impact of failed apoptotic cell removal (efferocytosis) on chronic inflammatory lung disease. Chest 129:1673–1682

Yamaryo T, Oishi K, Yoshimine H et al (2003) Fourteen-member macrolides promote the phosphatidylserine receptor-dependent phagocytosis of apoptotic neutrophils by alveolar macrophages. Antimicrob Agents Chemother 47:48–53

Yang IA, Seeney SL, Wolter JM et al (2003) Mannose-binding lectin gene polymorphism predicts hospital admissions for COPD infections. Genes Immun 4:269–274

Chapter 15
Environmental Factors Affecting Phagocytosis of Dying Cells: Smoking and Static Magnetic Fields

Luciana Dini and Cristian Vergallo

Abstract: Cell surface modifications are fundamental for the correct and fast removal of apoptotic cells. These changes include the appearance of tethering molecules on the surface of apoptotic cells, the externalization of PS, oxidation of phospholipids and qualitative and quantitative changes in surface sugars and ICAM-3. Phagocytes, both professional and non-professional, use specific receptors that bind to the apoptotic cells either directly or through bridging molecules. In non-pathological conditions, apoptotic cells are normally cleared via an anti-inflammatory pathway. In contrast, the uptake and removal of necrotic cells normally involves inflammation and an immune response. Besides the "eat me" signals on the dying cells, phagocytes can also recognize "leave-me-alone" signals on healthy cells. The correct repertoire of molecules exposed on the cell surfaces prevents the engulfment of living undamaged cells. Thus, any factors influencing cell surface molecule expression on both phagocytes and/or apoptotic cells can in turn affect recognition of living and/or apoptotic cells. One such factor is cigarette smoke, which contains highly reactive carbonyls, which can modify proteins that directly or indirectly affect cellular functions. Moreover cigarette smoke is a major etiological factor in the development of COPD, in which apoptosis and defective PACs play a fundamental role. Another environmental factor that may interfere with the normal correct exposure of molecules on cell surfaces is exposure to (S)MFs. Despite the multiplicity of experimental conditions (i.e. in vitro or in vivo models, intensity and type of field, time of exposure, metabolic state of the cells, etc), converging data indicate that the primary site of action of (S)MFs and (E)MFs is the plasma membrane: i.e. they affect the electrochemical balance of the membrane, the distribution of membrane proteins and membrane receptors, cell-cell and cell-matrix junctions, sugar residues

L. Dini (✉) · C. Vergallo
Department of Biological and Environmental Science
and Technology (Di.S.Te.B.A.)
University of the Salento
Via per Monteroni, Lecce, Italy
Tel.: +390832298614
Fax: +390832298937
E-mail: luciana.dini@unile.it

D. V. Krysko, P. Vandenabeele (eds.), *Phagocytosis of Dying Cells*,
DOI: 10.1007/978-1-4020-9292-3_15, © Springer Science+Business Media B.V. 2009

on cell membranes and trans-membrane fluxes of various ions, especially calcium. The effects of cigarette smoke and SMF exposure on the phagocytosis of apoptotic cells will be discussed in this chapter.

Keywords: Apoptosis • Cigarette smoke • Magnetic fields • Calcium • Phagocytes

15.1 Introduction

The best fate of dying cells is their efficient clearance by neighbouring or specialized cells. The recognition and the engulfment of apoptotic cells is regulated by a well-defined sequence of events, conserved during evolution, in many of the signal transduction pathways implicated in the uptake of dying cells. To this purpose, studies conducted on *C. elegans* and on *D. melanogaster* have played an important role identifying the genetic pathways implicated in the phagocytosis of dead cells (Gronski and Ravichandran, Chap. 5, this Vol.). These studies led to the isolation of several mutations affecting engulfment. To date, seven genes that function in two genetic phagocytosis pathways (Ellis et al. 1991; Gumienny and Hengartner 2001; Zhou et al. 2001; deBakker et al. 2004; Reddien and Horvitz 2004; Henson 2005) have been identified. In addition, many receptors, adaptors and chemotactic molecules are reported to be involved in the removal of apoptotic cells (Peter et al., Chap. 3; Napirei and Mannherz, Chap. 4; Gregory and Pound, Chap. 9, this Vol.; Emoto et al. 1997; Arur et al. 2003; Fadeel 2003; Grimsley and Ravichandran 2003; deAlmeida and Linden 2005). Indeed, the cell surface of apoptotic cells is of great interest regarding their removal, and recent observations have highlighted the complex interaction between apoptotic cells and phagocytes (Krysko et al. 2006).

In its lifespan, an organism produces millions of cells that die during development and in other particular physiological conditions (Krysko et al., Chap. 1; Diez-Fraile et al., Chap. 2, this Vol.; Baehrecke 2002). Thus the clearance of dying cells is a fundamental process designed to protect the organism as a whole and is not just a simple anti- or pro-inflammatory response. Recognition of apoptotic cells induces a host of additional effects in normal tissue, each of which, if disrupted, could contribute to pathogenesis. For example, apoptotic cell recognition induces anti-inflammatory and anti-immunogenic responses, which may stimulate the production of anti-proteases, and in particular, may promote cell replacement and tissue repair (Ucker, Chap. 6; Lacy-Hulbert, Chap. 7, this Vol.). Phagocytosis is an extremely important process in many conditions in which the resolution of an inflammatory state is needed. Interestingly, some of the recognition mechanisms for apoptotic cells are related to innate immune system processes, suggesting that cell deletion and innate immunity may have evolved together from the earliest metazoa. Defective clearance can therefore lead to chronic inflammatory conditions or to the onset of autoimmune diseases. Impairment of phagocytosis of apoptotic cells (PACs) may be implicated in several diseases, including Systemic Lupus Erythematosus (SLE) and Chronic Obstructive Pulmonary Disease (COPD), and in atherosclerosis-related lesions (Mevorach, Chap. 10; Martinet et al., Chap. 13; Reynolds and Hodge, Chap. 14, this Vol.; Haslett 1999; Ward et al. 1999; Schrijvers et al. 2005; Gaipl et al. 2007).

The development of modern society has been accompanied by a dramatic increase in the numbers of environmental contaminants in the air, water and soil. It is generally difficult to recognize them, since they are frequently invisible, colourless, odourless and tasteless. However, these harmful factors may affect our health even at molecular levels. Apoptosis of various target cells appears to be induced by many environmental factors, including cadmium, methylmercury, bisphenol A, dicyclohexylphthalate, p-octylphenol, and manganese ethylene-bisdithiocarbamate (maneb; Ishido 2006). If the environment interferes with the production of apoptotic cells, it is possible that it may play also a role in their clearance, but studies in this field are still in their infancy. The few data in the literature report the effects of exposure to static magnetic fields (SMFs) produced by magnets and electrical devices and the consequences of smoking for macrophage recognition of apoptotic cells (Kirkham et al. 2004; Dini and Abbro 2005). These environmental factors could interfere with the clearance of apoptotic cells in two ways: from the perspective of the apoptotic cells, by both modifying the rate at which apoptotic cells are produced and disrupting characteristics such as the plasma membrane, and from the perspective of the phagocytes, by both modifying phagocytic activity and surface receptor systems.

In this chapter we will discuss the effects of these two environmental factors: smoking and exposure to SMFs. They are extremely different from each other; smoking entails a mixture of chemical compounds in the gaseous and particulate states. Magnetic fields (MFs) are invisible lines of force created by magnets or as a consequence of the movement of electric charges. SMFs do not vary with time (zero frequency) but vary in intensity, allowing classification of SMFs as weak (< 1 mT), moderate (1 mT to 1 T), strong (1 T to 5 T) and ultrastrong (> 5 T). Growing evidence suggests that the geomagnetic field (GMF), to which all living organisms have adapted, modulates the biological effects of artificial MFs (Gmitrov 2007). This opens up a further scenario, not discussed in the present review, entailing the potential adaptation of living organisms to unexpected fields or artificial magnetic and electromagnetic fields, to which they have became subject only in the 20th century.

Specifically, in this chapter we will discuss (1) the effect of cigarette smoking on PACs; (2) the effect of SMFs on apoptosis, cell membranes, organelles and macrophage defence; (3) the effect of SMFs on phagocytosis of apoptotic human lymphocytes; (4) immunological and health implications.

15.2   Tobacco Smoke

15.2.1   Cigarette Smoke: Health Implications

Tobacco smoke is considered a dangerous air pollutant. Due to their inherently deleterious nature and ingredients, when lit they emit dangerous substances. Tobacco smoke is composed of a great variety of constituents. Some of them, including polycyclic hydrocarbons, nitrosamines and aromatic amines, are known to contribute to the carcinogenic risk of tobacco smoke (Hoffmann and Hoffmann 1997; Smith et al. 2003). Nicotine, the major tobacco alkaloid, is held to be responsible for the addictive potential of smoking, mediated by neuronal nicotinic acetylcholine receptors in the central nervous system (Dajas-Bailador and Wonnacott 2004). With regard to the possible carcinogenic risk, nicotine may stimulate tumour development by both genotoxic and non-genotoxic mechanisms (Arabi 2004; Argentin and Cicchetti 2004; Cooke and Bitterman 2004; Kleinsasser et al. 2005). Furthermore, nicotine may also stimulate cellular growth, particularly when receptor-regulated (Schuller 1994; Shin et al. 2004). It also has an anti-apoptotic effect (Argentin and Cicchetti 2004).

Cigarette smoke is a major etiological factor in the development of COPD (Reynolds and Hodge, Chap. 14, this Vol.; Barnes 2000). COPD is a highly prevalent airway disease characterized by an abnormal inflammatory response of the lungs to noxious particles and gases. COPD is characterized within the airways by a dramatic increase in neutrophils (Pesci et al. 1998), macrophages (Shapiro 1999), and CD8$^+$ T cells (Saetta et al. 1998). Accumulating evidence suggests that apoptosis may play an important role in the pathogenesis of this disease. Apoptotic cells are rarely seen in the non-diseased human lung, due to both low rates of cell death and a rapid removal process. The increased apoptosis of alveolar structural cells may be responsible for the alveolar wall destruction associated with emphysema. The increased apoptosis associated with COPD is paralleled by defective alveolar macrophage phagocytic function, which, in turn, may lead to secondary necrosis and tissue damage (Ferrara et al. 1996; Vandivier et al. 2002, 2006).

15.2.2   Cigarette Smoke and Phagocytosis of Apoptotic Cells

Increased apoptosis in COPD may imply defects in the normal physiological clearance of apoptotic cells. Indeed, one hypothesis regarding pathogenesis in COPD is that the accumulation of apoptotic airway epithelial cells could be due to defective phagocytic clearance by alveolar macrophages. The imbalance in the effective clearance of apoptotic neutrophils, resulting in vivo in secondary necrosis, exacerbates the inflammatory response and establishes a chronic inflammatory state. In support of this notion, the intratracheal instillation of apoptotic Jurkat T cell in the presence of TNF-α (an inhibitor of PACs) results in the development of a transient emphysema-like injury pattern (Morimoto et al. 2006; Vandivier et al. 2006).

Cigarette smoke extract can dramatically suppress neutrophil caspase-3-like activity, which correlates with reduced cleavage of the glutamate-L-cysteine ligase catalytic subunit, a known target of active caspase-3 (Stringer et al. 2007). Cigarette smoke extract has a direct effect on active but not on pro-caspase-3, and caspase-3 activity is reduced in a concentration-dependent manner. Despite near-complete suppression of caspase-3 activity, spontaneous apoptosis is not altered. Cigarette smoke extract does not suppress caspase-3 activity to the extent of actually preventing spontaneous apoptosis, but the level of inhibition is sufficient to impair neutrophil phagocytic activity. Interestingly, the rate of cigarette smoke-induced apoptosis is not diminished after cessation of cigarette smoking, indicating that other mechanisms perpetuate apoptosis in COPD, including protease/anti-protease imbalance, inflammation and oxidative stress, which may synergically promote apoptosis or contribute to impaired clearance of apoptotic cells (Park et al. 2007). Some direct effect of cigarette smoke on PACs is likely but does not easily explain the persistent effects even after smoking cessation, or its effect on circulating blood cells. There are a significant number of explanations for the defective uptake. Increased lung proteolytic activity probably contributes by digesting surface apoptotic cell recognition receptors and ligands on apoptotic cells (Vandivier et al. 2002). Macrophages matured from blood monocytes of patients with COPD also showed decreased uptake of *Escherichia coli* (Prieto et al. 2001; Finney-Hayward et al. 2005).

To gain more insight into the effects of cigarette smoke on phagocytic activity, studies have investigated ex vivo alveolar macrophage phagocytic ability (with emphasis on the recognition molecules CD36, integrin $\alpha_v\beta_3$, CD31, CD91 and CD44), in current smokers with and without COPD and after exposure of alveolar macrophage to cigarette smoke in vitro. Alveolar macrophages in current smokers with and without COPD exhibited reduced CD31, CD91, CD44, and CD71 and enhanced Ki-67 compared with healthy never-smoker control subjects. The phagocytic ability of alveolar macrophages was significantly decreased in smokers with and without COPD compared with control subjects. In addition, the suppressive effects of cigarette smoke on alveolar macrophage recognition molecules associated with an increase in cyclic Adenosine monophosphate (cAMP) has been demonstrated in vitro. Smoking-related reduction of alveolar macrophage phagocytic ability and reduced expression of several important recognition molecules may be at least partially normalized in those subjects with COPD who cease smoking. There is evidence that the 15-member macrolide antibiotic azithromycin may mitigate the effects of smoking by virtue of its anti-inflammatory properties (Hodge et al. 2007). Its effects may be increased in the lung due to its ability to reach high concentrations in alveolar macrophages. Low-doses (500 ng x mL^{-1}) of azithromycin were found to significantly improve the phagocytosis of epithelial cells or neutrophils by alveolar macrophages in COPD subjects, often reaching levels comparable with controls (Hodge et al. 2006). The increase in phagocytosis was partially inhibited by phosphatidylserine, implicating the phosphatidylserine pathway in the pro-phagocytic effects of azithromycin. Azithromycin had no effect on other recognition molecules (granulocyte-macrophage colony-stimulating factor, CD44, CD31, CD36, CD91, $\alpha_v\beta_3$ integrin).

Importantly, defective phagocytic clearance by alveolar macrophages was not observed using polystyrene beads, suggesting that the failure to resolve epithelial damage in COPD may result, at least partially, from specific defects in the phagocytic ability of alveolar macrophages to ingest apoptotic airway epithelial cells (Reynolds and Hodge, Chap. 14, this Vol.; Hodge et al. 2003).

What are the mechanisms by which cigarette smoke impairs the clearance of apoptotic cells? The complex mixture of chemicals and oxidants of which cigarette smoke is composed includes the highly reactive aldehydes acrolein and 4-hydroxynonenal (4-HNE; Esterbauer et al. 1991), which can also be formed in vivo as a result of lipid peroxidation. These highly reactive water-soluble carbonyls can

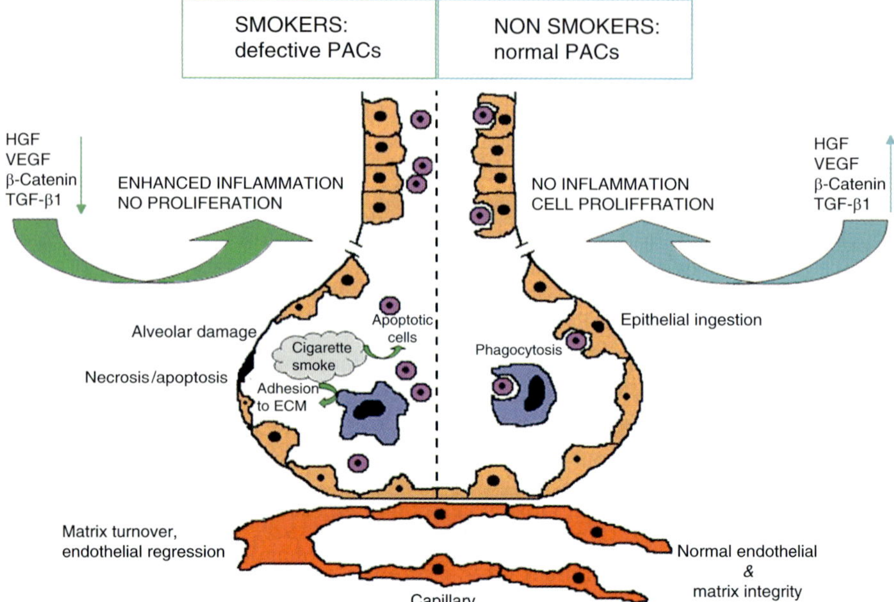

Fig. 15.1   Possible role of defective apoptotic cell recognition processes in inflammatory pulmonary diseases. An efficient phagocytosis promotes the maintenance of normal airway and alveolar structures of lung in the physiological state (right half of the figure) while a defective PACs is found in the inflamed lung (left half of the figure). The efficient phagocytosis promotes the release of HGF and VEGF, and the activation of the β-catenin pathway, leading to the proliferation of epithelial and endothelial cells and to the maintenance of normal lung structure. In contrast, the release of TGF-β1 when PACs is defective (e.g. COPD), suppresses a variety of inflammatory mediators in autocrine/paracrine manner. As consequence, the impaired release of anti-inflammatory and pro-growth mediators results in sustained inflammation and inadequate tissue repair.

modify proteins (by attacking residues such as lysine, arginine, cysteine or histidine) which directly or indirectly affect cellular function. Macrophages can adhere to these carbonyl modified proteins via the macrophage SR-A (Kirkham et al. 2003), which plays a functional role in the phagocytosis of apoptotic thymocytes in vitro (Platt et al. 1996). It has been shown in vitro that surfaces coated with extracellular matrix (ECM) proteins modified by cigarette smoke carbonyls or the products of lipid peroxidation (such as acrolein and 4-HNE), can trigger macrophage adhesion. This in turn results in a down-regulation of the ability of these adherent macrophages to phagocytose apoptotic neutrophils (Kirkham et al. 2004). These findings may at least in part account for the increased macrophage retention and adhesion observed by several authors in the lungs of smokers with and without COPD. Various alveolar macrophage receptors are involved in the recognition of carbonyl-modified proteins, including SR-A, CD36 and the hyaluron receptor CD44. Kirkham et al. (2004) suggest that it is the sequestration of receptors as a result of the adhesion of macrophages to post-translationally modified ECM proteins, which prevents recognition of apoptotic neutrophils and consequently reduces their phagocytosis. In addition, this adhesion prevents scavenger receptor phagocytosis of modified low-density lipoprotein (LDL) particles but not Fc-mediated phagocytosis of IgG-coated red blood cells. Consequently, the clearance of apoptotic cells most likely involves the coordinated response of numerous receptors and a dysfunction in any one can have a detrimental effect. In addition, the defective airway clearance of apoptotic cells described in COPD may also be due to elastase-mediated cleavage of PS receptors in phagocytes that, in turn, may lead to an ongoing inflammation degree and progressive airway damage (Vandivier et al. 2002).

15.3 SMF Exposure

SMFs are naturally present everywhere as the Earth is surrounded by fields varying between 25 and 65 μT. Superimposed on the Earth's magnetic field there may also be artificial SMFs. The widespread use of medical diagnostic instrumentation, such as nuclear magnetic resonance (NMR), and the generation of extremely low frequency electromagnetic fields (ELF-EMFs) by common electrical devices have introduced many sources of SMFs and "quasi"-SMFs into our living environment.

Exposure of workers in many industrial sectors (for example aluminium smelting, chloralkali processes, welding, railways, NMR) and of the general population to artificial SMFs is commonplace (Feychting 2005). Whereas fields over 2 T can induce acute effects, e.g. nausea, vertigo, metallic taste, there is still much debate exists concerning the biological effects of SMFs of weak or moderate intensity. The weak MF of the Earth is used by many organisms (from bacteria to fish and birds) for spatial orientation and navigation (Belyaev et al. 1997). These organisms have developed systems that employ biogenic magnetite (Fe_3O_4) as field sensors (Hong 1995). These observations suggest that at least some cells or cellular structures are sensitive to weak changes in SMF (Semm et al. 1980). Zero magnetic field has been used as a reference in studies of the effects of geomagnetic fields on animals and humans (Beischer 1971).

Living cells and organisms are able to respond to a wide range of environmental stimuli and stresses (including MFs), leading to intracellular and extracellular changes (Saffer and Phillips 1996). Thus, the influence of SMFs is becoming a topic of considerable interest, due to the recent substantial evidence of their influence on living organisms, tissues and cells (Rosen 2003b; Dini and Abbro 2005), as well as their health implications. The cellular and molecular modifications induced when MFs interact with biological materials depend on the duration of exposure, their intensity, tissue penetration and the type of cells (Walleczek and Liburdy 1990; Tenuzzo et al. 2006). However, difficulties remain in resolving the contradictory results that arise from the multiplicity of experimental conditions, including the magnetic flux density of the SMFs, exposure times (from minutes to months), temperature of exposure, cell types, etc. (Walleczek and Liburdy 1990; Chionna et al. 2005; Tenuzzo et al. 2006). In addition, for the same cell line the response to exposure is strictly dependent on field intensity (Ghibelli et al. 2006). While exposure to SMFs alone has little or no effect on cell growth and genetic toxicity regardless of the magnetic density (Ikehata et al. 1999; Wiskirchen et al. 2000; Nakahara et al. 2002; Schiffer et al. 2003), its bioeffects are stronger when applied in combination with other external factors such as ionizing radiation and some chemicals (Zhang et al. 2003; Zmyślony et al. 2000).

15.3.1 Bioeffects of Magnetic Fields

15.3.1.1 Plasma Membrane

The plasma membrane is considered a primary site of SMF action. More than a decade of research into the effects of SMFs on biological systems has yielded compelling evidence for the involvement of the Ca^{2+} signalling pathway as a major target of MFs. As first demonstrated by Liburdy and colleagues (1993) and subsequently by a number of other authors, the Ca^{2+} influx pathways of isolated or cultured cells are severely affected by low-intensity MFs. To better understand membrane-related responses to SMFs, Rosen (1996) directly measured membrane calcium currents using the whole-cell patch clamp technique. Exposure to a 120 mT SMF resulted in a slight reduction in the peak calcium current amplitude and shifts in the current–voltage relationship in cultured GH3 cells. Patch-clamp studies of calcium channels have also provided support for temperature dependency, which may be explained with reference to membrane thermotropic phase transition (Rosen 1996). There is a slight shift in the current–voltage relationship and a less than 5% reduction in peak current during MF exposure to 125 mT in voltage-activated Na^+ channels in proliferating GH3 cells (Rosen 2003a). More pronounced is the increase in the activation time constant, tm, during exposure to the field and for at least 100 s afterwards. The temperature dependence of this phenomenon is probably due to the greater ease with which the liquid crystal membrane is deformed. The experimental threshold gradient and the calculated threshold field intensity required for blockade of action potentials are estimated to be approximately 0.02 mT/mm and 0.02 mT, respectively (Cavopol et al. 1995).

To explain the action of SMFs on biological systems it also is important to determine the physical properties of the membrane's molecular structure that are influenced by these fields. Certain proteins and other macro-molecules are anisotropic, and this leads to torques that cause them to adopt a particular orientation with respect to the applied field, a phenomenon known as diamagnetic anisotropy. Magnetic anisotropy in proteins and polypeptides permits orientation of small structures under MFs of several kiloG, and can be attributed to the diamagnetic anisotropy of the planar peptide bonds. The α helix in particular has large anisotropy due to the axial alignment of the peptide bonds. The regular arrangements of the peptide bonds in β-pleated sheets and collagen structures also produce substantial anisotropy, but less than the α helix (Worcester 1978).

At the beginning of the cascade of cellular changes, it is likely that SMFs influence the diamagnetic properties of the plasma membrane. The reorientation of diamagnetic

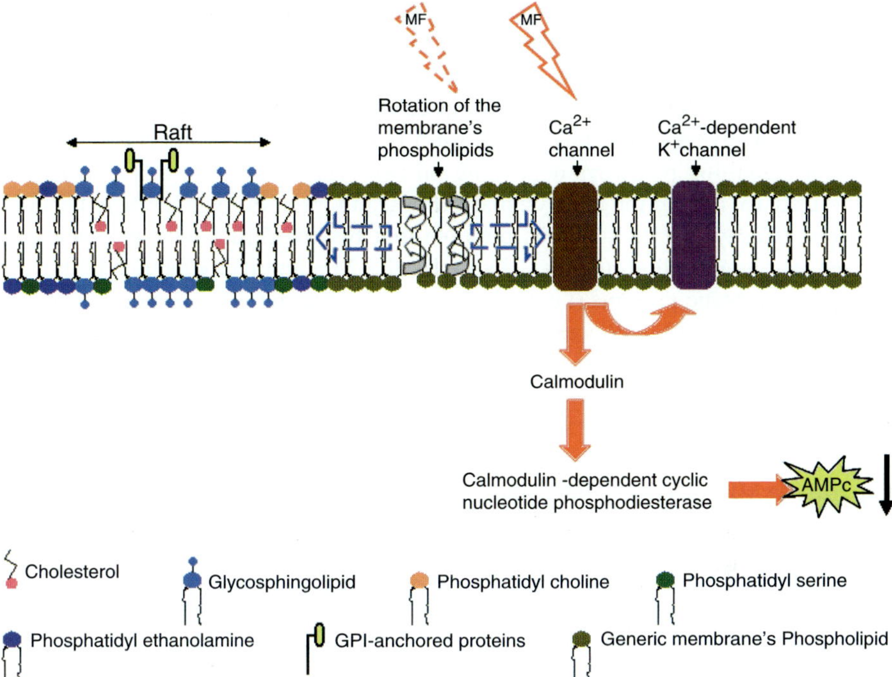

Fig. 15.2 SMFs and plasma membrane interaction. SMFs affect plasma membrane directly (by rotation of the membrane's phospholipids) or indirectly (by possible sphingolipid-cholesterol rafts disorder). Exposure to the SMFs increases Ca^{2+} influx, inhibits K^+ influx via Ca^{2+}-dependent K^+ channels and alters calmodulin-dependent cyclic nucleotide phosphodiesterase activity.

anisotropic molecules in the cell membrane distorts embedded ion channels, those of Ca^{2+} in particular, to the point of altering their function (St Pierre and Dobson 2000; Rosen 2003a). Thus, reorientation of membrane phospholipids ultimately results in the alteration of ion channel activation kinetics and alters calmodulin-dependent cyclic nucleotide phosphodiesterase activity. However, the level of MF intensity at which this effect is observed (~20 μT) implies that calmodulin activation may be functionally sensitive to the geomagnetic field (Liboff et al. 2003).

In the prevailing view of cellular membrane structure, lipids in the bilayer function mainly as solvent for membrane proteins. In the fluid bilayer, numerous lipid species are asymmetrically distributed over the exoplasmic and cytoplasmic leaflets of the membrane (Van Meer 1989). MFs affect the reorientation of diamagnetic molecular domains and the rotation of the membrane's phospholipids, and change protein distribution (Chionna et al. 2003a and 2005; Pagliara et al. 2005; Silva et al. 2006; Tenuzzo et al. 2006). The lipids are organized in the lateral dimension and are responsible for more short- and long-range order that was previously recognized (Kusumi and Sako 1996). This lateral organization probably results from the preferential packing of sphingolipids and cholesterol into moving platforms or rafts to which specific proteins within the bilayer attach (Kai and Ikonen 1997).

Sphingolipid-cholesterol rafts are insoluble in the detergent Triton X-100 a 4°C, in which they form detergent-insoluble glycolipid-enriched complexes (DIGs; Brown and Rose 1992; Parton and Simons, 1995). Several signalling molecules also partition into DIGs (Anderson 1993; Lisanti et al. 1994; Parton and Simons 1995). Clustering of glycosyl phosphatidyl inositol (GPI)-anchored proteins can activate different signalling pathways depending on cell type. Clustering of GPI anchors might lead to interaction with signal-transducing proteins which then become activated (Casey 1995). It is thus possible that by interfering with phospholipid bilayer membranes (Tenforde and Liburdy 1988), MFs indirectly interact both with intrinsic membrane proteins (i.e. Ca^{2+} channels) and with (GPI)-anchored proteins present inside the raft, with various consequences.

15.3.1.2 Cell Surface

Exposure to MFs of varying intensities and types has been shown to cause modifications to cell surface morphology (Hamada et al. 1989; Paradisi et al. 1993; Santoro et al. 1997; Lisi et al. 2000; Chionna et al. 2003a; Rieti et al. 2004; Chionna et al. 2005; Dini and Abbro 2005; Tenuzzo et al. 2006) and the redistribution of plasma membrane proteins, receptors (Bersani et al. 1997) and sugar residues (Bordiushkov et al. 2000; Chionna et al. 2003a; Tenuzzo et al. 2006). The quality and quantity of the alterations is strictly related to the type of molecules and cells investigated (Rosen 2003b; Tenuzzo et al. 2006). Alterations to surface molecules may in turn lead to impairment of cell-specific activities regulated by the plasma membrane. The cell surface changes observed within the SMF-exposed cells may result from a number of causes, including the reported effects of SMFs on gene expression, particularly at the transcriptional level (Goodman et al. 1992; Hirose et al. 2003; De Mattei et al. 2005). Lipid peroxidation could be another cause of plasma membrane modification, since it has been observed in THP-1 leukemia cells as a consequence of exposure to SMFs (Cutolo et al. 2001; Amara et al. 2007). Preliminary data from a study currently being performed in our laboratory show high levels of lipid peroxidation in U937 cells and human lymphocytes exposed to SMFs.

15.3.1.3 Ca^{2+} Fluxes and Concentration

Ca^{2+} fluxes and modulation of intracellular Ca^{2+} concentration are of particular interest to many biological functions, due to the activation of signalling pathways by this ion. Changes in $[Ca^{2+}]$ have frequently been reported during SMF exposure (Flipo et al. 1998; Fanelli et al. 1999; Chionna et al. 2005; Tenuzzo et al. 2006). Nuccitelli et al. (2006) investigated the interference of MFs with cell metabolism by analysing cell parameters linked to apoptotic signalling and regulation of Ca^{2+} fluxes; they showed that different types of MFs (static and extremely low-frequency) at varying intensities alter plasma membrane potential. These findings suggest that alterations to membrane potential induced by

MFs are involved in the anti-apoptotic effects of MFs (Nuccitelli et al. 2006), which are discussed below.

Many attempts have been made to understand the cellular events that mediate the increase in Ca^{2+} at plasma membrane level resulting from SMF exposure, but none of them has yet provided a definitive explanation. Recently the theory of microvillar Ca^{2+} signalling has been put forward, and cable-like ion conductance in actin filaments has been discovered (Lange 1999 and 2000; Gartzke and Lange 2002). It has also been shown that U937 cells exposed to SMFs bear cell surface lamellar microvilli and de-arrange the F-actin network (Pagliara et al. 2005). F-actin is one of the Ca^{2+} storage compartments in the cells; removal of Ca^{2+} causes de-polymerisation of F-actin and the de-arrangement of the actin network, leading to modification of both cell shape and surface molecule distribution. Lange (2000) suggests that actin filaments create a diffusion barrier between membrane microvilli and the interior of the cell. Gelsolin, which de-polymerises the microvilli filaments, is activated by the influx of Ca^{2+} into the cell, which in turn allows additional influx of Ca^{2+} from the microvilli.

15.3.1.4 Cell Shape and Organelles

Living cells respond to SMFs exposure with both intracellular and extracellular changes (Miyakoshi 2005; Amara et al. 2007). The biophysical properties of biological membranes, such as membrane tension or bending stiffness, are critical for many aspects of cellular function, including cell shape, cell motility, surface area regulation, membrane repair, endocytosis, exocytosis and intracellular traffic (Raucher and Sheetz 1999, 2000; Togo et al. 2000; Morris and Homann 2001). Chionna et al. (2003a, 2005) conducted research into SMF-induced cell shape and cell surface modifications, providing evidence for time-related changes. As a general effect, cells growing in suspension lost their round shape and became irregularly elongated, while attached cells modified their shape and orientation or detached themselves, becoming freely suspended in the culture medium. All cells, irrespective of growth patterns, were found to undergo surface modifications (lamellar microvilli and blebs) and structural membrane changes. Interestingly, lamellar microvilli were observed well before the distortion of the cell shape, which was found at periods of exposure longer than 24 hours (Chionna et al. 2003a).The presence of many lamellar or bubble-like microvilli has been observed in various cell lines (K562, U937, human lymphocytes) exposed to fields of varying type (high-frequency electric field; 50 Hz magnetic field; moderate intensity SMFs) and intensity (from 500 MHz to 2,5 mT; Popov 1991; Paradisi et al. 1993; Chionna et al. 2003a), and can thus be taken as an index of exposure. The extent of cell shape modifications is determined, up to a point, by the duration of the exposure: longer periods of SMF exposure are associated with more extensive cell shape or plasma membrane modification (Rosen 1993; Chionna et al. 2003a). Cell type is also crucial for the extent of morphological modifications. Indeed, changes to the morphology of HepG2, U937, HeLa cells and lymphocytes after short or long exposure to a 6 mT SMF were found to be significantly greater than for 3DO, FRTL-5 cells and thymocytes (Tenuzzo et al. 2006). In addition, another factor significantly influencing cell response to SMF exposure is the age of the cells. The degree of cell shape modifications exerted by SMFs was observed to be amplified in senescent lymphocytes (Dini and Abbro 2005).

As a consequence of SMF exposure, many cellular and subcellular components are subjected to general subcellular redistribution; for example, smooth endoplasmic reticulum (SER), mitochondria and lysosomes were preferentially localized at the cell periphery and polarised (Somosy 2000; Testorf et al. 2002). For example, the flat irregular cell shape with many cytoplasmic protrusions and long pseudopodia observed in tetradecanoil-13-phorbol acetate (TPA)-differentiated monocytes shifts towards an irregular shape in which protrusions and pseudopodia are present only at one pole of the cells (personal laboratory observations). The question of whether and how these morphological changes induce more or less significant perturbations of some cellular activities remains to be resolved.

The increase and mobilization of $[Ca^{2+}]i$ during SMF exposure seems to be a crucial event in the maintenance of the morphology of the cell. Some of the available studies suggest that the reorganisation and breakdown of different cytoskeleton elements is related to modified $[Ca^{2+}]i$ homeostasis or to the altered phosphorylation/de-phosphorylation state of proteins in exposed cells (Popov et al. 1991; Santoro et al. 1997). The influence of SMFs on $[Ca^{2+}]i$ has also been reported in *Fusarium culmorum*, where it modified the Ca^{2+}-dependent signal transduction pathways involved in conidia germination and thus interfered with its normal morphology (Albertini et al. 2003).

15.3.2   *Apoptosis as Modulated by SMFs*

One of the first steps in investigating the relationship between MF exposure and the apoptotic program as a whole is to determine whether MFs affect apoptotic rates. Although MFs themslves have not been found to affect cell death (necrosis is not enhanced), several studies have shown that MFs alter apoptosis when it is induced by other factors, such as heat, chemicals and radiation (Fanelli et al. 1999; Teodori et al. 2002a, b; Chionna et al. 2003a, 2005; Volpe 2003; Ghibelli et al. 2006). SMFs of moderate intensity (1 mT (Tesla) to 1 T) have been found to reduce apoptosis when this is induced by various agents in human cell systems, in an intensity-dependent fashion (plateau reached at 60 mT; Ding et al. 2000; Ghibelli et al. 2006). This was not due to a change in the mode of cell death, i.e. necrosis, or to a delay in the process itself, but, rather, to the presence of MFs, which allowed the indefinite survival and replication of the cells struck by apoptogenic agents. Tian et al. (2002) found that exposure to MFs suppressed X-ray-induced apoptosis in Ku80-deficient cells (xrs5) and had no effect on apoptosis in Ku80-proficient cells (CHO-K1).

In a comparative study of the in vitro biological effects of moderate-intensity SMFs on various cell types, including primary cultures and transformed or stabilized cell lines derived from various tissues (e.g. U937 cells, human lymphocytes, thymocytes, FTRL-5, HepG2 cells, HeLa cells), it was found that SMF exposure influences the rate of spontaneous and induced apoptosis, whose quality and quantity was dependent on cell type and length of exposure (Tenuzzo et al. 2006). In addition, the type (increase and decrease) and extent of modifications was found to be independent of the apoptotic agent used (Tenuzzo et al. 2006).

Ca^{2+} as a mediator of intracellular signalling is crucial for the development of apoptosis. In cells committed to apoptosis, an increase in $[Ca^{2+}]i$ due to the emptying of intracellular Ca^{2+} stores and Ca^{2+} influx from the extracellular medium is a fairly general phenomenon, independent of the apoptotic stimuli (Bian et al. 1997). However, the role of $[Ca^{2+}]i$ increase during apoptosis is ambiguous because it has different effects in different cell systems (Magnelli et al. 1994; Teodori et al. 2002 a, b). Therefore, an increase in $[Ca^{2+}]i$ accompanies apoptosis both in cells where $[Ca^{2+}]i$ plays an anti-apoptotic role and in cells where $[Ca^{2+}]i$ induces apoptosis (Teodori et al. 2002 a, b). The choice of cell system may explain the many results reported by other researchers who in contrast have failed to detect any apoptotic effects of MFs (Buemi et al. 2001; Teodori et al. 2002 a, b). It is known that rapidly proliferating and transformed cells have differently de-polarized cell membranes compared with normal cells (Marino et al. 1994) and that epithelial cells lose their transepithelial potential during carcinogenesis (Capko et al. 1996). By acting on the movement of charged particles, SMFs may selectively modulate different cell signalling pathways in different cells, depending on their membrane potential, and thus exert differing effects on survival mechanisms (Rosen 1996; Tofani et al. 2003).

MFs alter plasma membrane potential; specifically, they induce plasma membrane hyper-polarization in cells sensitive to the anti-apoptotic effects of MFs, and plasma membrane depolarization in insensitive cells (Nuccitelli et al. 2006). These opposing effects suggest that protection against apoptosis and membrane potential modulation are correlated, plasma membrane hyper-polarization possibly being part of the signal transduction chain determining the anti-apoptotic effects of MFs. The anti-apoptotic effects of SMFs were found to be independent of the age of lymphocytes (Abbro et al. 2004; Dini 2006). This finding further supports the notion that SMFs, independently of the physiological condition of the cell, interfere directly with apoptotic intracellular signalling, as demonstrated by the use of apoptogenic substances that bypass drug uptake such as hydrogen peroxide (Dini and Abbro 2005).

The anti-apoptotic effects of SMFs are characterized by an immediate and generally reversible response; however, SMFs also entail long-term changes such as altered patterns of gene expression. The small number of studies of the effects of SMFs on gene expression that have been published to date have yielded contradictory data, and only a few of them refer to apoptotic genes. When young and aged human lymphocytes were exposed to a SMF for at least 24 h the expression of Bcl-2, Bax, p53 and heat shock protein (Hsp)70 was modulated and an altered rate of apoptosis was observed (submitted manuscript).

From a morphological point of view, SMFs interfere with the "typical" morphological features of apoptosis: roundness and smoothness of the cell surface is only partially achieved, and the distribution of some cell surface "eat me" molecules is modified (Dini and Abbro 2005).

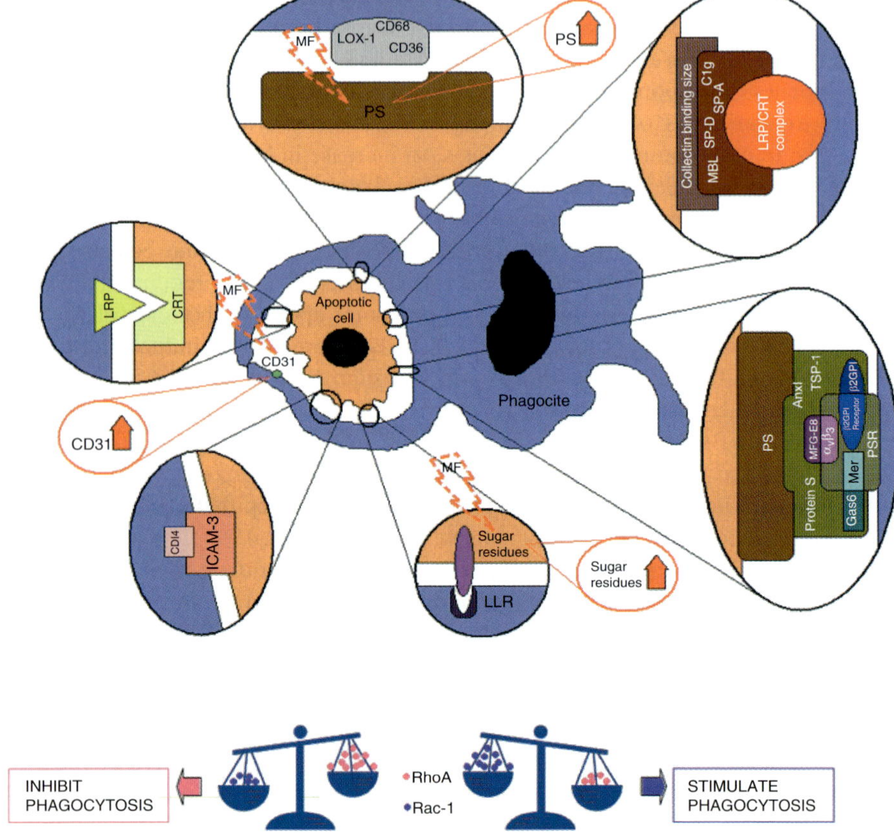

Fig. 15.3 Simplified scheme of interactions between apoptotic cell and phagocyte. "Eat me" signals on the apoptotic cell, such as calreticulin (CRT) and PS, bind directly to receptors on the phagocyte surface (LRP or CD91, phosphatidylserine receptor (PSR), CD44, CD14) or by soluble bridging proteins. PS binds multiple PS-bridging proteins (milk-fat-globule-factor E8, MFG-E8; thrombospondin 1, TSP-1; growth –arrest-specific factor 6, Gas6; β2-glycoprotein I; Annexin I, AnxI; and Protein S), which then engage specific receptors on the phagocyte surface, initiating ingestion. In most cases the receptors on phagocytes that recognize these PS-bridging proteins have not yet been defined, but it has been reported that Gas6 is a ligand for the tyrosine kinase receptor Mer and that MFG-E8 can bind to the vitronectin receptor $\alpha_v\beta_3$. Other molecules that bind PS with vary specificity are the oxidized low-density lipoprotein receptor-1 (LOX-1) and the scavenger receptor CD36 and CD68. Collectins (mannose-binding lectin, MBL; scavenger receptor class (SP)-A, SP-D) and the fist component of the complement cascade collectin-like C1q binds to an unknown collectin binding site on the apoptotic cell surface via their globular heads. Their collagenous tails then engage the LRP/CRT complex on the phagocyte, initiating ingestion. Apoptotic CRT may also directly engage phagocyte LRP in *cis* or in *trans* configuration by passing collectin intermediates. ICAM-3 interacts with the membrane glycoprotein CD14. During apoptosis, CD31 switches from a *don't-eat-me* (detachment) signal (on viable cells) to a tethering system that allows *eat-me* signals to engage. The sugar residues expressed on apoptotic cell are recognized by LLRs. The exposure to SMF of non-apoptotic cells enhances the cell surface expression of sugar residues, PS and CD31. The ingestion of apoptotic cells is governed by a balance between the Rho GTPases, Rac-1 and RhoA, whereby Rac-1 is the principal activator of phagocytes and RhoA is the principal inhibitor via the downstream effector, Rho kinase.

15.3.3   Macrophage Defence Is Modulated by MFs

Studies of the phagocytic activity of macrophages in the presence of MFs (static or pulsating) have also yielded contradictory data. While Simkó et al. (2001) reported that 50 Hertz (Hz) ELF-MF exposure is able to activate primary monocytes and macrophages from different species and cell lines, with an activation potential comparable to activation by certain chemicals, resulting in physiologically significant cellular responses. Flipo et al. (1998) detected decreased phagocytic activity in pre-stimulated peritoneal macrophages exposed to SMFs (24 h) with flux densities ranging from 25 to 106 mT. These contradictory data can be explained by differences in exposure conditions and by the fact that the cells were pre-stimulated. Our data also showed reduced phagocytic activity in phorbol ester-differentiated U937 cells after short SMF exposure times, while longer exposure times did not interfere with the internalization rate (Pagliara et al. 2005). Surprisingly, the ability to bind to but not to ingest latex beads was always higher under exposure to SMFs, thus confirming the modification of the cell surface (Pagliara et al. 2005). Under exposure to SMFs, the overall macrophage defence was found to be depressed (Cappelli et al. 2001; Volpe et al. 2002). SMFs, from very low intensities up to 670 mT, were found to exert a continuous effect on the expression of the CD14 and CD64 proteins of the human monocytes-derived macrophages (MDMs), on the interactions of MDM with *Mycobacterium tuberculosis* (MTB), and on the division of MDM inside the bacterial cytosol. It is possible that at certain intensities, MFs might facilitate or prevent the penetration of MDM into an MTB cell or, independently of the quantity of MDM entering an MTB cell, they might facilitate or prevent direct digestion of MTB cells. Without applying artificial fields, one MDM per MTB cell was observed entrapped in a very large phagolysosome. At 91 mT, about 10 MDMs per MTB cell were observed, but the phagolysosome was of a smaller size. At 670 mT the MTB cells were free to replicate because the MDMs were unable to construct a phagolysosome (Volpe 2003).

15.4   Effects of SMFs on Phagocytosis of Apoptotic Human Lymphocytes

15.4.1   SMF-induced Modification of Lymphocytes

Exposure of isolated human lymphocytes to SMFs induces the same morphological modifications (i.e. cell shape, cytoskeleton and cell surface) and the same anti-apoptotic effects that have been described in previous sections (Dini and Abbro 2005; Dini 2006). Decreased cell viability or altered cell growth rates are never observed. Surprisingly, the cell surface of non-exposed apoptotic lymphocytes and that of non-apoptotic SMF-exposed lymphocytes was shown to have many common modifications, including some proteins and sugar residues (Chionna et al. 2005; Dini and Abbro 2005; Tenuzzo et al. 2006). Conversely, the surface of lymphocytes induced to apoptosis under SMFs was a halfway stage between the surface of non-exposed apoptotic and non-apoptotic cells (Dini and Abbro 2005; Dini 2006). The expression and distribution of mannose/D-glucosammine and galactose/D-galactosamine were significantly enhanced after exposure. Apoptosis as well as exposure to SMFs induced modifications in the protein profile of cells. Indeed, analysis of membrane proteins and whole-cell lysate proteins (Dini et al. 2007) showed that there were many differences, not only between exposed and non-exposed apoptotic and non-apoptotic lymphocytes, but also between young and aged cells. Aged lymphocytes exposed to SMFs were always highly susceptible to exposure. In addition, SDS-PAGE tests show that the main modifications were found for proteins in the 130, 95 and 45–43 kiloDalton (Da) bands; non-apoptotic exposed cells accumulate proteins in the 60 kiloDa band (Dini et al. 2007). At least 60% of the modifications are shared by non-apoptotic SMF-exposed and non-exposed apoptotic lymphocytes. Western blots of membrane glycoproteins, performed with biotinylated lectins, show 70% similarity between non-exposed apoptotic and non-apoptotic exposed lymphocytes (Dini et al. 2007). The similarity in terms of α-mannose and α-glucose residues was 100%, while it was 30% for α-L-fucose, and 70% for N-acetyl-neuraminic acid, N-acetyl-glucosamine, and N-acetyl-galactosamine.

Analysis of the lysate protein patterns of non-apoptotic exposed and non-exposed apoptotic lymphocytes performed by 2D gel electrophoresis indicated that exposure to SMFs modulates both low- and high-weight acid proteins. Lipid peroxidation of the cell membrane, which is a possible cause of the glycoprotein rearrangement seen with fluorescent-lectin (Dini and Abbro 2005), was seen to occur at high levels in SMF-exposed lymphocytes.

The identification of those (glyco)proteins whose amount and distribution is modified by exposure to SMFs is in progress. Possible candidates include adhesion molecules and proteins that belong to transduction pathways: they comprise, therefore, molecules involved in recognition, cell-cell adhesion and intracellular pathways; processes that are all important for the recognition and engulfment of apoptotic cells. We found that continuous exposure to an SMF of 6 mT for up to 48 h induced an increase in the expression of CD5, an intrinsic plasma membrane glycoprotein typical of mature lymphocytes, and CD99 (MIC2), an extrinsic protein cited as a pro-apoptotic factor (Bernard et al. 1997; Alberti et al. 2002) and CD31. In contrast, the expression of CD102 (inter-cellular adhesion molecule-2, ICAM-2), expressed on the cell surface of mature lymphocytes, decreased (Maio and Del Vecchio 1992).

Indeed, data that might help to determine whether the modifications in cell proteins are due to conformational, transductional or transcriptional changes, or indeed to a combination of these factors, are currently lacking.

15.4.2 Liver Recognition and Internalization of Normal and Apoptotic SMF-exposed Lymphocytes

The initial event in phagocytosis of apoptotic cells is the recognition of the target. Successful engulfment requires that apoptotic cells expose "eat me" signals on their surface. Because the interaction between an apoptotic cell and a phagocyte involves an array of receptors and bridging molecules, it has been also termed "phagocytic synapse"(Stuart and Ezekowitz 2005). Receptors known to be associated with phagocytosis of apoptotic cells include lectin-like receptors (LLRs), lipoprotein receptor-related protein (LRP or CD91), Mer receptor tyrosine kinase, $\alpha_v\beta_3$- and $\alpha_v\beta_5$-integrins, scavenger receptors, CD44, CD14, and complement receptors 3 and 4 (Lauber et al. 2004). Adenosine triphosphate binding cassette proteins including ATP-binding cassette ABC-A1, ABC-A7, and cystic fibrosis transmembrane regulator (CFTR) also contribute to the phagocytosis of apoptotic cells (Jehle et al. 2006) although perhaps not as true receptors (Fig. 15.3).

One convenient model for the study of the PACs is the liver. Indeed, liver is suitable for both in vivo and in situ experiments and, since hepatocytes and non-parenchymal cells (i.e. Kupffer cells, endothelial cells, pit cells and liver NK cells) can be cultivated as isolated and purified cell populations, also for in vitro experiments. Recent studies have revealed that the liver is the main site for apoptosis and burying of blood cells (Crispe et al. 2000; Sun and Shi 2001). Liver phagocytosis is highly performed by endothelial and Kupffer cells (Dini and Carlà 1998; Dini et al. 2002), which preferentially accumulate apoptotic lymphocytes. As Kupffer cells account for the 80% of all macrophages in mammals, it is reasonable to assume that the liver is a specialized organ for clearing apoptotic cells. Liver cells recognize and ingest circulating apoptotic cells and/or neighbouring ones by using either the asialoglycoprotein receptor of hepatocytes and the carbohydrate-specific and scavenger receptors of sinusoidal cells (Dini et al. 2002).

The liver clearance of apoptotic lymphocytes can be influenced at two different levels: phagocyte receptors and apoptotic cells. In general, the efficiency of phagocytosis is proportional to the number of expressed receptors. The liver LLRs, that function in the recognition of apoptotic cells, are modulated (quantity and distribution) in relation to the type of the cells (i.e. Kupffer cells > endothelial cells > hepatocytes; Dini et al. 1995; Dini 2005), to the physiological or pathological status of the organ (adult > embryo; cancer > normal; Dini et al. 1995, 2002; Massimi et al. 1995) and to the cell localization inside the lobule (periportal > perivenous; Dini and Carlà 1998). With respect to this latter aspect, according to the number of carbohydrate specific receptors the efficiency of recognition and internalization is higher in the periportal tracts of the liver lobule, than in the perivenous ones. On the side of the edible cells, the quality of apoptotic cells can determine the efficiency of clearance by the liver (Chionna et al. 2003b; Pagliara et al. 2003). It is known that the steps in the process of apoptosis (i.e. early, mature or late apoptosis) influence the efficiency of removal since the modifications of the cell surface change accordingly with the progression through the apoptotic program (Pagliara et al. 2003). A hierarchy of the expression of specific epitopes on the cell surface of apoptotic lymphocytes has been reported (Pagliara et al. 2003). In particular, exposure of PS, mannose and N-acetylgalactosamine appears as soon as cell is triggered to apoptosis, and are considered very early changes. In the mature phase of lymphocyte's apoptosis an extensive cell surface exposure of PS and sugar residues (mannose and N-acetylgalactosamine for the most) is present. The late phase of lymphocytes apoptosis is characterized by a decreased exposure of PS and by the abundant expression of sugar residues (mainly fucose), that are absent from the surface of early apoptotic lymphocytes (Chionna et al. 2003a). Therefore, the surface of dying cells progressively changes leading to widely amplified possibilities of a safe recognition and to modulation in the rate of recognition and engulfment. The model of the acute liver intoxication with lead nitrate beautiful shows that the amount of cell surface exposed LLRs and the quantity and quality of "eat me"signals drives the efficiency of PACs (Ruzittu et al. 1999). Lead nitrate, intravenously administrated, induces first liver hyperplasia followed by the increase of apoptotic cells and by the increased expression of LLRs (Dini et al. 1993). Liver PACs increases accordingly with the number of apoptotic cells and LLRs.

The SMF-induced modifications of phagocytes and apoptotic cells influence liver PACs (Dini 2006). The presence of SMF during the drug-induced apoptosis and/or during the PACs influences this process at all steps, i.e. tethering, docking and engulfment. Tethering is influenced by the cell surface modifications (i.e. glycoproteins and glycolipids) of apoptotic and non-apoptotic cells and/or phagocytes exposed to SMFs. The second step of PACs (i.e. docking) is impaired by the de-arrangement of F-actin network. The engulfment of the apoptotic cells and the formation of phagolysosome in presence of SMFs is delayed.

We are currently studying the in situ liver recognition of non-apoptotic or apoptotic human lymphocytes exposed or not (from 24 to 48 h) to 6 mT SMF. Liver cells actively retain non-exposed apoptotic lymphocytes, but fail to bind non-exposed non-apoptotic cells (less than 5% of injected non-apoptotic cells bind to sinusoids). When internalization is allowed (i.e. in situ liver perfusion at 37°C), a great number of non-exposed apoptotic lymphocytes is engulfed by sinusoidal cells. Non-apoptotic lymphocytes are never observed inside sinusoidal cells. Unexpectedly, non-apoptotic young or aged lymphocytes, exposed to SMF, are bound to and engulfed by the sinusoidal liver cells at a rate similar to the liver clearance of apoptotic cells. Indeed more than 70% of the exposed non-apoptotic lymphocytes injected into the liver circulation are retained by liver cells. An increased number of apoptotic lymphocytes, drug-induced under SMF exposure, bind to sinusoidal cells. Conversely, their internalization is reduced. The amount of bound apoptotic cells and the delay of their engulfment is dependent on the duration of SMF exposure. The clearance of apoptotic and SMF-exposed apoptotic lymphocytes is, in all cases, receptor-mediated as indicated by inhibition experiments. The almost complete inhibition achieved with experiments of sugar competition indicates that this recognition is mediated by the LLRs. However, in the same experiments about 50% of inhibition of binding of non-apoptotic SMF-exposed lymphocytes is obtained, suggesting

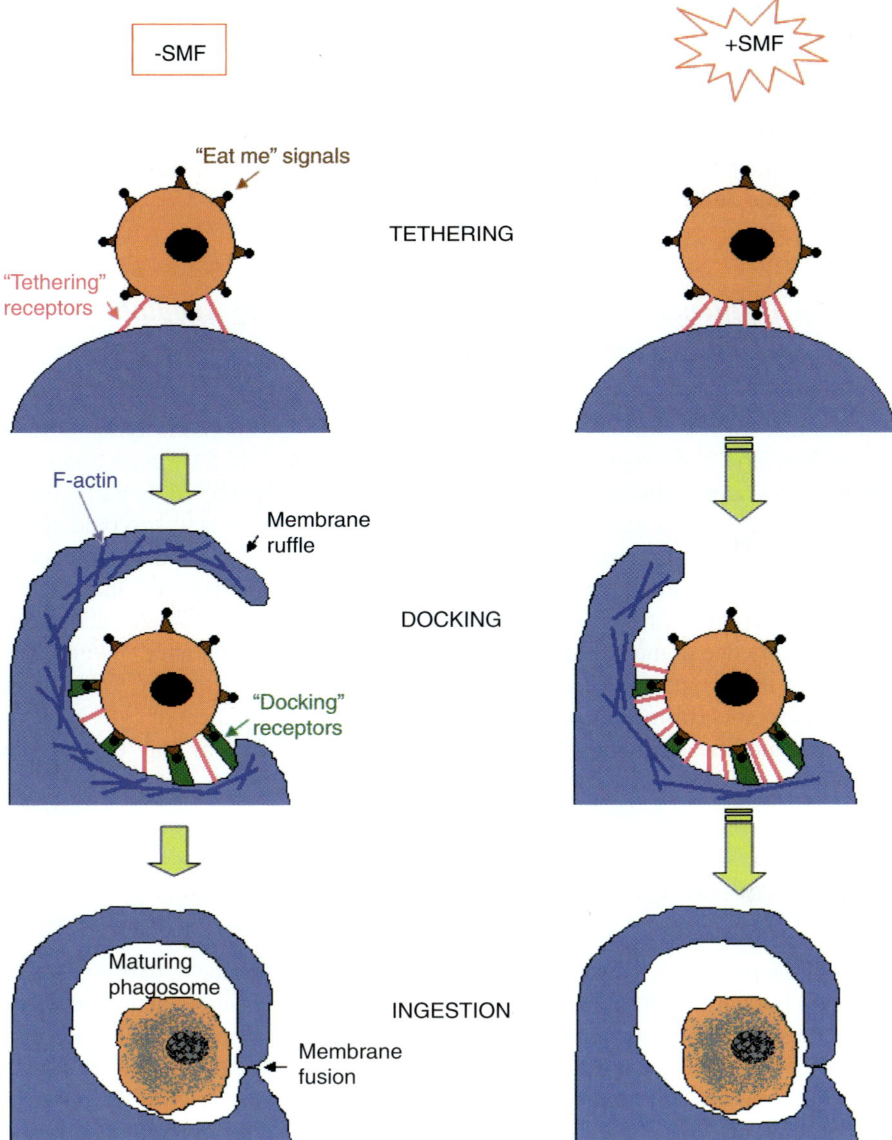

Fig. 15.4 Tethering and docking hypothesis for physiological and SMF-exposed phagocytosis. Tethering receptors (CD14, CD31, LLR) hold the apoptotic cell in close vicinity of the phagocyte. Tethering adhesions increase in presence of SMF. Docking receptors (PSRs, LRP or CD91) then engage *eat-me* signals on the apoptotic cell surface. This binding is followed by signal transduction that activates Rac-1 and results in the extension of membrane ruffles. SMF exposure delays the formation of membrane ruffles and de-arranges the F-actin network of the ruffles. The ingestion of the apoptotic cell is through a fluid-filled phagosome where is digested. The engulfment of the apoptotic cell, the formation of the phagosome and the digestion of the dead cell are significantly slowed down in presence of SMF.

that LLRs are not the unique receptors involved. In this context, the non-apoptotic SMF-exposed cells are recognized at the same extent in both the perivenous and periportal tract of the liver sinusoids (Abbro et al. 2004). Therefore, the liver clearance of SMF-exposed non-apoptotic and non-exposed apoptotic lymphocytes is, at least in part, through the same receptors. The ability of phagocytes to recognize non-apoptotic SMF-exposed lymphocytes is also confirmed in in vitro experiments by using Raw 264.7 macrophages (unpublished results).

15.4.3 Influence of SMF on the Phagocytosis of Apoptotic Lymphocytes

Exposure to SMFs can also impair the phagocytosis of apoptotic lymphocytes. When Raw 264.7 macrophages are challenged in presence of SMF with apoptotic lymphocytes, a perturbation of the phagocytosis activity is observed. In particular, a delayed engulfment is measured: the phagocytes became very active to bind apoptotic cells, but very sluggish to engulf them. In presence of SMFs, Raw 264.7 needs more time to internalize the same amount of apoptotic lymphocytes, that can be engulfed in absence of SMFs. The increased apoptotic cell/macrophage adhesions may be likely due to the modifications of the phagocyte's membranes and/or apoptotic cells, that happen quickly (within 30 min/1 hour) under exposure to SMFs. A variety of factors (from altered signal molecules to de-arrangement of cytoskeleton, lysosome modification, and delayed phagolysosome formation) may be listed to explain the slowdown of engulfment.

15.5 Immunological and Health Consequences

Many of the pollutants, nowadays present in our living environment, have a negative effect on human health by interfering with many metabolic and genetic pathways. Cigarette smoke and exposure to SMF impair the PACs. A disordering of apoptosis pathways is very likely to promote human disease, e.g. cancer, while disorders in the proper recognition and engulfment of apoptotic cells can lead to the onset of autoimmune diseases (Maderna and Godson 2003; Kim et al. 2005; Mahoney and Rosen 2005; Krysko et al. 2003, 2006). In non-pathological conditions, an efficient clearance is assured by the redundant interaction between apoptotic cell ligands and multiple molecules present on the engulfing cell surface. Although a substantial amount of work has produced significant accomplishment in identifying molecules involved in the engulfment of apoptotic cells, the exact role of individual molecules during this process remains unclear (Steinman et al. 2000; Duffield 2003; Sun et al. 2004). How certain receptors can promote different outcomes, remains an important question. PACs appears to be a key regulatory checkpoint for the innate immune system, the adaptive immune system, and cell proliferation. There is now overwhelming evidence that phagocyte function is profoundly altered following apoptotic cell uptake, with consequences for the ensuing innate and adaptive immune response (Erwig and Henson 2007). Thus PACs could be used to manipulate the immune response for therapeutic purpose.

Efficient engulfment of apoptotic cells is of paramount importance in vivo, in part because clearance of apoptotic cells prior to their lyses is critical for the resolution of inflammation, due to the suppression of the release of pro-inflammatory cytokines. The role played by cigarette smoke in the pathogenesis of chronic inflammatory diseases has been described in different researches. The myriad of compounds present in the cigarette smoke can either increase the number of apoptotic deaths and impair the phagocytic activity of alveolar macrophages. COPD appears to be associated with the delayed removal of dying cells, which may directly impact the natural ability of the injured organism to shut down inflammation and initiate tissue repair. The failure of PACs may contribute to disease pathogenesis by impeding both the resolution of inflammation and the maintenance of alveolar integrity. A central regulatory role for normal apoptotic cell recognition as the key homeostatic regulator for alveolar–capillary membrane integrity has been proposed (Vandivier et al. 2007). In this paradigm, continual surveillance and phagocytosis of senescent, apoptotic alveolar epithelial (or endothelial) cells by professional and non-professional phagocytes result in synthesis and secretion of growth factors (among them two key growth factors for alveolar cells, vascular endothelial growth factor, VEGF and hepat ocyte growth factor, HGF), cytokines, and chemokines that maintain alveolar epithelial and pulmonary vascular integrity (Botto et al. 1998; Douglas et al. 2006; Vandivier et al. 2007).

Lymphocytes have been widely used in the study of apoptosis due to their considerable importance during immunosystem maturation and certain pathologies (Rathmell and Thompson 2002). Indeed, a correct process of apoptosis is fundamental in lymphocyte physiology: it is the process needed for the maturation of the different lymphocyte populations and plays a critical role in lymphocyte development and homeostasis (Lenardo 1997; Krammer 2000). Impaired phagocytosis of apoptotic lymphocytes can lead to immunosuppresion when the clearance is enhanced, or to autoimmunity in the presence of defective clearance (Scott et al. 2001).

SMF exposure decreases the drug-induced apoptotic rate of lymphocytes, but increases the liver removal of control living SMF-exposed cells by using in part the same recognition system utilized by liver sinusoids for the clearance of circulating apoptotic cells. Evidences suggest that non-apoptotic SMF-exposed and apoptotic cells express similar cell surface changes. Indeed, morphological modifications are not only a simple mechanical process able to dismantle cell structures; they are necessary to create a dead cell, which is "edible" for the phagocytes. Interestingly, the morphological modifications of SMF-exposed non-apoptotic lymphocytes create "edible cells" too but "living".

The meaning of cell death is defined not only by the efficiency of the apoptotic cell signalling, but also by the efficiency of removal (Savill et al. 2002). The exposure to SMF decreases the efficiency of removal of apoptotic cells, delaying the engulfment. It could be hypothesized that a continuous in vivo exposure could provoke a series of events whose effects need to be investigated: 1) removal from the blood of the non-apoptotic lymphocytes, recognized as apoptotic cells for the SMF-induced cell surface modifications; 2) increase of the apoptotic circulating lymphocytes, whose clearance is delayed by SMF-induced modifications, 3) increase of circulating lymphocytes that survive the apoptotic struck.

The release of anti-inflammatory and immunosuppressive signals following the recognition and engulfment of the apoptotic cells, makes the process of PACs different from the phagocytosis of microrganisms or debries (Krysko et al. 2003). Thus, the modulation of phagocytic capacity for apoptotic cell clearance represents a potential therapeutic target in the control of inflammatory disease (Gupta et al. 2006). However, in the liver, the recognition and engulfment of apoptotic cells by Kupffer cells stimulates death ligands and cytokines, that have also profound effects on lipid metabolism (Cambay et al. 2003). The disposition of apoptotic cells is, therefore, an unexplored and potentially key mechanism linking apoptosis to inflammation and fibrosis in the liver. In this respect, consistent with a role for Kupffer cells in liver inflammation and fibrosis, gadolinium chloride attenuated both PACs and death ligand generation (Cambay et al. 2003). Although the liver provides a "tolerogenic" immune environment for antigen-specific T cells, activation of Kupffer cells, recruited macrophages and inflammatory cells results in the production of cytokines and chemokines that can lead to prolonged inflammation, hepatocyte damage, and/or cholestasis. Inflammation is a pathogenic component of various types of acute and chronic liver diseases, and it contributes to progressive liver damage and fibrosis. Kupffer cells are activated via toll-like receptor (TLR)4 complex on the cell surface, that, in turn, results in the release of a myriad of pro-inflammatory mediators, inducing hepatic injury and fibrosis (Szabo et al. 2007). TLRs are conserved membrane receptors of innate immunity whose activation represents a double-edged sword. On one end, it launches a quick anti-microbial reaction and directs adaptative immunity to mount a protective response. On the other hand, by ensuing cytokine and chemokine overproduction as well as inappropriate activation of antigen-presenting cells, TLR activation may have detrimental effects culminating in the induction, progression or exacerbation of disease. Thus TLR signalling must be tightly regulated. It is of note that TLR-2, -3 and -4 may be also involved in the pathogenesis of COPD (Papadimitraki et al. 2007).

Further investigations will be necessary to determine whether the clearance of circulating non-apoptotic SMF-exposed cells suppresses, as in PACs, the release of pro-inflammatory cytokines and allows the release of growth signals to surrounding cells, promoting their survival (Krysko et al. 2006). It has been recently reported that when cells dying through autophagy are phagocyted, a pro-inflammatory response is evoked in macrophages (Petrovski et al. 2007). Moreover, it should be also investigated if Kupffer cells removal of non-apoptotic SMF exposed cells can be linked to inflammation and fibrosis in the liver. Indeed, models, in which a perturbation in the clearance of apoptotic cells is present, are useful to study the consequences of inadequate PACs (Taylor et al. 2000; Scott et al. 2001).

15.6 Conclusion

The strong and reproducible effects of cigarette smoke and of SMF exposure on the process of PACs have been discussed. These two factors represent the first two environmental elements up to now investigated for their ability to modulate PACs by interfering with both the apoptotic cells and the phagocytes. Cell death, removal, and replenishment are essential both to maintain homeostasis in the non-diseased organism and to appropriately regulate inflammation and tissue repair during disease. Cigarette smoke increases the production of apoptotic deaths and impairs the PACs, leading to inflammation. Chronic inflammation in the lung appears to be associated with the delayed removal of dying cells, which may directly impact the natural ability of the injured organism to shut down inflammation and initiate tissue repair. Further investigations will be necessary to determine whether impaired PACs directly impacts disease pathogenesis (e.g. the development of emphysema) and whether novel biomarkers can be identified based on these mechanisms. However, the possibility that reversal of a normally immunosuppressive environment in the lung due to ineffective apoptotic cell recognition (Hoffmann et al. 2005) might contribute to local lymphocyte accumulation (increased number of T lymphocytes has been reported) and immune responses seen in severe COPD, is intriguing. The recognition of impaired PACs as a mechanism of disease may allow a new therapeutic approach based on fundamental processes that have remained highly conserved across the metazoa.

Conversely, the knowledge of the effects of SMFs on the whole apoptotic program is still in its infancy. This is likely due to the complexity of the experimental approaches (type of field, intensity, time, cell type, in vivo or in vitro model, etc.). However, anti-apoptotic effect and delayed PACs have been demonstrated. Ca^{2+}, which represents the most important intracellular signal governing almost all physiological functions in cells, seems to be responsible for the SMF's anti-apoptotic effects. The dramatic modifications of plasma membrane (glyco)proteins and lipids in cells exposed to SMF, that overlap for more than 70% with the apoptosis-induced surface changes, allow them to be recognized by the reticulo-endothelial system. Thus, it is likely that the elimination of SMF exposed cells, as "apoptotic dressed cells", through PACs instead of pathogens and debris phagocytosis, could allow a silent removal of these spoiled cells without eliciting an inflammatory tissue response.

References

Abbro L, Lanubile R, Dini L (2004) Liver recognition of young and aged lymphocytes exposed to magnetic pollution. Recent Res Dev Cell Sci 1:83–97

Alberti I, Bernard G, Rouquette-Jazdanian AK et al (2002) CD99 isoforms expression dictates T cell functional outcomes. FASEB J 16:1946–1948

Albertini MC, Accorsi A, Citterio B et al (2003) Morphological and biochemical modifications induced by a static magnetic field on *Fusarium culmorum*. Biochimie 85:963–970

Amara S, Douki T, Ravanat JL et al (2007) Influence of a static magnetic field (250 mT) on the antioxidant response and DNA integrity in THP1 cells. Phys Med Biol 52:889–898

Anderson RGW (1993) Caveolae: where incoming and outgoing messengers meet. Proc Natl Acad Sci USA 90:10909–10913

Arabi M (2004) Nicotinic infertility: assessing DNA and plasma membrane integrity of human spermatozoa. Andrologia 36:305–310

Argentin G, Cicchetti R (2004) Genotoxic and anti-apoptotic effect of nicotine on human gingival fibroblasts. Toxicol Sci 79:75–81

Arur S, Uche UE, Rezaul K et al (2003) Annexin I is an endogenous ligand that mediates apoptotic cell engulfment. Dev Cell 4:587–598

Baehrecke EH (2002) How death shapes life during development. Nat Rev Mol Cell Biol 3:779–787

Barnes P (2000) Mechanisms in COPD: differences from asthma. Chest 117:10S–14S

Beischer DE (1971) The null magnetic field as reference for the study of geomagnetic directional effects in animals and man. Ann N Y Acad Sci 188:324–330

Belyaev IYa, Alipov YD, Harms-Ringdahl M (1997) Effects of zero magnetic fieldon the conformation of chromatin in human cells. Biochim Biophys Acta 1336:465–473

Bernard G, Breittmayer JP, de Matteis M et al (1997) Apoptosis of immature thymocytes mediated by E2/CD99. J Immunol 158:2543–2550

Bersani F, Marinelli F, Ognibene A et al (1997) Intramembrane protein distribution in cell cultures is affected by 50 Hz pulsed magnetic fields. Bioelectromagnetics 18:463–469

Bian X, Hughes FM Jr, Huang Y et al (1997) Roles of cytoplasmic Ca2+ and intracellular Ca2+ stores in induction and suppression of apoptosis in S49 cells. Am J Physiol 272;C1241–1249

Bordiushkov IuN, Goroshinskaia IA, Frantsiiants EM et al (2000) Structural-functional changes in lymphocyte and erythrocyte membranes after exposure to alternating magnetic field. Vop Med Khim 46:72–80

Botto M, Dell'Agnola C, Bygrave AE et al (1998) Homozygous C1q deficiency causes glomerulonephritis associated with multiple apoptotic bodies. Nat Genet 19:56–59

Brown D, Rose J (1992) Sorting of GPI-anchored proteins to glycolipid-enriched membrane subdomains during transport to the apical cell surface. Cell 68:533–544

Buemi M, Marino D, Di Pasquale G et al (2001) Cell proliferation/cell death balance in renal cell cultures after exposure to a static magnetic field. Nephron 87:269–273

Cambay A, Feldstein AE, Higuchi H et al (2003) Kupffer cell engulfment of apoptotic bodies stimulates death ligand and cytokine expression. Hepatology 38:1188–1198

Capko D, Zhuravkov A, Davies RJ (1996) Transepithelial depolarisation in breast cancer. Breast Cancer Res 41:230

Cappelli G, Volpe P, Sanduzzi A et al (2001) Human macrophage gamma interferon decreases gene expression but not replication of *Mycobacterium tuberculosis*: analysis of the host-pathogen reciprocal influence on transcription in a comparison of strains H37Rv and CMT97. Infect Immun 69:7262–7270

Casey PJ (1995) Protein lipidation in cell signalling. Science 268:221–225

Cavopol AV, Wamil AW, Holcomb RR et al (1995) Measurement and analysis of static magnetic fields that block action potentials in cultured neurons. Bioelectromagnetics 16:197–206

Chionna A, Dwikat M, Panzarini E et al (2003a) Cell shape and plasma membrane alterations after static magnetic fields exposure. Eur J Histochem 47:299–308

Chionna A, Panzarini E, Pagliara P et al (2003b) Hepatic clearance of apoptotic lymphocytes: simply removal of waste cells? Eur J Histochem 47:97–104

Chionna A, Tenuzzo B, Panzarini E et al (2005) Time dependent modifications of Hep G2 cells during exposure to static magnetic fields. Bioelectromagnetics 26:275–286

Cooke JP, Bitterman H (2004) Nicotine and angiogenesis: A new paradigm for tobacco-related diseases. Ann Med 36:33–40

Crispe IN, Dao T, Klugewitz K et al (2000) The liver as a site of T-cell apoptosis: graveyard, or killing field? Immunol Rev 174:47–62

Cutolo M, Carruba G, Villaggio B et al (2001) Phorbol diester 12-O-tetradecanoylphorbol 13-acetate (TPA) up-regulates the expression of estrogen receptors in human THP-1 leukemia cells. J Cell Biochem 83:390–400

Dajas-Bailador F, Wonnacott S (2004) Nicotinic acetylcholine receptors and the regulation of neuronal signalling. Trends Pharmacol Sci 25:317–324

De Mattei M, Gagliano N, Moscheni C et al (2005) Changes in polyamines, c-myc and c-fos gene expression in osteoblast-like cells exposed to pulsed electromagnetic fields. Bioelectromagnetics 26:207–214

deAlmeida CJ, Linden R (2005) Phagocytosis of apoptotic cells: a matter of balance. Cell Mol Life Sci 62:1532–1546

deBakker CD, Haney LB, Kinchen JM et al (2004) Phagocytosis of apoptotic cells is regulated by a UNC-73/TRIO-MIG-2/RhoG signaling module and armadillo repeats of CED-12/ELMO. Curr Biol 14:2208–2216

Ding GR, Yaguchi H, Yoshida M et al (2000) Increase in X-ray-induced mutations by exposure to magnetic field (60 Hz, 5 mT) in NF-κB-inhibited cells. Biochem Biophys Res Commun 276:238–243

Dini L (2005) Phagocyte and apoptotic cell interplay. In: Scovassi AI (ed) Apoptosis, 2005. Research Signpost, Kerala, India pp 107–129

Dini L (2006) Recognition and engulfment of the apoptotic cells: influence of the environment. In: Shultz LB (ed) Cell Apoptosis: regulation and enviromental factors. Nova Science Publishers, New York pp. 85–111

Dini L, Abbro L (2005) Bioeffects of moderate-intensity static magnetic fields on cell cultures. Micron 36:195–217

Dini L, Carlà EC (1998) Hepatic sinusoidal endothelium heterogeneity with respect to the recognition of apoptotic cells. Exp Cell Res 240:388–393

Dini L, Falasca L, Lentini A et al (1993) Galactose-specific receptor modulation related to the onset of apoptosis in rat liver. Eur J Cell Biol 61:329–337

Dini L, Lentini A, Diez GD et al (1995) Phagocytosis of apoptotic bodies by liver endothelial cells. J Cell Sci 108:967–973

Dini L, Pagliara P, Carlà EC (2002) Phagocytosis of apoptotic cells by liver: a morphological study. Microsc Res Tech 57:530–540

Dini L, Tenuzzo B, Vergallo C et al (2007) Static magnetic field exposure labels human lymphocytes as apoptotic cells. 75th Gordon Research Conferences. "Apoptotic cell recognition & clearance". Bates College, Lewiston, Maine (ME)—USA

Douglas IS, Diaz Del Valle F, Winn RA et al (2006) Beta-catenin in the fibroproliferative response to acute lung injury. Am J Resp Cell Mol 34:274–285

Duffield JS (2003) The inflammatory macrophage: a story of Jekyll and Hyde. Clin Sci (Lond) 104:27–38

Ellis RE, Jacobson DM, Horvitz HR (1991) Genes required for the engulfment of cell corpses during programmed cell death in Caenorhabditis elegans. Genetics 129:79–94

Emoto K, Toyama-Sorimachi N, Karasuyama H et al (1997) Exposure of phosphatidylethanolamine on the surface of apoptotic cells. Exp Cell Res 232:430–434

Erwig LP, Henson PM (2007) Immunological consequences of apoptotic cell phagocytosis. Am J Pathol 171:2–8

Esterbauer H, Schaur RJ, Zollner H (1991) Chemistry and biochemistry of 4-hydroxynonenal, malonaldehyde and related aldehydes. Free Radic Biol Med 11:81–128

Fadeel B (2003) Programmed cell clearance. Cell Mol Life Sci 60:2575–2585

Fanelli C, Coppola S, Barone R et al (1999) Magnetic fields increase cell survival by inhibiting apoptosis via modulation of Ca2+ influx. FASEB J 13:95–102

Ferrara A, Bisetti A, Azzarà V (1996) The Rational basis of hyperreactivity of the integrated respiratory tract. Acta Otorhinolaryngol Ital 16:4–28

Feychting M (2005) Health effects of static magnetic fields - a review of the epidemiological evidence. Prog Biophys Mol Biol 87:241–246

Finney-Hayward TK, Russell REK, Kon OM et al (2005) Decreased phagocytotic activity of monocyte-derived macrophages in patients with COPD. Proc Am Thorac Soc 2:A17

Flipo D, Fournier M, Benquet C et al (1998) Increased apoptosis, changes in intracellular Ca2+, and functional alterations in lymphocytes and macrophages after in vitro exposure to static magnetic field. J Toxicol Environ Health A 54:63–76

Gaipl US, Munoz LE, Grossmayer G et al (2007) Clearance deficiency and systemic lupus erythematosus (SLE). J Autoimmun 28:114–121

Gartzke J, Lange K (2002) Cellular target of weak magnetic fields: ionic conduction along actin filaments of microvilli. Am J Physiol Cell Physiol 283:C1333–C1346

Ghibelli L, Cerella C, Cordisco S et al (2006) NMR exposure sensitizes tumor cells to apoptosis. Apoptosis 11:359–365

Gmitrov J (2007) Geomagnetic field modulates artificial static magnetic field effect on arterial baroreflex and on microcirculation. Int J Biometeorol 51:335–344

Goodman R, Wei LX, Bumann J et al (1992) Exposure of human cells to electromagnetic fields: effect of time and field strength on transcript levels. J Electro Magnetobiol 11:19–28

Grimsley C, Ravichandran KS (2003) Cues for apoptotic cell engulfment: eat-me, don't eat-me and come-get-me signals. Trends Cell Biol 13:648–656

Gumienny TL, Hengartner MO (2001) How the worm removes corpses: the nematode *C. elegans* as a model system to study engulfment. Cell Death & Differ 8:564–568

Gupta S, Agrawal A, Agrawal S et al (2006) A paradox of immunodeficiency and inflammation in human aging: lessons learned from apoptosis. Immun Ageing 19:3–5

Hamada SH, Witkus R, Griffith Jr R (1989) Cell surface changes during electromagnetic field exposure. Exp Cell Biol 57:1–10

Haslett C (1999) Granulocyte apoptosis and its role in the resolution and control of lung inflammation. Am J Respir Crit Care Med 160:S5–11

Henson PM (2005) Engulfment: ingestion and migration with Rac, Rho and TRIO. Curr Biol 15: R29–R30

Hirose H, Nakahara T, Zhang QM et al (2003) Static magnetic field with a strong magnetic field gradient (41.7 T/m) induces c-Jun expression in HL-60 cells. In Vitro Cell Dev Biol Anim 39:348–352

Hodge S, Hodge G, Ahern J et al (2007) Smoking alters alveolar macrophage recognition and phagocytic ability: implications in chronic obstructive pulmonary disease. Am J Resp Cell Mol Biol 37:748–755

Hodge S, Hodge G, Brozyna S et al (2006) Azithromycin increases phagocytosis of apoptotic bronchial epithelial cells by alveolar macrophages. Eur Respir J 28:486–495

Hodge S, Hodge G, Scicchitano R et al (2003) Alveolar macrophages from subjects with chronic obstructive pulmonary disease are deficient in their ability to phagocytose apoptotic airway epithelial cells. Immun Cell Biol 81:289–296

Hoffmann D, Hoffmann I (1997) The changing cigarette, 1950–,1995. J Toxicol Env Health 50:307–364

Hoffmann PR, Kench JA, Vondracek A et al (2005) Interaction between phosphatidylserine and the phosphatidylserine receptor inhibits immune responses in vivo. J Immunol 174:1393–1404

Hong FT (1995) Magnetic field effects on biomolecules, cells, and living organisms. Biosystems 36:187–229

Ikehata M, Koana T, Suzuki Y et al (1999) Mutagenicity and co-mutagenicity of static magnetic fields detected by bacterial mutation assay. Mutat Res 427:147–156

Ishido M (2006) Apoptosis induced by environmental factors. In: Shultz LB (ed) Cell Apoptosis: regulation and enviromental factors. Nova Science Publishers, New York pp141–156

Jehle AW, Gardai SJ, Li S et al (2006) ATP-binding cassette transporter A7 enhances phagocytosis of apoptotic cells and associated ERK signaling in macrophages. J Cell Biol 174:547–556

Kai S, Ikonen E (1997) Functional rafts in cell membrane. Nature 387:569–572

Kim S, Chung EY, Ma X (2005) Immunological consequences of macrophage-mediated clearance of apoptotic cells. Cell Cycle 4:231–234

Kirkham PA, Spooner G, Ffoulkes-Jones C et al (2003) Cigarette smoke triggers macrophage adhesion and activation: role of lipid peroxidation products and scavenger receptor. Free Rad Biol Med 35:697–710

Kirkham PA, Spooner G, Rahman I et al (2004) Macrophage phagocytosis of apoptotic neutrophils is compromised by matrix proteins modified by cigarette smoke and lipid peroxidation products. Biochem Biophys Res Commun 318:32–37

Kleinsasser NH, Sassen AW, Semmler MP et al (2005) The tobacco alkaloid nicotine demonstrates genotoxicity in human tonsillar tissue and lymphocytes. Toxicol Sci 86:309–317

Krammer PH (2000) CD95's deadly mission in the immune system. Nature 407:789–795

Krysko DV, Brouckaert G, Kalai M et al (2003) Mechanisms of internalization of apoptotic and necrotic L929 cells by a macrophage cell line studied by electron microscopy. J Morphol 258:336–345

Krysko DV, Denecker G, Festjens N et al (2006) Macrophages use different internalization mechanisms to clear apoptotic and necrotic cells. Cell Death Diff 13:2011–2022

Kusumi A, Sako Y (1996) Cell surface organization by the membrane skeleton. Curr Opin Cell Biol 8:566–574

Lange K (1999) Microvillar Ca^{2+} signaling: a new view of an old problem. J Cell Physiol 180:19–34

Lange K (2000) Microvillar ion channels: cytoskeletal modulation of ion fluxes. J Theor Biol 206:561–884

Lauber K, Blumenthal SG, Waibel M et al (2004) Clearance of apoptotic cells: getting rid of the corpses. Mol Cell 14:277–287

Lenardo MJ (1997) The molecular regulation of lymphocyte apoptosis. Semin Immunol 9:1–5

Liboff AR, Cherng S, Jenrow KA et al (2003) Calmodulin-dependent cyclic nucleotide phosphodiesterase activity is altered by 20 mT magnetostatic fields. Bioelectromagnetics 24:32–38

Liburdy RP, Sloma TR, Sokolic R et al (1993) ELF magnetic fields, breast cancer, and melatonin: 50 Hz fields block melatonin's oncostatic action on ER+ breast cancer cell proliferation. J Pineal Res 14:89–97

Lisanti MP, Scherer PE, Tang ZL et al (1994) Caveolaer, caveolin and caveolin-rich membrane domains: a signalling hypothesis. Trends Cell Biol 4:231–235

Lisi A, Pozzi D, Pasquali E et al (2000) Three dimensional (3D) analysis of the morphological changes induced by 50 Hz magnetic field exposure on human lymphoblastoid cells (Raji). Bioelectromagnetics 21:46–51

Maderna P, Godson C (2003) Phagocytosis of apoptotic cells and the resolution of inflammation. Biochim Biophys Acta 1639:141–151

Magnelli L, Cinelli M, Turchetti A et al (1994) Apoptosis induction in 32D cells by IL-3 withdrawal is preceded by a drop in the intracellular calcium level. Biochem Biophys Res Commun 194:1394–1397

Mahoney JA, Rosen A (2005) Apoptosis and autoimmunity. Curr Opin Immunol 17:583–588

Maio M, Del Vecchio L (1992) Expression and functional role of CD54/Intercellular Adhesion Molecule-1 (ICAM-1) on human blood cells. Leuk Lymphoma 8:23–33

Marino AA, Iliev IG, Schwalke MA et al (1994) Association between cell membrane potential and breast cancer. Tumour Biol 15:82–89

Massimi M, Devirgiliis LC, Kolb-Bachofen V et al (1995) Independent modulation of galactose-specific receptor expression in rat liver cells. Hepatology 22:1819–1828

Miyakoshi J (2005) Effects of static magnetic fields at the cellular level. Prog Biophys Mol Biol 87:213–223

Morimoto K, Janssen WJ, Fessler MB et al (2006) Lovastatin enhances clearance of apoptotic cells (efferocytosis) with implications for chronic obstructive pulmonary disease. J Immunol 176:7657–7665

Morris CE, Homann U (2001) Cell surface area regulation and membrane tension. J Membr Biol 179:79–102

Nakahara T, Yaguchi H, Yoshida M et al (2002) Effects of exposure of CHO-K1 cells to a 10-T static magnetic field. Radiology 224:817–822

Nuccitelli S, Cerella C, Cordisco S et al (2006) Hyperpolarization of plasma membrane of tumour cells sensitive to anti-apoptotic effects of magnetic fields. Ann N Y Acad Sci 1090:217–225

Pagliara P, Chionna A, Panzarini E et al (2003) Lymphocytes apoptosis: young vs. aged and humans vs. rats. Tissue Cell 35:29–36

Pagliara P, Lanubile R, Dwikat M et al (2005) Differentiation of monocytic U937 cells under static magnetic field exposure. Eur J Histochem 49:75–86

Paradisi S, Donelli G, Santini MT et al (1993) A 50-Hz magnetic field induces structural and biophysical changes in membranes. Bioelectromagnetics 14:247–255

Park JW, Ryter SW, Choi AMK (2007) Functional significance of apoptosis in chronic obstructive pulmonary disease. COPD 4:347–353

Parton RG, Simons K (1995) Digging into caveole. Science 269:1398–1399

Papadimitraki ED, Bertsias GK, Boumpas DT (2007) Toll like receptors and autoimmunity: a critical appraisal. J Autoimmun 29:310–318

Pesci A, Balbi B, Majori M et al (1998) Inflammatory cells and mediators in bronchial lavage of patients with chronic obstructive pulmonary disease. Eur Respir J 12:380–386

Petrovski G, Zahuczky G, Màjai G et al (2007) Phagocytosis of cells dying through autophagy evokes a pro-inflammatory response in macrophages. Autophagy 3: 509–511

Platt N, Suzuki H, Kurihara Y et al (1996) Role for the class A macrophage scavenger receptor in the phagocytosis of apoptotic thymocytes in vitro. Proc Natl Acad Sci USA 93:12456–12460

Popov SV, Svitkina TM, Margolis LB et al (1991) Mechanism of cell protrudion formation in electrical field: the role of actin. Biochem Biophys Acta 1066:151–158

Prieto A, Reyes E, Bernstein ED et al (2001) Defective natural killer and phagocytic activities in chronic obstructive pulmonary disease are restored by glycophosphopeptical (inmunoferon). Am J Respir Crit Care Med 163:1578–1583

Rathmell JC, Thompson CB (2002) Pathways of apoptosis in lymphocyte development, omeostasis, and disease. Cell 109:S97–S107

Raucher D, Sheetz MP (1999) Characteristics of a membrane reservoir buffering membrane tension. Biophys J 77:1992–2002

Raucher D, Sheetz MP (2000) Cell spreading and lamellipodial extension rate is regulated by membrane tension. J Cell Biol 148:127–136

Reddien PW, Horvitz HR (2004) The engulfment process of programmed cell death in *Caenorhabditis elegans*. Annu Rev Cell Dev Biol 20:193–221

Rieti S, Manni V, Lisi A et al (2004) SNOM and AFM microscopy techniques to study the effect of non-ionizing radiation on the morphological and biochemical properties of human keratinocytes cell line (HaCaT). J Micros 213:20–28

Rosen AD (1993) Membrane response to static magnetic fields: effect of exposure duration. Biochim Biophys Acta 1148:317–320

Rosen AD (1996) Inhibition of calcium channel activation in GH3 cells by static magnetic fields. Biochim Biophys Acta 1282:149–155

Rosen AD (2003a) Effect of a 125 mT static magnetic field on the kinetics of voltage activated Na+ channels in GH3 cells. Bioelectromagnetics 24:517–523

Rosen AD (2003b) Mechanism of action of moderate-intensity static magnetic fields on biological systems. Cell Biochem Biophys 39:163–173

Ruzittu M, Carlá EC, Montinari MR et al (1999) Modulation of cell surface expression of liver carbohydrate receptors during in vivo induction of apoptosis with lead nitrate. Cell Tissue Res 298:105–112

Saetta M, Di Stefano A, Turato G et al (1998) CD8+ T-lymphocytes in peripheral airways of smokers with chronic bstructive pulmonary disease. Am J Respir Crit Care Med 157:822–826

Saffer JD, Phillips JL (1996) Evaluating the biological aspects of in vitro studies in bioelectromagnetics. Bioelectrochem. Bioenergetics 40:1–7

Santoro N, Lisi A, Pozzi D et al (1997) Effect of extremely low frequency (ELF) magnetic field exposure on morphological and biophysical properties of human lymphoid cell line (Raji). Biochem Biophys Acta 1357:281–290

Savill J, Dransfield I, Gregory C et al (2002) A blast from the past: clearance of apoptotic cells regulates immune responses. Nat Rev Immunol 2:965–975

Schiffer IB, Schreiber WG, Graf R et al (2003) No influence of magnetic fields on cell cycle progression using conditions relevant for patients during MRI. Bioelectromagnetics 4:241–250

Schrijvers DM, De Meyer GRY, Kockx MM et al (2005) Phagocytosis of apoptotic cells by macrophages is impaired in atherosclerosis. Arterioscler Thromb Vasc Biol25:1256

Schuller HM (1994) Carbon dioxide potentiates the mitogenic effects of nicotine and its carcinogenic derivative, NNK, in normal and neoplastic neuroendocrine lung cells via stimulation of autocrine and protein kinase C–dependent mitogenic pathways. Neurotoxicology 15:877–886

Scott RS, McMahon EJ, Pop SM et al (2001) Phagocytosis and clearance of apoptotic cells is mediated by MER. Nature 411:207– 211

Semm P, Schneider T, Vollrath L (1980) Effects of an Earth-strength magnetic field on electrical activity of pineal cells. Nature 288:607–608

Shapiro SD (1999) The macrophage in chronic obstructive pulmonary disease. Am J Respir Crit Care Med 160:S29–S32

Shin VY, Wu WK, Ye YN et al (2004) Nicotine promotes gastric tumor growth and neovascularization by activating extracellular signal-regulated kinase and cyclooxygenase-2. Carcinogenesis 25:2487–2495

Schuller HM (1994) Carbon dioxide potentiates the mitogenic effects of nicotine and its carcinogenic derivative, NNK, in normal and neoplastic neuroendocrine lung cells via stimulation of autocrine and protein kinase C–dependent mitogenic pathways. Neurotoxicology 15:877–886

Silva AKA, Silva EL, Egito EST et al (2006) Safety concerns related to magnetic field exposure. Radiat Environ Biophys 45:245–252

Simkó M, Droste S, Kriehuber R et al (2001) Stimulation of phagocytosis and free radical production in murine macrophages by 50 Hz electromagnetic fields. Eur J Cell Biol 80:562–566

Smith CJ, Perfetti TA, Garg R et al (2003) IARC carcinogens reported in cigarette mainstream smoke and their calculated log P values. Food Chem Toxicol 41:807–817

Somosy Z (2000) Radiation response of cell organelles. Micron 31:165–181

Sonnier H, Kolomytkin O, Marino A (2003) Action potential from human neuroblastoma cells in magnetic fields. Neurosci Lett 337:163–166

St Pierre TG, Dobson J (2000) Theoretical evaluation of cell membrane ion channel activation by applied magnetic fields. Eur Biophys J 29:455–456

Steinman RM, Turley S, Mellman I et al (2000) The induction of tolerance by dendritic cells that have captured apoptotic cells. J Exp Med 191:411–416

Stringer KA, Tobias M, O'Neill HC et al (2007) Cigarette smoke extract-induced suppression of caspase-3-like activity impairs human neutrophil phagocytosis. Am J Physiol Lung Cell Mol Physiol 292:L1572–1579

Stuart LM, Ezekowitz RA (2005) Phagocytosis: elegant complexity. Immunity 22:539–550

Sun EW, Shi YF (2001) Apoptosis: the quiet death silences the immune system. Pharmacol Ther 92:135–145

Szabo G, Mandrekar P, Dolganiuc A (2007) Innate immune response and hepatic inflammation. Semin Liver Dis 27:339–350

Taylor PR, Carugati A, Fadok VA et al (2000) A hierarchical role for classical pathway complement proteins in the clearance of apoptotic cells. J Exp Med 192:359–366

Tenforde TS, Liburdy RP (1988) Magnetic deformation of phospholipid bilayers: Effects on lysosome shape and solute permeability at prephase transition temperatures. J Theor Biol 133:385–396

Tenuzzo B, Chionna A, Panzarini E et al (2006) Biological effects of 6 mT static magnetic fields: a comparative study in different cell types. Bioelectromagnetics 27:560–577

Teodori L, Göhde W, Valente MG et al (2002a) Static magnetic fields affect calcium fluxes and inhibit stress-induced apoptosis in human glioblastoma cells. Cytometry 49:143–149

Teodori L, Grabarek J, Smolewski P et al (2002b) Exposure of cells to static magnetic field accelerates loss of integrity of plasma membrane during apoptosis. Cytometry 49:113–118

Testorf MF, Oberg PA, Iwasaka M et al (2002) Melanophore aggregation in strong static magnetic fields. Bioelectromagnetics 23:444–449

Tian F, Nakahara T, Yoshida M et al (2002) Exposure to power frequency magnetic fields suppresses X-ray-induced apoptosis transiently in Ku80-deficient xrs5 cells. Biochem Biophys Res Commun 292:355–361

Tofani S, Barone D, Berardelli M et al (2003) Static and ELF magnetic fields enhance the in vivo anti-tumor efficacy of cis-platin against lewis lung carcinoma, but not of cyclophosphamide against B16 melanotic melanoma. Pharmacol Res 48:83–90

Togo T, Krasieva TB, Steinhardt RA (2000) A decrease in membrane tension precedes successful cell-membrane repair. Mol Biol Cell 11:4339–4346

Van Meer G (1989) Lipid traffic in animal cells. Annu Rev Cell Biol 5:247-275

Vandivier RW, Fadok VA, Hoffmann PR et al (2002) Elastase-mediated phosphatidylserine receptor cleavage impairs apoptotic cell clearance in cystic fibrosis and bronchiectasis. J Clin Invest 109:661–670

Vandivier RW, Henson PM, Douglas IS (2006) Burying the dead: the impact of failed apoptotic cell removal (efferocytosis) on chronic inflammatory lung disease. Chest 129:1673–1682

Volpe P (2003) Interactions of zero-frequency and oscillating magnetic fields with biostructures and biosystems. Photochem Photobiol Sci 2:637–648

Volpe P, Cappelli G, Mariani F et al (2002) Macrophage sensitivity to static magnetic fields. In: Kostarakis P, Demokritos Publishers (ed) Biological Effects of EMFs, vol I. Rhodes, pp 374–381

Walleczek J, Liburdy RP (1990) Nonthermal 60-Hz sinusoidal magnetic field exposure enhances 45CaCC uptake in rat thymocytes: dependence on mitogen activation. FEBS Lett 271:157–160

Ward C, Dransfield I, Chilvers ER et al (1999) Pharmacological manipulation of granulocyte apoptosis: potential therapeutic targerts. Trends Pharmacol Sci 20:503–509

Wiskirchen J, Grönewäller EF, Heinzelmann F et al (2000) Human fetal lung fibroblasts: in vitro study of repetitive magnetic field exposure at 0.2, 1.0, and 1.5 T. Radiol 215:858–862

Worcester DL (1978) Structural origins of diamagnetic anisotropy in proteins. Biophys J 75:5475–5477

Zhang QM, Tokiwa M, Doi T et al (2003) Strong static magnetic field and the induction of mutations through elevated production of reactive oxygen species in *E. coli soxR*. Int J Radiat Biol 79:281–286

Zhou Z, Hartwieg E, Horvitz HR (2001) CED-1 is a transmembrane receptor that mediates cell corpse engulfment in *C. elegans*. Cell 104:43–56

Zmyślony M, Palus J, Jajte J et al (2000) DNA damage in rat lymphocytes treated in vitro with iron cations and exposed to 7 mT magnetic fields (static or 50 Hz). Mutat Res 453:89–96

Index

Complete Solutions - from classics to novel technologies

BD Biosciences, a segment of Becton, Dickinson and Company, is one of the world's leading businesses focused on bringing innovative tools to life science researchers and clinicians. Principal product lines include flow cytometers and cell sorters, cell imaging systems, monoclonal antibodies and kits, reagent systems, tools to aid in drug discovery and growing tissue and cells, and diagnostic assays.

BD Biosciences customers are involved in basic research, the discovery and development of drugs and vaccines, clinical trials, diagnostic testing, and disease management. This diverse customer base includes academic and government institutions, pharmaceutical and biotech companies, and clinical laboratories.

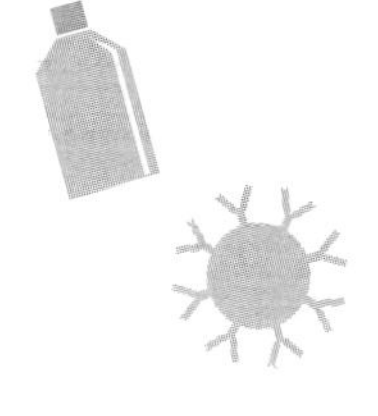

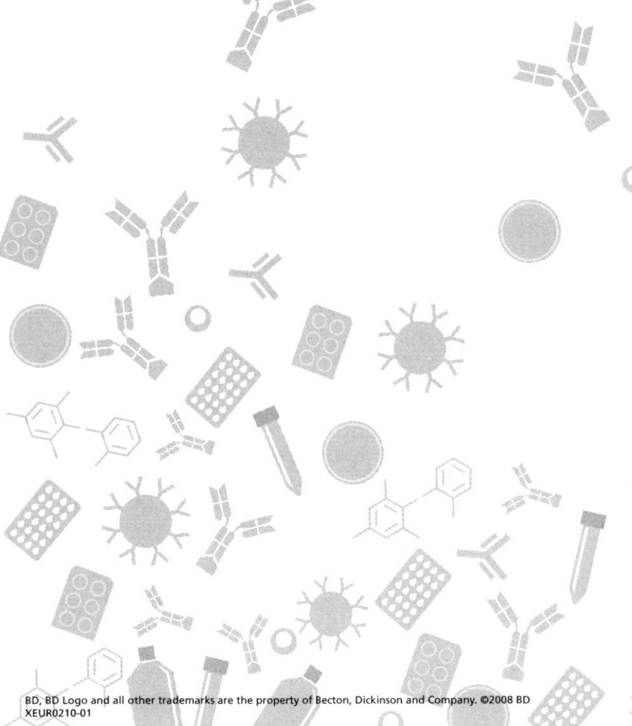

BD Biosciences
Europe
POB 13, Erembodegem-Dorp 86
B-9320 Erembodegem
Belgium
Tel.: (32) 2 400 98 95
Fax: (32) 2 401 70 94
help.biosciences@europe.bd.com
www.biosciences.com